Annals of Mathematics Studies

Number 176

Let $\tau \colon \mathbb{N} \to \mathbb{Z}$ be defined by

$$\sum_{n \geq 0} \tau(n)q^n = q \prod_{n \geq 1} (1 - q^n)^{24} \quad \text{in } \mathbb{Z}[[q]].$$

Then we have, in $\mathbb{Z}/19\mathbb{Z}$,

$$\tau(10^{1000} + 1357) = \pm 4,$$
$$\tau(10^{1000} + 7383) = \pm 2,$$
$$\tau(10^{1000} + 21567) = \pm 3,$$
$$\tau(10^{1000} + 27057) = 0,$$
$$\tau(10^{1000} + 46227) = 0,$$
$$\tau(10^{1000} + 57867) = 0,$$
$$\tau(10^{1000} + 64749) = \pm 7,$$
$$\tau(10^{1000} + 68367) = 0,$$
$$\tau(10^{1000} + 78199) = \pm 8,$$
$$\tau(10^{1000} + 128647) = 0.$$

Section 7.5, Lemma 7.4.1, and [Magma], [SAGE] and [PARI].

Computational Aspects of Modular Forms and Galois Representations

How One Can Compute in Polynomial Time the Value of Ramanujan's Tau at a Prime

Edited by Jean-Marc Couveignes and Bas Edixhoven

PRINCETON UNIVERSITY PRESS

PRINCETON AND OXFORD

Published by Princeton University Press,
41 William Street, Princeton, New Jersey 08540
In the United Kingdom: Princeton University Press,
6 Oxford Street, Woodstock, Oxfordshire OX20 1TW

press.princeton.edu

Library of Congress Cataloging-in-Publication Data

Computational aspects of modular forms and Galois representations :
how one can compute in polynomial time the value of Ramanujan's tau at a prime /
edited by Bas Edixhoven, Jean-Marc Couveignes.
    p. cm. -- (Annals of mathematics studies ; 176)
   ISBN 978-0-691-14201-2 (hardback) -- ISBN 978-0-691-14202-9 (pbk.)
    1. Galois modules (Algebra) 2. Class field theory. I. Edixhoven, B. (Bas), 1962-
        II. Couveignes, Jean-Marc.
  QA247.C638 2011
  512'.32--dc22
                                2010053185

British Library Cataloging-in-Publication Data is available

The publisher would like to acknowledge the authors of this volume
for providing the camera-ready copy from which this book was printed

10 9 8 7 6 5 4 3 2 1

Contents

Chapter 5. Computing complex zeros of polynomials and power series
by *J.-M. Couveignes*   95

Chapter 6. Computations with modular forms and Galois representations
by *J. Bosman*   129

Chapter 7. Polynomials for projective representations of level one forms
by *J. Bosman*   159

Chapter 8. Description of $X_1(5l)$
by *B. Edixhoven*   173

Chapter 9. Applying Arakelov theory
by *B. Edixhoven and R. de Jong*   187

Chapter 10. An upper bound for Green functions on Riemann surfaces
by *F. Merkl*   203

Chapter 11. Bounds for Arakelov invariants of modular curves

Chapter 15. Computing coefficients of modular forms
by *B. Edixhoven*

Epilogue

Bibliography

Index

Preface

This is a book about computational aspects of modular forms and the Galois representations attached to them. The main result is the following: Galois representations over finite fields attached to modular forms of level one can, in almost all cases, be computed in polynomial time in the weight and the size of the finite field. As a consequence, coefficients of modular forms can be computed fast via congruences, as in Schoof's algorithm for the number of points of elliptic curves over finite fields. The most important feature of the proof of the main result is that exact computations involving systems of polynomial equations in many variables are avoided by approximations and height bounds, that is, bounds for the accuracy that is necessary to derive exact values from the approximations.

The book's authors are the two editors, Jean-Marc Couveignes and Bas Edixhoven, together with Johan Bosman, Robin de Jong, and Franz Merkl. Each chapter has its own group of authors.

Chapter 1 gives an introduction to the subject, precise statements of the main results, and places these in a somewhat wider context. Chapter 2 provides the necessary background concerning modular curves and modular forms. Chapter 3 gives a first, informal description of the algorithms. These first three chapters should allow readers without much background in arithmetic geometry to still get a good idea of what happens in the book, skipping, if necessary, some parts of Chapter 2.

Chapters 4 and 5 provide the necessary background on heights, Arakelov theory, and on algorithmic aspects of the computation with a desired accuracy of the roots of complex polynomials and power series.

Chapters 6 and 7 are concerned with some real computations of Galois representations attached to modular forms, and end with a table dealing with all cases of forms of level one and of weight at most 22, and finite fields of characteristic at most 23.

The main ingredients for the proof of the main result are established in Chapters 8, 9, 10, 11, 12, and 13. The topics dealt with are, respectively, construction of suitable divisors on modular curves, bounding heights using Arakelov theory, bounding Arakelov invariants of certain modular curves, approximation of divisors using complex numbers, and using finite fields.

The main result on the computation of Galois representations is proved in Chapter 14, where one finds a detailed description of the algorithm and a rigorous proof of the complexity bound.

Chapter 15 contains the application of the main result to the computation of coefficients of modular forms.

The Epilogue describes some work on generalisations and applications, as well as a direction of further research outside the context of modular forms.

ACKNOWLEDGMENTS

This book started as a report, written in the context of a contract (Contrat d'Études 04.42.217) between the University of Leiden and the French CELAR (Centre Électronique de l'Armement). We thank the CELAR, and, in particular, David Lubicz and Reynald Lercier, for this financial support.

From 2005 until 2010, our project was supported in the Netherlands by NWO, the Dutch organization for scientific research (Nederlandse organisatie voor Wetenschappelijk Onderzoek) in the form of a VICI-grant to Bas Edixhoven.

As the contract with the CELAR stipulated, the report was published on the Internet (arxiv, homepage), in May 2006, and our aim was to extract a research article from it. We thank Eyal Goren for his suggestion to expand it into this book instead.

Jean-Marc Couveignes is supported by the Agence Nationale de la Recherche through the ALGOL project.

We thank René Schoof for asking the question, in 1995, that this book answers. We thank Peter Bruin for carefully reading preliminary versions and pointing out some errors, and Hendrik Lenstra for some suggestions and references.

AUTHOR INFORMATION

Johan Bosman obtained his PhD in Leiden, where he worked in the arithmetic geometry group of Bas Edixhoven. He has held a postdoc position at the Institut für Experimentelle Mathematik in Essen, and is currently at the University of Warwick as a Marie Curie postdoc.

Jean-Marc Couveignes is professor at the Université de Toulouse le Mirail and member of the Institut de Mathématiques de Toulouse, a joint laboratory of the Centre National de la Recherche Scientifique and the Université de Toulouse.

Bas Edixhoven is professor of mathematics at the University of Leiden. He obtained his PhD in Utrecht, and has worked in Berkeley, Utrecht, and Rennes.

Robin de Jong is assistant professor at the University of Leiden. He obtained his PhD at the University of Amsterdam.

Franz Merkl is professor of applied mathematics in München. He obtained his PhD at the ETH in Zürich, and has worked in Eindhoven, Bielefeld, and Leiden.

DEPENDENCIES BETWEEN THE CHAPTERS

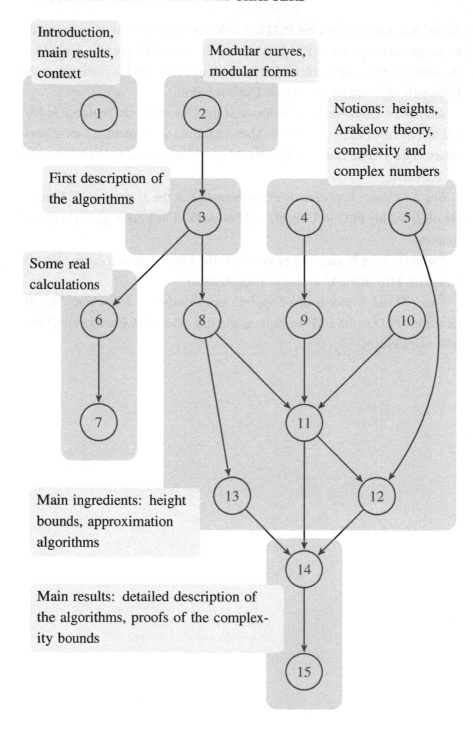

Introduction, main results, context

Modular curves, modular forms

Notions: heights, Arakelov theory, complexity and complex numbers

First description of the algorithms

Some real calculations

Main ingredients: height bounds, approximation algorithms

Main results: detailed description of the algorithms, proofs of the complexity bounds

Chapter One

Introduction, main results, context

by Bas Edixhoven

1.1 STATEMENT OF THE MAIN RESULTS

As the final results in this book are about fast computation of coefficients of modular forms, we start by describing the state of the art in this subject.

A convenient way to view modular forms and their coefficients in this context is as follows, in terms of Hecke algebras. For N and k positive integers, let $S_k(\Gamma_1(N))$ be the finite dimensional complex vector space of cusp forms of weight k on the congruence subgroup $\Gamma_1(N)$ of $\mathrm{SL}_2(\mathbb{Z})$. Each f in $S_k(\Gamma_1(N))$ has a power series expansion $f = \sum_{n \geq 1} a_n(f)q^n$, a complex power series converging on the open unit disk. These $a_n(f)$ are the coefficients of f that we want to compute, in particular for large n. For each positive integer n we have an endomorphism T_n of $S_k(\Gamma_1(N))$, and we let $\mathbb{T}(N, k)$ denote the sub-$\mathbb{Z}$-algebra of $\mathrm{End}(S_k(\Gamma_1(N))$ generated by them. The $\mathbb{T}(N, k)$ are commutative, and free $\mathbb{Z}$-modules of rank the dimension of $S_k(\Gamma_1(N))$, which is of polynomially bounded growth in N and k. For each N, k, and n one has the identity $a_n(f) = a_1(T_n f)$. The $\mathbb{C}$-valued pairing between $S_k(\Gamma_1(N))$ and $\mathbb{T}(N, k)$ given by $(f, t) \mapsto a_1(tf)$ identifies $S_k(\Gamma_1(N))$ with the space of $\mathbb{Z}$-linear maps from $\mathbb{T}(N, k)$ to $\mathbb{C}$, and we can write $f(T_n)$ for $a_n(f)$. All together this means that the key to the computation of coefficients of modular forms is the computation of the Hecke algebras $\mathbb{T}(N, k)$ and their elements T_n. A modular form f in $S_k(\Gamma_1(N))$ is determined by the $f(T_i)$ with $i \leq k \cdot [\mathrm{SL}_2(\mathbb{Z}) : \Gamma_1(N)]/12$, hence if T_n is known as a $\mathbb{Z}$-linear combination of these T_i, then $f(T_n)$ can be computed as the same $\mathbb{Z}$-linear combination of the $f(T_i)$.

The state of the art in computing the algebras $\mathbb{T}(N, k)$ can now be summarized as follows.

> There is a deterministic algorithm that on input positive integers N and $k \geq 2$, computes $\mathbb{T}(N, k)$: it gives a $\mathbb{Z}$-basis and the multiplication table for this basis, in running time polynomial in N and k. Moreover, the Hecke operator T_n can be expressed in this $\mathbb{Z}$-basis in deterministic polynomial time in N, k and n.

We do not know a precise reference for this statement, but it is rather obvious from the literature on calculations with modular forms for which we refer to William Stein's book [Ste2], and in particular to Section 8.10.2 of it. The algorithms alluded to above use that $S_k(\Gamma_1(N))$, viewed as $\mathbb{R}$-vector space, is naturally isomorphic to the $\mathbb{R}$-vector space obtained from the so-called "cuspidal subspace":

$$\mathrm{H}^1(\Gamma_1(N), \mathbb{Z}[x, y]_{k-2})_{\mathrm{cusp}} \subset \mathrm{H}^1(\Gamma_1(N), \mathbb{Z}[x, y]_{k-2})$$

in group cohomology. Here, $\mathbb{Z}[x, y]_{k-2}$ is the homogeneous part of degree $k-2$ of the polynomial ring $\mathbb{Z}[x, y]$ on which $\mathrm{SL}_2(\mathbb{Z})$ acts via its standard representation on $\mathbb{Z}[x, y]_1$. In this way, $\mathrm{H}^1(\Gamma_1(N), \mathbb{Z}[x, y]_{k-2})_{\mathrm{cusp}}$, modulo its torsion subgroup, is a free $\mathbb{Z}$-module of finite rank that is a faithful $\mathbb{T}(N, k)$-module, and the action of the T_n is described explicitly. The algorithms then use a presentation of $\mathrm{H}^1(\Gamma_1(N), \mathbb{Z}[x, y]_{k-2})_{\mathrm{cusp}}$ in terms of so-called "modular symbols" and we call them therefore modular symbols algorithms. The theory of modular symbols was developed by Birch, Manin, Shokurov, Cremona, Merel, ... It has led to many algorithms, implementations, and calculations, which together form the point of departure for this book.

The computation of the element T_n of $\mathbb{T}(N, k)$, using modular symbols algorithms, involves sums in which the number of terms grows at least linearly in n. If one computes such sums by evaluating and adding the terms one by one, the computation of T_n, for N and k fixed, will take time at least linear in n, and hence exponential in $\log n$. The same is true for other methods for computing T_n that we know of: computations with q-expansions that involve multiplication of power series, using linear combinations of

theta series, the "graph method" of Mestre and Oesterlé, and the Lefschetz
trace formula for correspondences, holomorphic or not. Efforts to evaluate
the encountered sums more quickly seem to lead, in each case, to the problem of computing coefficients of modular forms. For example, the graph
method leads to the problem of quickly computing representation numbers
of integer quadratic forms in 4 variables. In the case of the trace formula,
there are maybe only $O(\sqrt{n})$ terms, but they contain class numbers of imaginary quadratic orders, these numbers being themselves directly related to
coefficients of modular forms of half integral weight.

Let us now state one of the main results in this book, Theorem 15.2.1.

> *Assume that the generalized Riemann hypothesis (GRH) for
> zeta functions of number fields holds. There exists a determin-
> istic algorithm that on input positive integers n and k, together
> with the factorization of n into prime factors, computes the ele-
> ment T_n of $\mathbb{T}(1, k)$ in running time polynomial in k and $\log n$.*

The restriction to modular forms of level 1 in this result is there for a technical reason. The result will certainly be generalized to much more general
levels; see the Epilogue. The condition that the factorization of n into primes
must be part of the input is necessary because we do not have a polynomial
time algorithm for factoring integers. Conversely, see Remark 2.2.4 for evidence that factoring is not harder than computing coefficients of modular
forms.

Let us describe how the computation of Galois representations is used
for the computation of T_n. Standard identities express T_n in terms of the
T_p for p dividing n. These T_p are computed, via the LLL basis reduction algorithm, from sufficiently many of their images under morphisms f
from $\mathbb{T}(1, k)$ to finite fields, analogously to Schoof's algorithm for counting points of an elliptic curve over a finite field. Indeed, for such an f from
$\mathbb{T}(1, k)$ to $\mathbb{F}$, with p not the characteristic, l, say, of $\mathbb{F}$, the image $f(T_p)$ is
equal to the trace of $\rho_f(\text{Frob}_p)$, where $\rho_f \colon \text{Gal}(\overline{\mathbb{Q}}/\mathbb{Q}) \to \text{GL}_2(\mathbb{F})$ is the Galois representation attached to f, and $\rho_f(\text{Frob}_p)$ a Frobenius element at p.
The representation ρ_f is characterized by the following three conditions: it
is semisimple, it is unramified outside l, and for all prime numbers $p \neq l$

one has

$$\text{trace}(\rho_f(\text{Frob}_p)) = f(T_p) \quad \text{and} \quad \det(\rho_f(\text{Frob}_p)) = p^{k-1} \quad \text{in } \mathbb{F}.$$

It is *the* main result of this book, Theorem 14.1.1, plus some standard computational number theory, that enables us to compute $\rho_f(\text{Frob}_p)$ in time polynomial in k, #$\mathbb{F}$, and $\log p$ (note the log!). Under GRH, existence of sufficiently many maximal ideals of small enough index is guaranteed. We partly quote Theorem 14.1.1:

> There is a deterministic algorithm that on input a positive integer k, a finite field $\mathbb{F}$, and a surjective ring morphism f from $\mathbb{T}(1,k)$ to $\mathbb{F}$ such that the associated Galois representation $\rho_f \colon \text{Gal}(\overline{\mathbb{Q}}/\mathbb{Q}) \to \text{GL}_2(\mathbb{F})$ is reducible or has image containing $\text{SL}_2(\mathbb{F})$, computes ρ_f in time polynomial in k and #$\mathbb{F}$.

By "computing ρ_f" we mean the following. Let $K_f \subset \overline{\mathbb{Q}}$ be the finite Galois extension such that ρ_f factors as the natural surjection from $\text{Gal}(\overline{\mathbb{Q}}/\mathbb{Q})$ to $\text{Gal}(K_f/\mathbb{Q})$, followed by an injection into $\text{GL}_2(\mathbb{F})$. Then to give ρ_f means to give K_f as $\mathbb{Q}$-algebra, in terms of a multiplication table with respect to a $\mathbb{Q}$-basis, together with a list of all elements of $\text{Gal}(K_f/\mathbb{Q})$, as matrices with coefficients in $\mathbb{Q}$, and, for each σ in $\text{Gal}(K_f/\mathbb{Q})$, to give the corresponding element $\rho_f(\sigma)$ of $\text{GL}_2(\mathbb{F})$.

Before we describe in more detail, in the next sections, some history and context concerning our main results, we give one example and make some brief remarks. Many of these remarks are treated with more detail farther on.

The first nontrivial example is given by $k = 12$. The space of cuspidal modular forms of level one and weight 12 is one-dimensional, generated by the *discriminant modular form* Δ, whose coefficients are given by Ramanujan's τ-function:

$$\Delta = q \prod_{n \geq 1}(1-q^n)^{24} = \sum_{n \geq 1} \tau(n)q^n = q - 24q^2 + 252q^3 + \cdots \quad \text{in } \mathbb{Z}[[q]].$$

In this case, the Hecke algebra $\mathbb{T}(1,12)$ is the ring $\mathbb{Z}$, and, for each n in $\mathbb{Z}_{>0}$, we have $T_n = \tau(n)$. The results above mean that

> for p prime, Ramanujan's $\tau(p)$ can be computed in time polynomial in $\log p$.

For l prime, let ρ_l denote the Galois representation to $\mathrm{GL}_2(\mathbb{F}_l)$ attached to Δ. It was proved by Swinnerton-Dyer that for l not in $\{2, 3, 5, 7, 23, 691\}$ the image of ρ_l contains $\mathrm{SL}_2(\mathbb{F}_l)$. This means that for all l not in this short list the representation ρ_l has nonsolvable image, and so cannot be computed using computational class field theory. The classical congruences for Ramanujan's τ-function correspond to the l in the list above. Our results provide a generalization of these congruences in the sense that the number fields K_l that give the ρ_l "encode" the $\tau(p) \bmod l$ in such a way that $\tau(p)$ mod l can be computed in time polynomial in l and $\log p$, that is, just the same complexity as in the case where one has explicit congruences.

More generally, we hope that nonsolvable global field extensions whose existence and local properties are implied by the Langlands program can be made accessible to computation and so become even more useful members of the society of mathematical objects. Explicit descriptions of these fields make the study of global properties such as class groups and groups of units possible. Certainly, if we only knew the maximal Abelian extension of $\mathbb{Q}$ as described by general class field theory, then roots of unity would be very much welcomed.

The natural habitat for Galois representations such as the ρ_f above is that of higher degree étale cohomology with $\mathbb{F}_\ell$-coefficients of algebraic varieties over $\overline{\mathbb{Q}}$, together with the action of $\mathrm{Gal}(\overline{\mathbb{Q}}/\mathbb{Q})$. Our results provide some evidence that, also in interesting cases, such objects can be computed in reasonable time. We stress that this question is not restricted to varieties related to modular forms or automorphic forms. In fact, thinking of elliptic curves, over $\mathbb{Q}$, say, knowing that these are modular does not help for computing their number of points over finite fields: Schoof's algorithm uses algebraic geometry, not modularity.

The problem of computing étale cohomology with Galois action is clearly related to the question of the existence of polynomial time algorithms for computing the number of solutions in $\mathbb{F}_p$ of a fixed system of polynomial equations over $\mathbb{Z}$, when p varies. Our results treat this problem for the 11-dimensional variety that gives rise to Δ; see Section 1.5 for more details and also for an explicit variety of dimension 19 related to this.

The Epilogue describes a striking application of a generalization of our results to the problem of computing representation numbers of the $\mathbb{Z}^{2k}$

equipped with their standard inner products. This again is an example where there are explicit formulas only for small k, but where in general there (surely) exists an algorithm that computes such numbers as quickly as if such formulas did exist. Hence, from a computational perspective, such algorithms form a natural generalization of the finite series of formulas.

We very briefly describe the method by which we compute the ρ_f. Their duals occur in the higher degree étale cohomology of certain higher dimensional varieties, but no-one seems to know how to compute with this directly.

Via some standard methods in étale cohomology (the Leray spectral sequence, and passing to a finite cover to trivialize a locally constant sheaf of finite dimensional $\mathbb{F}_l$-vector spaces), or from the theory of congruences between modular forms, it is well known that the ρ_f are realized by subspaces V_f in the l-torsion $J_l(\overline{\mathbb{Q}})[l]$ of the Jacobian variety J_l of some modular curve X_l defined over $\mathbb{Q}$. The field K_f is then the field generated by suitable "coordinates" of the points $x \in V_f \subset J_l(\overline{\mathbb{Q}})[l]$. We are now in the more familiar situation of torsion points on Abelian varieties. But the price that we have paid for this is that the Abelian variety J_l depends on l, and that its dimension, equal to the genus of X_l, that is, equal to $(l - 5)(l - 7)/24$, grows quadratically with l. This makes it impossible to directly compute the $x \in V_l$ using computer algebra: known algorithms for solving systems of nonlinear polynomial equations take time exponential in the dimension, that is, exponential in l.

Instead of using computer algebra directly, Jean-Marc Couveignes suggested that we use approximations and height bounds. In its simplest form, this works as follows. Suppose that x is a rational number, $x = a/b$, with a and b in $\mathbb{Z}$ coprime. Suppose that we have an upper bound M for $\max(|a|, |b|)$. Then x is determined by any approximation $y \in \mathbb{R}$ of x such that $|y - x| < 1/2M^2$, simply because for all $x' \neq x$ with $x' = a'/b'$, where a' and b' in $\mathbb{Z}$ satisfy $\max(|a'|, |b'|) < M$, we have $|x' - x| = |(a'b - ab')/bb'| \geq 1/M^2$.

For the computation of K_f, we consider the minimal polynomial P_f in $\mathbb{Q}[T]$ of a carefully theoretically constructed generator α of K_f. We use approximations of all Galois conjugates of α, that is, of all roots of P_f. Instead of working directly with torsion points of J_l, we work with divisors on the curve X_l. Using this strategy, the problem of showing that P_f can be computed in time polynomial in k and #$\mathbb{F}$ is divided into two different tasks. First, to show that the number of digits necessary for a good enough approximation of P_f is bounded by a fixed power of #$\mathbb{F}$. Second, to show that, given f and n, the coefficients of P_f can be approximated with a precision of n digits in time polynomial in $n \cdot$ #$\mathbb{F}$. The first problem is dealt with in Chapters 9, 10, and 11, using Arakelov geometry. The second problem is solved in Chapters 12 and 13, in two ways: complex approximations (numerical analysis), and approximations in the sense of reductions modulo many small primes, using exact computations in Jacobians of modular curves over finite fields. These five chapters form the technical heart of this book. The preceding chapters are meant as an introduction to them, or motivation for them, and the two chapters following them give the main results as relatively straightforward applications.

Chapters 6 and 7 stand a bit apart, as they are concerned with some real computations of Galois representations attached to modular forms. They use the method by complex approximations, but do not use a rigorously proved bound for sufficient accuracy. Instead, the approximations provide good *candidates* for polynomials P_f. These candidates have the correct Galois group and the right ramification properties. Recent modularity results by Khare, Wintenberger and Kisin (see [Kh-Wi1], [Kh-Wi2], [Kis1], and [Kis2]) are then applied to *prove* that the candidates do indeed give the right Galois representations.

1.2 HISTORICAL CONTEXT: SCHOOF'S ALGORITHM

The computation of Hecke operators from Galois representations and congruences can be viewed as a generalization of Schoof's method to count points on elliptic curves over finite fields; see [Sch2] and [Sch3]. René Schoof gave an algorithm to compute, for E an elliptic curve over a finite field $\mathbb{F}_q$, the number #$E(\mathbb{F}_q)$ of $\mathbb{F}_q$-rational points in a time $O((\log q)^{5+\varepsilon})$. His algorithm works as follows.

The elliptic curve is embedded, as usual, in the projective plane $\mathbb{P}^2_{\mathbb{F}_q}$ as the zero locus of a Weierstrass equation, which, in inhomogeneous coordinates, is of the form

$$y^2 + a_1 xy + a_3 y = x^3 + a_2 x^2 + a_4 x + a_6,$$

with the a_i in $\mathbb{F}_q$. We let $\mathbb{F}_q \rightarrow \overline{\mathbb{F}}_q$ be an algebraic closure. We let $F_q \colon E \rightarrow E$ denote the so-called q-Frobenius. It is the endomorphism of E with the property that for all (a, b) in the affine part of $E(\overline{\mathbb{F}}_q)$ given by the Weierstrass equation above, we have $F_q((a, b)) = (a^q, b^q)$. The theory of elliptic curves over finite fields says:

1. there is a unique integer a, called the *trace* of F_q, such that in the endomorphism ring of E one has $F_q^2 - aF_q + q = 0$;

2. $\#E(\mathbb{F}_q) = 1 - a + q$;

3. $|a| \leq 2q^{1/2}$.

So, computing $\#E(\mathbb{F}_q)$ is equivalent to computing this integer a. Schoof's idea is now to compute a modulo l for small prime numbers l. If the product of the prime numbers l exceeds $4q^{1/2}$, the length of the interval in which we know a to lie, then the congruences modulo these l determine a uniquely. Analytic number theory tells us that it will be sufficient to take all primes l up to approximately $(\log q)/2$.

Then the question is how one computes a modulo l. This should be done in time polynomial in $\log q$ and l. The idea is to use the elements of order dividing l in $E(\overline{\mathbb{F}}_q)$. We assume now that l does not divide q, that is, we avoid the characteristic of $\mathbb{F}_q$. For each l, the kernel $E(\overline{\mathbb{F}}_q)[l]$ of multiplication by l on $E(\overline{\mathbb{F}}_q)$ is a two-dimensional vector space over $\mathbb{F}_l$. The map F_q gives an endomorphism of $E(\overline{\mathbb{F}}_q)[l]$, and it follows that the image of a in $\mathbb{F}_l$ is the unique element of $\mathbb{F}_l$, also denoted a, such that for each v in $E(\overline{\mathbb{F}}_q)[l]$ we have $aF_q v = F_q^2 v + qv$. We remark that the image of a in $\mathbb{F}_l$ is the trace of the endomorphism of $E(\overline{\mathbb{F}}_q)[l]$ given by F_q, but this is not really used at this point.

To find this element a of $\mathbb{F}_l$, one proceeds as follows. We suppose that $l \neq 2$. There is a unique monic element ψ_l of $\mathbb{F}_q[x]$ of degree $(l^2-1)/2$, whose roots in $\overline{\mathbb{F}}_q$ are precisely the x-coordinates of the l^2-1 nonzero elements in $E(\overline{\mathbb{F}}_q)[l]$ (the rational function x on E is a degree two map to $\mathbb{P}^1_{\overline{\mathbb{F}}_q}$, which as such is the quotient for the multiplication by -1 map on E). One then lets A_l be the $\mathbb{F}_q$-algebra obtained as

$$A_l := \mathbb{F}_q[x, y]/(y^2 + a_1 xy + a_3 y - x^3 - a_2 x^2 - a_4 x - a_6, \psi_l(x)).$$

The dimension of A_l as $\mathbb{F}_q$-vector space is $l^2 - 1$. An equivalent description of A_l is to say that it is the affine coordinate ring of the subscheme of points of order l of E. By construction of A_l, there is a tautological A_l-valued point v in $E(A_l)$ (its coordinates are the images of x and y in A_l). Now to find the element a of $\mathbb{F}_l$ that we are looking for, we then try one by one the elements i in $0, \pm 1, \ldots, \pm(l-1)/2$ until $i \mathbb{F}_q v = \mathbb{F}_q^2 v + qv$; then $i = a$ mod l.

It is easy to see that the computation of the integer a can be done in time $O((\log q)^{5+\varepsilon})$ (using fast arithmetic for the elementary operations, for example, a multiplication in A_l costs about $(l^2(\log q))^{1+\varepsilon}$ time; $l^2(\log q)$ is the number of bits needed to store one element of A_l).

For the sake of completeness, let us mention that shortly after the appearance of Schoof's algorithm, Atkin and Elkies added some improvements to it, making it possible in certain cases to reduce the dimension of the $\mathbb{F}_q$-algebra from $l^2 - 1$ to linear in $l+1$ or $l-1$. This improvement, called the Schoof-Atkin-Elkies (SEA) algorithm, is important mainly for implementations. Its (average) complexity is $O((\log q)^{4+\varepsilon})$; for details, the reader is referred to [Sch3].

1.3 SCHOOF'S ALGORITHM DESCRIBED IN TERMS OF ÉTALE CO-HOMOLOGY

In order to describe Schoof's algorithm in the previous section, we referred to the theory of elliptic curves over finite fields. But there is a more general framework for getting information on the number of rational points of algebraic varieties over finite fields: cohomology, and Lefschetz's trace formula. Cohomology exists in many versions. The version directly related to Schoof's algorithm is étale cohomology with coefficients in $\mathbb{F}_l$. Standard references for étale cohomology are [SGA4], [SGA4.5], [SGA5], [Mil1], [Fr-Ki]. The reader is referred to these references for the notions that we will use below. We also recommend Appendix C of [Hart].

For the sake of precision, let us say that we define the notion of *algebraic variety over a field k* to mean *k-scheme that is separated and of finite type.* Attached to an algebraic variety X over a field k there are *étale cohomology groups with compact supports* $H^i_c(X_{et}, \mathbb{F}_l)$, for all $i \geq 0$ and for all prime numbers l. Actually, the coefficients $\mathbb{F}_l$ can be replaced by more general objects, sheaves of Abelian groups on the étale site X_{et} of X, but we do not need this now. If X is a proper k-scheme, then the $H^i_c(X_{et}, \mathbb{F}_l)$ are equal to the étale cohomology groups $H^i(X_{et}, \mathbb{F}_l)$ without condition on supports.

If k is separably closed, then the $H^i_c(X_{et}, \mathbb{F}_l)$ are finite dimensional $\mathbb{F}_l$-vector spaces, zero for $i > 2\dim(X)$. In that case, they are the analog of the more easily defined cohomology groups $H^i_c(X, \mathcal{F})$ for complex analytic varieties: the derived functors of the functor that associates to a sheaf $\mathcal{F}$ of $\mathbb{Z}$-modules on X equipped with its Archimedean topology its $\mathbb{Z}$-module of global sections whose support is compact.

The construction of the $H^i_c(X_{et}, \mathbb{F}_l)$ is functorial for proper morphisms: a proper morphism $f: X \to Y$ of algebraic varieties over k induces a pullback morphism f^* from $H^i_c(Y_{et}, \mathbb{F}_l)$ to $H^i_c(X_{et}, \mathbb{F}_l)$.

Let now X be an algebraic variety over $\mathbb{F}_q$. Then we have the q-Frobenius morphism F_q from X to itself, and, by extending the base field from $\mathbb{F}_q$ to $\overline{\mathbb{F}}_q$, from $X_{\overline{\mathbb{F}}_q}$ to itself. This morphism F_q is proper, hence induces maps

$$F_q^*: H^i_c(X_{\overline{\mathbb{F}}_q, et}, \mathbb{F}_l) \longrightarrow H^i_c(X_{\overline{\mathbb{F}}_q, et}, \mathbb{F}_l).$$

Hence, for each i in $\mathbb{Z}$, the trace $\mathrm{trace}(F_q^*, H^i_c(X_{\overline{\mathbb{F}}_q, et}, \mathbb{F}_l))$ of the map is defined, and it is zero for $i < 0$ and $i > 2\dim(X)$. The set of fixed points of F_q on $X(\overline{\mathbb{F}}_q)$ is precisely the subset $X(\mathbb{F}_q)$. The *Lefschetz trace formula* then gives the following identity in $\mathbb{F}_l$:

$$(1.3.1) \qquad \#X(\mathbb{F}_q) = \sum_i (-1)^i \mathrm{trace}(F_q^*, H^i_c(X_{\overline{\mathbb{F}}_q, et}, \mathbb{F}_l)).$$

We can now say how Schoof's algorithm is related to étale cohomology. We consider again an elliptic curve E over a finite field $\mathbb{F}_q$. We assume that l does not divide q. Then, as for any smooth proper geometrically connected curve, $H^0(E_{\overline{\mathbb{F}}_q, et}, \mathbb{F}_l) = \mathbb{F}_l$ and F_q^* acts on it as the identity, and $H^2(E_{\overline{\mathbb{F}}_q, et}, \mathbb{F}_l)$ is one-dimensional and F_q^* acts on it by multiplication by q, the degree of F_q. According to the trace formula (1.3.1), we have

$$\#E(\mathbb{F}_q) = 1 - \text{trace}(F_q^*, H^1(E_{\overline{\mathbb{F}}_q,\text{et}}, \mathbb{F}_l)) + q.$$

It follows that for the integer a of the previous section, the trace of Frobenius, we have, for all l not dividing p the identity in $\mathbb{F}_l$,

$$a = \text{trace}(F_q^*, H^1(E_{\overline{\mathbb{F}}_q,\text{et}}, \mathbb{F}_l)).$$

This identity is explained by the fact that there is a natural isomorphism, compatible with the action of F_q,

$$H^1(E_{\overline{\mathbb{F}}_q,\text{et}}, \mathbb{F}_l) = E(\overline{\mathbb{F}}_q)[l].$$

Let us describe how one constructs this isomorphism. On E_{et} we have the short exact sequence of sheaves called the Kummer sequence:

$$0 \longrightarrow \mu_l \longrightarrow \mathbb{G}_m \longrightarrow \mathbb{G}_m \longrightarrow 0$$

where the map on $\mathbb{G}_m$ is multiplication by l in the group law of $\mathbb{G}_m$, that is, taking lth powers. This short exact sequence gives an exact sequence of cohomology groups after pullback to $E_{\overline{\mathbb{F}}_q,\text{et}}$:

$$\{1\} \longrightarrow \mu_l(\overline{\mathbb{F}}_q) \longrightarrow \overline{\mathbb{F}}_q^\times \longrightarrow \overline{\mathbb{F}}_q^\times \longrightarrow H^1(E_{\overline{\mathbb{F}}_q,\text{et}}, \mu_l)$$
$$\longrightarrow H^1(E_{\overline{\mathbb{F}}_q,\text{et}}, \mathbb{G}_m) \longrightarrow H^1(E_{\overline{\mathbb{F}}_q,\text{et}}, \mathbb{G}_m) \longrightarrow \cdots$$

Just as for any scheme, one has

$$H^1(E_{\overline{\mathbb{F}}_q,\text{et}}, \mathbb{G}_m) = \text{Pic}(E_{\overline{\mathbb{F}}_q}).$$

It follows that

$$H^1(E_{\overline{\mathbb{F}}_q,\text{et}}, \mu_l) = \text{Pic}(E_{\overline{\mathbb{F}}_q})[l].$$

Finally, using the exact sequence

$$0 \longrightarrow \text{Pic}^0(E_{\overline{\mathbb{F}}_q}) \longrightarrow \text{Pic}(E_{\overline{\mathbb{F}}_q}) \xrightarrow{\text{deg}} \mathbb{Z} \longrightarrow 0$$

and the fact that E is its own Jacobian variety, that is, $\text{Pic}^0(E_{\overline{\mathbb{F}}_q}) = E(\overline{\mathbb{F}}_q)$, we obtain

$$H^1(E_{\overline{\mathbb{F}}_q,\text{et}}, \mu_l) = \text{Pic}(E_{\overline{\mathbb{F}}_q})[l] = \text{Pic}^0(E_{\overline{\mathbb{F}}_q})[l] = E(\overline{\mathbb{F}}_q)[l].$$

The choice of an isomorphism between $\mu_l(\overline{\mathbb{F}}_q)$ and $\mathbb{F}_l$ gives us the desired isomorphism between $\mathrm{H}^1(E_{\overline{\mathbb{F}}_q,\mathrm{et}}, \mathbb{F}_l)$ and $E(\overline{\mathbb{F}}_q)[l]$. In fact, we note that by using the Weil pairing from $E(\overline{\mathbb{F}}_q)[l] \times E(\overline{\mathbb{F}}_q)[l]$ to $\mu_l(\overline{\mathbb{F}}_q)$, we get an isomorphism,

$$\mathrm{H}^1(E_{\overline{\mathbb{F}}_q,\mathrm{et}}, \mathbb{F}_l) = E(\overline{\mathbb{F}}_q)[l]^{\vee},$$

that is more natural than the one used above; in particular, it does not depend on the choice of an isomorphism $\mathbb{F}_l \twoheadrightarrow \mu_l(\overline{\mathbb{F}}_q)$.

1.4 SOME NATURAL NEW DIRECTIONS

We have seen that the two-dimensional $\mathbb{F}_l$-vector spaces that are used in Schoof's algorithm for elliptic curves can also be seen as étale cohomology groups. A natural question that arises is then the following.

> Are there other interesting cases where étale cohomology can be used to construct polynomial time algorithms for counting rational points of varieties over finite fields?

A more precise question is the following.

> Let n and m be in $\mathbb{Z}_{\geq 0}$, and let $f_1, \ldots, f_m$ be in $\mathbb{Z}[x_1, \ldots, x_n]$. Is there an algorithm that on input a prime number p computes $\#\{a \in \mathbb{F}_p^n \mid \forall i : f_i(a) = 0\}$ in time polynomial in $\log p$?

We believe that the answer to this question is yes, and that the way in which such an algorithm can work is to compute étale cohomology.

1.4.1 Curves of higher genus

The first step in the direction of this question was taken by Jonathan Pila. In [Pil] he considered principally polarized Abelian varieties of a fixed dimension, and curves of a fixed genus, and showed that in those cases polynomial time algorithms for computing the number of rational points over finite fields exist. In these cases, the only relevant cohomology groups are in degree one, that is, they are of the form $\mathrm{H}^1(X_{\overline{\mathbb{F}}_q,\mathrm{et}}, \mathbb{F}_l)$ with X a smooth proper curve, or an Abelian variety, over the field $\mathbb{F}_q$. As in Schoof's algorithm, the way to deal with these cohomology groups is to view them as

$J(\overline{\mathbb{F}}_q)[l]$, the kernel of multiplication by l on the Abelian variety J. In the case where X is a curve, one lets J be the Jacobian variety of X.

As Pila makes use of explicit systems of equations for Abelian varieties, his algorithm has a running time that is at least exponential in the dimension of the Abelian variety, and hence, in the case of curves, as a function of the genus of the curve.

The current state of the art concerning the question of counting the rational points of curves over finite fields seems still to be the same: algorithms have a running time that is exponential in the genus. As an illustration, let us mention that in [Ad-Hu] Adleman and Huang give an algorithm that computes $\#X(\mathbb{F}_q)$ in time $(\log q)^{O(g^2 \log g)}$, where X is a hyperelliptic curve over $\mathbb{F}_q$ and g is the genus of X.

Recent progress in the case where the characteristic of the finite fields $\mathbb{F}_q$ is fixed, using so-called p-adic methods, will be discussed in Section 1.6 below. In that case, there are algorithms whose running time is polynomial in g and $\log q$.

1.4.2 Higher degree cohomology, modular forms

Another direction in which one can try to generalize Schoof's algorithm is to varieties of higher dimension, where nontrivial cohomology groups of degree higher than one are needed. In this context, we would call the degree two cohomology group of a curve trivial, because the trace of F_q on it is q.

More generally speaking, cohomology groups, but now with l-adic coefficients, that are of dimension one are expected to have the property that the trace of F_q can only be of the form $q^n \zeta$, with n an integer greater than or equal to zero, and ζ a root of unity. This means that one-dimensional cohomology groups are not so challenging. Indeed, it is the fact that for elliptic curves over $\mathbb{F}_p$ all integers in the interval $[p+1-2p^{1/2}, p+1+2p^{1/2}]$ can occur that makes the problem of point counting very different from point counting on nonsingular quadric surfaces in $\mathbb{P}^3_{\mathbb{F}_q}$, for example, where the outcome can only be $q^2 + 2q + 1$ or $q^2 + 1$.

It follows that the simplest case to consider is cohomology groups of dimension two, in degree at least two, on which the action of F_q is not given by a simple rule as in the one-dimensional case. Such cohomology groups are provided by modular forms, as we will explain later in Section 2.2. Let us just say for the moment that there is a direct relation with elliptic curves, via the concept of *modularity* of elliptic curves over $\mathbb{Q}$, that we will now sketch.

Let E be an elliptic curve over $\mathbb{Q}$, given by some Weierstrass equation. Such a Weierstrass equation can be chosen to have its coefficients in $\mathbb{Z}$. A Weierstrass equation for E with coefficients in $\mathbb{Z}$ is called *minimal* if the absolute value of its *discriminant* is minimal among all Weierstrass equations for E with coefficients in $\mathbb{Z}$; this discriminant then only depends on E and will be denoted discr(E). In fact, two minimal Weierstrass equations define isomorphic curves in $\mathbb{P}^2_{\mathbb{Z}}$, the projective plane over $\mathbb{Z}$. In other words, E has a Weierstrass minimal model over $\mathbb{Z}$, which will be denoted by $E_{\mathbb{Z}}$. For each prime number p, we let $E_{\mathbb{F}_p}$ denote the curve over $\mathbb{F}_p$ given by reducing a minimal Weierstrass equation modulo p; it is the fiber of $E_{\mathbb{Z}}$ over $\mathbb{F}_p$. The curve $E_{\mathbb{F}_p}$ is smooth if and only if p does not divide discr(E). The possible singular fibers have exactly one singular point: an ordinary double point with rational tangents, or with conjugate tangents, or an ordinary cusp. The three types of reduction are called split multiplicative, nonsplit multiplicative, and additive, respectively, after the type of group law that one gets on the complement of the singular point. For each p we then get an integer a_p by requiring the following identity:

$$p + 1 - a_p = \#E(\mathbb{F}_p).$$

This means that for all p, a_p is the trace of F_p on the degree one étale cohomology of $E_{\overline{\mathbb{F}}_p}$, with coefficients in $\mathbb{F}_l$, or in $\mathbb{Z}/l^n\mathbb{Z}$ or in the l-adic numbers $\mathbb{Z}_l$. For p not dividing discr(E) we know that $|a_p| \leq 2p^{1/2}$. If $E_{\mathbb{F}_p}$ is multiplicative, then $a_p = 1$ or -1 in the split and nonsplit case. If $E_{\mathbb{F}_p}$ is additive, then $a_p = 0$. We also define for each p an element $\varepsilon(p)$ in $\{0, 1\}$ by setting $\varepsilon(p) = 1$ for p not dividing discr(E) and setting $\varepsilon(p) = 0$ for p dividing discr(E). The *Hasse-Weil L-function* of E is then defined as

$$L_E(s) = \prod_p L_{E,p}(s), \qquad L_{E,p}(s) = \left(1 - a_p p^{-s} + \varepsilon(p) p p^{-2s}\right)^{-1},$$

for s in $\mathbb{C}$ with $\Re(s) > 3/2$ (indeed, the fact that $|a_p| \leq 2p^{1/2}$ implies that the product converges for such s). To explain this function more conceptually, we note that for all p and for all $l \neq p$ we have the identity

$$1 - a_p t + \varepsilon(p) p t^2 = \det(1 - t F_p^*, H^1(E_{\overline{\mathbb{F}}, \text{et}}, \mathbb{Q}_l)).$$

The reader should notice that now we use étale cohomology with coefficients in $\mathbb{Q}_l$, the field of l-adic numbers, and not in $\mathbb{F}_l$. The reason for this is that we want the last identity above to be an identity between polynomials with integer coefficients, and not with coefficients in $\mathbb{F}_l$.

The function L_E was conjectured to have a holomorphic continuation over all of $\mathbb{C}$, and to satisfy a certain precisely given functional equation relating the values at s and $2-s$. In that functional equation appears a certain positive integer N_E called the *conductor* of E, composed of the primes p dividing $\text{discr}(E)$ with exponents that depend on the behavior of E at p, that is, on $E_{\mathbb{Z}_p}$. This conjecture on continuation and functional equations was proved for semistable E (that is, E such that there is no p where E has additive reduction) by Wiles and Taylor-Wiles, and in the general case by Breuil, Conrad, Diamond, and Taylor; see [Edi2] for an overview of this. In fact, the continuation and functional equation are direct consequences of the modularity of E that was proved by Wiles, Taylor-Wiles and so on (see below). The weak Birch and Swinnerton-Dyer conjecture says that the dimension of the $\mathbb{Q}$-vector space $\mathbb{Q} \otimes E(\mathbb{Q})$ is equal to the order of vanishing of L_E at 1. Anyway, the function L_E gives us integers a_n for all $n \geq 1$ as follows:

$$L_E(s) = \sum_{n \geq 1} a_n n^{-s} \quad \text{for } \Re(s) > 3/2.$$

From these a_n one can then consider the function

$$f_E : \mathbb{H} = \{z \in \mathbb{C} \mid \Im(z) > 0\} \to \mathbb{C}, \quad z \mapsto \sum_{n \geq 1} a_n e^{2\pi i n z}.$$

Equivalently, we have

$$f_E = \sum_{n \geq 1} a_n q^n, \quad \text{with} \quad q : \mathbb{H} \to \mathbb{C}, \quad z \mapsto e^{2\pi i z}.$$

A more conceptual way to state the relation between L_E and f_E is to say that L_E is obtained, up to elementary factors, as the *Mellin transform* of f_E:

$$\int_0^\infty f_E(it) t^s \frac{dt}{t} = (2\pi)^{-s} \Gamma(s) L_E(s) \quad \text{for } \Re(s) > 3/2.$$

After all these preparations, we can finally state what the modularity of E means:

> f_E is a modular form of weight two for the congruence sub-group $\Gamma_0(N_E)$ of $SL_2(\mathbb{Z})$.

For some more details on the concept of modular forms we refer to Section 2.2. At this moment, we just want to say that the last statement means that f_E has, as Mazur says in Singh's BBC documentary on Wiles's proof of Fermat's Last Theorem, an enormous amount of symmetry. This symmetry is with respect to the action of $GL_2(\mathbb{Q})^+$, the group of invertible 2-by-2 matrices with coefficients in $\mathbb{Q}$ whose determinant is positive, on the upper half-plane $\mathbb{H}$. This symmetry gives, by Mellin transformation, the functional equation of L_E. Conversely, it had been proved in [Wei1] by Weil that if sufficiently many twists of L_E by Dirichlet characters satisfy the conjectured holomorphic continuation and functional equation, then f_E is a modular form of the type mentioned.

We now remark that Schoof's algorithm implies that, for p prime, the coefficient a_p in the q-expansion of $f_E = \sum_{n \geq 1} a_n q^n$ can be computed in time polynomial in $\log p$. One of the aims of this book is to generalize this last fact to certain modular forms of higher weight. Before we give precise definitions in Section 2.2, we will discuss a typical case in the next section.

1.5 MORE HISTORICAL CONTEXT: CONGRUENCES FOR RAMANU-JAN'S τ-FUNCTION

References for this section are the articles [Ser2], [Swi], and [Del1] by Serre, Swinnerton-Dyer, and Deligne.

A typical example of a modular form of weight higher than two is the discriminant modular form, usually denoted Δ. One way to view Δ is as the holomorphic function on the upper half-plane $\mathbb{H}$ given by

$$(1.5.1) \qquad \Delta = q \prod_{n \geq 1} (1 - q^n)^{24},$$

where q is the function from $\mathbb{H}$ to $\mathbb{C}$ given by $z \mapsto \exp(2\pi i z)$. The coefficients in the power series expansion

$$(1.5.2) \qquad \Delta = \sum_{n \geq 1} \tau(n) q^n$$

define the famous *Ramanujan τ-function*.

To say that Δ is a modular form of weight 12 for the group $SL_2(\mathbb{Z})$ means that for all elements $\left(\begin{smallmatrix} a & b \\ c & d \end{smallmatrix}\right)$ of $SL_2(\mathbb{Z})$ the following identity holds for all z in $\mathbb{H}$:

$$(1.5.3) \qquad \Delta\left(\frac{az+b}{cz+d}\right) = (cz+d)^{12} \Delta(z),$$

which is equivalent to saying that the multidifferential form $\Delta(z)(dz)^{\otimes 6}$ is invariant under the action of $SL_2(\mathbb{Z})$. As $SL_2(\mathbb{Z})$ is generated by the elements $\left(\begin{smallmatrix} 1 & 1 \\ 0 & 1 \end{smallmatrix}\right)$ and $\left(\begin{smallmatrix} 0 & -1 \\ 1 & 0 \end{smallmatrix}\right)$, it suffices to check the identity in (1.5.3) for these two elements. The fact that Δ is q times a power series in q means that Δ is a *cusp form*: it vanishes at "$q = 0$". It is a fact that Δ is the first example of a nonzero cusp form for $SL_2(\mathbb{Z})$: there is no nonzero cusp form for $SL_2(\mathbb{Z})$ of weight smaller than 12, that is, there are no nonzero holomorphic functions on $\mathbb{H}$ satisfying (1.5.3) with the exponent 12 replaced by a smaller integer, whose Laurent series expansion in q is q times a power series. Moreover, the $\mathbb{C}$-vector space of such functions of weight 12 is one-dimensional, and hence Δ is a basis of it.

The one-dimensionality of this space has as a consequence that Δ is an eigenform for certain operators on this space, called *Hecke operators*, that arise from the action on $\mathbb{H}$ of $GL_2(\mathbb{Q})^+$, the subgroup of $GL_2(\mathbb{Q})$ of elements whose determinant is positive. This fact explains that the coefficients $\tau(n)$ satisfy certain relations that are summarized by the following identity of Dirichlet series (converging for $\Re(s) \gg 0$, for the moment, or just formal series, if one prefers that):

$$(1.5.4) \qquad L_\Delta(s) := \sum_{n \geq 1} \tau(n) n^{-s} = \prod_p (1 - \tau(p) p^{-s} + p^{11} p^{-2s})^{-1},$$

where the product is over all prime numbers. These relations

$$(1.5.5) \quad \begin{array}{ll} \tau(mn) = \tau(m)\tau(n) & \text{if } \gcd(m,n) = 1, \\ \tau(p^n) = \tau(p^{n-1})\tau(p) - p^{11}\tau(p^{n-2}) & \text{if } p \text{ is prime and } n \geq 2, \end{array}$$

were conjectured by Ramanujan, and proved by Mordell. Using these identities, $\tau(n)$ can be expressed in terms of the $\tau(p)$ for p dividing n.

As L_Δ is the Mellin transform of Δ, L_Δ is holomorphic on $\mathbb{C}$, and satisfies the functional equation (Hecke)

$$(2\pi)^{-(12-s)}\Gamma(12-s)L_\Delta(12-s) = (2\pi)^{-s}\Gamma(s)L_\Delta(s).$$

The famous *Ramanujan conjecture* states that for all primes p one has the inequality

$$(1.5.6) \qquad\qquad |\tau(p)| < 2p^{11/2},$$

or, equivalently, that the complex roots of the polynomial $x^2 - \tau(p)x + p^{11}$ are complex conjugates of each other, and hence are of absolute value $p^{11/2}$. This conjecture was proved by Deligne as a consequence of [Del1] and his proof of the analog of the Riemann hypothesis in the Weil conjectures in [Del2].

Finally, Ramanujan conjectured congruences for the integers $\tau(p)$ with p prime, modulo certain powers of certain small prime numbers. In order to state these congruences we define, for $n \geq 1$ and $r \geq 0$:

$$\sigma_r(n) := \sum_{1 \leq d \mid n} d^r;$$

that is, $\sigma_r(n)$ is the sum of the rth powers of the positive divisors of n. We will now list the congruences that are given in the first pages of [Swi]:

$$\tau(n) \equiv \sigma_{11}(n) \qquad\qquad \text{mod } 2^{11} \qquad \text{if } n \equiv 1 \quad \text{mod } 8,$$
$$\tau(n) \equiv 1217\sigma_{11}(n) \qquad \text{mod } 2^{13} \qquad \text{if } n \equiv 3 \quad \text{mod } 8,$$
$$\tau(n) \equiv 1537\sigma_{11}(n) \qquad \text{mod } 2^{12} \qquad \text{if } n \equiv 5 \quad \text{mod } 8,$$
$$\tau(n) \equiv 705\sigma_{11}(n) \qquad\ \text{mod } 2^{14} \qquad \text{if } n \equiv 7 \quad \text{mod } 8,$$

$$\tau(n) \equiv n^{-610}\sigma_{1231}(n) \qquad \text{mod } 3^6 \qquad \text{if } n \equiv 1 \quad \text{mod } 3,$$
$$\tau(n) \equiv n^{-610}\sigma_{1231}(n) \qquad \text{mod } 3^7 \qquad \text{if } n \equiv 2 \quad \text{mod } 3,$$

$$\tau(n) \equiv n^{-30}\sigma_{71}(n) \qquad\qquad \text{mod } 5^3 \qquad \text{if } n \text{ is prime to } 5,$$

$$\tau(n) \equiv n\sigma_9(n) \qquad \text{mod } 7 \qquad \text{if } n \equiv 0, 1, 2 \text{ or } 4 \quad \text{mod } 7,$$

$$\tau(n) \equiv n\sigma_9(n) \qquad \text{mod } 7^2 \qquad \text{if } n \equiv 3, 5 \text{ or } 6 \quad \text{mod } 7,$$

$$\tau(p) \equiv 0 \qquad \text{mod } 23 \quad \text{if } p \text{ is prime and not a square mod 23,}$$

$$\tau(p) \equiv 2 \qquad \text{mod } 23 \quad \text{if } p \neq 23 \text{ is a prime of the form } u^2 + 23v^2,$$

$$\tau(p) \equiv -1 \qquad \text{mod } 23 \quad \text{for other primes } p \neq 23,$$

$$\tau(n) \equiv \sigma_{11}(n) \quad \text{mod } 691.$$

The reader is referred to [Swi] for the origin and for proofs of these congruences. There, Swinnerton-Dyer remarks that the proofs do little to explain why such congruences occur. Serre conjectured an explanation in [Ser2]. First of all, Serre conjectured the existence, for each prime number l, of a continuous representation

$$(1.5.7) \qquad \rho_l \colon \mathrm{Gal}(\overline{\mathbb{Q}}/\mathbb{Q}) \longrightarrow \mathrm{Aut}(V_l),$$

with V_l a two-dimensional $\mathbb{Q}_l$-vector space, such that ρ_l is unramified at all primes $p \neq l$, and such that for all $p \neq l$ the characteristic polynomial of $\rho_l(\mathrm{Frob}_p)$ is given by

$$(1.5.8) \qquad \det(1 - x\mathrm{Frob}_p, V_l) = 1 - \tau(p)x + p^{11}x^2.$$

To help the reader, let us explain what unramified at p means, and what the Frobenius elements Frob_p are. For p prime, we let $\mathbb{Q}_p$ denote the topological field of p-adic numbers, and $\mathbb{Q}_p \to \overline{\mathbb{Q}}_p$ an algebraic closure. The action of $\mathrm{Gal}(\overline{\mathbb{Q}}/\mathbb{Q})$ on the set $\mathrm{Hom}(\overline{\mathbb{Q}}, \overline{\mathbb{Q}}_p)$ of embeddings of $\overline{\mathbb{Q}}$ into $\overline{\mathbb{Q}}_p$ is transitive, and each embedding induces an injection from $\mathrm{Gal}(\overline{\mathbb{Q}}_p/\mathbb{Q}_p)$ into $\mathrm{Gal}(\overline{\mathbb{Q}}/\mathbb{Q})$, the image of which is called a decomposition group of $\mathrm{Gal}(\overline{\mathbb{Q}}/\mathbb{Q})$ at p. The injections from $\mathrm{Gal}(\overline{\mathbb{Q}}_p/\mathbb{Q}_p)$ into $\mathrm{Gal}(\overline{\mathbb{Q}}/\mathbb{Q})$ and the corresponding decomposition groups at p obtained like this are all conjugated by the action of $\mathrm{Gal}(\overline{\mathbb{Q}}/\mathbb{Q})$. In order to go farther we need to say a bit about the structure of $\mathrm{Gal}(\overline{\mathbb{Q}}_p/\mathbb{Q}_p)$. We let $\mathbb{Q}_p^{\mathrm{unr}}$ be the maximal unramified extension of $\mathbb{Q}_p$ in $\overline{\mathbb{Q}}_p$, that is, the composite of all finite extensions K of $\mathbb{Q}_p$ in $\overline{\mathbb{Q}}_p$ such that p is a uniformizer for the integral closure O_K of $\mathbb{Z}_p$ in K. We let $\mathbb{Z}_p^{\mathrm{unr}}$ be the integral closure of $\mathbb{Z}_p$ in $\mathbb{Q}_p^{\mathrm{unr}}$; it is a local ring, and its residue field is an algebraic closure $\overline{\mathbb{F}}_p$ of $\mathbb{F}_p$. The subextension $\mathbb{Q}_p^{\mathrm{unr}}$ gives a short exact sequence:

$$(1.5.9) \qquad I_p \hookrightarrow \mathrm{Gal}(\overline{\mathbb{Q}}_p/\mathbb{Q}_p) \twoheadrightarrow \mathrm{Gal}(\overline{\mathbb{F}}_p/\mathbb{F}_p).$$

The subgroup I_p of $\mathrm{Gal}(\overline{\mathbb{Q}}_p/\mathbb{Q}_p)$ is called the inertia subgroup. The quotient $\mathrm{Gal}(\overline{\mathbb{F}}_p/\mathbb{F}_p)$ is canonically isomorphic to $\hat{\mathbb{Z}}$, the profinite completion of $\mathbb{Z}$, by demanding that the element 1 of $\hat{\mathbb{Z}}$ corresponds to the Frobenius element Frob_p of $\mathrm{Gal}(\overline{\mathbb{F}}_p/\mathbb{F}_p)$ that sends x to x^p for each x in $\overline{\mathbb{F}}_p$.

Let now ρ_l be a continuous representation from $\mathrm{Gal}(\overline{\mathbb{Q}}/\mathbb{Q})$ to $\mathrm{GL}(V_l)$ with V_l a finite dimensional $\mathbb{Q}_l$-vector space. Each embedding of $\overline{\mathbb{Q}}$ into $\overline{\mathbb{Q}}_p$ then gives a representation of $\mathrm{Gal}(\overline{\mathbb{Q}}_p/\mathbb{Q}_p)$ on V_l. Different embeddings give isomorphic representations because they are conjugated by an element in the image of $\mathrm{Gal}(\overline{\mathbb{Q}}/\mathbb{Q})$ under ρ_l. We now choose one embedding, and call the representation $\rho_{l,p}$ of $\mathrm{Gal}(\overline{\mathbb{Q}}_p/\mathbb{Q}_p)$ on V_l obtained like this the local representation at p attached to ρ_l. This being defined, ρ_l is then said to be unramified at a prime p if $\rho_{l,p}$ factors through the quotient $\mathrm{Gal}(\overline{\mathbb{Q}}_p/\mathbb{Q}_p) \to \mathrm{Gal}(\overline{\mathbb{F}}_p/\mathbb{F}_p)$, that is, if I_p acts trivially on V_l. If ρ_l is unramified at p, then we get an element $\rho_l(\mathrm{Frob}_p)$ in $\mathrm{GL}(V_l)$. This element depends on our chosen embedding of $\overline{\mathbb{Q}}$ into $\overline{\mathbb{Q}}_p$, but its conjugacy class under $\rho_l(\mathrm{Gal}(\overline{\mathbb{Q}}/\mathbb{Q}))$ does not. In particular, we get a well defined conjugacy class in $\mathrm{GL}(V_l)$, and so the characteristic polynomial of $\rho_l(\mathrm{Frob}_p)$ is now defined if ρ_l is unramified at p.

Continuous representations such as ρ_l can be reduced modulo powers of l as follows. The compactness of $\mathrm{Gal}(\overline{\mathbb{Q}}/\mathbb{Q})$ implies that with respect to a suitable basis of V_l the representation ρ_l lands in $\mathrm{GL}_2(\mathbb{Z}_l)$, and hence gives representations to $\mathrm{GL}_2(\mathbb{Z}/l^n\mathbb{Z})$ for all $n \geq 0$. This reduction of ρ_l modulo powers of l is not unique, but the semisimplification of the reduction modulo l is well defined, that is, two reductions lead to the same Jordan-Hölder constituents. According to Serre, the congruences above would then be explained by properties of the image of ρ_l.

For example, if the image of the reduction modulo l of ρ_l is reducible, say an extension of two characters α and β from $\mathrm{Gal}(\overline{\mathbb{Q}}/\mathbb{Q})$ to $\mathbb{F}_l^\times$, then one has the identity in $\mathbb{F}_l$, for all $p \neq l$,

$$(1.5.10) \qquad\qquad \tau(p) \equiv \alpha(\mathrm{Frob}_p) + \beta(\mathrm{Frob}_p).$$

The characters α and β are unramified outside l. By the Kronecker-Weber theorem, the maximal Abelian subextension of $\mathbb{Q} \to \overline{\mathbb{Q}}$ that is unramified outside l is the cyclotomic extension generated by all l-power roots of unity, with Galois group $\mathbb{Z}_l^\times$. It then follows that $\alpha = \chi_l^n$ and $\beta = \chi_l^m$ for suitable n and m, where χ_l is the character giving the action on the lth roots of unity in $\overline{\mathbb{Q}}$: for all σ in $\mathrm{Gal}(\overline{\mathbb{Q}}/\mathbb{Q})$ and for all ζ in $\overline{\mathbb{Q}}^\times$ with $\zeta^l = 1$ one has $\sigma(\zeta) = \zeta^{\chi_l(\sigma)}$. The identity (1.5.10) in $\mathbb{F}_l$ above then takes the form

$$(1.5.11) \qquad\qquad \tau(p) = p^n + p^m \quad \mathrm{mod}\ l, \quad \text{for all } p \neq l,$$

which indeed is of the same form as the congruences mod l for $\tau(p)$ listed above. For example, the congruence mod 691 corresponds to the statement that the reduction modulo l of ρ_l gives the two characters 1 and χ_l^{11}.

Deligne, in [Del1], proved the existence of the ρ_l, as conjectured by Serre, by showing that they occur in the degree one l-adic étale cohomology of certain sheaves on certain curves, and in the degree 11 étale cohomology with $\mathbb{Q}_l$-coefficients of a variety of dimension 11. This last variety is, loosely speaking, the 10-fold fibered product of the universal elliptic curve. Deligne's constructions will be discussed in detail in Sections 2.2 and 2.4. It should be said that Shimura had already shown how to construct Galois representations in the case of modular forms of weight two; in that case one does not need étale cohomology, because torsion points of Jacobians of modular curves suffice, see [Shi1]. Blasius proved in [Bla] that the ρ_l cannot be obtained from the cohomology of Abelian varieties, using tensor constructions and subquotients.

At this point we give the following precise and simple statement, relating Ramanujan's τ-function to point counting on an algebraic variety C_{10} (more precisely, a quasi-projective scheme over $\mathbb{Z}$), for which one easily writes down a system of equations. Moreover, the statement relates the weight of Δ to the classical question in geometry on cubic plane curves passing through a given set of points: up to 9 points the situation is easy and the count is given by a polynomial. See also [Ber], especially Section 15.

1.5.12 Proposition *For $n \in \mathbb{Z}_{\geq 0}$, q a prime power and $\mathbb{F}_q$ a finite field with q elements, let $C_n(\mathbb{F}_q)$ be the set of $(C, P_1, \ldots, P_n)$, where C is a smooth cubic in $\mathbb{P}^2_{\mathbb{F}_q}$ and $P_i \in C(\mathbb{F}_q)$. Then there are $f_0, \ldots, f_{10} \in \mathbb{Z}[x]$ such that for all $\mathbb{F}_q$ and $n \leq 9$ one has $\#C_n(\mathbb{F}_q)/\#\mathrm{PGL}_3(\mathbb{F}_q) = f_n(q)$, and for all prime numbers p,*

$$\#C_{10}(\mathbb{F}_p) \, / \, \#\mathrm{PGL}_3(\mathbb{F}_p) = -\tau(p) + f_{10}(p).$$

Proof For n in $\mathbb{Z}_{\geq 0}$ and $\mathbb{F}_q$ a field with q elements, let $\mathcal{E}_n(\mathbb{F}_q)$ denote the category, and also its set of objects, of $(E/\mathbb{F}_q, P_1, \ldots, P_n)$, where $E/\mathbb{F}_q$ is an elliptic curve and $P_i \in E(\mathbb{F}_q)$; the morphisms are the isomorphisms $\phi \colon E \to E'$ such that $\phi(P_i) = P_i'$. For each n, the category $\mathcal{E}_n(\mathbb{F}_q)$ has only finitely many objects up to isomorphism, and one defines

$$\#\mathcal{E}_n(\mathbb{F}_q) = \sum_{x \in \mathcal{E}_n(\mathbb{F}_q)} \frac{1}{\#\operatorname{Aut}(x)},$$

where, in the sum, one takes one x per isomorphism class. It is well known (see [Del1], [Beh]) that for $n \leq 9$ the functions $q \mapsto \#\mathcal{E}_n(\mathbb{F}_q)$ are given by certain elements f_n in $\mathbb{Z}[x]$, and that there is an f_{10} in $\mathbb{Z}[x]$ such that for all prime numbers p one has $\#\mathcal{E}_{10}(\mathbb{F}_p) = -\tau(p) + f_{10}(p)$. In view of this, the claims in Proposition 1.5.12 are a consequence of the following equality, for all $n \in \mathbb{Z}_{\geq 0}$ and all prime powers q:

(1.5.13) for all $n \in \mathbb{Z}_{\geq 0}$ and all $\mathbb{F}_q$: $\#C_n(\mathbb{F}_q) = \#\mathrm{PGL}_3(\mathbb{F}_q) \cdot \#\mathcal{E}_n(\mathbb{F}_q)$.

We prove (1.5.13) by comparing the subsets on both sides in which the underlying curves are fixed.

Let $n \in \mathbb{Z}_{\geq 0}$ and q a prime power. Let F be a nonsingular projective geometrically irreducible curve of genus one over $\mathbb{F}_q$, and let E_0 be its Jacobian. Then F is an E_0-torsor. By Lang's theorem, Theorem 2 of [Lan2], $F(\mathbb{F}_q)$ is not empty.

Let $C_n(\mathbb{F}_q)_F$ be the subset of $C_n(\mathbb{F}_q)$ consisting of the $(C, P_1, \ldots, P_n)$ with C isomorphic to F. The number of C in $\mathbb{P}^2_{\mathbb{F}_q}$ that are isomorphic to F is the number of embeddings $i \colon F \to \mathbb{P}^2_{\mathbb{F}_q}$, divided by $\#\operatorname{Aut}(F)$. Such embeddings are obtained from line bundles $\mathcal{L}$ of degree three on F, together with a basis, up to $\mathbb{F}_q^\times$, of $\mathcal{L}(F)$. Hence the number of embeddings is

$\#\mathrm{PGL}_3(\mathbb{F}_q) \cdot \#E_0(\mathbb{F}_q)$. The group $\mathrm{Aut}(F)$ has the subgroup of translations $E_0(\mathbb{F}_q)$, with quotient $\mathrm{Aut}(E_0)$. So we find

$$\#C_n(\mathbb{F}_q)_F = \#\mathrm{PGL}_3(\mathbb{F}_q) \cdot (\#E_0(\mathbb{F}_q))^n / \#\mathrm{Aut}(E_0).$$

On the other hand, let $\mathcal{E}_n(\mathbb{F}_q)_{E_0}$ be the full subcategory of $\mathcal{E}_n(\mathbb{F}_q)$ with objects the $(E_0, P_1, \ldots, P_n)$, with P_i in $E_0(\mathbb{F}_q)$. The group $\mathrm{Aut}(E_0)$ acts on the set of objects of $\mathcal{E}_n(\mathbb{F}_q)_{E_0}$, and this action *is* the set of morphisms in $\mathcal{E}_n(\mathbb{F}_q)_{E_0}$. This means that:

$$\#\mathcal{E}_n(\mathbb{F}_q)_{E_0} = (\#E_0(\mathbb{F}_q))^n / \#\mathrm{Aut}(E_0).$$

Summing over the isomorphism classes of F gives (1.5.13). $\qquad\square$

1.5.14 Remark For those who like more technical statements, we mention (without further explanation) that, for all $n \in \mathbb{Z}_{\geq 1}$, we have a Zariski PGL_3-torsor $C_n \to \mathcal{E}_n$, sending $(C \subset \mathbb{P}_S^2, P_1, \ldots, P_n)$ to $((C, P_0), P_1, \ldots, P_n)$, where $P_0 \in C(S)$ is determined uniquely by the condition that, locally on S, $\mathcal{O}_{\mathbb{P}^2}(1)|_C$ is isomorphic to $\mathcal{O}_C(P_0 + 2P_1)$.

1.5.15 Remark The polynomials f_n mentioned in Proposition 1.5.12 have been computed by Carel Faber and Gerard van der Geer. Their result is

$$
\begin{aligned}
f_0 &= x, \\
f_1 &= x^2 + x, \\
f_2 &= x^3 + 3x^2 + x - 1, \\
f_3 &= x^4 + 6x^3 + 6x^2 - 2x - 3, \\
f_4 &= x^5 + 10x^4 + 20x^3 + 4x^2 - 14x - 7, \\
f_5 &= x^6 + 15x^5 + 50x^4 + 40x^3 - 30x^2 - 49x - 15, \\
f_6 &= x^7 + 21x^6 + 105x^5 + 160x^4 - 183x^2 - 139x - 31,
\end{aligned}
$$

$$f_7 = x^8 + 28x^7 + 196x^6 + 469x^5 + 280x^4 - 427x^3 - 700x^2$$
$$- 356x - 63,$$
$$f_8 = x^9 + 36x^8 + 336x^7 + 1148x^6 + 1386x^5 - 406x^4 - 2436x^3$$
$$- 2224x^2 - 860x - 127,$$
$$f_9 = x^{10} + 45x^9 + 540x^8 + 2484x^7 + 4662x^6 + 1764x^5 - 6090x^4$$
$$- 9804x^3 - 6372x^2 - 2003x - 255,$$
$$f_{10} = x^{11} + 55x^{10} + 825x^9 + 4905x^8 + 12870x^7 + 12264x^6$$
$$- 9240x^5 - 33210x^4 - 33495x^3 - 17095x^2 - 4553x - 511.$$

We refer to Birch [Bir] for results on the distribution of the number of rational points on elliptic curves over finite fields, that also make $\tau(p)$ appear.

In [Swi], Swinnerton-Dyer gives results, partly resulting from his correspondence with Serre, in which the consequences of the existence of the ρ_l for congruences of $\tau(p)$ modulo l are explored. A natural question to ask is if there are primes l other than 2, 3, 5, 7, 23, and 691 modulo which there are similar congruences for $\tau(p)$.

For each $p \neq l$, $\tau(p)$ is the trace of $\rho_l(\mathrm{Frob}_p)$, and the determinant of $\rho_l(\mathrm{Frob}_p)$ equals p^{11}. Hence, a polynomial relation between $\tau(p)$ and p^{11}, valid modulo some l^n for all $p \neq l$, is a relation between the determinant and the trace of all $\rho_l(\mathrm{Frob}_p)$ in $\mathrm{GL}_2(\mathbb{Z}/l^n\mathbb{Z})$. But Chebotarev's theorem (see [Lan6] or [Ca-Fr], for example) implies that every element of the image of $\mathrm{Gal}(\overline{\mathbb{Q}}/\mathbb{Q})$ in $\mathrm{GL}_2(\mathbb{Z}/l^n\mathbb{Z})$ is of the form Frob_p for infinitely many p. Hence, such a polynomial relation is valid for all elements in the image of $\mathrm{Gal}(\overline{\mathbb{Q}}/\mathbb{Q})$ in $\mathrm{GL}_2(\mathbb{Z}/l^n\mathbb{Z})$. For this reason, the existence of nontrivial congruences modulo l^n as above for $\tau(p)$ depends on this image.

The image of $\mathrm{Gal}(\overline{\mathbb{Q}}/\mathbb{Q})$ in $\mathbb{Z}_l^\times$ under $\det \circ \rho_l$ is equal to the subgroup of 11th powers in $\mathbb{Z}_l^\times$. To explain this, we note that $\det \circ \rho_l$ is a continuous character from $\mathrm{Gal}(\overline{\mathbb{Q}}/\mathbb{Q})$ to $\mathbb{Z}_l^\times$, unramified outside l, and such that Frob_p is mapped to p^{11} for all $p \neq l$; this implies that $\det \circ \rho_l$ is the 11th power of the l-adic cyclotomic character $\chi_l \colon \mathrm{Gal}(\overline{\mathbb{Q}}/\mathbb{Q}) \to \mathbb{Z}_l^\times$, defined by $\sigma(z) = z^{\chi_l(\sigma)}$, for all σ in $\mathrm{Gal}(\overline{\mathbb{Q}}/\mathbb{Q})$ and all z in $\overline{\mathbb{Q}}^\times$ of l-power order.

In order to state the results in [Swi], one calls a prime number l *exceptional (for Δ)* if the image of ρ_l, taking values in $\mathrm{GL}_2(\mathbb{Z}_l)$, does *not* contain $\mathrm{SL}_2(\mathbb{Z}_l)$. For l not exceptional, that is, such that the image of ρ_l contains $SL_2(\mathbb{Z}_l)$, the image of $\text{Gal}(\overline{\mathbb{Q}}/\mathbb{Q})$ in $GL_2(\mathbb{Z}_l) \times \mathbb{Z}_l^\times$, under (ρ_l, χ_l), is the subgroup H of elements (g, t) such that $\det(g) = t^{11}$. This subgroup H maps surjectively to $\mathbb{F}_l \times \mathbb{F}_l^\times$ under $(g, t) \mapsto (\text{trace}(g), t)$, and therefore there can be no congruence for $\tau(p)$ modulo l as above.

The Corollary to Theorem 4 in [Swi] states, among other results, that the list of primes which are exceptional for Δ is $\{2, 3, 5, 7, 23, 691\}$. The main tool that is used that we have not discussed, is the theory of modular forms modulo l, or, equivalently, the theory of congruences modulo l between modular forms. As a consequence, there are no similar congruences for $\tau(p)$ modulo primes other than the ones listed above. The special form of the congruences modulo 23 is explained by the fact that in that case the image of $\text{Gal}(\overline{\mathbb{Q}}/\mathbb{Q})$ in $GL_2(\mathbb{F}_{23})$ is dihedral; in the other cases the residual representation, that is, the representation to $GL_2(\mathbb{F}_l)$, is reducible. In the case $l = 2$, Swinnerton-Dyer has determined the image of $\text{Gal}(\overline{\mathbb{Q}}/\mathbb{Q})$ in $GL_2(\mathbb{Z}_2)$ exactly: see the appendix in [Swi].

The direction in which we generalize Schoof's algorithm is to give an algorithm that computes for prime numbers l that are not exceptional for Δ the field extension $\mathbb{Q} \to K_l$ that corresponds to the representation of $\text{Gal}(\overline{\mathbb{Q}}/\mathbb{Q})$ to $GL_2(\mathbb{F}_l)$ that comes from Δ. The field K_l is given in the form $\mathbb{Q}[x]/(f_l)$. The computation has a running time that is polynomial in l. It is fair to say that this algorithm makes the mod l Galois representations attached to Δ accessible to computation, at least theoretically. As the field extensions that are involved are nonsolvable, this should be seen as a step beyond computational class field theory, and beyond the case of elliptic curves, in the direction to make the results of the Langlands program accessible to computations.

As a consequence, one can compute $\tau(p)$ mod l in time polynomial in $\log p$ and l, by reducing f_l as above mod p and some more computations that will be described later (see Section 15.1). By doing this for sufficiently many l, just as in Schoof's algorithm, one then gets an algorithm that computes $\tau(p)$ in time polynomial in $\log p$.

In Section 15.2 the method used here is generalized to the case of modular forms for $SL_2(\mathbb{Z})$ of arbitrary weight. The main result there is Theorem 15.2.1.

1.6 COMPARISON WITH p-ADIC METHODS

Before we start seriously with the theory of modular forms and the Galois
representations attached to them in the next chapter, we compare our generalization of Schoof's algorithm with the so-called p-adic methods that
have been developed since 2000 by Satoh [Sat], Kedlaya [Ked] (see also
[Edi3]), Hubrechts, [Hub], Lauder and Wan [La-Wa1], [La-Wa2], [Lau1]
and [Lau2], Fouquet, Gaudry, Gürel and Harley [Fo-Ga-Ha], [Ga-Gu], Denef and Vercauteren and Castryk [De-Ve], [Ca-De-Ve], Mestre, Lercier and
Lubicz [Le-Lu], Carls, Kohel and Lubicz, [Ca-Ko-Lu], [Ca-Lu], and Gerkmann, [Ger1] and [Ger2]. Actually, we should notice that such a method
was already introduced in [Ka-Lu] in 1982, but that this article seems to
have been forgotten (we thank Fre Vercauteren for having drawn our attention to this article). An overview of the p-adic approach is given in [Ch].

   In all these methods, one works with fields of small characteristic p,
hence of the form $\mathbb{F}_q$ with $q = p^m$ and p fixed. All articles cited in the
previous paragraph have the common property that they give algorithms for
computing the number of $\mathbb{F}_q$-rational points on certain varieties X over $\mathbb{F}_q$,
using, sometimes indirectly, cohomology groups with p-adic coefficients,
whence the terminology "p-adic methods."

   For example, Satoh [Sat] uses the canonical lift of ordinary elliptic curves
and the action of the lifted Frobenius endomorphism on the tangent space,
which can be interpreted in terms of the algebraic de Rham cohomology of
the lifted curve. Kedlaya [Ked] uses Monsky-Washnitzer cohomology of
certain affine pieces of hyperelliptic curves. In fact, all cohomology groups
used here are de Rham type cohomology groups, given by complexes of
differential forms on certain p-adic lifts of the varieties in question. Just as
an example, let us mention that Kedlaya [Ked] gives an algorithm that for
fixed $p \neq 2$ computes the zeta functions of hyperelliptic curves given by
the equations

$$y^2 = f(x),$$

where f has arbitrary degree, in time $m^3 \deg(f)^4$. The running times of
the other algorithms are similar, but all have in common that the running
time grows at least linearly in p (or linear in $O(p^{1/2})$, in [Harv]), hence

exponentially in log p. The explanation for this is that somehow in each case nonsparse polynomials of degree at least linear in p have to be manipulated.

Summarizing this recent progress, one can say that, at least from a theoretical point of view, the problem of counting the solutions of systems of polynomial equations over finite fields of a fixed characteristic p and in a fixed number of variables has been solved. If p is not bounded, then almost nothing is known about the existence of polynomial time algorithms.

A very important difference between this book, using étale cohomology with coefficients in $\mathbb{F}_l$, and the p-adic methods, is that the Galois representations on $\mathbb{F}_l$-vector spaces that we obtain are *global* in the sense that they are representations of the absolute Galois group of the global field $\mathbb{Q}$. The field extensions such as the $K_l = \mathbb{Q}[x]/(f_l)$ arising from Δ discussed in the previous section have the advantage that one can choose to do the required computations over the complex numbers, approximating f_l, or p-adically at some suitable prime p, or in $\mathbb{F}_p$ for sufficiently many small p. Also, as we have said already, being able to compute such field extensions K_l, that give mod l information on the Frobenius elements at all primes $p \neq l$, is very interesting. On the other hand, the p-adic methods force one to compute with p-adic numbers, or, actually, modulo some sufficiently high power of p, and it gives information only on the Frobenius at p. The main drawback of the étale cohomology with $\mathbb{F}_l$-coefficients seems to be that the degree of the field extensions as K_l to be dealt with grows exponentially in the dimension of the cohomology groups; for that reason, we do not know how to use étale cohomology to compute $\#X(\mathbb{F}_q)$ for X a curve of arbitrary genus in a time polynomial in log q and the genus of X. Nevertheless, for modular curves, see the end of Section 15.3.

Chapter Two

Modular curves, modular forms, lattices, Galois representations

by Bas Edixhoven

2.1 MODULAR CURVES

As a good reference for getting an overview of the theory of modular curves and modular forms we recommend the article [Di-Im] by Fred Diamond and John Im. This reference is quite complete as results are concerned, and gives good references for the proofs of those results. Moreover, it is one of the few references that treats the various approaches to the theory of modular forms, from the classical analytic theory on the upper half-plane to the more modern representation theory of adelic groups. Another good first introduction is the book [Di-Sh]. Let us also mention the book [Conr] by Brian Conrad, and also the information in the *wikipedia* is getting more and more detailed.

In this section our aim is just to give the necessary definitions and results for what we need later (and we need at least to fix our notation). Readers who want more details, or more conceptual explanations are encouraged to consult [Di-Im].

2.1.1 Definition For n an integer greater than or equal to one we let $\Gamma(n)$ be the kernel of the surjective morphism of groups $\mathrm{SL}_2(\mathbb{Z}) \to \mathrm{SL}_2(\mathbb{Z}/n\mathbb{Z})$ given by reduction of the coefficients modulo n, and we let $\Gamma_1(n)$ be the inverse image of the subgroup of $\mathrm{SL}_2(\mathbb{Z}/n\mathbb{Z})$ that fixes the element $(1, 0)$ of $(\mathbb{Z}/n\mathbb{Z})^2$. Similarly, we let $\Gamma_0(n)$ be the inverse image of the subgroup of $SL_2(\mathbb{Z}/n\mathbb{Z})$ that fixes the subgroup $\mathbb{Z}/n\mathbb{Z}\cdot(1,0)$ of $(\mathbb{Z}/n\mathbb{Z})^2$. Hence the elements of $\Gamma_0(n)$ are the $\left(\begin{smallmatrix} a & b \\ c & d \end{smallmatrix}\right)$ of $SL_2(\mathbb{Z})$ such that $c \equiv 0 \mod n$, those of $\Gamma_1(n)$ are the ones that satisfy the extra conditions $a \equiv 1 \mod n$ and $d \equiv 1 \mod n$, and those of $\Gamma(n)$ are the ones that satisfy the extra condition $b \equiv 0 \mod n$.

The group $SL_2(\mathbb{R})$ acts on the upper half-plane $\mathbb{H}$ by fractional linear transformations:

$$\begin{pmatrix} a & b \\ c & d \end{pmatrix} \cdot z = \frac{az+b}{cz+d}.$$

The subgroup $SL_2(\mathbb{Z})$ of $SL_2(\mathbb{R})$ acts discontinuously in the sense that for each z in $\mathbb{H}$ the stabilizer $SL_2(\mathbb{Z})_z$ is finite and there is an open neighborhood U of z such that each translate γU with γ in $SL_2(\mathbb{Z})$ contains exactly one element of the orbit $SL_2(\mathbb{Z})\cdot z$ and any two translates γU and $\gamma' U$ with γ and γ' in $SL_2(\mathbb{Z})$ are either equal or disjoint. This property implies that the quotient $SL_2(\mathbb{Z})\backslash\mathbb{H}$, equipped with the quotient topology and with, on each open subset U, the $SL_2(\mathbb{Z})$-invariant holomorphic functions on the inverse image of U, is a complex analytic manifold of dimension one, that is, each point of the quotient has an open neighborhood that is isomorphic to the complex unit disk. Globally, the well-known j-function from $\mathbb{H}$ to $\mathbb{C}$ is in fact the quotient map for this action. One way to see this is to associate to each z in $\mathbb{H}$ the elliptic curve $E_z := \mathbb{C}/(\mathbb{Z} + \mathbb{Z}z)$, and to note that for z and z' in $\mathbb{H}$ the elliptic curves E_z and $E_{z'}$ are isomorphic if and only if z and z' are in the same $SL_2(\mathbb{Z})$-orbit, and to use the fact that two complex elliptic curves are isomorphic if and only if their j-invariants are equal.

The quotient set $\Gamma(n)\backslash\mathbb{H}$ can be identified with the set of isomorphism classes of pairs (E, ϕ), where E is a complex elliptic curve and where $\phi: (\mathbb{Z}/n\mathbb{Z})^2 \to E[n]$ is an isomorphism of groups, compatible with the Weil pairing $E[n] \times E[n] \to \mu_n(\mathbb{C})$ and the $\mu_n(\mathbb{C})$-valued pairing on $(\mathbb{Z}/n\mathbb{Z})^2$ that sends $((a_1, a_2), (b_1, b_2))$ to $\zeta_n^{a_1 b_2 - a_2 b_1}$, where $\zeta_n = e^{2\pi i/n}$.

The quotient $\Gamma_0(n)\backslash\mathbb{H}$ is then identified with the set of pairs (E, G), where E is a complex elliptic curve and $G \subset E$ a subgroup that is isomorphic to $\mathbb{Z}/n\mathbb{Z}$. Equivalently, we may view $\Gamma_0(n)\backslash\mathbb{H}$ as the set of isomorphism classes of $E_1 \xrightarrow{\phi} E_2$, where ϕ is a morphism of complex elliptic curves, and $\ker(\phi)$ is isomorphic to $\mathbb{Z}/n\mathbb{Z}$.

Finally, the quotient $\Gamma_1(n)\backslash\mathbb{H}$ is then identified with the set of pairs (E, P), where E is a complex elliptic curve and P is a point of order n of E. Explicitly, to each z in $\mathbb{H}$ corresponds the pair $(\mathbb{C}/(\mathbb{Z}z + \mathbb{Z}), [1/n])$, where $[1/n]$ denotes the image of $1/n$ in $\mathbb{C}/(\mathbb{Z}z + \mathbb{Z})$.

In order to understand that the quotients considered above are in fact the complex analytic varieties associated with affine complex algebraic curves, it is necessary (and sufficient!) to show that these quotients can be compactified to compact Riemann surfaces by adding a finite number of points, called *the cusps*. As the quotient by $\mathrm{SL}_2(\mathbb{Z})$ is given by $j: \mathbb{H} \to \mathbb{C}$, it can be compactified easily by embedding $\mathbb{C}$ into $\mathbb{P}^1(\mathbb{C})$; the point ∞ of $\mathbb{P}^1(\mathbb{C})$ is called the cusp. Another way to view this is to note that the equivalence relation on $\mathbb{H}$ given by the action of $\mathrm{SL}_2(\mathbb{Z})$ identifies two elements z and z' with $\Im(z) > 1$ and $\Im(z') > 1$ if and only if $z' = z + n$ for some n in $\mathbb{Z}$; this follows from the identity, for all $\left(\begin{smallmatrix} a & b \\ c & d \end{smallmatrix}\right)$ in $\mathrm{SL}_2(\mathbb{R})$ and z in $\mathbb{H}$,

$$(2.1.2) \qquad \Im\left(\frac{az + b}{cz + d}\right) = \frac{\Im(z)}{|cz + d|^2}.$$

Indeed, if moreover $c \neq 0$, then

$$(2.1.3) \qquad \frac{\Im(z)}{|cz + d|^2} \leq \frac{\Im(z)}{(\Im(cz))^2} = \frac{1}{c^2 \Im(z)}.$$

Hence on the part "$\Im(z) > 1$" of $\mathbb{H}$ the equivalence relation given by $\mathrm{SL}_2(\mathbb{Z})$ is given by the action of $\mathbb{Z}$ by translation. As the quotient for that action is given by the map $z \mapsto \exp(2\pi i z)$ onto $D(0, e^{-2\pi})^*$, the open disk of radius $e^{-2\pi}$, centered at 0, and with 0 removed, we get an open immersion of $D(0, e^{-2\pi})^*$ into $\mathrm{SL}_2(\mathbb{Z})\backslash\mathbb{H}$. The compactification is then obtained by replacing $D(0, e^{-2\pi})^*$ with $D(0, e^{-2\pi})$, that is, by adding the center back into the punctured disk.

Let us now consider the problem of compactifying the other quotients above. Let Γ be one of the groups considered above, or, in fact, any subgroup of finite index in $\mathrm{SL}_2(\mathbb{Z})$. We consider our compactification $\mathbb{P}^1(\mathbb{C})$ of $\mathbb{C}$ and the natural morphism $f: \Gamma\backslash\mathbb{H} \to \mathrm{SL}_2(\mathbb{Z})\backslash\mathbb{H} = \mathbb{C}$. By construction, f is proper (that is, the inverse image of a compact subset of $\mathbb{C}$ is compact). Also, we know that ramification can only occur at points with j-invariant 0 or 1728. Let D^* be the punctured disk described above. Then $f: f^{-1}D^* \to D^*$ is an unramified covering of degree $\#(\mathrm{SL}_2(\mathbb{Z})/\Gamma)$ if Γ

does not contain -1, and of degree $\#(\mathrm{SL}_2(\mathbb{Z})/\Gamma)/2$ if -1 is in Γ. Up to isomorphism, the only connected unramified covering of degree n, with $n \geq 1$, of D^* is the map $D_n^* \to D^*$, with $D_n^* = \{z \in \mathbb{C} \mid 0 < |z| < e^{-2\pi/n}\}$, sending $z \mapsto z^n$. It follows that $f^{-1}D^*$ is, as a covering of D^*, a disjoint union of copies of such $D_n^* \to D^*$. Each D_n^* has the natural compactification $D_n := \{z \in \mathbb{C} \mid |z| < e^{-2\pi/n}\}$. We compactify $\Gamma \backslash \mathbb{H}$ by adding the origin to each punctured disk in $f^{-1}D^*$. The points we have added are called the cusps. By construction, the morphism $f: \Gamma \backslash \mathbb{H} \to \mathrm{SL}_2(\mathbb{Z}) \backslash \mathbb{H}$ extends to the compactifications. It is a fact that a compact Riemann surface can be embedded into some projective space, using the theorem of Riemann-Roch, and that the image of such an embedding is a complex algebraic curve. This means that our quotients are, canonically, the Riemann surfaces associated with smooth complex algebraic curves.

2.1.4 Definition For $n \in \mathbb{Z}_{\geq 1}$ we define $X(n)$, $X_1(n)$ and $X_0(n)$ to be the proper smooth complex algebraic curves obtained via the compactifications of $\Gamma(n) \backslash \mathbb{H}$, $\Gamma_1(n) \backslash \mathbb{H}$, and $\Gamma_0(n) \backslash \mathbb{H}$, respectively. The affine parts obtained by removing the cusps are denoted $Y(n)$, $Y_1(n)$ and $Y_0(n)$.

The next step in the theory is to show that these complex algebraic curves are naturally defined over certain number fields. Let us start with the $X_0(n)$ and $X_1(n)$, which are defined over $\mathbb{Q}$. A simple way to produce a model of $X_0(n)$ over $\mathbb{Q}$, that is, an algebraic curve $X_0(n)_{\mathbb{Q}}$ over $\mathbb{Q}$ that gives $X_0(n)$ via extension of scalars via $\mathbb{Q} \to \mathbb{C}$, is to use the map:

$$(j, j'): \mathbb{H} \longrightarrow \mathbb{C} \times \mathbb{C}, \quad z \mapsto (j(z), j(nz)).$$

This map factors through the action of $\Gamma_0(n)$, and induces a map from $X_0(n)$ to $\mathbb{P}^1 \times \mathbb{P}^1$ that is birational to its image. This image is a curve in $\mathbb{P}^1 \times \mathbb{P}^1$, hence the zero locus of a bihomogeneous polynomial often denoted Φ_n, the minimal polynomial of j' over $\mathbb{C}(j)$. One can then check, using some properties of the j-function, that Φ_n has integer coefficients. The normalization of the curve in $\mathbb{P}^1_{\mathbb{Q}} \times \mathbb{P}^1_{\mathbb{Q}}$ defined by Φ_n is then the desired curve $X_0(n)_{\mathbb{Q}}$. As Φ_n has coefficients in $\mathbb{Z}$, it even defines a curve in $\mathbb{P}^1_{\mathbb{Z}} \times \mathbb{P}^1_{\mathbb{Z}}$ (here, one has to work with schemes), whose normalization $X_0(n)_{\mathbb{Z}}$ can be characterized as a so-called *coarse moduli space*. For this notion, and for the necessary proofs, the reader is referred to [Di-Im], Section II.8, to [De-Ra], and to [Ka-Ma].

One consequence of this statement is that for any algebraically closed field k in which n is invertible, the k-points of $Y_0(n)_{\mathbb{Z}}$ (the complement of the cusps) correspond bijectively to isomorphism classes of $E_1 \overset{\phi}{\to} E_2$ where ϕ is a morphism of elliptic curves over k of which the kernel is cyclic of order n.

The notion of moduli space also gives natural models over $\mathbb{Z}[1/n]$ of $X_1(n)$ and $Y_1(n)$. For $n \geq 4$ the defining property of $Y_1(n)_{\mathbb{Z}[1/n]}$ is not hard to state. There is an elliptic curve $\mathbb{E}$ over $Y_1(n)_{\mathbb{Z}[1/n]}$ with a point $\mathbb{P}$ in $E(Y_1(n)_{\mathbb{Z}[1/n]})$ that has order n in every fiber, such that *any* pair $(E/S, P)$ with S a $\mathbb{Z}[1/n]$-scheme and P in $E(S)$ of order n in all fibers arises by a *unique* base change

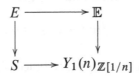

that is compatible with the sections P and $\mathbb{P}$. The pair $(\mathbb{E}/Y_1(n)_{\mathbb{Z}[1/n]}, \mathbb{P})$ is therefore called *universal*.

The moduli interpretation of $X(n)$ is a bit more complicated, because of the occurrence of the Weil pairing on $E[n]$ that we have seen above. The curve $X(n)$ has a natural model $X(n)_{\mathbb{Z}[1/n, \zeta_n]}$ over $\mathbb{Z}[1/n, \zeta_n]$. For $n \in \mathbb{Z}_{\geq 3}$, the complement of the cusps $Y(n)_{\mathbb{Z}[1/n, \zeta_n]}$ then has an elliptic curve $\mathbb{E}$ over it, and an isomorphism ϕ between the constant group scheme $(\mathbb{Z}/n\mathbb{Z})^2$ and $\mathbb{E}[n]$ that respects the pairings on each side. The pair $(\mathbb{E}/Y(n)_{\mathbb{Z}[1/n, \zeta_n]}, \phi)$ is universal in the same sense as above. We warn the reader that the notation $X(n)$ is also used sometimes for the moduli scheme for pairs (E, ϕ), where ϕ does not necessarily respect the pairings on the two sides.

For n and m in $\mathbb{Z}_{>1}$ that are relatively prime we will sometimes view $Y_1(nm)_{\mathbb{Z}[1/nm]}$ as the moduli space of triples $(E/S, P_n, P_m)$, where S is a scheme over $\mathbb{Z}[1/nm]$, E/S an elliptic curve, P_n and P_m in $E(S)$ that are everywhere (that is, in every geometric fiber of E/S) of orders n and m, respectively. Indeed, for such a triple, $P_n + P_m$ is everywhere of order nm, and the inverse construction starting with a point P_{nm} that is everywhere of order nm is given by multiplying with the two idempotents of $\mathbb{Z}/nm\mathbb{Z}$

corresponding to the isomorphism of rings $\mathbb{Z}/nm\mathbb{Z} \to \mathbb{Z}/n\mathbb{Z} \times \mathbb{Z}/m\mathbb{Z}$.

2.2 MODULAR FORMS

Let us now turn our attention to modular forms. It will be enough for us to work with modular forms for the congruence subgroups $\Gamma_1(n)$. Therefore, we restrict ourselves to that case.

2.2.1 Definition Let $n \geq 1$ and k an integer. A (holomorphic) *modular form* for $\Gamma_1(n)$ is a holomorphic function $f : \mathbb{H} \to \mathbb{C}$ that satisfies the following properties:

1. for all $\left(\begin{smallmatrix} a & b \\ c & d \end{smallmatrix}\right) \in \Gamma_1(n)$ and for all $z \in \mathbb{H}$,

$$f((az+b)/(cz+d)) = (cz+d)^k f(z);$$

2. f is holomorphic at the cusps (see below for an explanation).

A modular form is called a *cusp form* if it vanishes at the cusps.

We still need to explain the condition that f is holomorphic at the cusps. In order to do that, we first explain what this means at the cusp ∞. That cusp is the point that was added to the punctured disk obtained by taking the quotient of $\mathbb{H}$ by the unipotent subgroup $\left(\begin{smallmatrix} 1 & * \\ 0 & 1 \end{smallmatrix}\right)$, which acts on $\mathbb{H}$ by translations by integers. The coordinate of that disk is q, the map that sends z to $\exp(2\pi i z)$. Therefore, f has a Laurent series expansion in q,

$$(2.2.2) \qquad f = \sum_{n \in \mathbb{Z}} a_n(f) q^n, \quad \text{called the } q\text{-expansion at } \infty.$$

With this notation, f is called holomorphic at ∞ if $a_n(f)$ is zero for all $n < 0$, and f is said to vanish at ∞ if $a_n(f)$ is zero for all $n \leq 0$.

To state this condition at the other cusps, we need some description of the set of cusps. First, we note that $\mathbb{P}^1(\mathbb{C}) - \mathbb{P}^1(\mathbb{R})$ is the same as $\mathbb{C} - \mathbb{R}$, and therefore the disjoint union of $\mathbb{H}$ and its complex conjugate (which explains, by the way, that $\mathrm{GL}_2(\mathbb{R})^+$ acts by fractional linear transformations on $\mathbb{H}$). We can then consider $\mathbb{H} \cup \mathbb{P}^1(\mathbb{Q})$ inside $\mathbb{P}^1(\mathbb{C})$, with the $\mathrm{SL}_2(\mathbb{Z})$ action on it. Then the subgroup $\left(\begin{smallmatrix} 1 & * \\ 0 & 1 \end{smallmatrix}\right)$ stabilizes the point $\infty = (1:0)$ of $\mathbb{P}^1(\mathbb{Q})$, and ∞ can be naturally identified with the origin that we added to the disk D^*

above, because ∞ is the unique element of $\mathbb{P}^1(\mathbb{Q})$ that lies in the closure in $\mathbb{P}^1(\mathbb{C})$ of the inverse image "$\Im(z) > 1$" of D^* in $\mathbb{H}$. Then, the images of the region "$\Im(z) > 1$" under the action of elements of $\mathrm{SL}_2(\mathbb{Z})$ correspond bijectively to the elements of $\mathbb{P}^1(\mathbb{Q})$ (note that $\mathrm{SL}_2(\mathbb{Z})$ acts transitively on $\mathbb{P}^1(\mathbb{Q}) = \mathbb{P}^1(\mathbb{Z})$), and also to the maximal unipotent subgroups of $\mathrm{SL}_2(\mathbb{Z})$ (that is, the subgroups that consist of elements whose eigenvalues are 1). It follows that we can identify the set of cusps of $X_1(n)$ with $\Gamma_1(n)\backslash\mathbb{P}^1(\mathbb{Q})$, and that the images of the region "$\Im(z) > 1$" under $\mathrm{SL}_2(\mathbb{Z})$ give us punctured disks around the other cusps. Let $\gamma = \left(\begin{smallmatrix} a & b \\ c & d \end{smallmatrix}\right)$ be an element of $\mathrm{SL}_2(\mathbb{Z})$. The conditions of holomorphy and vanishing at the cusp $\gamma\infty = (a : c)$ are then given in terms of the q-expansion of $z \mapsto (cz + d)^{-k} f(\gamma z)$ at ∞. The group $\gamma^{-1}\Gamma_1(n)\gamma$ contains the group $\left(\begin{smallmatrix} 1 & n* \\ 0 & 1 \end{smallmatrix}\right)$ (indeed, $\Gamma_1(n)$ contains $\Gamma(n)$ and that one is normal in $\mathrm{SL}_2(\mathbb{Z})$). Therefore, putting $q_n : \mathbb{H} \to \mathbb{C}$, $z \mapsto \exp(2\pi i z/n)$, the function $z \mapsto (cz + d)^{-k} f(\gamma z)$ has a Laurent series expansion in q_n, and one asks that this Laurent series is a power series (for holomorphy) or a power series with constant term zero (for vanishing).

We let $M_k(\Gamma_1(n))$ denote the space of modular forms of weight k on $\Gamma_1(n)$, and the subspace of cusp forms by $S_k(\Gamma_1(n))$. We define $M(\Gamma_1(n))$ to be the direct sum over the k in $\mathbb{Z}$ of the $M_k(\Gamma_1(n))$; it is a $\mathbb{Z}$-graded commutative $\mathbb{C}$-algebra under pointwise multiplication.

2.2.3 Example Some simple examples of modular forms for $\mathrm{SL}_2(\mathbb{Z})$ are given by Eisenstein series. For each even $k \geq 4$ one has the function E_k:

$$E_k : \mathbb{H} \to \mathbb{C}, \quad z \mapsto \frac{1}{2\zeta(k)} \sum_{\substack{(n,m)\in\mathbb{Z}^2 \\ (n,m)\neq(0,0)}} \frac{1}{(n + mz)^k}.$$

The q-expansions of these E_k are given by

$$E_k = 1 - \frac{2k}{B_k} \sum_{n\geq 1} \sigma_{k-1}(n)q^n,$$

where the B_k are the Bernoulli numbers defined by

$$\frac{te^t}{e^t - 1} = \sum_{k\geq 0} B_k \frac{t^k}{k!},$$

and where, as before, $\sigma_r(n)$ denotes the sum of the rth powers of the positive divisors of n. In particular, one has the formulas

$$E_4 = 1 + 240 \sum_{n \geq 1} \sigma_3(n) q^n, \quad E_6 = 1 - 504 \sum_{n \geq 1} \sigma_5(n) q^n,$$

$$\Delta = \frac{E_4^3 - E_6^2}{1728}.$$

2.2.4 Remark We note that, from a computational point of view, the coefficients of q^p with p prime of the E_k are very easy to compute, namely, up to a constant factor they are $1 + p^{k-1}$, but that computing the $\sigma_{k-1}(n)$ for composite n seems equivalent to factoring n, at least when the number of prime factors of n is bounded. This is a strong indication that, for computing coefficients $a_n(f)$ of a modular form f, there is a real difference between the case where n is prime and the case where n is composite.

Indeed, Denis Charles and Eric Bach have shown that for $n = pq$ a product of two distinct primes such that $\tau(n) \neq 0$, one can compute p and q from n, $\tau(n)$, and $\tau(n^2)$ in time polynomial in $\log n$; see [Ba-Ch]. Their argument is very simple: one uses the identities in (1.5.5) to compute the rational number $\tau(p)^2/q^{11}$ and notes that the denominator is of the form q^r with r odd. Lehmer has conjectured that for all $n \in \mathbb{Z}_{\geq 1}$, $\tau(n) \neq 0$. See Corollary 7.1.4.

The Eisenstein series E_4 and E_6 generate the $\mathbb{C}$-algebra $M(\mathrm{SL}_2(\mathbb{Z}))$, and are algebraically independent:

$$M(\mathrm{SL}_2(\mathbb{Z})) = \mathbb{C}[E_4, E_6].$$

In particular, we have:

$$\dim_{\mathbb{C}} M_k(\mathrm{SL}_2(\mathbb{Z})) = \#\{(a, b) \in \mathbb{Z}_{\geq 0}^2 \mid 4a + 6b = k\}.$$

The space $S_k(\Gamma_1(n))$ can be interpreted as the space of sections of some holomorphic line bundle $\underline{\omega}^{\otimes k}(-\mathrm{Cusps})$ on $X_1(n)$, if $n \geq 5$ (for $n < 4$ the action of $\Gamma_1(n)$ on $\mathbb{H}$ is not free, and for $n = 4$ there is a cusp whose stabilizer is not unipotent):

$$(2.2.5) \qquad S_k(\Gamma_1(n)) = H^0(X_1(n), \underline{\omega}^{\otimes k}(-\mathrm{Cusps})) \quad \text{if } n \geq 5.$$

This implies that the spaces $S_k(\Gamma_1(n))$ are finite dimensional, and in fact zero if $k \leq 0$ because the line bundle in question then has negative degree. The restriction to $Y_1(n)$ of the line bundle giving the weight k forms is given by dividing out the action of $\Gamma_1(n)$ on $\mathbb{C} \times \mathbb{H}$ given by

$$(2.2.6) \qquad \begin{pmatrix} a & b \\ c & d \end{pmatrix} : (x, z) \mapsto \left((cz + d)^k x, \frac{az + b}{cz + d} \right).$$

The extension of this line bundle over the cusps is then given by decreeing that, at the cusp ∞, the constant section 1 (which is indeed invariant under the translations $z \mapsto z + n$) is a generator for the bundle of holomorphic forms, and q times 1 is a generator for the bundle of cusp forms.

The moduli interpretation for $Y_1(n)$ can be extended to the holomorphic line bundles giving the modular forms as follows. Recall that a point on $Y_1(n)$ is an isomorphism class of a pair (E, P) with E a complex elliptic curve and P a point of order n of E. The complex line at (E, P) of the bundle of forms of weight k is then $\underline{\omega}_E^{\otimes k}$, the kth tensor power of the dual of the tangent space at 0 of E. In this way, a modular form f of weight k for $\Gamma_1(n)$ can be described as follows: it is a function that assigns to each (E, P) an element $f(E, P)$ of $\underline{\omega}_E^{\otimes k}$, varying holomorphically with (E, P), and such that it has the right property at the cusps (being holomorphic or vanishing). The function f has to be compatible with isomorphisms: if $\phi : E \to E'$ is an isomorphism, and $\phi(P) = P'$, then $f(E, P)$ has to be equal to $(\phi^*)^{\otimes k} f(E', P')$. In what follows we will simply write ϕ^* for $(\phi^*)^{\otimes k}$. In particular, for $\phi = -1$, we have $f(E, -P) = (-1)^k f(E, P)$.

The fact that f should be holomorphic can be stated by evaluating it on the family of elliptic curves that we have over $\mathbb{H}$. Recall that to z in $\mathbb{H}$ we attached the pair $(\mathbb{C}/(\mathbb{Z}z + \mathbb{Z}), [1/n])$. Let us denote by x the coordinate of $\mathbb{C}$; then dx is a generator of the cotangent space at 0 of this elliptic curve. Then for f a function as above, we can write:

$$(2.2.7) \qquad f((\mathbb{C}/(\mathbb{Z}z + \mathbb{Z}), [1/n])) = F_f(z) \cdot (dx)^{\otimes k}, \qquad F_f : \mathbb{H} \to \mathbb{C}.$$

The function F_f is then required to be holomorphic. The requirement that f is compatible with isomorphisms means precisely that F_f transforms under $\Gamma_1(n)$ as in Definition 2.2.1. The requirement that f vanishes at the cusps is equivalent to the statement that the Laurent expansions in

$$q^{1/n} : z \mapsto \exp(2\pi i z / n)$$

obtained by evaluating f on all pairs $(\mathbb{C}/(\mathbb{Z}z + \mathbb{Z}), (az + b)/n)$, with a and b in $\mathbb{Z}$ such that $(az + b)/n$ is of order n, are in fact power series with constant term zero.

The spaces $S_k(\Gamma_1(n))$ are equipped with certain operators, called Hecke operators and diamond operators. These operators arise from the fact that for every element γ of $\mathrm{GL}_2(\mathbb{Q})^+$ the subgroups $\Gamma_1(n)$ and $\gamma\Gamma_1(n)\gamma^{-1}$ are commensurable, that is, their intersection has finite index in each of them. The diamond operators are then the simplest to describe. For each a in $(\mathbb{Z}/n\mathbb{Z})^\times$, $Y_1(n)$ has the automorphism $\langle a \rangle$ given by the property that it sends (E, P) to (E, aP). This action is then extended on modular forms by

$$(2.2.8) \qquad\qquad (\langle a \rangle f)(E, P) = f(E, aP).$$

For $a = -1$ and f in $S_k(\Gamma_1(n))$ we have

$$(2.2.9) \qquad (\langle -1 \rangle f)(E, P) = f(E, -P) = (-1)^k f(E, P),$$

hence $\langle -1 \rangle f = (-1)^k f$.

Similarly, there are Hecke operators T_m on $S_k(\Gamma_1(n))$ for all integers $m \geq 1$, defined by

$$(2.2.10) \qquad (T_m f)(E, P) = \frac{1}{m} \sum_\phi \phi^* f(E_\phi, \phi(P)),$$

where the sum runs over all quotients $\phi \colon E \to E_\phi$ of degree m such that $\phi(P)$ is of order n. Intuitively, the operator T_m is to be understood as a kind of averaging operator over all possible isogenies of degree m. However, the normalizing factor $1/m$ is not equal to the inverse of the number of such isogenies. Instead, this factor is there to make the Eichler-Shimura isomorphism (see (2.4.5)) T_m-equivariant.

Of course, each element f of $S_k(\Gamma_1(n))$ is determined by its q-expansion $\sum_{m \geq 1} a_m(f)q^m$ at the cusp ∞. The action of the Hecke operators can be expressed in terms of these q-expansions (see [Di-Im], equation (12.4.1)) as

$$(2.2.11) \qquad a_m(T_r f) = \sum_{\substack{0 < d \mid (r,m) \\ (d,n)=1}} d^{k-1} a_{rm/d^2}(\langle d \rangle f),$$

for f in $S_k(\Gamma_1(n))$, r and m positive integers.

From this formula, a lot can be deduced. It can be seen that the T_r commute with each other (but there are better ways to understand this). The $\mathbb{Z}$-algebra generated by the T_m for $m \geq 1$ and the $\langle a \rangle$ for a in $(\mathbb{Z}/n\mathbb{Z})^\times$ is in fact generated by the T_m with $m \geq 1$, that is, one does not need the diamond operators, and also by the T_p for p prime and the $\langle a \rangle$ with a in $(\mathbb{Z}/n\mathbb{Z})^\times$ (see Section 3.5 in [Di-Im]). The multiplication rules for the T_m acting on $S_k(\Gamma_1(n))$ can be read off from the formal identity (Section 3.4 in [Di-Im]):

$$(2.2.12) \qquad \sum_{m \geq 1} T_m m^{-s} = \prod_p (1 - T_p p^{-s} + p^{k-1} \langle p \rangle p^{-2s})^{-1},$$

where $\langle p \rangle$ is to be interpreted as zero when p divides n. The fact that the Hecke and diamond operators commute means that they have common eigenspaces. Taking $m = 1$ in (2.2.11) gives

$$(2.2.13) \qquad a_1(T_r f) = a_r(f).$$

It follows that if f is a nonzero eigenvector for all T_r, then $a_1(f) \neq 0$, so that we can scale f uniquely such that that $a_1(f) = 1$. Then, for all $r \geq 1$, $a_r(f)$ is the eigenvalue for T_r. In particular, this means that the common eigenspaces for the T_r are one-dimensional, and automatically eigenspaces for the diamond operators. Eigenforms with $a_1(f) = 1$ are called *normalized eigenforms*.

From (2.2.12) it follows that for a normalized eigenform f one has

$$(2.2.14) \qquad \begin{aligned} L_f(s) &:= \sum_{m \geq 1} a_m(f) m^{-s} \\ &= \prod_p \left(1 - a_p(f) p^{-s} + p^{k-1} \varepsilon_f(p) p^{-2s} \right)^{-1}, \end{aligned}$$

where $\varepsilon_f \colon (\mathbb{Z}/n\mathbb{Z})^\times \to \mathbb{C}^\times$ is the character via which the diamond operators act on f, with the convention that $\varepsilon_f(p) = 0$ if p divides n. In particular, the L-function of a modular form has such an Euler product expansion if and only if the modular form is an eigenform for all Hecke operators.

An element of $S_k(\Gamma_1(n))$ that is a normalized eigenform for all Hecke operators is called a *newform* if the system of eigenvalues $a_p(f)$, with p not dividing n, does not occur in a level strictly smaller than n, that is, in some $S_k(\Gamma_1(m))$ with $m < n$ (actually, we will see in a moment that one only needs to consider the m's dividing n). The set of newforms in $S_k(\Gamma_1(n))$ will be denoted $S_k(\Gamma_1(n))^{\text{new}}$.

Now we want to recall briefly how one obtains a basis of $S_k(\Gamma_1(n))$ in terms of the sets of newforms $S_k(\Gamma_1(m))^{\text{new}}$ for m dividing n. For details and references to proofs, see [Di-Im], Section I.6. First of all, for each n, $S_k(\Gamma_1(n))^{\text{new}}$ is a linearly independent subset of $S_k(\Gamma_1(n))$, hence finite. For m dividing n and for d dividing n/m, we have a map $B_{n,m,d}$ from $X_1(n)$ to $X_1(m)$, whose moduli interpretation is that it maps (E, P) to $(E/\langle (n/d)P \rangle, d'P)$, where $dd' = n/m$. For example, this means

$$(2.2.15) \qquad B_{n,m,d} : (\mathbb{C}/(\mathbb{Z}z + \mathbb{Z}), 1/n) \mapsto (\mathbb{C}/(\mathbb{Z}zd + \mathbb{Z}), 1/m),$$

which means that the cusp ∞ of $X_1(n)$ is mapped to the cusp ∞ of $X_1(m)$. Each such map $B_{n,m,d}$ induces by pullback a map:

$$(2.2.16) \qquad B_{n,m,d}^* : S_k(\Gamma_1(m)) \to S_k(\Gamma_1(n)).$$

In terms of q-expansions at the cusp ∞ we have, for f in $S_k(\Gamma_1(m))$,

$$(2.2.17) \qquad B_{n,m,d}^* f = \sum_{r \geq 1} a_r(f) q^{dr},$$

that is, the effect is just substitution of q by q^d. With these definitions, we can describe a basis for $S_k(\Gamma_1(n))$:

$$(2.2.18) \qquad \coprod_{m|n} \coprod_{d|(n/m)} B_{n,m,d}^* S_k(\Gamma_1(m))^{\text{new}} \quad \text{is a basis of } S_k(\Gamma_1(n)).$$

In the case where $\Gamma_1(n)$ is replaced by $\Gamma_0(n)$, this kind of basis is due to Atkin and Lehner.

In the sequel, we will also make use of a (Hermitian) inner product on the $S_k(\Gamma_1(n))$: the Petersson scalar product. It is defined as follows. For f and g in $S_k(\Gamma_1(n))$, viewed as functions on $\mathbb{H}$ as in Definition 2.2.1 one has

$$(2.2.19) \qquad \langle f, g \rangle = \int_{\Gamma_1(n) \backslash \mathbb{H}} f(z) \overline{g(z)} \, y^k \, \frac{dx \, dy}{y^2},$$

where the integral over $\Gamma_1(n) \backslash \mathbb{H}$ means that one can perform it over any fundamental domain. Indeed, (2.1.2) shows that the function $\mathbb{H} \to \mathbb{C}$, $z \mapsto f(z)\overline{g(z)}y^k$ is invariant under $\Gamma_1(n)$.

We also want to explain the definition of $\langle f, g \rangle$ in terms of the moduli interpretation of $S_k(\Gamma_1(n))$, if $k \geq 2$. For simplicity, let us suppose $n \geq 5$ now. Then $S_k(\Gamma_1(n))$ is the space of global sections of $\underline{\omega}^{\otimes k}(-\text{Cusps})$ on $X_1(n)$. Now we let $\Omega^1 := \Omega^1_{X_1(n)}$ denote the line bundle of holomorphic differentials on $X_1(n)$. Then there is an isomorphism, named after Kodaira and Spencer,

(2.2.20)   KS: $\underline{\omega}^{\otimes 2}(-\text{Cusps}) \xrightarrow{\sim} \Omega^1$,   Kodaira-Spencer isomorphism.

Explicitly, for $f: \mathbb{H} \to \mathbb{C}$ in $S_2(\Gamma_1(n))$ we have the $\Gamma_1(n)$-invariant section $f(dx)^{\otimes 2}$ of $\underline{\omega}^{\otimes 2}$ for the family of elliptic curves over $\mathbb{H}$ whose fiber at z is $\mathbb{C}/(\mathbb{Z}z + \mathbb{Z})$, and

(2.2.21) $$\text{KS}: f(dx)^{\otimes 2} \mapsto (2\pi i)^{-2} f \frac{dq}{q}.$$

Equivalently, for this family of elliptic curves the Kodaira-Spencer isomorphism sends $(dx)^{\otimes 2}$ to $(2\pi i)^{-2}(dq)/q$. Note that indeed $(dx)^{\otimes 2}$ and $(dq)/q$ transform in the same way under the action of $SL_2(\mathbb{R})$. We note that without f being required to vanish at the cusps, $\text{KS}(f)$ could have poles of order one at the cusps. The factor $(2\pi i)^{-2}$ is to make the isomorphism compatible with the coordinates $t = \exp(2\pi i x)$ on $\mathbb{C}^\times/\langle \exp(2\pi i z) \rangle$ (which is another way to write $\mathbb{C}/(\mathbb{Z}z + \mathbb{Z})$), and the coordinate $q = \exp(2\pi i z)$ on the unit disk. In those coordinates that have a meaning "over $\mathbb{Z}$," which means that formulas relating them are power series (or Laurent series) with integer coefficients, KS sends $((dt)/t)^{\otimes 2}$ to $(dq)/q$.

For every complex elliptic curve, the one-dimensional complex vector space $\underline{\omega}_E$ has the inner product given by

(2.2.22) $$\langle \alpha, \beta \rangle = \frac{i}{2} \int_E \alpha \, \overline{\beta},$$

where we interpret α and β as translation invariant differential forms on E. The factor $i/2$ is explained by the fact that, when writing $z = x + iy$, one has $dx \, dy = (i/2)dz \, d\overline{z}$. Applying this to the family of elliptic curves $\mathbb{C}/(\mathbb{Z}z + \mathbb{Z})$ over $\mathbb{H}$ gives an inner product on the line bundle $\underline{\omega}$ on $\mathbb{H}$, and also on the line bundle $\underline{\omega}$ on $Y_1(n)$ (recall that we are supposing that $n \geq 5$). Taking tensor powers and duals, this induces inner products on $\underline{\omega}^{\otimes k}$ for all k. The Kodaira-Spencer isomorphism (2.2.21) gives isomorphisms

$$(2.2.23) \qquad \mathrm{KS} \colon \underline{\omega}^{\otimes k}(-\mathrm{Cusps}) \xrightarrow{\sim} \Omega^1 \otimes \underline{\omega}^{\otimes(k-2)}.$$

For f and g in $S_k(\Gamma_1(n))$, now viewed as sections of $\underline{\omega}^{\otimes k}(-\mathrm{Cusps})$ over $X_1(n)$, one has

$$(2.2.24) \qquad \langle f, g \rangle = \frac{i}{2} \int_{X_1(n)} \langle \mathrm{KS}(f), \mathrm{KS}(g) \rangle,$$

where the inner product on the left-hand side is that of (2.2.19), and where for two local sections $\omega_1 \otimes \alpha_1^{\otimes(k-2)}$ and $\omega_2 \otimes \alpha_2^{\otimes(k-2)}$ of $\Omega^1 \otimes \underline{\omega}^{\otimes(k-2)}$ we have defined

$$(2.2.25) \qquad \langle \omega_1 \otimes \alpha_1^{\otimes(k-2)}, \omega_2 \otimes \alpha_2^{\otimes(k-2)} \rangle = \langle \alpha_1, \alpha_2 \rangle^{k-2} \omega_1 \overline{\omega_2}.$$

The operators T_m on $S_k(\Gamma_1(n))$ with m relatively prime to n are normal: they commute with their adjoint. As a consequence, distinct newforms in $S_k(\Gamma_1(n))$ are orthogonal to each other. On the other hand, the basis (2.2.18) above of $S_k(\Gamma_1(n))$ is not orthogonal if it consists of more than only newforms.

2.3 LATTICES AND MODULAR FORMS

Before we move on to Galois representations attached to modular forms, we briefly discuss the relation between modular forms and lattices.

Let us consider a free $\mathbb{Z}$-module L of finite rank n, equipped with a positive definite symmetric bilinear form $b \colon L \times L \to \mathbb{Z}$. Then $L_{\mathbb{R}} := \mathbb{R} \otimes L$ is an $\mathbb{R}$-vector space of dimension n on which b gives an inner product, and hence L is a lattice in the Euclidean space $L_{\mathbb{R}}$. For m in $\mathbb{Z}$ the *representation numbers* of (L, b) are defined as

$$(2.3.1) \qquad r_L(m) = r_{L,b}(m) := \#\{x \in L \mid b(x, x) = m\}.$$

In this situation, one considers the *theta-function* attached to (L, b):

$$(2.3.2) \qquad \theta_L = \theta_{L,b} = \sum_{x \in L} q^{b(x,x)/2} = \sum_{m \geq 0} r_L(m) q^{m/2}, \qquad \mathbb{H} \to \mathbb{C},$$

where $q^{1/2}: \mathbb{H} \to \mathbb{C}$ is the function $q^{1/2}: z \mapsto \exp(\pi i z)$.

If (L, b) is the orthogonal direct sum of (L_1, b_1) and (L_2, b_2) then we have

(2.3.3)
$$
\begin{aligned}
\theta_L &= \sum_{x \in L} q^{b(x,x)/2} \\
&= \sum_{(x_1,x_2) \in L_1 \times L_2} q^{b_1(x_1,x_1)/2 + b_2(x_2,x_2)/2} = \theta_{L_1} \cdot \theta_{L_2}.
\end{aligned}
$$

The *discriminant* of b is $\det(B)$, where B is the matrix of b with respect to some basis of L (indeed, this determinant does not depend on the choice of basis); we denote it by $\mathrm{discr}(b)$.

We define the positive integer N_L to be the exponent of the cokernel of the map $\phi_b: L \to L^\vee$ given by b, or, equivalently, to be the denominator of B^{-1}, where B is the matrix of b with respect to some basis e of L. The map $N_L \phi_b^{-1}: L_{\mathbb{Q}}^\vee \to L_{\mathbb{Q}}$ restricts to a map $N_L \phi_b^{-1}: L^\vee \to L$. Viewing L as $(L^\vee)^\vee$ in the usual way, this gives a positive definite symmetric bilinear form $b': L^\vee \times L^\vee \to \mathbb{Z}$. The matrix of this form with respect to the basis $e^\vee$ dual to e is $N_L B^{-1}$. Applying this same construction to b' gives a $b'': L \times L \to \mathbb{Z}$ that is not necessarily equal to b: one has $b = mb''$, with $m \in \mathbb{Z}_{>0}$ and b'' primitive (that is, the $\mathbb{Z}$-linear map $L \otimes L \to \mathbb{Z}$ induced by b'' is surjective). Poisson's summation formula gives the following functional equation; see [Ser5], Chapter VII, Section 6, Proposition 16.

2.3.4 Theorem *Let L be a free $\mathbb{Z}$-module, of finite rank n, equipped with a positive definite symmetric bilinear form $b: L \times L \to \mathbb{Z}$. We have, with the notation as above, for all $z \in \mathbb{H}$:*

$$
\theta_{L,b}(-1/N_L z) = \frac{(-N_L i z)^{n/2}}{\mathrm{discr}(b)^{1/2}} \theta_{L^\vee, b'}(z),
$$

where the square root of $-N_L i z$ is holomorphic in z and positive for $z \in \mathbb{R} i$.

The form b is called *even* if $b(x, x)$ is even for all x in L. Equivalently, b is even if and only if the matrix B of b with respect to some basis of L has only even numbers on the diagonal.

The form b is called *unimodular* if $\phi_b: L \to L^\vee$ is an isomorphism, or, equivalently, if $N_L = 1$. In this case, ϕ_b is an isomorphism from (L, b) to $(L^\vee, b')$.

With this terminology, one has the following result, see [Miy], Corollary 4.9.5, the proof of which has as main ingredient the functional equation of Theorem 2.3.4.

2.3.5 Theorem *Let L be a free $\mathbb{Z}$-module of finite rank n, equipped with a positive definite symmetric bilinear form $b: L \times L \to \mathbb{Z}$. Assume that n is even. Let N_L be as defined above, and let χ_L be the character given by*

$$\chi_L: (\mathbb{Z}/N_L\mathbb{Z})^\times \to \mathbb{C}^\times, \quad (a \bmod N_L) \mapsto \left(\frac{(-1)^{n/2}\mathrm{discr}(b)}{a} \right),$$

where the fraction denotes the Kronecker symbol.

1. *The function $z \mapsto \theta_L(2z)$ is a modular form on $\Gamma_1(4N_L)$ of weight $n/2$ and with character χ_L.*

2. *If b is even then the function θ_L is a modular form on $\Gamma_1(2N_L)$ of weight $n/2$ and with character χ_L.*

3. *If both b and b' (see above for its definition) are even, then the function θ_L is a modular form on $\Gamma_1(N_L)$ of weight $n/2$ and with character χ_L.*

This theorem says nothing about the case where n is odd. In that case, θ_L is a modular form of half-integral weight $n/2$; see Corollary 4.9.7 in [Miy]. For even unimodular forms, we have the following corollary of Theorem 2.3.5.

2.3.6 Corollary *Let L be a free $\mathbb{Z}$-module of finite rank n, equipped with a bilinear form $b: L \times L \to \mathbb{Z}$ that is symmetric, positive definite, even, and unimodular. Then n is even, and θ_L is a modular form on $\mathrm{SL}_2(\mathbb{Z})$ of weight $n/2$.*

Proof As b is unimodular, we have $N_L = 1$. The fact that n is even follows from the fact that b induces a nondegenerate alternating bilinear form on $\mathbb{F}_2 \otimes L$. As $\phi_b: L \to L^\vee$ is an isomorphism between (L, b) and $(L^\vee, b')$, we have that b' is even as well. Theorem 2.3.5 gives the conclusion. $\square$

2.3.7 Remark In fact, the rank of an even unimodular lattice is a multiple of 8. This follows directly from Theorems 2.3.4 and 2.3.5 (see Corollary 4.9.6 in [Miy], or [Ser5], VII, Section 6, Theorem 8).

Let us consider some examples.

2.3.8 Example For $n \in \mathbb{Z}_{\geq 0}$ we consider $\mathbb{Z}^n$ with its standard inner product. For m in $\mathbb{Z}$ we have

$$r_{\mathbb{Z}^n}(m) = \#\{x \in \mathbb{Z}^n \mid x_1^2 + \cdots + x_n^2 = m\},$$

the number of ways in which m can be written as a sum of n squares of integers. Theorem 2.3.5 tells us that for even n the theta function $z \mapsto \theta_{\mathbb{Z}^n}(2z)$ is a modular form on $\Gamma_1(4)$ of weight $n/2$. According to (2.3.3) we have $\theta_{\mathbb{Z}^n} = \theta_{\mathbb{Z}}^n$, and so all the functions $z \mapsto \theta_{\mathbb{Z}^n}(2z)$ are powers of the modular form $\sum_{m \in \mathbb{Z}} q^{m^2}$ of weight $1/2$ on $\Gamma_1(4)$.

2.3.9 Example We consider the $E8$-lattice, that is, $E8 := \mathbb{Z}^8$ equipped with the inner product given by the Dynkin diagram $E8$ with numbered vertices:

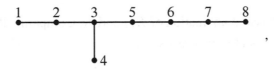

that is, whose matrix with respect to the standard basis is:

$$\begin{pmatrix} 2 & -1 & & & & & & \\ -1 & 2 & -1 & & & & & \\ & -1 & 2 & -1 & -1 & & & \\ & & -1 & 2 & & & & \\ & & -1 & & 2 & -1 & & \\ & & & & -1 & 2 & -1 & \\ & & & & & -1 & 2 & -1 \\ & & & & & & -1 & 2 \end{pmatrix}.$$

The lattice $E8$ is unimodular and even, hence, by Corollary 2.3.6, θ_{E8} is a modular form on $\mathrm{SL}_2(\mathbb{Z})$ of weight 4, that is, θ_{E8} is in $M_4(\mathrm{SL}_2(\mathbb{Z}))$. The dimension of $M_4(\mathrm{SL}_2(\mathbb{Z}))$ is one, with the Eisenstein series E_4 as basis. Therefore, θ_{E8} is a constant times E_4. Comparing constant terms, we get

$$\theta_{E8} = E_4 = 1 + 240 \sum_{n \geq 1} \sigma_3(n) q^n.$$

2.3.10 Example Let L be the *Leech lattice*. This lattice, of rank 24, even, and unimodular, is named after John Leech; see [Lee2] and [Lee1]. Apparently, it had already been discovered by Ernst Witt in 1940 (unpublished, see [Wit]). John Horton Conway showed in [Conw] that L is the only nonzero even unimodular lattice of rank less than 32 with $r_L(2) = 0$; this also follows from Hans-Volker Niemeier's classification of even unimodular lattices of rank 24 in [Nie]. According to Henry Cohn and Abhinav Kumar [Co-Ku], the Leech lattice gives the densest lattice sphere packing in dimension 24.

Corollary 2.3.6 above shows that the theta function θ_L of the Leech lattice is a modular form of level 1 and weight 12. The space of such modular forms is two-dimensional, with basis the Eisenstein series E_{12} and the discriminant form Δ, where

$$E_{12} = 1 + \frac{65520}{691} \sum_{m \geq 1} \sigma_{11}(m)q^m.$$

Hence θ_L is a linear combination of E_{12} and Δ. Comparing the coefficients of q^m for $m = 0$ and $m = 1$ gives

$$\theta_L = E_{12} - \frac{65520}{691}\Delta.$$

2.4 GALOIS REPRESENTATIONS ATTACHED TO EIGENFORMS

The aim of this section is to describe the construction of the Galois representations attached to modular forms, that came up in the case of Δ in Section 1.5. Before giving the construction, let us state the result, which is due to Eichler and Shimura [Shi1] for $k = 2$, to Deligne [Del1] for $k > 2$, and to Deligne and Serre [De-Se] for $k = 1$. See Section 12.5 in [Di-Im]. A long account of the construction in the case $k \geq 2$ is given in the book [Conr].

2.4.1 Theorem *Let f be a normalized newform, let n be its level, let k be its weight, and let $\varepsilon \colon (\mathbb{Z}/n\mathbb{Z})^\times \to \mathbb{C}^\times$ be its character. Then the subfield K of $\mathbb{C}$ generated over $\mathbb{Q}$ by the $a_n(f)$, $n \geq 1$, and the image of ε is finite over $\mathbb{Q}$. For every prime number l and for any embedding λ of K into $\overline{\mathbb{Q}}_l$, there is a continuous two-dimensional representation V_λ over $\overline{\mathbb{Q}}_l$ of $\mathrm{Gal}(\overline{\mathbb{Q}}/\mathbb{Q})$ that is unramified outside nl and such that for each prime* *number p not dividing nl the characteristic polynomial of the Frobenius at p acting on V_λ is*

$$\det(1 - x\mathrm{Frob}_p, V_\lambda) = 1 - a_p(f)x + \varepsilon(p)p^{k-1}x^2.$$

For $k \geq 2$ the representations V_λ can be found in the l-adic étale cohomology in degree $k-1$ of some variety of dimension $k-1$, or in the cohomology in degree one of some sheaf on a curve, as we will describe below. The determinant of the action of $\mathrm{Gal}(\overline{\mathbb{Q}}/\mathbb{Q})$ on V_λ is easily described. We let $\varepsilon \colon \mathrm{Gal}(\overline{\mathbb{Q}}/\mathbb{Q}) \to K^\times$ be the composition of the character $\varepsilon \colon (\mathbb{Z}/n\mathbb{Z})^\times \to K^\times$ with the mod n cyclotomic character from $\mathrm{Gal}(\overline{\mathbb{Q}}/\mathbb{Q})$ to $(\mathbb{Z}/n\mathbb{Z})^\times$ given by the action of $\mathrm{Gal}(\overline{\mathbb{Q}}/\mathbb{Q})$ on $\mu_n(\overline{\mathbb{Q}})$. With these definitions, the determinant of the action of $\mathrm{Gal}(\overline{\mathbb{Q}}/\mathbb{Q})$ on V_λ is given by the character $\varepsilon \chi_l^{k-1}$, where $\chi_l \colon \mathrm{Gal}(\overline{\mathbb{Q}}/\mathbb{Q}) \to \mathbb{Z}_l^\times$ is the l-adic cyclotomic character. As the image of $\mathrm{Gal}(\overline{\mathbb{Q}}/\mathbb{Q})$ under the determinant of V_λ is infinite, its image in $\mathrm{GL}(V_\lambda)$ is infinite. An important property of the $\rho_{f,\lambda}$ is that they are *odd*, meaning that the determinant of the image of a complex conjugation is -1, as $\varepsilon(-1) = (-1)^k$ by (2.2.9).

On the other hand, for $k = 1$, the image of $\mathrm{Gal}(\overline{\mathbb{Q}}/\mathbb{Q})$ in $\mathrm{GL}(V_\lambda)$ is finite, and in fact all these representations when λ varies can be realized over some fixed finite extension of $\mathbb{Q}$. The proof of Theorem 2.4.1 by Deligne and Serre in the case $k = 1$ is quite different from the case $k \geq 2$: the reductions to finite coefficient fields (see Section 2.5) can be constructed via congruences to forms of weight 2, and then it is shown that these representations can be lifted to characteristic zero. No direct construction of the characteristic zero Galois representations for forms of weight one is known. We remark that for $k = 2$ the V_λ occur in the first degree étale cohomology with constant coefficients $\mathbb{Q}_l$ of modular curves, hence can be constructed from l-power torsion points of Jacobians of modular curves (in fact, of the modular curve $X_1(n)$).

The representation V_λ is irreducible by a theorem of Ribet, see Theorem 2.3 of [Rib2], and hence it is characterized by its trace. As the Frobenius conjugacy classes at the primes not dividing nl are dense by Chebotarev's theorem, the representation V_λ is unique up to isomorphism. Noncuspidal eigenforms lead to Galois representations that are reducible; as our interest lies in going beyond class field theory, we do not discuss this case.

Let us now start the description of the construction, by Deligne, of the representation V_λ as in Theorem 2.4.1 above in the case where $k \geq 2$. First, if $n < 5$, we replace n by, say, $5n$ and f by a normalized Hecke eigenform in the two-dimensional $\mathbb{C}$-vector space generated by $f(q)$ and $f(q^5)$. Then f is no longer a newform, but it is an eigenform, which will be good enough, and as $n \geq 5$ we can view it as a section of the line bundle $\underline{\omega}^{\otimes k}(-\text{Cusps})$ on the smooth complex projective curve $X_1(n)$. The eigenvalues at primes other than 5 have not been changed by this operation. As one can compute from the formulas in the previous section, the two possible eigenvalues for T_5 on the space generated by $f(q)$ and $f(q^5)$ are the two roots of the polynomial $x^2 - a_5(f)x + \varepsilon(5)5^{k-1}$, that is, the two eigenvalues of the Frobenius element at 5 attached to f if λ does not divide 5. For a detailed computation for this, see Section 4 of [Co-Ed]; that article also explains why one should expect the two eigenvalues always to be distinct, and that this is a theorem if $k = 2$.

On $Y_1(n)$, we have a universal family $(\mathbb{E}/Y_1(n), \mathbb{P})$ of elliptic curves with a given point of order n. Taking fiber-wise the cohomology $H^1(\mathbb{E}_s, \mathbb{Z})$ gives us a locally constant sheaf on $Y_1(n)$, denoted $R^1 p_* \mathbb{Z}_\mathbb{E}$ because it is the first higher direct image of the constant sheaf $\mathbb{Z}_\mathbb{E}$ on $\mathbb{E}$ via the morphism $p : \mathbb{E} \to Y_1(n)$. The stalks of the locally constant sheaf $R^1 p_* \mathbb{Z}_\mathbb{E}$ on $Y_1(n)$ are free $\mathbb{Z}$-modules of rank 2. More concretely, the sheaf $R^1 p_* \mathbb{Z}_\mathbb{E}$ is obtained from the constant sheaf $\mathbb{Z}^2$ on $\mathbb{H}$ by dividing out the $\Gamma_1(n)$-action given by

$$(2.4.2) \quad \gamma \cdot \left(\begin{pmatrix} n \\ m \end{pmatrix}, \tau \right) = \left(\gamma \cdot \begin{pmatrix} n \\ m \end{pmatrix}, \gamma \cdot \tau \right) = \left(\begin{pmatrix} an + bm \\ cn + dm \end{pmatrix}, \frac{a\tau + b}{c\tau + d} \right),$$

where $\gamma = \begin{pmatrix} a & b \\ c & d \end{pmatrix}$.

We will also use other locally constant sheaves on $Y_1(n)$ that are obtained from $R^1 p_* \mathbb{Z}_\mathbb{E}$ by tensor constructions. The classification of the irreducible representations of the algebraic group GL_2 over $\mathbb{Q}$ implies that these tensor constructions are finite direct sums in which each term is a symmetric power of $R^1 p_* \mathbb{Z}_\mathbb{E}$, tensored with a power of the determinant of $R^1 p_* \mathbb{Z}_\mathbb{E}$. We define

$$(2.4.3) \qquad \qquad \mathcal{F}_k := \text{Sym}^{k-2}(R^1 p_* \mathbb{Z}_\mathbb{E}),$$

where Sym^{k-2} is the operation of taking the $(k-2)$th symmetric power. The sheaf $\mathcal{F}_k$ is then obtained by dividing out the $\Gamma_1(n)$-action on the constant sheaf $\mathrm{Sym}^{k-2}(\mathbb{Z}^2)$ on $\mathbb{H}$. It is useful to view $\mathbb{Z}^2$ as the $\mathbb{Z}$-submodule $\mathbb{Z}x \oplus \mathbb{Z}y$ of the polynomial ring $\mathbb{Z}[x,y]$. The grading $\mathbb{Z}[x,y] = \oplus_i \mathbb{Z}[x,y]_i$ by the degree then gives the symmetric powers of $\mathbb{Z}x \oplus \mathbb{Z}y$:

$$(2.4.4) \qquad \mathrm{Sym}^{k-2}(\mathbb{Z}^2) = \mathbb{Z}[x,y]_{k-2} = \bigoplus_{i+j=k-2} \mathbb{Z}x^i y^j.$$

We extend the sheaf $\mathcal{F}_k$ to $X_1(n)$ by taking the direct image via the open immersion $j : Y_1(n) \rightarrow X_1(n)$; this gives us $j_* \mathcal{F}_k$ on $X_1(n)$, again denoted $\mathcal{F}_k$. Outside the cusps, $\mathcal{F}_k$ is locally constant, with stalks free of rank $k-1$ as $\mathbb{Z}$-modules. At the cusps, the stalks of $\mathcal{F}_k$ are free of rank one. At the cusp ∞ this follows from the fact that the subring of invariants of $\mathbb{Z}[x,y]$ for the action of $\left(\begin{smallmatrix} 1 & * \\ 0 & 1 \end{smallmatrix}\right)$ is $\mathbb{Z}[x]$. At the other cusps it then follows by conjugating with a suitable element of $\mathrm{SL}_2(\mathbb{Z})$. We note that for $k=2$ the sheaf $\mathcal{F}_k$ is the constant sheaf $\mathbb{Z}$ on $X_1(n)$.

The *Eichler-Shimura isomorphism* gives a relation between the cohomology of $\mathcal{F}_k$ and modular forms. One way to view this, due to Deligne, is in terms of Hodge structures. More precisely, the complex vector space $\mathbb{C} \otimes \mathrm{H}^1(X_1(n), \mathcal{F}_k)$ carries a Hodge decomposition

$$(2.4.5) \qquad \mathbb{C} \otimes \mathrm{H}^1(X_1(n), \mathcal{F}_k) \overset{\sim}{\longrightarrow} S_k(\Gamma_1(n)) \oplus \overline{S_k(\Gamma_1(n))},$$

where the two terms on the right are of type $(k-1,0)$ and $(0,k-1)$, respectively. The complex conjugation on the second term on the right comes from the complex conjugation on the tensor factor $\mathbb{C}$ on the left. A good reference for this decomposition and its properties is [Ba-Ne]; we will not go into details here. For an account using group cohomology we refer to Section 12.2 of [Di-Im]. For $k=2$ all of this is quite easy. Via the Kodaira-Spencer isomorphism (2.2.21) it then is the decomposition

$$(2.4.6) \qquad \mathrm{H}^1(X_1(n), \mathbb{C}) = \mathrm{H}^0(X_1(n), \Omega^1) \oplus \overline{\mathrm{H}^0(X_1(n), \Omega^1)}.$$

We should mention that instead of working with the sheaf $\mathcal{F}_k$ on the curve $X_1(n)$, one can also work with a constant sheaf on a $(k-1)$-dimensional variety. As before, we let $(\mathbb{E}, \mathbb{P})$ denote the universal object over $Y_1(n)$. Then we let $\mathbb{E}^{k-2}$ denote the $(k-2)$-fold fiber power of $\mathbb{E}$ over $Y_1(n)$; these are the simplest cases of so-called Kuga-Sato varieties. The graded commutative algebra structure on cohomology gives, for s in $Y_1(n)$, a map, equivariant for the action of the symmetric group S_{k-2},

$$\mathbb{Z}(\varepsilon) \otimes \mathrm{H}^1(\mathbb{E}_s, \mathbb{Z}) \otimes \cdots \otimes \mathrm{H}^1(\mathbb{E}_s, \mathbb{Z}) \longrightarrow \mathrm{H}^{k-2}(\mathbb{E}_s^{k-2}, \mathbb{Z}),$$

where $\mathbb{Z}(\varepsilon)$ denotes the sign representation. Twisting this map by $\mathbb{Z}(\varepsilon)$ and taking co-invariants gives a map

$$(2.4.7) \qquad \mathcal{F}_{k,s} = \mathrm{Sym}^{k-2}(\mathrm{H}^1(\mathbb{E}_s, \mathbb{Z})) \longrightarrow \mathrm{H}^{k-2}(\mathbb{E}_s^{k-2}, \mathbb{Z})_{\varepsilon},$$

where the subscript ε means the largest quotient on which S_{k-2} acts via the sign representation. In view of the Leray spectral sequence for the cohomology $\mathrm{H}(\mathbb{E}^{k-2}, \mathbb{Z})$ of $\mathbb{E}^{k-2}$ in terms of the cohomology of the higher derived direct images $\mathrm{H}(Y_1(n), \mathrm{R}p_*\mathbb{Z}_{\mathbb{E}^{k-2}})$, it is then not so surprising that $S_k(\Gamma_1(n))$ can be identified with a piece of $\mathrm{H}^{k-1}(\overline{\mathbb{E}^{k-2}}, \mathbb{C})$, where $\overline{\mathbb{E}^{k-2}}$ is a certain smooth projective model of $\mathbb{E}^{k-2}$ over $X_1(n)$. Some details for this can be found in [Del1], and more of them in [Sch1], and still more in [Conr]. A very explicit way to describe this identification is the map

$$(2.4.8) \qquad \begin{aligned} S_k(\Gamma_1(n)) &\longrightarrow \mathrm{H}^{k-1}(\overline{\mathbb{E}^{k-2}}, \mathbb{C}), \\ f &\mapsto (2\pi i)^{k-1} f \, d\tau \, dz_1 \cdots dz_{k-2}, \end{aligned}$$

where τ is the coordinate on $\mathbb{H}$, and the z_j are the coordinates on the copies of $\mathbb{C}$ using $\mathbb{E}_\tau = \mathbb{C}/(\mathbb{Z}\tau + \mathbb{Z})$. It is indeed easy to verify that the differential form on the right is invariant under the actions of $\mathbb{Z}^{2(k-2)}$ and $\mathrm{SL}_2(\mathbb{Z})$, precisely because f is a modular form of weight k for $\Gamma_1(n)$. The claim (proved in the references above) is that it extends without poles over $\overline{\mathbb{E}^{k-2}}$. As it is a holomorphic form of top-degree, it is automatically closed, and hence defines a class in the de Rham cohomology of $\overline{\mathbb{E}^{k-2}}$, hence in $\mathrm{H}^{k-1}(\overline{\mathbb{E}^{k-2}}, \mathbb{C})$.

There are natural Hecke correspondences on $\mathbb{C} \otimes \mathrm{H}^1(X_1(n), \mathcal{F}_k)$ and on $\mathrm{H}^{k-1}(\overline{\mathbb{E}^{k-2}}, \mathbb{C})$, and the identification of $S_k(\Gamma_1(n))$ as a piece of these cohomology groups is compatible with these correspondences. Let now f be our eigenform in $S_k(\Gamma_1(n))$ as above. Then the Hecke eigenspace in $\mathbb{C} \otimes \mathrm{H}^1(X_1(n), \mathcal{F}_k)$ with the eigenvalues $a_m(f)$ for T_m is two-dimensional: the sum of the one-dimensional subspace $\mathbb{C}f$ in $S_k(\Gamma_1(n))$ and the one-dimensional subspace $\mathbb{C}\overline{f'}$ in $\overline{S_k(\Gamma_1(n))}$, where $f' = \sum_{m \geq 1} \overline{a_m(f)} q^m$, the Galois conjugate of f obtained by letting complex conjugation act on the coefficients of f. This element f' has eigenvalue $\overline{a_m(f)}$ for T_m, hence $\overline{f'}$ has eigenvalue $a_m(f)$ again. The $(k-1)$-form corresponding to $\overline{f'}$ is $\overline{f'} \, d\overline{\tau} \, d\overline{z_1} \cdots d\overline{z_{k-2}}$, indeed a form of type $(0, k-1)$.

We let $\mathbb{T}(n, k)$ or just $\mathbb{T}$ denote the $\mathbb{Z}$-algebra in $\mathrm{End}_{\mathbb{C}}(S_k(\Gamma_1(n)))$ generated by the T_m ($m \geq 1$) and the $\langle a \rangle$ (a in $(\mathbb{Z}/n\mathbb{Z})^\times$). The fact that the Eichler-Shimura isomorphism (2.4.5) is equivariant for the Hecke correspondences acting on both sides implies that the image of $\mathrm{H}^1(X_1(n), \mathcal{F}_k)$ in $\mathbb{C} \otimes \mathrm{H}^1(X_1(n), \mathcal{F}_k)$ is a faithful $\mathbb{T}(n, k)$-module. As this image is free of finite rank as $\mathbb{Z}$-module, $\mathbb{T}(n, k)$ is free of finite rank as $\mathbb{Z}$-module.

Let us for a moment drop the assumption that $n \geq 5$. For A a subring of $\mathbb{C}$ and for $n \geq 1$, we let $M_k(\Gamma_1(n), A)$ be the sub-A-module of $M_k(\Gamma_1(n))$ consisting of elements g such that $a_m(g) \in A$ for all $m \geq 0$. In particular, $M_k(\Gamma_1(n), \mathbb{Z})$ is the submodule of forms whose q-expansion has all its coefficients in $\mathbb{Z}$. Similarly, for A a subring of $\mathbb{C}$ and for $n \geq 1$, we let $S_k(\Gamma_1(n), A)$ be the sub-A-module of $S_k(\Gamma_1(n))$ consisting of elements g such that $a_m(g) \in A$ for all $m \geq 1$. For example, Δ belongs to $S_{12}(\mathrm{SL}_2(\mathbb{Z}), \mathbb{Z})$. The $S_k(\Gamma_1(n), A)$ are $\mathbb{T}(n, k)$-submodules of $S_k(\Gamma_1(n))$; see Propositions 12.3.11 and 12.4.1 of [Di-Im].

We have the following pairing between $\mathbb{T}(n, k)$ and $S_k(\Gamma_1(n), \mathbb{Z})$:

$$(2.4.9) \qquad S_k(\Gamma_1(n), \mathbb{Z}) \times \mathbb{T}(n, k) \longrightarrow \mathbb{Z}, \quad (g, t) \mapsto a_1(tg).$$

This pairing is perfect, in the sense that it identifies each side with the $\mathbb{Z}$-linear dual of the other; this follows easily from the identity (2.2.13). It follows that the $\mathbb{Z}$-dual $S_k(\Gamma_1(n), \mathbb{Z})^\vee$ of $S_k(\Gamma_1(n), \mathbb{Z})$ is free of rank one as $\mathbb{T}(n, k)$-module. See [Di-Im], Proposition 12.4.13.

For any $\mathbb{Z}$-algebra A we let $\mathbb{T}_A = \mathbb{T}(n, k)_A$ denote $A \otimes \mathbb{T}(n, k)$, and $\mathbb{T}_A^\vee$ will denote the A-linear dual of $\mathbb{T}_A$. It can be proved that $\mathbb{T}_{\mathbb{Q}}^\vee$ is free of rank one as $\mathbb{T}_{\mathbb{Q}}$-module, that is, that $\mathbb{T}_{\mathbb{Q}}$ is *Gorenstein*. One proof is by explicit computation; see Theorem 3.5 and Corollary 3.6 of [Par]. Another, more conceptual proof uses the Petersson inner product, and a so-called Atkin-Lehner pseudo-involution w_{ζ_n}, to show that $S_k(\Gamma_1(n))^\vee$ is isomorphic as $\mathbb{T}_{\mathbb{C}}$-module to $S_k(\Gamma_1(n))$ itself; see [Di-Im], Section 12.4.14. It follows that $S_k(\Gamma_1(n))$ is free of rank one as $\mathbb{T}_{\mathbb{C}}$-module, and, if $n \geq 5$, that $\mathbb{Q} \otimes \mathrm{H}^1(X_1(n), \mathcal{F}_k)$ and its dual $\mathbb{Q} \otimes \mathrm{H}^1(X_1(n), \mathcal{F}_k)^\vee$ are free of rank two as $\mathbb{T}_{\mathbb{Q}}$-module. It is this freeness result that will lead to the fact that the Galois representations we get are two-dimensional.

We assume again that $n \geq 5$. The step from the cohomological interpretation of modular forms, given, over the complex numbers, by the Eichler-Shimura isomorphism (2.4.5), to two-dimensional l-adic Galois representations is made by comparing the cohomology groups above to their l-adic counterparts for the étale topology, and noting that $p \colon \mathbb{E} \to X_1(n)$ is naturally defined over $\mathbb{Z}[1/n]$ as we have seen at the end of Section 2.1. From now on we will denote by $X_1(n)$ this model over $\mathbb{Z}[1/n]$, and by $X_1(n)(\mathbb{C})$ the Riemann surface given by $X_1(n)$. For any $\mathbb{Z}[1/n]$-algebra A, $X_1(n)_A$ will denote the A-scheme obtained from $X_1(n)$ by extending scalars via $\mathbb{Z}[1/n] \to A$.

We let $\mathcal{F}_{k,l}$ denote the sheaf of $\mathbb{Q}_l$-vector spaces $\mathbb{Q}_l \otimes \mathcal{F}_k$ on $X_1(n)(\mathbb{C})$. Then we have a canonical isomorphism

$$(2.4.10) \qquad \mathrm{H}^1(X_1(n)(\mathbb{C}), \mathcal{F}_{k,l}) = \mathbb{Q}_l \otimes \mathrm{H}^1(X_1(n)(\mathbb{C}), \mathcal{F}_k).$$

The sheaves $\mathcal{F}_{k,l}$ can also be constructed on the étale site $X_1(n)_{\mathrm{et}}$, by taking the first derived direct image of the constant sheaf $\mathbb{Q}_l$ on $\mathbb{E}_{\mathrm{et}}$ under $p \colon \mathbb{E} \to Y_1(n)$, then the $(k-2)$th symmetric power of that, and finally the pushforward from $Y_1(n)$ to $X_1(n)$.

The usual comparison theorems (comparing cohomology for étale and Archimedean topology, and étale cohomology over various algebraically closed fields) give

$$(2.4.11) \qquad \begin{aligned} \mathrm{H}^1(X_1(n)(\mathbb{C}), \mathcal{F}_{k,l}) &= \mathrm{H}^1(X_1(n)_{\mathbb{C},\mathrm{et}}, \mathcal{F}_{k,l}) \\ &= \mathrm{H}^1(X_1(n)_{\overline{\mathbb{Q}},\mathrm{et}}, \mathcal{F}_{k,l}). \end{aligned}$$

We put

$$(2.4.12) \qquad W_l := \mathrm{H}^1(X_1(n)_{\overline{\mathbb{Q}},\mathrm{et}}, \mathcal{F}_{k,l})^{\vee}.$$

By the results and the comparisons above, W_l is, as $\mathbb{T}_{\mathbb{Q}_l}$-module, free of rank two, and $\mathrm{Gal}(\overline{\mathbb{Q}}/\mathbb{Q})$ acts continuously on it. To be precise: an element σ of $\mathrm{Gal}(\overline{\mathbb{Q}}/\mathbb{Q})$ acts as $((\mathrm{id} \times \mathrm{Spec}(\sigma^{-1}))^*)^{\vee}$, which is indeed covariant in σ. The fact that the Hecke correspondences exist over $\mathbb{Q}$ makes that the $\mathrm{Gal}(\overline{\mathbb{Q}}/\mathbb{Q})$-action on W_l commutes with the Hecke operators. The choice of a $\mathbb{T}_{\mathbb{Q}_l}$-basis of W_l gives us a representation:

$$(2.4.13) \qquad \rho_l: \mathrm{Gal}(\overline{\mathbb{Q}}/\mathbb{Q}) \longrightarrow \mathrm{GL}_2(\mathbb{T}_{\mathbb{Q}_l}).$$

Recall that we have fixed an eigenform f in $S_k(\Gamma_1(n), \mathbb{C})$. Sending a Hecke operator to its eigenvalue for f then gives us a morphism of rings:

$$(2.4.14) \qquad \phi_f: \mathbb{T} \longrightarrow \mathbb{C}.$$

We let $K(f)$ be the image of $\mathbb{T}_{\mathbb{Q}}$ under ϕ_f; it is the finite extension of $\mathbb{Q}$ obtained by adjoining all coefficients $a_m(f)$ of the q-expansion of f. We now view ϕ_f as a morphism from $\mathbb{T}$ to $K(f)$. The tensor product $\mathbb{Q}_l \otimes K(f)$ is the product of the completions $K(f)_\lambda$, with λ ranging through the finite places of $K(f)$ that divide l. For each such λ we then get a morphism $\phi_{f,\lambda}: \mathbb{T}_{\mathbb{Q}_l} \to K(f)_\lambda$, and a representation

$$(2.4.15) \qquad \rho_{f,\lambda}: \mathrm{Gal}(\overline{\mathbb{Q}}/\mathbb{Q}) \longrightarrow \mathrm{GL}_2(K(f)_\lambda).$$

These are the representations mentioned in Theorem 2.4.1. It may be useful to note that the space on which the representation is realized is

$$(2.4.16) \qquad V_{f,\lambda} := K(f)_\lambda \otimes_{\mathbb{T}_{\mathbb{Q}_l}} W_l.$$

The representations $\rho_{f,\lambda}$ are continuous by construction. The sheaves $\mathcal{F}_{k,l}$ on $X_1(n)_{\mathbb{Z}[1/nl]}$ are "lisse" away from the cusps, and tamely ramified at the cusps; hence, by Proposition 2.1.9 of [SGA7], XIII, Section 2, $\rho_{f,\lambda}$ is unramified at all p not dividing nl.

In the case where $k = 2$ the construction of the $\rho_{f,\lambda}$ is much simpler, because then the sheaf $\mathcal{F}_k$ is the constant sheaf $\mathbb{Z}$ on $X_1(n)(\mathbb{C})$. The use of étale cohomology can then be replaced by Tate modules of the Jacobian variety of $X_1(n)$. We let $J := J_1(n)$ be this Jacobian variety, actually an Abelian scheme over $\mathbb{Z}[1/n]$. Then we have

$$(2.4.17) \qquad W_l = \mathbb{Q} \otimes \varprojlim_m J(\overline{\mathbb{Q}})[l^m].$$

The fact that for p a prime not dividing nl, the characteristic polynomial of $\rho_{f,\lambda}(\mathrm{Frob}_p)$ is as stated in Theorem 2.4.1 is obtained by studying the reduction modulo p of the Hecke correspondence T_p, that is, as a correspondence on $X_1(nl)_{\mathbb{F}_p}$, compatibly with the sheaf $\mathcal{F}_{k,l}$. For details we refer to Conrad's book [Conr] and Deligne's article [Del1]. In the case $k = 2$ this result is known as the *Eichler-Shimura congruence relation*, expressing the endomorphism T_p of $J_{\mathbb{F}_p}$ as $F + \langle p \rangle V$, where F denotes the Frobenius endomorphism, and V its dual, that is, the endomorphism satisfying $FV = p = VF$ in $\text{End}(J_{\mathbb{F}_p})$. For details in the case $k = 2$ we refer to Section 12.5 of [Di-Im].

Now that we have sketched the construction of the l-adic Galois representations attached to modular forms, we mention some more of their properties that are not mentioned in Theorem 2.4.1 and in the remarks directly following that theorem.

The fact that Deligne proved the Riemann hypothesis part of Weil's conjectures in [Del2] implies very precise bounds on the coefficients of modular forms. The reason for that is that the roots of $x^2 - a_p(f)x + \varepsilon_f(p)p^{k-1}$ are eigenvalues of the Frobenius at p on the space $H^{k-1}(\overline{\mathbb{E}^{k-2}}_{\mathbb{F}_p, \text{et}}, \mathbb{Q}_l)$. We state these bounds, called *Ramanujan bounds*, in a theorem, due to Deligne in the case $k \geq 2$, and to Deligne-Serre ([De-Se]) in the case $k = 1$.

2.4.18 Theorem *Let f be a normalized newform, let n be its level, and let k be its weight. Then for p not dividing n, we have*

$$(2.4.19) \qquad |a_p(f)| \leq 2 \cdot p^{(k-1)/2}.$$

The slightly weaker result that for f as above, $|a_m(f)| = O(m^{k/2})$, can be obtained in a very elementary way; see Corollary 2.1.6 of [Miy] (the idea is to use that the function $z \mapsto |f(z)|(\Im(z))^k$ is bounded on $\mathbb{H}$ and to view $a_m(f)$ as a residue).

Theorem 2.4.1 gives us information on the restriction $\rho_{f,\lambda,p}$ of $\rho_{f,\lambda}$ to decomposition groups $\text{Gal}(\overline{\mathbb{Q}}_p/\mathbb{Q}_p)$ for p not dividing nl. Namely, the theorem says that such restrictions $\rho_{f,\lambda,p}$ are unramified, and it gives the eigenvalues of $\rho_{f,\lambda,p}(\text{Frob}_p)$. It is not known whether $\rho_{f,\lambda,p}(\text{Frob}_p)$ is semisimple; see [Co-Ed] for information on this.

We should note that also in the case that p divides nl almost everything is known about $\rho_{f,\lambda,p}$. For p not dividing l, this is the very general statement that the "Frobenius semisimplification" of $\rho_{f,\lambda,p}$ corresponds, via a suitably normalized local Langlands correspondence, to a certain representation $\pi_{f,p}$ of $GL_2(\mathbb{Q}_p)$ attached to f. This result is due, in increasing order of generality, to Langlands, Deligne, and Carayol. For details on this the reader is referred to [Car], which gives this result in the more general context of Hilbert modular forms (that is, $\mathbb{Q}$ is replaced by a totally real number field). The result for $p = l$ uses Fontaine's p-adic Hodge theory, and is due to Saito ([Sai1] for the case of modular forms, and [Sai2] for the case of Hilbert modular forms).

2.5 GALOIS REPRESENTATIONS OVER FINITE FIELDS, AND REDUCTION TO TORSION IN JACOBIANS

We start this section by explaining how to pass from l-adic Galois representations to Galois representations over finite fields.

Let $f = \sum a_m q^m$ be a (complex) normalized cuspidal eigenform for all Hecke operators T_m, $m \geq 1$, of some level $n \geq 1$ and of some weight $k \geq 2$. As in Theorem 2.4.1 we have the Galois representations $\rho_{f,\lambda}$, from $\mathrm{Gal}(\overline{\mathbb{Q}}/\mathbb{Q})$ to $\mathrm{GL}_2(\overline{\mathbb{Q}}_l)$. It follows from the construction of those representations that there is a finite subextension $\mathbb{Q}_l \to E$ of $\mathbb{Q}_l \to \overline{\mathbb{Q}}_l$ such that $\rho_{f,\lambda}$ takes its values in $\mathrm{GL}_2(E)$. (Actually, this can also be deduced from the continuity alone; see the proof of Corollary 5 in [Dic1] for an argument.) The question as to what the smallest possible E is can be easily answered. Such an E must contain the traces $a_p(f)$ of the $\rho_{f,\lambda}(\mathrm{Frob}_p)$ for all p not dividing nl. So let K be the extension of $\mathbb{Q}$ generated by the $a_p(f)$ with p not dividing n, that is, K is the field of definition of the newform corresponding to f. Then E can be taken to be K_λ, the l-adic completion of K specified by the embedding λ of K into $\overline{\mathbb{Q}}_l$ (see Section 12.5 in [Di-Im]).

Let now $\rho_{f,\lambda}\colon \mathrm{Gal}(\overline{\mathbb{Q}}/\mathbb{Q}) \to \mathrm{GL}_2(E)$ be a realization of $\rho_{f,\lambda}$ over E as above. As $\rho_{f,\lambda}$ is semisimple (it is even irreducible), such a realization is unique up to isomorphism (because it is determined by the traces). Let O_E be the ring of integers in E, that is, the integral closure of $\mathbb{Z}_l$ in E. As $\mathrm{Gal}(\overline{\mathbb{Q}}/\mathbb{Q})$ is compact, it stabilizes some O_E-lattice in E^2 (in the set of lattices, the orbits under $\mathrm{Gal}(\overline{\mathbb{Q}}/\mathbb{Q})$ are finite, so take the intersection, or the sum, of the lattices in one orbit). This means that, after suitable conjugation (choose such a lattice, and an O_E-basis of it), $\rho_{f,\lambda}$ takes values in $\mathrm{GL}_2(O_E)$. We let $O_E \to \overline{\mathbb{F}}_l$ denote the morphism induced by the given embedding of E into $\overline{\mathbb{Q}}_l$ (we view $\overline{\mathbb{F}}_l$ as the residue field of the subring of integers $\overline{\mathbb{Z}}_l$ of $\overline{\mathbb{Q}}_l$). We can then define the *residual* Galois representation $\overline{\rho}_{f,\lambda}$ to be the *semisimplification* of the composed representation $\mathrm{Gal}(\overline{\mathbb{Q}}/\mathbb{Q}) \to \mathrm{GL}_2(O_E) \to \mathrm{GL}_2(\overline{\mathbb{F}}_l)$. Another choice of E or of lattice or basis leads to an isomorphic $\overline{\rho}_{f,\lambda}$, but we note that without the operation of semisimplification this would not be true (see Chapter III of [Ser7]).

Given f, all but finitely many of the $\overline{\rho}_{f,\lambda}$ are irreducible. This was proved for f of level one and with coefficients in $\mathbb{Z}$ in Theorem 4 of [Swi]. The general case follows easily from Theorem 2.3 of [Fa-Jo], which says that if $\overline{\rho}_{f,\lambda}$ is reducible with $l > k$ not dividing n, then $\overline{\rho}_{f,\lambda} = \alpha \oplus \beta \overline{\chi}_l^{k-1}$ with α and β unramified outside n, and $\overline{\chi}_l \colon \mathrm{Gal}(\overline{\mathbb{Q}}/\mathbb{Q}) \to \mathbb{F}_l^\times$ the mod l cyclotomic character. Moreover, the proof shows that the set of l such that some $\overline{\rho}_{f,\lambda}$ is reducible can be bounded explicitly in terms of n and k.

The next question that we want to answer is the following: what is the smallest subfield of $\overline{\mathbb{F}}_l$ over which $\overline{\rho}_{f,\lambda}$ can be realized? Just as for $\rho_{f,\lambda}$ itself, that subfield must contain the traces of the $\overline{\rho}_{f,\lambda}(\mathrm{Frob}_p)$ for all p not dividing nl. That condition turns out to be sufficient, as we will now show. So we let, in this paragraph, $\mathbb{F}$ be the subfield of $\overline{\mathbb{F}}_l$ that is generated by the images $\overline{a_p(f)}$ in $\overline{\mathbb{F}}_l$ of the $a_p(f)$ in $\overline{\mathbb{Z}}_l$. Then for any σ in $\mathrm{Gal}(\overline{\mathbb{F}}_l/\mathbb{F})$ the conjugate $\overline{\rho}_{f,\lambda}^\sigma$ of $\overline{\rho}_{f,\lambda}$ and $\overline{\rho}_{f,\lambda}$ itself are both semisimple and give the same characteristic polynomials as functions on $\mathrm{Gal}(\overline{\mathbb{Q}}/\mathbb{Q})$. Therefore, by a theorem of Brauer-Nesbitt (see Theorem 30.16 of [Cu-Re]), $\overline{\rho}_{f,\lambda}$ is isomorphic to all its conjugates over $\mathbb{F}$. (A more general statement of this kind is given in Exercise 1 of Section 18.2 of [Ser7].) The fact that $\mathrm{Gal}(\overline{\mathbb{F}}_l/\mathbb{F})$ is equal to $\hat{\mathbb{Z}}$ then implies that $\overline{\rho}_{f,\lambda}$ can be realized over $\mathbb{F}$. Let us give an argument for that in terms of matrices, although a much more conceptual argument would be to say that a "gerbe over a finite field is trivial". Let σ be the Frobenius element of $\mathrm{Gal}(\overline{\mathbb{F}}_l/\mathbb{F})$, and let s be an element of $\mathrm{GL}_2(\overline{\mathbb{F}}_l)$ such that for all g in the image of $\overline{\rho}_{f,\lambda}$ we have $\sigma(g) = sgs^{-1}$. Then take a t in $\mathrm{GL}_2(\overline{\mathbb{F}}_l)$ such that $s = \sigma(t)^{-1}t$. Then all tgt^{-1} are in $\mathrm{GL}_2(\mathbb{F})$. By Brauer-Nesbitt, the realization over $\mathbb{F}$ is unique.

For a discussion on possible images of $\overline{\rho}_{f,\lambda}$ we refer the reader to the introduction of [Ki-Ve] (we note however that for f a "CM-form," that is, a form for which all l-adic Galois representations have dihedral image infinitely many of the $\overline{\rho}_{f,\lambda}$ can have dihedral image in $\mathrm{PGL}_2(\overline{\mathbb{F}}_l)$). In particular, Theorem 2.1 of [Rib3] states that for f not a CM-form only finitely many of the images of the $\overline{\rho}_{f,\lambda}$ are *exceptional* in the sense that they are of order prime to l. See also Theorem 2.5.20 for the case where $n = 1$. For $l > 3$ such that $\overline{\rho}_{f,\lambda}$ is irreducible and not exceptional, a result of Dickson (see Chapter XII of [Dic2], or the proof of Theorem 2.5 of [Rib4]) says that the image of $\overline{\rho}_{f,\lambda}$ in $\mathrm{PGL}_2(\overline{\mathbb{F}}_l)$ is, after suitable conjugation, equal to $\mathrm{PGL}_2(\mathbb{F})$ or $\mathrm{SL}_2(\mathbb{F})/\{1, -1\}$ for some finite extension $\mathbb{F}$ of $\mathbb{F}_l$. We note that this field $\mathbb{F}$ can be smaller than the extension of $\mathbb{F}_l$ generated by the traces of $\overline{\rho}_{f,\lambda}$ (indeed, twisting does not change the projective image, but it can make the field generated by the traces bigger).

2.5.1 Lemma *Let l be a prime number, let V be a two-dimensional $\mathbb{F}_l$-vector space, and let G be a subgroup of $\mathrm{Aut}(V)$ of order a multiple of l, and such that V is irreducible as a representation of G. Then G contains $\mathrm{SL}(V)$, and acts transitively on $V - \{0\}$.*

Proof Let g_1 be an element of G of order l. Then the kernel L_1 of $g_1 - \mathrm{id}_V$ is a line. As V is irreducible, L_1 is not G-invariant; hence we can take an element h in G such that $L_2 := hL_1$ is not L_1. Then $g_2 := hg_1h^{-1}$ is of order l and fixes L_2. Let e_1 and e_2 be nonzero elements of L_1 and L_2, respectively. Then with respect to the basis $e = (e_1, e_2)$ of V, g_1 and g_2 are given by elementary matrices of the form $\left(\begin{smallmatrix} 1 & a \\ 0 & 1 \end{smallmatrix}\right)$ and $\left(\begin{smallmatrix} 1 & 0 \\ b & 1 \end{smallmatrix}\right)$, respectively, with a and b nonzero, and hence generate $\mathrm{SL}(V)$. $\qquad\square$

It follows that if $\overline{\rho}_{f,\lambda}$ takes values in $\mathrm{GL}_2(\mathbb{F}_l)$, and is irreducible and not exceptional, then $\mathrm{im}\overline{\rho}_{f,\lambda}$ contains $\mathrm{SL}_2(\mathbb{F}_l)$, and therefore is the subgroup of elements of $\mathrm{GL}_2(\mathbb{F}_l)$ whose determinant is in the image of the character $\overline{\varepsilon_f}\cdot\overline{\chi}_l^{k-1}$. In that case, $\mathrm{im}\overline{\rho}_\lambda$ acts transitively on $\mathbb{F}_l^2 - \{0\}$.

The properties of residual Galois representations that we have seen above show that we do not need to define them via l-adic Galois representations, but that we can start from maximal ideals in Hecke algebras.

2.5.2 Theorem *Let n and k be positive integers. Let $\mathbb{F}$ be a finite field, and $f: \mathbb{T}(n, k) \to \mathbb{F}$ a surjective morphism of rings. Then there is a continuous semisimple representation*

$$\rho_f: \mathrm{Gal}(\overline{\mathbb{Q}}/\mathbb{Q}) \longrightarrow \mathrm{GL}_2(\mathbb{F})$$

that is unramified outside nl, where l is the characteristic of $\mathbb{F}$, such that for all p not dividing nl we have, in $\mathbb{F}$:

$$\text{trace}(\rho(\text{Frob}_p)) = f(T_p) \quad \text{and} \quad \det(\rho(\text{Frob}_p)) = f(\langle p \rangle) p^{k-1}.$$

Such a ρ_f is unique up to isomorphism (that is, up to conjugation).

Proof Let n, k, $\mathbb{F}$, and f be given. As $\mathbb{T} := \mathbb{T}(n,k)$ is free of finite rank as $\mathbb{Z}$-module, $\text{Spec}(\mathbb{T})$ has only finitely many irreducible components, each of which is one-dimensional and finite over $\text{Spec}(\mathbb{Z})$. Therefore, the maximal ideal $\ker(f)$ of $\mathbb{T}_{\mathbb{F}_l}$ is the specialization of a maximal ideal m of $\mathbb{T}_{\mathbb{Q}}$. Let K be the quotient $\mathbb{T}_{\mathbb{Q}}/m$. Then the quotient morphism $\mathbb{T}_{\mathbb{Q}} \to K$ is a normalized eigenform $\tilde{f}$ in $S_k(\Gamma_1(n))_K$, and ρ_f is the realization over $\mathbb{F}$ of the reduction of some $\rho_{\tilde{f},\lambda}$. $\square$

Let now f be as in Theorem 2.5.2, and let us suppose now that ρ_f is irreducible. The construction of l-adic Galois representations that we recalled in Section 2.4 implies that the dual of ρ_f occurs in $\text{H}^{k-1}(\overline{\mathbb{E}^{k-2}}_{\mathbb{Q},\text{et}}, \mathbb{F}_l)$, as well as in $\text{H}^1(X_1(n)_{\overline{\mathbb{Q}},\text{et}}, \overline{\mathcal{F}}_{k,l})$, where $\overline{\mathcal{F}}_{k,l}$ is defined as $\mathcal{F}_{k,l}$ but with $\mathbb{Q}_l$ replaced by $\mathbb{F}_l$. Let us now assume that $k > 2$. Then both these realizations are difficult to deal with computationally. In the first realization the difficulty arises from the degree $k-1$ étale cohomology; it seems to be unknown how to deal explicitly with elements of such cohomology groups. In the second realization, the elements of the cohomology group are isomorphism classes of $\overline{\mathcal{F}}_{k,l}$-torsors on $X_1(n)_{\overline{\mathbb{Q}},\text{et}}$. Such torsors can be described explicitly, as certain covers of $X_1(n)_{\overline{\mathbb{Q}}}$ with certain extra data. The set of such torsors can probably be described by a system of polynomial equations that can be written down in time polynomial in nl (think of the variables as coefficients of certain equations for the torsors). But the problem is that, apparently, there are no good methods known to solve these systems of equations (the number of variables grows too fast with l and the equations are not linear). In fact, the *satisfiability problem* SAT, which is known to be NP-complete (Cook's theorem, see for example [Mor1], or wikipedia), is a special case of the problem of deciding whether a polynomial system of equations over $\mathbb{F}_2$ has a solution over $\mathbb{F}_2$. We note that the description of the set of torsors by a system of polynomial equations should also work over suitable finite extensions of finite fields $\mathbb{F}_p$, in time polynomial in $l \log p$.

Another place where ρ_f occurs is in $J_1(nl)(\overline{\mathbb{Q}})[l]$, that is, in the l-torsion of the Jacobian of the modular curve with level nl, if $l+1 \geq k$ and $l \nmid n$. This means that at the cost of increasing the level by a factor l, we are reduced to dealing with torsion points on Abelian varieties. Of course, the l-adic representations $\rho_{\tilde{f},\lambda}$ attached to lifts of f do not occur in the Jacobian of any curve, simply because the Frobenius eigenvalues are Weil numbers of the wrong weight. What happens here for ρ_f is a "mod l phenomenon" having to do with "congruences" between modular forms. Before we give a detailed statement, let us explain why this happens (such explanations date back at least to the 1960s; Shimura, Igusa, Serre, ...).

For simplicity, and only during this explanation, we assume that $n \geq 5$. Then we have a universal elliptic curve with a given point of order n over $\mathbb{Z}[1/nl]$-schemes: $(\mathbb{E}/Y_1(n), \mathbb{P})$. We let $p: \mathbb{E} \to Y_1(n)$ denote the structure morphism. By definition, we have

$$(2.5.3) \qquad \overline{\mathcal{F}}_{k,l} = \mathrm{Sym}^{k-2}\mathrm{R}^1 p_* \mathbb{F}_l.$$

As explained at the end of Section 1.3, we have a natural isomorphism

$$(2.5.4) \qquad \mathrm{R}^1 p_* \mathbb{F}_l = \mathbb{E}[l]^\vee.$$

And by the definition of $Y_1(nl)$ and the Weil pairing, we have an exact sequence on $Y_1(nl)_{\mathrm{et}}$,

$$(2.5.5) \qquad 0 \longrightarrow \mathbb{F}_l \longrightarrow \mathbb{E}[l] \longrightarrow \mu_l \longrightarrow 0,$$

where $\mathbb{F}_l$ and μ_l denote the corresponding constant sheaves. It follows that the pullback of $\mathrm{R}^1 p_* \mathbb{F}_l$ to $Y_1(nl)_{\mathrm{et}}$ has a 2-step filtration with successive quotients $\mathbb{F}_l$ and $\mu_l^\vee$. Therefore, $\overline{\mathcal{F}}_{k,l}$ has a filtration in $k-1$ steps, with successive quotients $\mathbb{F}_l^{\otimes i} \otimes (\mu_l^\vee)^{\otimes j} = \mu_l^{\otimes -j}$, with $i+j = k-2$, $i \geq 0$, $j \geq 0$. In particular, we get a map:

$$(2.5.6) \qquad \begin{aligned} \mathrm{H}^1(X_1(n)_{\overline{\mathbb{Q}},\mathrm{et}}, \overline{\mathcal{F}}_{k,l}) &\longrightarrow \mathrm{H}^1(X_1(nl)_{\overline{\mathbb{Q}},\mathrm{et}}, \overline{\mathcal{F}}_{k,l}) \\ &\longrightarrow \mathrm{H}^1(X_1(nl)_{\overline{\mathbb{Q}},\mathrm{et}}, \mathbb{F}_l) = J_1(nl)(\overline{\mathbb{Q}})[l]^\vee. \end{aligned}$$

This map explains that ρ_f is likely to occur in $J_1(nl)(\overline{\mathbb{Q}})[l]$. A better way to analyze this map is in fact by studying the direct image of the constant sheaf $\mathbb{F}_l$ via the map $X_1(nl) \to X_1(n)$. A recent detailed treatment of this method, and precise results can be found in [Wie1].

Another way to show that ρ_f occurs in $J_1(nl)(\overline{\mathbb{Q}})[l]$ is to study modular forms mod l of level nl and weight 2. This is more complicated than the modular forms that we have seen before, as it uses the study of the reduction mod l of the modular curve $X_1(nl)$, which is not smooth. The study of these reductions has its roots in Kronecker's congruence relation. The most complete modern accounts of such material are given in the article [De-Ra] by Deligne and Rapoport and in the book [Ka-Ma] by Katz and Mazur. A construction of ρ_f in $J_1(nl)(\overline{\mathbb{Q}})[l]$, following suggestions from Serre, was given by Gross in [Gro].

We are now in a position to state the following theorem, which, combining Gross's result with a so-called multiplicity one theorem, gives us a useful realization of ρ_f. As it is nowadays customary to say, it is a result due to "many people" (mainly Mazur, Ribet, Gross (and Edixhoven for the multiplicity one part)).

2.5.7 Theorem *Let n and k be positive integers, $\mathbb{F}$ a finite field and l its characteristic, and $f\colon \mathbb{T}(n,k) \to \mathbb{F}$ a surjective ring morphism. Assume that $2 < k \leq l+1$ and that the associated Galois representation ρ_f from $\mathrm{Gal}(\overline{\mathbb{Q}}/\mathbb{Q})$ to $\mathrm{GL}_2(\mathbb{F})$ is absolutely irreducible. Then there is a unique ring morphism $f_2\colon \mathbb{T}(nl,2) \to \mathbb{F}$ such that for all $i \geq 1$ one has $f_2(T_i) = f(T_i)$. The morphism f_2 is surjective. Let $m_f = \ker(f_2)$, and let $V_f \subset J_1(nl)(\overline{\mathbb{Q}})$ denote the kernel of m_f, that is, the $\mathbb{F}$-vector space of elements x in $J_1(nl)(\overline{\mathbb{Q}})$ such that $tx = 0$ for all t in m_f. Then V_f is a finite, nonzero, direct sum of copies of ρ_f. If $k < l$ then the multiplicity of ρ_f in V_f is one, that is, V_f realizes ρ_f. For all $a \in (\mathbb{Z}/nl\mathbb{Z})^\times$, one has $f_2(\langle a \rangle) = f(\langle a \rangle)a^{k-2}$, where we still denote by a its images in $\mathbb{Z}/n\mathbb{Z}$ and in $\mathbb{F}_l$.*

Proof The existence of f_2 and the statement that V_f is a successive extension of copies of ρ_f are given in [Gro] (see his Proposition 11.8). In Section 6 of [Edi1] it is proved, applying results from [Bo-Le-Ri], that V_f is a direct sum of copies of ρ_f. Case 1 of Theorem 9.2 of [Edi1] gives the multiplicity one result. $\square$

2.5.8 Remark Let $l > 2$ be prime, $\mathbb{F}$ a finite extension of $\mathbb{F}_l$, and ρ from $\mathrm{Gal}(\overline{\mathbb{Q}}/\mathbb{Q})$ to $\mathrm{GL}_2(\mathbb{F})$ an odd representation. Then ρ is irreducible if and only if it is absolutely irreducible, simply because the image of a complex conjugation under ρ has distinct eigenvalues 1 and -1.

2.5.9 Remark See [Wie2], Corollary 4.5, for a complete result on the multiplicity one question for weights k with $2 \leq k \leq l+1$. In particular, if $k = l$ and ρ_f is unramified at l and $\rho_f(\mathrm{Frob}_l)$ is scalar, then this multiplicity is not one.

As we want to describe V_f explicitly, we will need a bound on the amount of Hecke operators needed to describe $\mathbb{T}(nl, 2)$ and its ideal m_f. We start by quoting a result of Jacob Sturm (see [Stu]).

2.5.10 Theorem (Sturm) *Let $N \geq 1$ be an integer and let Γ be a subgroup of $\mathrm{SL}_2(\mathbb{Z})$ containing $\Gamma(N)$. Let N' be the "width" of the cusp ∞ for Γ, that is, the positive integer defined by $\Gamma \cap \left(\begin{smallmatrix} 1 & \mathbb{Z} \\ 0 & 1 \end{smallmatrix}\right) = \left(\begin{smallmatrix} 1 & N'\mathbb{Z} \\ 0 & 1 \end{smallmatrix}\right)$. Let f be a modular form on Γ of weight k, with coefficients in a discrete valuation ring R contained in $\mathbb{C}$. Let F be the residue field of R, and suppose that the image $\sum a_n q^{n/N'}$ in $F[[q^{1/N'}]]$ of the q-expansion of f has $a_n = 0$ for all $n \leq k[\mathrm{SL}_2(\mathbb{Z}) : \Gamma]/12$. Then $a_n = 0$ for all n, that is, f is congruent to 0 modulo the maximal ideal of R.*

This result of Sturm gives as a direct consequence a bound for up to where one has to take T_i so that one gets a system of generators of the Hecke algebra as $\mathbb{Z}$-module, for a given level and weight. See Section 9.4 of [Ste2] for a detailed proof of Sturm's result, and of this consequence. For convenience we also state and prove this result in the precise context where we use it.

2.5.11 Theorem *Let $N \geq 1$ and $k \geq 1$ be integers. Then $\mathbb{T}(N, k)$ is generated, as $\mathbb{Z}$-module, by the T_i with $1 \leq i \leq k[\mathrm{SL}_2(\mathbb{Z}) : \Gamma_1(N)]/12$.*

Proof Let S be the $\mathbb{Z}$-module $S_k(\Gamma_1(N), \mathbb{Z})$. Then by (2.4.9) we have isomorphisms of $\mathbb{T}(N, k)$-modules: $S = \mathbb{T}(N, k)^\vee$, and $\mathbb{T}(N, k) = S^\vee$. Now the result of Sturm above says that for each prime number p, the elements T_i, $1 \leq i \leq k[\mathrm{SL}_2(\mathbb{Z}) : \Gamma_1(N)]/12$, generate the $\mathbb{F}_p$-vector space $\mathbb{F}_p \otimes S^\vee$, and hence they generate $\mathbb{F}_p \otimes \mathbb{T}(N, k)$. So, indeed, these T_i generate $\mathbb{T}(N, k)$ as $\mathbb{Z}$-module. $\qquad\square$

We can now state a complement to Theorem 2.5.7.

2.5.12 Proposition *In the situation of Theorem 2.5.7, the Hecke algebra* $\mathbb{T}(nl, 2)$ *is generated, as $\mathbb{Z}$-module, by the Hecke operators T_i with i satisfying* $1 \leq i \leq [\mathrm{SL}_2(\mathbb{Z}) : \Gamma_1(nl)]/6$.

We remark that, still in the same situation, giving generators of m_f is then a matter of simple linear algebra over $\mathbb{F}_l$ in a vector space of suitably bounded dimension.

We consider the particular case of a mod l eigenform of level one and of weight k, viewed as a ring morphism f from $\mathbb{T}(1, k)$ to a finite extension of $\mathbb{F}_l$. Then we have the following result, which states explicitly how the Galois representation attached to f is realized in the Jacobian $J_1(l)(\overline{\mathbb{Q}})$. Recall that $\mathbb{T}(l, 2)$, the Hecke algebra acting on weight two cusp forms on $\Gamma_1(l)$, is generated as $\mathbb{Z}$-module by the T_j with $1 \leq j \leq (l^2-1)/6$.

2.5.13 Theorem *Let l be a prime number, let k be an integer such that* $2 < k \leq l+1$, *and let $f : \mathbb{F}_l \otimes \mathbb{T}(1, k) \to \mathbb{F}$ be a surjective ring morphism with $\mathbb{F}$ a finite field of characteristic l, such that the associated Galois representation $\rho : \mathrm{Gal}(\overline{\mathbb{Q}}/\mathbb{Q}) \to \mathrm{GL}_2(\mathbb{F})$ is absolutely irreducible. Let f_2 be the morphism of rings $\mathbb{F}_l \otimes \mathbb{T}(l, 2) \to \mathbb{F}$ such that for all $m \in \mathbb{Z}_{\geq 1}$ we have $f_2(T_m) = f(T_m)$ (see Theorem 2.5.7). Let $(t_1, \ldots, t_r)$ be a system of generators for $\ker(f_2)$. Let*

$$V_f := \bigcap_{1 \leq i \leq r} \ker(t_i, J_1(l)(\overline{\mathbb{Q}})[l]).$$

Then V_f is a two-dimensional $\mathbb{F}$-vector space realizing ρ. For $p \neq l$ prime, $\langle p \rangle$ acts on V_f as multiplication by p^{k-2}.

One can obtain a system of generators of $\ker(f_2)$ as follows. For i in $\{1, \ldots, (l^2 - 1)/6\}$, either $f(T_i)$ is an $\mathbb{F}_l$-linear combination of the $f(T_j)$ with $j < i$, or it is not. If it is not, then let $t_i = 0$. If it is, then pick one, $f(T_i) = \sum_{j < i} a_{i,j} f(T_j)$, and let $t_i = T_i - \sum_{j < i} a_{i,j} T_j$.

Proof Just as in the proof of Theorem 2.5.7, we use Theorem 9.2 of [Edi1], but this time in the case of level one. Case 1 of that theorem deals with the k that satisfy $2 < k < l$. Case 3 deals with the case $k = l$, because ρ, being unramified outside $\{l\}$, and being irreducible of dimension two, *is* ramified at l. Case 4 deals with the case $k = l+1$, because there are no nonzero cusp forms of weight two and level one. $\square$

We also state the following definition and theorem here, because the result, to be used later, is directly related to Theorem 2.5.7. The theorem is due, again, to "many people," just as Theorem 2.5.7 itself.

2.5.14 Definition Let $N \geq 1$, and let $\mathbb{Z}[\zeta_N]$ be the subring of $\mathbb{C}$ generated by a root of unity of order N. To a pair $(E/S/\mathbb{Z}[1/N, \zeta_N], P)$ consisting of an elliptic curve E over a $\mathbb{Z}[1/N, \zeta_N]$-scheme S, together with a point P in $E(S)$ that is of order N everywhere on S, we associate another such pair $(E'/S'/\mathbb{Z}[1/N, \zeta_N], P')$ as follows. Let $\beta: E \rightarrow E'$ be the isogeny whose kernel is the subgroup of E generated by P. Let $\beta^\vee: E' \rightarrow E$ be the dual of β (see Section 2.5 of [Ka-Ma]). Let P' be the unique element of $\ker(\beta^\vee)(S)$ such that $e_\beta(P, P') = \zeta_N$, where e_β is the perfect μ_N-valued pairing between $\ker(\beta)$ and $\ker(\beta^\vee)$ as described in Section 2.8 of [Ka-Ma]. This construction induces an automorphism w_{ζ_N} of the modular curve $X_1(N)_{\mathbb{Z}[1/N, \zeta_N]}$, called an "Atkin-Lehner pseudo-involution."

2.5.15 Theorem *In the situation of Theorem 2.5.7 the completion* $\mathbb{T}_{m_{f,\lambda}}$ *of* $\mathbb{T}$ *at* $m_{f,\lambda}$ *is Gorenstein, that is, the* $\mathbb{Z}_l$-*linear dual of* $\mathbb{T}_{m_{f,\lambda}}$ *is free of rank one as* $\mathbb{T}_{m_{f,\lambda}}$-*module. For all* $r \geq 1$, *the* $(\mathbb{Z}/l^r\mathbb{Z}) \otimes \mathbb{T}_{m_{f,\lambda}}$-*module* $J_1(nl)(\overline{\mathbb{Q}})[l^r]_{m_{f,\lambda}}$ *is free of rank two.*

For any t *in* $\mathbb{T}$ *we have* $t^\vee = wtw^{-1}$, *where* $t^\vee$ *is the dual of* t *as endomorphism of the self-dual Abelian variety* $J_1(nl)_{\mathbb{Q}(\zeta_{nl})}$, *and* w *is the endomorphism of* $J_1(nl)_{\mathbb{Q}(\zeta_{nl})}$ *induced via Picard functoriality by the automorphism* $w_{\zeta_{nl}}$ *of* $X_1(nl)_{\mathbb{Q}(\zeta_{nl})}$.

For $r \geq 0$, *let* $(\cdot, \cdot)_r$ *denote the Weil pairing on* $J_1(nl)(\overline{\mathbb{Q}})[l^r]$, *and let* $\langle \cdot, \cdot \rangle_r$ *denote the pairing defined by*

$$\langle x, y \rangle_r = (x, w(y))_r.$$

Then $\langle \cdot, \cdot \rangle_r$ *is a perfect pairing on* $J_1(nl)(\overline{\mathbb{Q}})[l^r]$ *for which the action of* $\mathbb{T}$ *is self-adjoint. As a consequence,* $\langle \cdot, \cdot \rangle_r$ *induces a perfect pairing on the localization* $J_1(nl)(\overline{\mathbb{Q}})[l^r]_{m_{f,\lambda}}$ *of* $J_1(nl)(\overline{\mathbb{Q}})[l^r]$ *at* $m_{f,\lambda}$.

Proof See Sections 6.4 and 6.8 of [Edi1]. □

2.5.16 Remark See Corollary 4.2 of [Wie2] for a proof that $\mathbb{T}_{m_{f,\lambda}}$ is *not* Gorenstein if the multiplicity of ρ_f in V_f is not one.

Our next goal is to give an effective criterion for whether two modular forms give isomorphic residual Galois representations. We will use some theory on "Katz modular forms"; see Sections 2 and 3 of [Edi1] for a short account. For a in $\mathbb{Z}$, we denote by $M_a(1, \overline{\mathbb{F}}_l)$ the space of Katz modular forms of level one and weight a over $\overline{\mathbb{F}}_l$, and by $S_a(1, \overline{\mathbb{F}}_l)$ its subspace of cuspidal forms. Our reason to use Katz modular forms over $\overline{\mathbb{F}}_l$ is that this gives us the Hasse invariant A in $M_{l-1}(1, \overline{\mathbb{F}}_l)$ and the operators $\theta \colon M_a(1, \overline{\mathbb{F}}_l) \to S_{a+l+1}(1, \overline{\mathbb{F}}_l)$, for all $a \in \mathbb{Z}_{\geq 0}$, that, on q-expansions, act as the differential operator $q \cdot d/dq$. See [Edi1], Section 3, for the properties of θ that we will use.

2.5.17 Lemma *Let $l \geq 5$ be a prime number, k in $\mathbb{Z}_{\geq 0}$, and f a morphism of rings from $\mathbb{T}(1, k)$ to $\overline{\mathbb{F}}_l$. Then there are a k' in $\mathbb{Z}_{\geq 0}$ with $k' \leq l^2 + l$, and a normalized eigenform f' in $S_{k'}(1, \overline{\mathbb{F}}_l)$ with $T_l(f') = 0$, such that $\rho_f \cong \rho_{f'}$.*

Proof Theorem 3.4 of [Edi1] shows that there exist an a in $\{0, \ldots, l+1\}$, and a nonzero g in $M_a(1, \mathbb{F}_l)$ that is an eigenform for *all* Hecke operators, including T_l, and an i in $\{1, \ldots, l-1\}$, such that $\rho_f \cong \rho_g \otimes \chi_l^i$. We choose such a g.

Assume that $a_1(g) \neq 0$. Let $f' := (a_1(g))^{-1} \cdot \theta^i g$. Then f' has all the desired properties.

Assume now that $a_1(g) = 0$. Then $a_n(g) = 0$ for all $n \in \mathbb{Z}_{>0}$, hence the q-expansion of g is the constant power series $a_0(g)$. We replace g by $a_0(g)^{-1} \cdot g$. Then the q-expansion of g is 1, hence a is a multiple of $l-1$ and $g = A^{a/(l-1)}$. This implies that $\rho_g \cong 1 \oplus \chi_l^{-1} \cong (1 \oplus \chi_l) \otimes \chi_l^{-1}$. Let j be the element of $\{1, \ldots, l-1\}$ such that $j = i-1$ in $\mathbb{Z}/(l-1)\mathbb{Z}$. Let $\overline{E}_{l+1} \neq 0$ in $M_{l+1}(1, \overline{\mathbb{F}}_l)$ be the reduction of the Eisenstein series E_{l+1}, normalized by $a_1(\overline{E}_{l+1}) = 1$. Then $f' := \theta^j \overline{E}_{l+1}$ has the required properties. $\square$

2.5.18 Proposition *Let $l \geq 5$ be a prime number, let k_1 and k_2 be in $\mathbb{Z}_{\geq 0}$, let $f_1 \colon \mathbb{T}(1, k_1) \to \overline{\mathbb{F}}_l$ and $f_2 \colon \mathbb{T}(1, k_2) \to \overline{\mathbb{F}}_l$ be two morphisms of rings, and let i be in $\{0, \ldots, l-2\}$. Then ρ_{f_1} and $\rho_{f_2} \otimes \chi_l^i$ are isomorphic if and only if $k_1 = k_2 + 2i$ in $\mathbb{Z}/(l-1)\mathbb{Z}$ and for all primes $p \neq l$ with $p \leq (l^2+l)/12$ we have $f_1(T_p) = p^i f_2(T_p)$.*

Proof Assume first that ρ_{f_1} and $\rho_{f_2} \otimes \chi_l^i$ are isomorphic. Then we have $\det \rho_{f_1} = \det(\rho_{f_2} \otimes \chi_l^i)$, hence $\chi_l^{k_1-1} = \chi_l^{k_2-1+2i}$, hence $k_1 = k_2+2i$ in $\mathbb{Z}/(l-1)\mathbb{Z}$. For all primes $p \neq l$, we have $f_1(T_p) = p^i f_2(T_p)$ because they are the traces of the images under ρ_{f_1} and $\rho_{f_2} \otimes \chi_l^i$ of the Frobenius at p.

Assume now that $k_1 = k_2+2i$ in $\mathbb{Z}/(l-1)\mathbb{Z}$ and that for all primes $p \neq l$ with $p \leq (l^2+l)/12$ we have $f_1(T_p) = p^i f_2(T_p)$. Then $\det \rho_{f_1}$ and $\det(\rho_{f_2} \otimes \chi_l^i)$ are equal, hence it suffices to prove that the traces of ρ_{f_1} and ρ_{f_2} are equal, that is, that for all primes $p \neq l$ we have $f_1(T_p) = p^i f_2(T_p)$.

By Lemma 2.5.17 we can take k_1' in $\mathbb{Z}_{\geq 0}$ and f_1' in $S_{k_1'}(1, \overline{\mathbb{F}}_l)$ a normalized eigenform such that $k_1' \leq l^2+l$, $T_l(f_1') = 0$ and $\rho_{f_1} \cong \rho_{f_1'}$. Similarly, we can take k_2' in $\mathbb{Z}_{\geq 0}$ and f_2' in $S_{k_2'}(1, \overline{\mathbb{F}}_l)$ a normalized eigenform such that $k_2' \leq l^2+l$, $T_l(f_2') = 0$ and $\rho_{f_2} \otimes \chi_l^i \cong \rho_{f_2'}$.

Then $k_1' = k_2'$ in $\mathbb{Z}/(l-1)\mathbb{Z}$, f_1' and f_2' are normalized eigenforms, annihilated by T_l, and with the same eigenvalues for all T_p with $p \leq (l^2+1)/12$ prime. This implies that for all $m \leq (l^2+1)/12$, f_1' and f_2' have the same eigenvalue for T_m. If $k_1' \geq k_2'$, then $f_1' = A^{(k_1'-k_2')/(l-1)} f_2'$ by Sturm's bound in this case: if the difference were nonzero, then the order of vanishing at ∞ contradicts the degree of the line bundle of which it is a section. If $k_2' \geq k_1'$, then $f_2' = A^{(k_2'-k_1')/(l-1)} f_1'$ for the same reason. We conclude that for all primes $p \neq l$ we have $f_1(T_p) = p^i f_2(T_p)$. $\square$

2.5.19 Remark Proposition 2.5.18 can be generalized to forms of higher level, by the similar trick of passing to a higher level n at which one has forms that gave the same Galois representation, but with eigenvalue 0 for all T_p with p dividing n.

The next result gives some conditions under which the Galois representation ρ attached to a surjective ring morphism $f: \mathbb{T}(1, k) \to \mathbb{F}$ has a large image in the sense that it contains $\mathrm{SL}_2(\mathbb{F})$. It is an effective version of Theorem 5.1 of [Rib1]. We will need such a result later on.

2.5.20 Theorem *Let k be a positive integer, l a prime number such that $l > 6(k-1)$, $\mathbb{F}$ a finite field and l its characteristic, and $f: \mathbb{T}(1, k) \to \mathbb{F}$ a surjective morphism of rings such that the associated Galois representation $\rho: \mathrm{Gal}(\overline{\mathbb{Q}}/\mathbb{Q}) \to \mathrm{GL}_2(\mathbb{F})$ is irreducible. Then the image of ρ con-*

tains $SL_2(\mathbb{F})$, *and is equal to the subgroup of* $GL_2(\mathbb{F})$ *of elements g whose determinant is in the subgroup of* $(k-1)$*th powers in* $\mathbb{F}_l^\times$.

Proof As $\mathbb{T}(1,k) = 0$ for $k < 12$, we have $k \geq 12$, and hence $l > 66$. As $l > 2$ and ρ is odd, ρ is absolutely irreducible. We also have $k \leq l+1$.

We apply what is known about the restriction of ρ to an inertia subgroup I at l. We denote by ψ and $\psi' := \psi^l$ the two fundamental characters from I to $\mathbb{F}_{l^2}^\times$ of level 2 (the tame quotient of I is the projective limit of the $\mathbb{F}_{l^n}^\times$ and the fundamental characters of level n to $\overline{\mathbb{F}}_l^\times$ are those that are induced by ring morphisms $\mathbb{F}_{l^n} \to \overline{\mathbb{F}}_l$). By Theorems 2.5 (due to Deligne) and 2.6 (due to Fontaine) in [Edi1], we have

$$\rho|_I = \begin{pmatrix} \chi_l^{k-1} & * \\ 0 & 1 \end{pmatrix} \quad \text{if } f(T_l) \neq 0, \text{ and}$$

$$\overline{\mathbb{F}}_l \otimes_{\mathbb{F}} \rho|_I = \begin{pmatrix} \psi^{k-1} & 0 \\ 0 & \psi'^{k-1} \end{pmatrix} \quad \text{if } f(T_l) = 0.$$

The classification of subgroups of $GL_2(\mathbb{F})$ of order prime to l (see, for example, Proposition 16, Section 2.5 in [Ser3]) says that the image of such a subgroup in $PGL_2(\mathbb{F})$ is either cyclic, dihedral, or isomorphic to A_4, S_4 or A_5.

As ρ is absolutely irreducible, its image in $PGL_2(\mathbb{F})$ cannot be cyclic (note that the kernel of $GL_2(\mathbb{F}) \to PGL_2(\mathbb{F})$ is the center of $GL_2(\mathbb{F})$).

Let us show that the projective image of ρ cannot be A_4, S_4, or A_5. Assume that it is. Then the image of $\rho(I)$ in $PGL_2(\mathbb{F})$ is cyclic and has order at least $(l-1)/(k-1)$ (the order of ψ/ψ' is $l+1$). As we assume that $l-1$ is at least $6(k-1)$, this image has an element of order at least 6, a contradiction.

Let us show that the projective image of ρ is not dihedral. Assume that it is. Then the image of ρ is contained in the normalizer of a Cartan subgroup (that is, the group of points of a split or nonsplit maximal torus), and there is a quadratic extension K of $\mathbb{Q}$ such that ρ is the induction from $\mathrm{Gal}(\overline{\mathbb{Q}}/K)$ to $\mathrm{Gal}(\overline{\mathbb{Q}}/\mathbb{Q})$ of a character of $\mathrm{Gal}(\overline{\mathbb{Q}}/K)$ that is not equal to its conjugate under $\mathrm{Gal}(K/\mathbb{Q})$. As ρ is unramified outside l, K must be the quadratic extension of $\mathbb{Q}$ that is ramified precisely at l. As $l-1 > 2(k-1)$, the description above of $\rho|_I$ shows that there are precisely two lines in $\overline{\mathbb{F}}_l^2$ whose orbit under ρ in $\mathbb{P}^1(\overline{\mathbb{F}}_l)$ has order at most two: these are the coordinate axes (in the first case, the extension must be split). But in the first case the characters on these two lines are not conjugate under $\mathrm{Gal}(K/\mathbb{Q})$, and in the second case the action of ρ on the set of these two lines is not ramified. These contradictions show that the image of ρ cannot be dihedral.

We conclude that the order of the image of ρ is divisible by l. As $l > 3$ a result of Dickson (see [Dic2], Chapter XII, or rather the proof of Theorem 2.5 in [Rib4]) says that the image of ρ in $\mathrm{PGL}_2(\mathbb{F})$ is, after suitable conjugation, equal to $\mathrm{PGL}_2(\mathbb{F}')$ or $\mathrm{SL}_2(\mathbb{F}')/\{1,-1\}$ for some subfield $\mathbb{F}'$ of $\mathbb{F}$.

We claim that $\mathbb{F}' = \mathbb{F}$. Assume that it is not. We let f' and ρ' be the conjugates of f and ρ by the Frobenius automorphism of $\mathbb{F}$ over $\mathbb{F}'$. Then ρ and ρ' are not isomorphic because the traces of the image of ρ generate $\mathbb{F}$ (use that $\mathbb{T}(1,k)$ is generated as $\mathbb{Z}$-module by the T_i with $i \leq k/12 < l$). But their projective representations to $\mathrm{PGL}_2(\mathbb{F})$ are equal. Hence ρ' is a twist of ρ by some character $\chi\colon \mathrm{Gal}(\overline{\mathbb{Q}}/\mathbb{Q}) \to \mathbb{F}^\times$. As ρ and ρ' are unramified outside l, χ is unramified outside l and hence equal to χ_l^i for some i in $\{0,\dots,l-2\}$. But then we have $\theta^{i+1} f = A^j \theta f'$ for some $j \in \mathbb{Z}_{\geq 0}$, with θ and A as in the proof of Proposition 2.5.18. A look at the theta cycles in Section 3 of [Edi1] or Section 7 of [Joc] shows that then $k = (l+3)/2$ if $f(T_l) = 0$, and $k = (l+1)/2$ if $f(T_l) \neq 0$ (in the terminology of [Gro], k is its own "companion weight"). This contradicts our assumption that $l > 6(k-1)$.

So the image G of ρ in $\mathrm{PGL}_2(\mathbb{F})$ contains the image of $\mathrm{SL}_2(\mathbb{F})$. Then, for each $a \in \mathbb{F}$, G contains elements of the form $t\cdot(\begin{smallmatrix}1 & a \\ 0 & 1\end{smallmatrix})$ and $s\cdot(\begin{smallmatrix}1 & 0 \\ a & 1\end{smallmatrix})$, for some t and s in $\mathbb{F}^\times$. Taking suitable powers, we conclude that G contains all $(\begin{smallmatrix}1 & a \\ 0 & 1\end{smallmatrix})$ and $(\begin{smallmatrix}1 & 0 \\ a & 1\end{smallmatrix})$, where a ranges through $\mathbb{F}$. These generate $\mathrm{SL}_2(\mathbb{F})$. As $\det \rho = \chi_l^{k-1}$ the last claim in the theorem follows. $\square$

2.5.21 Remark Eigenforms $f\colon \mathbb{T}(1,k) \to \mathbb{F}$ such that the projective image of ρ is A_4, S_4 or A_5 are related to complex modular forms of weight one and level l or l^2; see [Kh-Wi1], Theorem 10.1. There are tables of such forms. For example, in [Ba-Ki], p. 110, one finds an A_5-example with $l = 2083$ and an S_4-example with $l = 751$. See also Section 4.3 of [Ki-Wi]. We note that for $f = \Delta$ the prime 23 with $23-1 = 2(12-1)$ nicely illustrates one of the arguments used in the proof above: ρ is then dihedral. More generally, f with ρ dihedral come from class groups of imaginary quadratic orders that are unramified outside l.

Chapter Three

First description of the algorithms

by Jean-Marc Couveignes and Bas Edixhoven

We put ourselves in the situation of Theorem 2.5.7, and we ask how we can compute the Galois representation. More explicitly, let n and k be positive integers, $\mathbb{F}$ a finite field and l its characteristic, and $f \colon \mathbb{T}(n,k) \to \mathbb{F}$ a surjective ring morphism. Assume that $2 < k \le l+1$, and that the associated Galois representation $\rho \colon \mathrm{Gal}(\overline{\mathbb{Q}}/\mathbb{Q}) \to \mathrm{GL}_2(\mathbb{F})$ is absolutely irreducible. Let $f_2 \colon \mathbb{T}(nl,2) \to \mathbb{F}$ be the weight two eigenform as in Theorem 2.5.7 and let $m = \ker(f_2)$. Assume that the multiplicity of ρ in $V := J_1(nl)(\overline{\mathbb{Q}})[m]$ is one, that is, that ρ is realized by V.

We let $K \subset \overline{\mathbb{Q}}$ be the field "cut out by ρ", that is, the finite Galois extension of $\mathbb{Q}$ contained in $\overline{\mathbb{Q}}$ consisting of the elements of $\overline{\mathbb{Q}}$ that are fixed by all elements in $\ker(\rho)$. Then we have, by definition, the following factorization of ρ:

$$\rho \colon \mathrm{Gal}(\overline{\mathbb{Q}}/\mathbb{Q}) \twoheadrightarrow \mathrm{Gal}(K/\mathbb{Q}) \hookrightarrow \mathrm{GL}_2(\mathbb{F}).$$

Our aim is then to compute such residual representations ρ, in time polynomial in n, k, and $\#\mathbb{F}$. By computing ρ we mean giving K as a $\mathbb{Q}$-algebra, in the form of a monic polynomial in $\mathbb{Q}[T]$ that is the minimal polynomial of some generator t of K, and giving the elements σ of $\mathrm{Gal}(K/\mathbb{Q})$ by giving their matrices with respect to the $\mathbb{Q}$-basis of K consisting of the first so many powers of t, together with the element $\rho(\sigma)$ of $\mathrm{GL}_2(\mathbb{F})$. Once given such an explicit description of ρ, it becomes possible to compute $f(T_p) \in \mathbb{F}$ in deterministic polynomial time in $\log p$. Indeed this boils down to computing the Frobenius conjugacy class at p in $\mathrm{Gal}(K/\mathbb{Q})$. Chapter 15 explains how to do this.

It will be convenient for us to use the modern version of Galois theory that says that the functor $A \mapsto \mathrm{Hom}_{\mathbb{Q}}(A, \overline{\mathbb{Q}})$ is an anti-equivalence from the category of finite separable $\mathbb{Q}$-algebras to that of finite discrete (continuous) $\mathrm{Gal}(\overline{\mathbb{Q}}/\mathbb{Q})$-sets. An inverse is given by the functor that sends X to $\mathrm{Hom}_{\mathrm{Gal}(\overline{\mathbb{Q}}/\mathbb{Q})}(X, \overline{\mathbb{Q}})$, the $\mathbb{Q}$-algebra of functions f from X to $\overline{\mathbb{Q}}$ such that $f(gx) = g(f(x))$ for all g in $\mathrm{Gal}(\overline{\mathbb{Q}}/\mathbb{Q})$ and all x in X. Under this correspondence, fields correspond to transitive $\mathrm{Gal}(\overline{\mathbb{Q}}/\mathbb{Q})$-sets.

As a first step toward the computation of ρ we let A be the $\mathbb{Q}$-algebra corresponding to the $\mathrm{Gal}(\overline{\mathbb{Q}}/\mathbb{Q})$-set V. Before we explain our strategy to compute A, we sketch how one gets from A to K and ρ. The $\mathbb{Q}$-algebra corresponding to $V \times V$ is $A \otimes A$. The addition map $V \times V \to V$ corresponds to a morphism $A \to A \otimes A$, the co-addition. The $\mathbb{F}^{\times}$-action on V corresponds to an $\mathbb{F}^{\times}$-action on A. We will see later that the co-addition and the $\mathbb{F}^{\times}$-action on A can be computed by the same method by which A will be computed. Viewing $V \times V$ as $\mathrm{Hom}_{\mathbb{F}}(\mathbb{F}^2, V)$ gives a right-action by $\mathrm{GL}_2(\mathbb{F})$ on $V \times V$, hence a left-action on $A \otimes A$. This action can be expressed in the co-addition and the $\mathbb{F}^{\times}$-action.

Let B be the $\mathbb{Q}$-algebra corresponding to the subset $\mathrm{Isom}_{\mathbb{F}}(\mathbb{F}^2, V)$ of $\mathrm{Hom}_{\mathbb{F}}(\mathbb{F}^2, V)$. This factor B of $A \otimes A$ can be computed by linear algebra over $\mathbb{Q}$, using the $\mathrm{GL}_2(\mathbb{F})$-action on $A \otimes A$. In terms of $V \times V$, one removes the subset of (v_1, v_2) that are linearly dependent, that is, the point $(0, 0)$ and the $\mathrm{GL}_2(\mathbb{F})$-orbit of $(V - \{0\}) \times \{0\}$. The field K then corresponds to a $\mathrm{Gal}(\overline{\mathbb{Q}}/\mathbb{Q})$-orbit in $\mathrm{Isom}(\mathbb{F}^2, V)$, hence is obtained by factoring B as a product of fields, using factoring algorithms, and choosing one of the factors. See [Le-Le-Lo], [Len1], and [Lan1] for the fact that such factoring can be done in polynomial time. The equivalence between factoring algebras and polynomials is given in [Len2].

Let $G \subset \mathrm{GL}_2(\mathbb{F})$ be the stabilizer of the chosen factor K. Then G is $\mathrm{Gal}(K/\mathbb{Q})$ and the inclusion is a representation from G to $\mathrm{GL}_2(\mathbb{F})$. Let ϕ be in the chosen $\mathrm{Gal}(\overline{\mathbb{Q}}/\mathbb{Q})$-orbit in $\mathrm{Isom}_{\mathbb{F}}(\mathbb{F}^2, V)$. As this orbit is a right G-torsor on which $\mathrm{Gal}(\overline{\mathbb{Q}}/\mathbb{Q})$ acts, there is, for every σ in $\mathrm{Gal}(\overline{\mathbb{Q}}/\mathbb{Q})$, a unique $g(\sigma)$ in G such that $\rho(\sigma) \circ \phi = \phi \circ g(\sigma)$. Note also that evaluation at ϕ is an embedding of K in $\overline{\mathbb{Q}}$, such that σ in $\mathrm{Gal}(\overline{\mathbb{Q}}/\mathbb{Q})$ induces $g(\sigma)$ on K. It follows that ϕ is an isomorphism between ρ and the representation $g \colon \mathrm{Gal}(\overline{\mathbb{Q}}/\mathbb{Q}) \twoheadrightarrow \mathrm{Gal}(K/\mathbb{Q}) = G \subset \mathrm{GL}_2(\mathbb{F})$.

We now turn to the question of how to compute the $\mathbb{Q}$-algebra A corresponding to V. We wish to produce a generator of A, and its minimal polynomial over $\mathbb{Q}$. This means that we must produce a $\overline{\mathbb{Q}}$-valued function a on V such that $a(\sigma x) = \sigma(a(x))$ for all x in V and all σ in $\mathrm{Gal}(\overline{\mathbb{Q}}/\mathbb{Q})$. Such a function is a generator of A if and only if it does not arise from a strictly smaller quotient of V as $\mathrm{Gal}(\overline{\mathbb{Q}}/\mathbb{Q})$-set (such quotients correspond to subalgebras), hence, equivalently, if and only if a is injective. The minimal polynomial over $\mathbb{Q}$ of such a generator a is given as follows:

$$(3.1) \qquad P(T) = \prod_{x \in V} (T - a(x)).$$

The question now is how to produce such a generator. A direct way would be to compute the elements of V in $J_1(nl)(\overline{\mathbb{Q}})$, by writing down polynomial equations in a suitable coordinate system that is defined over $\mathbb{Q}$, and solving them, using computer algebra. This is essentially how Schoof's algorithm deals with elliptic curves. However, the dimension of $J_1(nl)$ is quadratic in l. Writing down equations in polynomial time still seems possible. But we do not know of a way of solving the equations in a time that is not exponential in the dimension.

The decisive idea is to use numerical computations to approximate the coefficients of a minimal polynomial P as above, in combination with a bound on the *height* of those coefficients. We recall that the (standard, logarithmic) height of a rational number a/b, with a and b integers that are relatively prime, is $\log \max\{|a|, |b|\}$ (a variant would be $\log(a^2 + b^2)$). This rational number $x = a/b$ is known if we know an upper bound h for its height and an approximation y of it (in $\mathbb{R}$, say), with $|x - y| < e^{-2h}/2$. Indeed, if $x' = a'/b'$ also has height at most h, and $x' \neq x$, then

$$|x - x'| = \left| \frac{a}{b} - \frac{a'}{b'} \right| = \left| \frac{ab' - ba'}{bb'} \right| \geq \frac{1}{|bb'|} \geq e^{-2h}.$$

We also note that there are good algorithms to deduce x from such a pair of an approximation y and a bound h, for example, by using continued fractions, as we will now explain.

In practice we will use rational approximations y of x. Every rational number y can be written uniquely as

$$[a_0, a_1, \ldots, a_n] = a_0 + \cfrac{1}{a_1 + \cfrac{1}{\ddots + \cfrac{1}{a_{n-1} + \cfrac{1}{a_n}}}},$$

where $n \in \mathbb{Z}_{\geq 0}$, $a_0 \in \mathbb{Z}$, $a_i \in \mathbb{Z}_{>0}$ for all $i > 0$, and $a_n > 1$ if $n > 0$. To find these a_i, one defines $a_0 := \lfloor y \rfloor$ and puts $n = 0$ if $y = a_0$; otherwise, one puts $y_1 := 1/(y - a_0)$ and $a_1 = \lfloor y_1 \rfloor$ and $n = 1$ if $y_1 = a_1$, and so on. The rational numbers $[a_0, a_1, \ldots, a_i]$ with $0 \leq i \leq n$ are called the *convergents* of the continued fraction of y. Then one has the following well-known result (see Theorem 184 from [Ha-Wr]).

3.2 Proposition *Let y be in $\mathbb{Q}$, a and b in $\mathbb{Z}$ with $b \neq 0$, and*

$$\left| \frac{a}{b} - y \right| < \frac{1}{2b^2}.$$

Then a/b is a convergent of the continued fraction of y.

The question is now: how we are going to implement this method? The basic idea in doing this is to work not on the Abelian variety $J_1(nl)$ but rather on the product $X_1(nl)^g$ of copies of $X_1(nl)$, where g is the genus of $X_1(nl)$. To compare the two, we first choose an effective divisor

$$D_0 = P_1 + \cdots + P_g$$

on $X_1(nl)_{\mathbb{Q}}$, and we consider the well-known map

(3.3)
$$\begin{aligned} X_1(nl)^g &\longrightarrow J_1(nl), \\ (Q_1, \ldots, Q_g) &\mapsto [Q_1 + \cdots + Q_g - D_0]. \end{aligned}$$

To understand the definition of this map rigorously, one must use the interpretation of $X_1(nl)$ as its functor of points with values in $\mathbb{Z}[1/nl]$-schemes, and that of $J_1(nl)$ as the degree zero part of the relative Picard functor $\mathrm{Pic}^0_{X_1(nl)/\mathbb{Z}[1/nl]}$. For the necessary background on this, see Chapters 8 and 9 of [Bo-Lu-Ra]. The divisor D_0 lives on $X_1(nl)_{\mathbb{Q}}$, and it extends uniquely over $\mathbb{Z}[1/nl]$ to an effective relative Cartier divisor of degree

g on $X_1(nl)$. The points P_i of which D_0 is the sum need not be rational over $\mathbb{Q}$.

The inverse image of a point x in $J_1(nl)(\overline{\mathbb{Q}})$ under the map (3.3) can be described as follows. Let $\mathcal{L}_x$ denote a line bundle of degree zero on $X_1(nl)_{\overline{\mathbb{Q}}}$ that corresponds to x (x is an isomorphism class of such line bundles). Then the inverse image of x is the set of $(Q_1, \ldots, Q_g)$, such that $\mathcal{L}_x$ has a rational section whose divisor is $Q_1 + \cdots + Q_g - D_0$, or, equivalently, the set of $(Q_1, \ldots, Q_g)$ such that there is a nonzero section of $\mathcal{L}_x(D_0)$ with divisor $Q_1 + \cdots + Q_g$.

When x ranges over $J_1(nl)(\overline{\mathbb{Q}})$, the class of $\mathcal{L}_x(D_0)$ ranges over the set $\mathrm{Pic}^g(X_1(nl)_{\overline{\mathbb{Q}}})$. The function on $J_1(nl)(\overline{\mathbb{Q}})$ that assigns to x the dimension $h^0(\mathcal{L}_x(D_0))$ of the space of global sections of $\mathcal{L}_x(D_0)$ is semicontinuous in the sense that for each i the locus of x where $h^0(\mathcal{L}_x(D_0)) \geq i$ is closed (the condition $h^0(\mathcal{L}_x(D_0)) \leq i$ need not be closed). On a dense open subset of $J_1(nl)(\overline{\mathbb{Q}})$ this value is one, as can be seen using the theorem of Riemann-Roch, and Serre duality. This means that for x outside a proper closed subset of $J_1(nl)(\overline{\mathbb{Q}})$, the inverse image in $X_1(nl)^g(\overline{\mathbb{Q}})$ of x consists of the g-tuples obtained by permutation of coordinates of a single $(Q_1, \ldots, Q_g)$. Another way to express this is to say that the map (3.3) factors through the symmetric product $X_1(nl)^{(g)}$ and that the map from $X_1(nl)^{(g)}$ to $J_1(nl)$ is birational (that is, an isomorphism on suitable nonempty open parts).

It is then reasonable to assume that we can take D_0 such that for all x in V there is, up to permutation of the coordinates, a unique $Q = (Q_1, \ldots, Q_g)$ in $X_1(nl)^g(\overline{\mathbb{Q}})$ that is mapped to x via the map (3.3). On the other hand, on a curve of high genus such as $X_1(nl)$ it is not clear how to make a large supply of inequivalent effective divisors D_0 on $X_1(nl)_{\mathbb{Q}}$. We will see later, in Theorem 8.1.7, that we can indeed find a suitable divisor, supported on the cusps, and defined over $\mathbb{Q}(\zeta_l)$, on the $X_1(5l)$, which will suffice for treating almost all modular forms of level one.

Remark. In situations where such a cuspidal divisor cannot be found, one could try to use points $P_1, \ldots, P_g$ in $X_1(nl)(L)$, corresponding to elliptic curves lying in one isogeny class, with complex multiplications, for example, by $\mathbb{Q}(i)$. Then L is a solvable Galois extension of $\mathbb{Q}$, so that K can be reconstructed from the compositum KL. If one chooses the P_i reasonably, the degree of L and the logarithm of the discriminant of L are polynomial in l. Another possibility is to try to work with a divisor D_0 of degree smaller than g, for example, a multiple of a rational cusp.

Let us now assume that we have a divisor D_0 as described above. Then we choose a nonconstant function

$$f: X_1(nl)_{\mathbb{Q}} \twoheadrightarrow \mathbb{P}^1_{\mathbb{Q}}$$

that will have to satisfy some conditions that will be given in a moment.

With these two choices, D_0 and f, and a choice of an integer m, we get an element $a_{D_0,f,m}$ of the $\mathbb{Q}$-algebra A corresponding to V as follows. For x in V we let $D_x = Q_{x,1} + \cdots + Q_{x,g}$ be the unique effective divisor of degree g such that

$$(3.4) \qquad\qquad\qquad x = [D_x - D_0].$$

Note that indeed for $x = 0$ we have $D_x = D_0$. We assume that for all $x \in V$ the divisor D_x is disjoint from the poles of f. Then, for each x in V, we define:

$$P_{D_0,f,x} = \prod_{i=1}^{g} \left(t - f(Q_{x,i}) \right) \quad \text{in } \overline{\mathbb{Q}}[t].$$

We then get an element $a_{D_0,f,m}$ of A by evaluating the $P_{D_0,f,x}$ at m:

$$(3.5) \quad a_{D_0,f,m}: V \longrightarrow \overline{\mathbb{Q}}, \quad x \mapsto P_{D_0,f,x}(m) = \prod_{i=1}^{g} \left(m - f(Q_{x,i}) \right).$$

The condition that all the D_x are disjoint from the locus of poles of f will not be guaranteed to hold later when we treat forms of level one, but then it will be possible to omit the $Q_{x,i}$ at which f has a pole from the sum in (3.5) (f will have its poles at certain cusps). For the moment, let us just assume that this condition is satisfied. Then the $f_* D_x$, for x in V, are effective divisors of degree g on $\mathbb{A}^1_{\mathbb{Q}}$.

We will choose f in such a way that the $f_* D_x$ are distinct; we assume now that this is so. Then there is an integer $m \geq 0$ with $m \leq g \cdot (\#\mathbb{F})^4$, such that $a_{D_0,f,m}$ is injective, and hence a generator of A: the polynomials $P_{D_0,f,x}$ are distinct when x varies, and m must not be a root of any difference of two of them.

Finally, we want to have control on the heights of the coefficients of the minimal polynomial of $a_{D_0,f,m}$, because these heights determine the required precision of the approximations of those coefficients that we must compute. The whole strategy depends on the possibility to choose a divisor D_0 and a function f such that, when n, k, and $\mathbb{F}$ vary, those heights grow at most polynomially in n, k, and $\#\mathbb{F}$. Using a great deal of machinery from Arakelov theory, we will show (at least in the case $n = 1$) that any reasonable choices of D_0 and f will lead to an at most polynomial growth of those heights. Intuitively, and completely nonrigorously, one can believe that this should work, because of the following argument. Our x are torsion points, so that their Néron-Tate height is zero. As x and D_0 determine D_x, the height of D_x should be not much bigger than the height of D_0. As we choose D_0 ourselves, it should have small height. Finally, the height of $a_{D_0,f,m}$ should be not much bigger than the sum of those of f and m and the D_x. Turning these optimistic arguments into rigorous statements implies a lot of work that will be done in Chapters 9–11. An important problem here is that in Arakelov theory many results are available that deal with a single curve over $\mathbb{Q}$, but in our situation we are dealing with the infinitely many curves $X_1(nl)$ as l varies.

A few words about the numerical computations involved. What we need is that these can be done in a time that is polynomial in n, $\#\mathbb{F}$, and the number of significant digits that one wants for the coefficients of the minimal polynomial $P_{D_0,f,m}$ of $a_{D_0,f,m}$. It is not at all obvious that this can be done, as the genus of $X_1(nl)$ and hence the dimension of $J_1(nl)$ are quadratic in l.

One way to do the computations is to use the complex uniformizations of $X_1(nl)(\mathbb{C})$ and $J_1(nl)(\mathbb{C})$. The Riemann surface $X_1(nl)(\mathbb{C})$ can be obtained by adding finitely many cusps (the set $\Gamma_1(nl)\backslash\mathbb{P}^1(\mathbb{Z})$) to the quotient $\Gamma_1(nl)\backslash\mathbb{H}$ (see Section 2.1). This means that $X_1(nl)(\mathbb{C})$ is covered by disks around the cusps, which are well suited for computations (functions have q-expansions, for example). In order to describe $J_1(nl)(\mathbb{C})$ as $\mathbb{C}^g$ modulo a lattice, we need a basis of the space of holomorphic differential forms

$H^0(X_1(nl)(\mathbb{C}), \Omega^1)$. The basis that we work with is the one provided by Atkin-Lehner theory, as given in (2.2.18); we write it as $\omega = (\omega_1, \ldots, \omega_g)$. Then we have the following complex description of the map (3.3):

$$(3.6) \qquad X_1(nl)(\mathbb{C})^g \longrightarrow J_1(nl)(\mathbb{C}) = \mathbb{C}^g/\Lambda,$$

$$(Q_1, \ldots, Q_g) \longmapsto [\textstyle\sum_{i=1}^{g} Q_i - \sum_{i=1}^{g} P_i] = \sum_{i=1}^{g} \int_{P_i}^{Q_i} \omega,$$

where Λ is the period lattice with respect to this basis, that is, the image of $H_1(X_1(nl)(\mathbb{C}), \mathbb{Z})$ under integration of the ω_i. This map can be computed up to any desired precision by formal integration of power series on the disks mentioned above. The coefficients needed from the power series expansions of the ω_i can be computed using the method of modular symbols, as has been implemented by William Stein in Magma (see his book [Ste2]). We note that modular symbols algorithms can be used very well to locate V inside $l^{-1}\Lambda/\Lambda$, hence in $J_1(nl)(\mathbb{C})$. A strategy to approximate a point $Q_x = (Q_{x,1}, \ldots, Q_{x,g})$ as above for a nonzero x in V is to lift the straight line that one can draw in $\mathbb{C}^g/\Lambda$ from 0 to x (within a suitable fundamental domain for Λ) to a path in $X_1(nl)(\mathbb{C})^g$ starting at $(P_1, \ldots, P_g)$. In practice this seems to work reasonably well; see Bosman's Chapters 6 and 7. A theoretical difficulty with this approach is that one needs to bound from below the distance to the ramification locus of $X_1(nl)^{(g)} \to J_1(nl)$. Chapter 12 gets around this difficulty and provides a proved algorithm for inverting the Jacobi map (3.3). The starting idea is to set $y = x/N$ for N a large enough integer. This y is no longer an l-torsion point but it is close to the origin in the torus $J_1(nl)(\mathbb{C})$, and this helps finding a preimage Q_y of y, because the behavior of the Jacobi map (3.3) is well understood at least in the neighborhood of the origin. The divisor Q_x we are looking for is such that $Q_x - D_0$ and $N(Q_y - D_0)$ are linearly equivalent. So Q_x can be computed from Q_y by repeated application of an explicit form of the Riemann-Roch theorem. The resulting algorithm reduces to computing approximations of the complex zeros of a great number of modular forms with level $5l$ and weight 4. Chapter 5 explains how to approximate the complex zeros of power series. It also contains a reminder of the necessary notions from computational complexity theory.

Another way to do the "approximation" is to compute the minimal polynomial $P_{D_0,f,m}$ of $a_{D_0,f,m}$ modulo many small primes p. Indeed, the map (3.3) can be reduced mod p. In this case one has no analytic description of the curve and its Jacobian, but one can make random points in $J_1(nl)(\mathbb{F}_q)$ for a suitable finite extension $\mathbb{F}_p \to \mathbb{F}_q$. Such random points can then be projected, using Hecke operators, into V. Elements of $J_1(nl)(\mathbb{F}_q)$ can be represented by divisors on $X_1(nl)_{\mathbb{F}_q}$, and all necessary operations can be done in polynomial time. This approach is explained in detail in Chapter 13. In order to deduce a rational number $x = a/b$ from the knowledge of sufficiently many of its reductions modulo primes p not dividing b we have the following well-known result.

3.7 Proposition Let $x = a/b$ be in $\mathbb{Q}$, with a and b in $\mathbb{Z}$, relatively prime. Let $M = \max\{|a|, |b|\}$. Let S be a finite set of prime numbers p with p not dividing b, such that $\prod_{p \in S} p > 2M^2$. For each p in S, let x_p in $\mathbb{F}_p$ be the reduction of x, and let $L \subset \mathbb{Z}^2$ be the submodule of (n, m) with the property that for all p in S: $n - x_p m = 0$ in $\mathbb{F}_p$. Then (a, b) and $(-a, -b)$ are the shortest nonzero elements of L with respect to the standard inner product on $\mathbb{R}^2$, and the lattice reduction algorithm in dimension two, Algorithm 1.3.14 in [Coh], finds these in time polynomial in $\log M$.

Proof The lattice reduction gives a shortest nonzero element, so it suffices to show that, under the assumptions in the Proposition, the two shortest nonzero elements of L are precisely $\pm(a, b)$. The volume of $\mathbb{R}^2/L$ is the index of L in $\mathbb{Z}^2$, hence equals $\prod_{p \in S} p$. Let l_1 be a shortest nonzero element of L. Then $\|l_1\| \leq \|(a, b)\| \leq \sqrt{2}M$. Let l_2 in L be linearly independent of l_1. Then

$$2M^2 < \mathrm{Vol}(\mathbb{R}^2/L) \leq \mathrm{Vol}(\mathbb{R}^2/(\mathbb{Z} \cdot l_1 + \mathbb{Z} \cdot l_2)) \leq \|l_1\| \cdot \|l_2\| \leq \sqrt{2}M \|l_2\|.$$

Hence $\|l_2\| > \sqrt{2}M \geq \|(a, b)\|$. It follows that (a, b) and l_1 are linearly dependent, and hence $l_1 = \pm(a, b)$. $\qquad\square$

Remark. In case one has a natural rigid analytic uniformization at some prime p, one may want to use that. For the modular curves that we are dealing with this is not the case, but the closely related Shimura curves attached to quaternion algebras over $\mathbb{Q}$ do admit such uniformizations at the primes where the quaternion algebra is ramified (as was proved by Cerednik, Drinfeld, see [Bo-Ca]). More generally, Shimura curves attached to quaternion algebras over totally real number fields admit such rigid uniformizations.

Chapter Four

Short introduction to heights and Arakelov theory

by Bas Edixhoven and Robin de Jong

Chapter 3 explained how the computation of the Galois representations V attached to modular forms over finite fields should proceed. The essential step is to approximate the minimal polynomial P of (3.1) with sufficient precision so that P itself can be obtained. The topic to be addressed now is to bound from above the precision that is needed for this. This means that we must bound the heights of the coefficients of P. As was hinted at in Chapter 3, we get such bounds using Arakelov theory, a tool we discuss in this section. It is not at all excluded that a direct approach to bound the coefficients of P exists, thus avoiding the complicated theory that we use. On the other hand, it is clear that the use of Arakelov theory provides a way to split the work into smaller steps, and that the quantities occurring in each step are intrinsic in the sense that they do not depend on coordinate systems or other choices one could make. We also want to point out that our method does not depend on cancellations of terms in the estimates we will do; all contributions encountered can be bounded appropriately.

   A good reference for a more detailed introduction to heights is Chapter 6 of [Co-Si]. Good references for the Arakelov theory we will use are [Fal1] and [Mor2]. A general reference for heights in the context of Diophantine geometry is [Bo-Gu].

4.1 HEIGHTS ON $\mathbb{Q}$ AND $\overline{\mathbb{Q}}$

The definition of the height of an element of $\mathbb{Q}$ has already been given in Chapter 3; for $x = a/b$ with a and $b \neq 0$ relatively prime integers, we have $h(x) = \log \max\{|a|, |b|\}$. We will now give an equivalent definition in terms of absolute values $|\cdot|_v$ on $\mathbb{Q}$ attached to all places v of $\mathbb{Q}$, the finite places, indexed by the prime numbers, and the infinite place denoted ∞.

The absolute value $|\cdot|_\infty$ is just the usual absolute value on $\mathbb{R}$, restricted to $\mathbb{Q}$. We note that $\mathbb{R}$ is the completion of $\mathbb{Q}$ for $|\cdot|_\infty$. For p prime, we let v_p be the p-adic valuation

$$(4.1.1) \qquad\qquad v_p : \mathbb{Z} \longrightarrow \mathbb{Z} \cup \{\infty\},$$

sending an integer to the maximal number of times it can be divided by p. This valuation v_p on $\mathbb{Z}$ extends uniquely to $\mathbb{Q}$ subject to the condition that $v_p(xy) = v_p(x) + v_p(y)$; we have $v_p(a/b) = v_p(a) - v_p(b)$ for integers a and $b \neq 0$. We let $|\cdot|_p$ denote the absolute value on $\mathbb{Q}$ defined by

$$(4.1.2) \qquad\qquad |x|_p = p^{-v_p(x)}, \quad |0|_p = 0.$$

The completion of $\mathbb{Q}$ with respect to $|\cdot|_p$ is the locally compact topological field $\mathbb{Q}_p$. An important property of these absolute values is that all together they satisfy the product formula

$$(4.1.3) \qquad\qquad \prod_v |x|_v = 1 \quad \text{for all } x \text{ in } \mathbb{Q}^\times.$$

With these definitions, we have

$$(4.1.4) \qquad h(x) = \sum_v \log \max\{1, |x|_v\} \quad \text{for all } x \in \mathbb{Q},$$

where v ranges over the set of all places of $\mathbb{Q}$ (note that almost all terms in the sum are equal to 0).

The height function on $\mathbb{Q}$ generalizes as follows to number fields. First of all, for a local field F we define the natural absolute value $|\cdot|$ on it by letting, for x in $F^\times$, $|x|_F$ be the factor by which all Haar measures on F are scaled by the homothecy $y \mapsto xy$ on F. For example, for $F = \mathbb{C}$ we have $|z|_\mathbb{C} = z\bar{z} = |z|^2$, the square of the usual absolute value. Let now K be a number field. By a finite place of K we mean a maximal ideal of O_K. An infinite place of K is an embedding of K into $\mathbb{C}$, up to complex conjugation. For each place v of K, let K_v be its completion at v; as K_v is a local field, we have the natural absolute value $|\cdot|_v := |\cdot|_{K_v}$ on K_v and K. In this case, the product formula is true (this can be shown easily by considering the adèles of K; see Chapter IV, Section 4, Theorem 5 of [Wei2]). The height function on $\mathbb{Q}$ also generalizes to K. For all x in K we define

$$
\begin{aligned}
h_K(x) &:= \sum_v \log \max\{1, |x|_v\} \\
&= \sum_{v \text{ finite}} \log \max\{1, |x|_v\} + \sum_{\sigma: K \to \mathbb{C}} \log \max\{1, |\sigma(x)|\}.
\end{aligned}
$$

(4.1.5)

This function h_K is called the *height function of K*. For $K \to K'$ an extension of number fields, and for x in K, we have $h_{K'}(x) = (\dim_K K') \cdot h_K(x)$. Therefore we have the *absolute* height function h on $\overline{\mathbb{Q}}$ defined by

$$
(4.1.6) \qquad h: \overline{\mathbb{Q}} \to \mathbb{R}, \quad h(x) = \frac{h_K(x)}{\dim_{\mathbb{Q}} K},
$$

where $K \subset \overline{\mathbb{Q}}$ is any number field that contains x.

4.2 HEIGHTS ON PROJECTIVE SPACES AND ON VARIETIES

For $n \geq 0$ and for K a number field, we define a height function on the projective space $\mathbb{P}^n(K)$ by

$$
\begin{aligned}
h_K((x_0 : \cdots : x_n)) &:= \sum_v \log \max\{|x_0|_v, \ldots, |x_n|_v\}, \\
h(x) &:= \frac{h_K(x)}{\dim_{\mathbb{Q}} K},
\end{aligned}
$$

(4.2.1)

where v ranges through the set of all places of K. We note that it is because of the product formula that $h_K(x)$ is well defined, and that this definition is compatible with our earlier definition of the height and absolute height on K if we view K as the complement of ∞ in $\mathbb{P}^1(K)$. The functions h on $\mathbb{P}^n(K)$ for varying K naturally induce the absolute height function on $\mathbb{P}^n(\overline{\mathbb{Q}})$.

A fundamental result, not difficult to prove, but too important to omit here (even though we will not use it), is Northcott's finiteness theorem.

4.2.2 Theorem (Northcott) *Let n, d, and C be integers. Then*

$$
\{x \in \mathbb{P}^n(\overline{\mathbb{Q}}) \mid h(x) \leq C \text{ and } \dim_{\mathbb{Q}}(\mathbb{Q}(x)) \leq d\}
$$

is a finite set.

For a proof the reader is referred to Chapter 6 of [Co-Si], or to Section 2.4 of [Ser10].

For any algebraic variety X embedded in a projective space $\mathbb{P}^n_K$ over some number field K, we get height functions h_K on $X(K)$ and h on $X(\overline{\mathbb{Q}})$ by restricting those from $\mathbb{P}^n$ to X.

For later use, we include here some simple facts. The height functions on the projective spaces $\mathbb{P}^n(\overline{\mathbb{Q}})$ are compatible with embeddings as coordinate planes, for example, by sending $(x_0 : \cdots : x_n)$ to $(x_0 : \cdots : x_n : 0)$ or to $(0 : x_0 : \cdots : x_n)$.

For all $n \in \mathbb{N}$, we view $\mathbb{A}^n(\overline{\mathbb{Q}})$ as a subvariety of $\mathbb{P}^n(\overline{\mathbb{Q}})$, embedded in one of the $n + 1$ standard ways by sticking in a 1 at the extra coordinate, for example, by sending $(x_1, \ldots, x_n)$ to $(1 : x_1 : \cdots : x_n)$. This gives us, for each n, a height function $h \colon \mathbb{A}^n(\overline{\mathbb{Q}}) \to \mathbb{R}$. These height functions are also compatible with embeddings as coordinate planes. For $n = 1$ the height function on $\mathbb{A}^1(\overline{\mathbb{Q}}) = \overline{\mathbb{Q}}$ is the function in (4.1.6).

4.2.3 Lemma *Let $n \in \mathbb{Z}_{\geq 1}$, and $x_1, \ldots, x_n$ in $\overline{\mathbb{Q}}$. Then*

$$h(x_1 \cdots x_n) \leq \sum_{i=1}^n h(x_i), \quad h(x_1 + \cdots + x_n) \leq \log n + \sum_{i=1}^n h(x_i),$$

$$h(x_i) \leq h((x_1, \ldots, x_n)) \leq h(x_1) + \cdots + h(x_n) \quad \text{for each } i.$$

Proof Let $n \in \mathbb{Z}_{\geq 1}$, and $x_1, \ldots, x_n$ in $\overline{\mathbb{Q}}$. Let $K \subset \overline{\mathbb{Q}}$ be a finite extension of $\mathbb{Q}$ containing the x_i. For the first inequality, we have

$$h_K(x_1 \cdots x_n) = \sum_v \log \max\{1, |x_1 \cdots x_n|_v\}$$

$$\leq \sum_v \log(\max\{1, |x_1|_v\} \cdots \max\{1, |x_n|_v\})$$

$$= h_K(x_1) + \cdots + h_K(x_n).$$

For the second inequality, let $x = x_1 + \cdots + x_n$. Then we have

$$h_K(x) = \sum_{v \text{ finite}} \log \max \left\{ 1, \left| \sum_i x_i \right|_v \right\} + \sum_{\sigma : K \to \overline{\mathbb{Q}}} \log \max \left\{ 1, \left| \sigma \sum_i x_i \right| \right\}$$

$$\leq \sum_{v \text{ finite}} \log \max\{1, \max_i |x_i|_v\} + \sum_\sigma \log \max \left\{ 1, \sum_i |\sigma(x_i)| \right\}$$

$$\leq \sum_{v \text{ finite}} \sum_i \log \max\{1, |x_i|_v\} + \sum_\sigma \max\{0, \log(n \cdot \max_i |\sigma(x_i)|)\}$$

$$\leq \sum_{v \text{ finite},i} \log \max\{1, |x_i|_v\} + \sum_\sigma \log n + \sum_{\sigma,i} \max\{0, \log |\sigma(x_i)|\}$$

$$= (\dim_{\mathbb{Q}} K) \cdot \log n + \sum_i h_K(x_i).$$

For the third inequality, let i be in $\{1, \ldots, n\}$. We have

$$h_K(x_1, \ldots, x_n) = h_K(1 : x_1 : \cdots : x_n)$$

$$= \sum_v \log \max\{1, |x_1|_v, \ldots, |x_n|_v\}$$

$$\geq \sum_v \log \max\{1, |x_i|_v\} = h_K(x_i).$$

Finally, for the last inequality

$$h_K(x_1, \ldots, x_n) = h_K(1 : x_1 : \cdots : x_n)$$

$$= \sum_v \log \max\{1, |x_1|_v, \ldots, |x_n|_v\}$$

$$\leq \sum_v \sum_i \log \max\{1, |x_i|_v\} = h_K(x_1) + \cdots + h_K(x_n).$$

$\square$

4.2.4 Lemma Let $d \geq 1$ and $n \geq d$ be integers. Let Σ_d denote the elementary symmetric polynomial of degree d in n variables. Let $y_1, \ldots, y_n$ be in $\overline{\mathbb{Q}}$. Then we have:

$$h(\Sigma_d(y_1, \ldots, y_n)) \leq n \log 2 + \sum_{1 \leq i \leq n} h(y_i).$$

Proof Let K be the compositum of the fields $\mathbb{Q}(y_i)$ for $i = 1, \ldots, n$. For each place v of K, we let $|\cdot|_v$ be the natural absolute value on K_v and on K

as at the end of Section 4.1. By the triangle inequality we obtain, for each place v of K,

$$|\Sigma_d(y_1, \ldots, y_n)|_v \leq c(v, n) \max_{1 \leq i_1 < \cdots < i_d \leq n} |y_{i_1} \cdots y_{i_d}|_v$$

$$\leq c(v, n) \prod_{1 \leq i \leq n} \max\{1, |y_i|_v\},$$

where $c(v, n) = 2^n$ if v is Archimedean, and $c(v, n) = 1$ if v is non-Archimedean. It follows that

$$\max\{1, |\Sigma_d(y_1, \ldots, y_n)|_v\} \leq c(v, n) \prod_{i=1}^{n} \max\{1, |y_i|_v\}.$$

The proof of the lemma is finished by taking logarithms, summing over the places v, and dividing by $\dim_\mathbb{Q} K$. $\square$

4.2.5 Lemma *Let $x \neq 0$ be in $\overline{\mathbb{Q}}$, let $K = \mathbb{Q}(x)$, and let $\sigma \colon K \to \mathbb{C}$. Then*

$$|\sigma(x)| \geq e^{-(\dim_\mathbb{Q} K) \cdot h(x)}.$$

Proof We have

$$-\log|\sigma(x)| = \log\left(|\sigma(x)|^{-1}\right)$$

$$\leq \log\max\{1, |\sigma(x)|^{-1}\}$$

$$\leq \sum_v \log\max\{1, |x|_v^{-1}\}$$

$$= (\dim_\mathbb{Q} K) \cdot h(x^{-1}) = (\dim_\mathbb{Q} K) \cdot h(x),$$

where the sum is over all places of K. $\square$

4.2.6 Lemma *Let K be a number field, let n be in $\mathbb{Z}_{\geq 1}$, and let a be an n-by-n matrix with coefficients in K. Then*

$$h(\det(a)) \leq \sum_{i,j} h(a_{i,j}) + \frac{1}{2}n \log n.$$

Proof Let v be a finite place of K. Then we have

$$|\det(a)|_v = \left| \sum_{s \in S_n} a_{1,s(1)} \cdots a_{n,s(n)} \right|_v \leq \max_s |a_{1,s(1)}|_v \cdots |a_{n,s(n)}|_v$$

$$\leq \prod_{i,j} \max\{1, |a_{i,j}|_v\}.$$

For $\sigma \colon K \to \mathbb{C}$ we have, by Hadamard's inequality and the comparison $\|\cdot\| \leq n^{1/2}\|\cdot\|_{\max}$ in $\mathbb{C}^n$ of the Euclidean norm and the max-norm,

$$
\begin{aligned}
|\det \sigma(a)| &\leq \prod_j \|\sigma(a_j)\| \\
&\leq \prod_j \left(n^{1/2} \|\sigma(a_j)\|_{\max} \right) \\
&\leq n^{n/2} \prod_{i,j} \max\{1, |\sigma(a_{i,j})|\},
\end{aligned}
$$

where a_j is the jth column of a. Then we have

$$
\begin{aligned}
h_K(\det(a)) &= \sum_v \log \max\{1, |\det(a)|_v\} + \sum_\sigma \log \max\{1, |\det \sigma(a)|\} \\
&\leq \sum_v \sum_{i,j} \log \max\{1, |a_{i,j}|_v\} \\
&\quad + \sum_\sigma \left(\frac{n}{2} \log n + \sum_{i,j} \log \max\{1, |\sigma(a_{i,j})|\} \right) \\
&= \frac{1}{2} (\dim_{\mathbb{Q}} K) \cdot n \log n + \sum_{i,j} h_K(a_{i,j}).
\end{aligned}
$$

Dividing by $\dim_{\mathbb{Q}} K$ gives the result. $\qquad\square$

4.2.7 Lemma *Let K be a number field, and let n be in $\mathbb{Z}_{\geq 1}$, y be in K^n, and a in $\mathrm{GL}_n(K)$. Let x be the unique element in K^n such that $ax = y$. Let b be the maximum of all $h(a_{i,j})$ and $h(y_i)$. Then we have, for all i,*

$$
h(x_i) \leq 2n^2 b + n \log n.
$$

Proof We apply Cramer's rule: $x_i = \det(a(i))/\det(a)$, where the n-by-n matrix $a(i)$ with coefficients in K is obtained by replacing the ith column by y. Lemma 4.2.6 gives us

$$
h(\det(a)) \leq n^2 b + \frac{1}{2} n \log n, \qquad h(\det(a(i))) \leq n^2 b + \frac{1}{2} n \log n.
$$

Therefore: $h(x_i) = h(\det(a(i))/\det(a)) \leq 2n^2 b + n \log n$. $\qquad\square$

4.3 THE ARAKELOV PERSPECTIVE ON HEIGHT FUNCTIONS

We have just defined height functions h_K and h on a variety X over a number field K, embedded into some projective space $\mathbb{P}_K^n$. Such an embedding determines a line bundle $\mathcal{L}$ on X: the restriction of the line bundle $\mathcal{O}(1)$ of $\mathbb{P}_K^n$ that corresponds to homogeneous forms of degree 1, in the variables $x_0, \ldots, x_n$, say. The embedding of X into $\mathbb{P}_K^n$ is given by the global sections $s_0, \ldots, s_n$ of $\mathcal{L}$ obtained by restricting the global sections $x_0, \ldots, x_n$ to X. Now any finite set of generating global sections $t_0, \ldots, t_m$ of $\mathcal{L}$ determines a morphism $f \colon X \to \mathbb{P}_K^m$, inducing height functions $h_{K,f}$ and h_f via pullback along f. For f and f' two such morphisms, the difference $|h_f - h_{f'}|$ is bounded on $X(\overline{\mathbb{Q}})$ (see Theorem 3.1 of Chapter 6 of [Co-Si]). For this reason, one usually associates to a line bundle $\mathcal{L}$ on a variety X a class of height functions $f_{\mathcal{L}}$, that is, an element in the set of functions $X(\overline{\mathbb{Q}}) \to \mathbb{R}$ modulo bounded functions; this map is then a morphism of groups on $\mathrm{Pic}(X)$: $f_{\mathcal{L}_1 \otimes \mathcal{L}_2} \equiv f_{\mathcal{L}_1} + f_{\mathcal{L}_2}$. However, in our situation, we cannot permit ourselves to work modulo bounded functions on each variety, as we have infinitely many curves $X_1(l)$ to deal with.

There is a geometric way to associate to a line bundle a specific height function, not just a class of functions modulo bounded functions. For this, the contributions from the finite as well as the infinite places must be provided. Those from the finite places come from a model of X over the ring of integers O_K of K, that is, an O_K-scheme X_{O_K} whose fiber over K is X, together with a line bundle $\mathcal{L}$ on X_{O_K} whose restriction to X is the line bundle that we had. The O_K-scheme X_{O_K} is required to be proper (for example, projective). The contributions from the infinite places are provided by a Hermitian metric (or inner product) on $\mathcal{L}$, a notion that we will briefly explain.

A Hermitian metric on a locally free $\mathcal{O}_X$-module of finite rank $\mathcal{E}$ consists of a Hermitian metric $\langle \cdot, \cdot \rangle_x$ on all $\mathbb{C}$-vector spaces $x^* \mathcal{E}$, where x runs through $X(\mathbb{C})$, the set of $x \colon \mathrm{Spec}(\mathbb{C}) \to X$. Each x in $X(\mathbb{C})$ induces a morphism $\mathrm{Spec}(\mathbb{C}) \to \mathrm{Spec}(K)$, that is, an embedding of K into $\mathbb{C}$. Therefore, $X(\mathbb{C})$ is the disjoint union of the complex analytic varieties X_σ, indexed by the $\sigma \colon K \to \mathbb{C}$. A Hermitian metric on $\mathcal{E}$ consists of Hermitian metrics on all the holomorphic vector bundles $\mathcal{E}_\sigma$ that $\mathcal{E}$ induces on the X_σ. The metrics to be used are required to be continuous, that is, for U open in X and s and t in $\mathcal{E}(U)$, the function $x \mapsto \langle s(x), t(x) \rangle_x$ on $U(\mathbb{C})$ must be continuous. Actually, the metrics that we will use live on nonsingular X, and will be required to be smooth (infinitely differentiable). Another condition that is usually imposed is a certain compatibility between the metrics at a point x in $X(\mathbb{C})$ and its complex conjugate $\overline{x}$. We do not give this condition in detail, but note that it will be fulfilled by the metrics that we will use. It is also customary to denote a Hermitian metric $\langle \cdot, \cdot \rangle$ by its norm $\|\cdot\|$, given by $\|s\|^2 = \langle s, s \rangle$. Indeed, a suitable polarization identity expresses the Hermitian metric in terms of its norm. A pair $(\mathcal{E}, \|\cdot\|)$ of a locally free $\mathcal{O}_{X_{O_K}}$-module with a Hermitian metric $\|\cdot\|$ is called a metrized vector bundle on X_{O_K}. Metrized vector bundles can be pulled back via morphisms $f: W_{O_K} \to X_{O_K}$ between O_K-schemes of the type considered.

An important example of the above is the case where $X = \mathrm{Spec}(K)$, just a point, and $X_{O_K} = \mathrm{Spec}(O_K)$. A metrized line bundle $(\mathcal{L}, \|\cdot\|)$ then corresponds to an invertible O_K-module, L, say, with Hermitian metrics on the $L_\sigma := \mathbb{C} \otimes_{\sigma, O_K} L$. The *Arakelov degree* of $(\mathcal{L}, \|\cdot\|)$ is the real number defined by

$$(4.3.1) \qquad \deg(\mathcal{L}, \|\cdot\|) = \log \#(L/O_K s) - \sum_{\sigma: K \to \mathbb{C}} \log \|s\|_\sigma,$$

where s is any nonzero element of L (independence of the choice of s follows from the product formula). This definition should be compared to that of the degree of a line bundle on a smooth projective curve over a field: there one takes a rational section, and counts zeros and poles. The first term in (4.3.1) counts the zeros of s at the finite places. Interpreting this term in terms of valuations, and then norms, at the finite places, leads to the second term, which "counts" the "zeros" (or minus the "poles," for that matter) at the infinite places. For a finite extension $K \to K'$, and $(\mathcal{L}, \|\cdot\|)$ on $\mathrm{Spec}(O_K)$ as above, the pullback $(\mathcal{L}', \|\cdot\|)$ to $\mathrm{Spec}(O_{K'})$ has degree $\dim_K K'$ times that on $\mathrm{Spec}(O_K)$.

We can now give the definition of the height given by a proper O_K-scheme X together with a Hermitian line bundle $(\mathcal{L}, \|\cdot\|)$. Let x be in $X(K)$. Then, by the properness of X over O_K, x extends uniquely to an O_K-valued point, also denoted x, and one defines

$$(4.3.2) \qquad\qquad h_K(x) := \deg x^*(\mathcal{L}, \|\cdot\|).$$

The same method as the one used above can be applied to get an absolute height $h: X(\overline{\mathbb{Q}}) \to \mathbb{R}$. For $K \to K'$ a finite extension, each x in $X(K')$ extends uniquely to an x in $X(O_{K'})$, and one defines

$$(4.3.3) \qquad\qquad h(x) := \frac{\deg x^*(\mathcal{L}, \|\cdot\|)}{\dim_{\mathbb{Q}} K'}.$$

It is not hard to verify that this height function h is in the class (modulo bounded functions) that is attached to X_K and $\mathcal{L}_K$ (without metric); see Proposition 7.2 of Chapter 6 of [Co-Si], or Theorem 4.5 of Chapter V in [Ed-Ev]. In fact, for $X = \mathbb{P}^n_{O_K}$, and $\mathcal{L} = \mathcal{O}(1)$ with a suitable metric, the height h just defined is *equal* to the one defined in (4.2.1).

4.4 ARITHMETIC SURFACES, INTERSECTION THEORY, AND ARITHMETIC RIEMANN-ROCH

The context in which we are going to apply Arakelov theory is that of smooth projective curves X over number fields K. In [Ara] Arakelov defined an intersection theory on the *arithmetic surfaces* attached to such curves, with the aim of proving certain results, known in the case of function fields, in the case of number fields. The idea is to take a regular projective model $\mathcal{X}$ over $B := \mathrm{Spec}(O_K)$ of X, and try to develop an intersection theory on the surface $\mathcal{X}$, analogous to the theory one has when K is a function field. If K is a function field over a finite field k, say, one gets a projective surface $\mathcal{X}$ over k, fibered over the nonsingular projective curve B over k that corresponds to K. On such a projective surface, intersecting with principal divisors gives zero, hence the intersection pairing factors through the Picard group of $\mathcal{X}$, the group of isomorphism classes of invertible $\mathcal{O}_{\mathcal{X}}$-modules. In the number field case one "compactifies" B by formally adding the infinite places of K; the product formula then means that principal divisors have degree zero. Instead of the Picard group of $\mathcal{X}$, one considers the group of isomorphism classes of certain metrized line bundles on $\mathcal{X}$, as defined above. In [Fal1] (see also Chapters II, III and I of [Szp]), Faltings extended

Arakelov's work by establishing results such as a Grothendieck-Riemann-Roch theorem in this context. Since then, Arakelov theory has been generalized by Gillet and Soulé (see [Sou] and [Fal2]). Below, we will use the theory as given in [Fal1] and Chapter II of [Szp]. We start with some preparations concerning Riemann surfaces. The aim of this subsection is to give the arithmetic Riemann-Roch theorem as stated and proved by Faltings.

Let X be a compact Riemann surface of genus $g > 0$. The space of holomorphic differentials $H^0(X, \Omega^1_X)$ carries a natural Hermitian inner product:

$$(4.4.1) \qquad (\omega, \eta) \mapsto \frac{i}{2} \int_X \omega \wedge \bar{\eta}.$$

Let $(\omega_1, \ldots, \omega_g)$ be an orthonormal basis with respect to this inner product. This leads to a positive $(1, 1)$-form μ on X given by

$$(4.4.2) \qquad \mu = \frac{i}{2g} \sum_{k=1}^{g} \omega_k \wedge \overline{\omega_k},$$

independent of the choice of orthonormal basis. Note that $\int_X \mu = 1$. We refer to [Ara] for a proof of the following proposition. Denote by C^∞ the sheaf of complex-valued C^∞-functions on X, and by $\mathcal{A}^1$ the sheaf of complex C^∞ 1-forms on X. Recall that we have a tautological differential operator $d: C^\infty \to \mathcal{A}^1$. It decomposes as $d = \partial + \bar{\partial}$ where, for any local C^∞ function f and any holomorphic local coordinate z, with real and imaginary parts x and y, one has $\partial f = \frac{1}{2}(\frac{\partial f}{\partial x} - i \frac{\partial f}{\partial y}) \cdot dz$ and $\bar{\partial} f = \frac{1}{2}(\frac{\partial f}{\partial x} + i \frac{\partial f}{\partial y}) \cdot d\bar{z}$.

4.4.3 Proposition *For each a in X, there exists a unique real-valued $g_{a,\mu}$ in $C^\infty(X - \{a\})$ such that the following properties hold:*

1. *we can write $g_{a,\mu} = \log|z - z(a)| + h$ in an open neighborhood of a, where z is a local holomorphic coordinate and h is a C^∞-function;*

2. *$\partial\bar{\partial} g_{a,\mu} = \pi i \mu$ on $X - \{a\}$;*

3. *$\int_X g_{a,\mu}\mu = 0$.*

We refer to μ and the $g_{a,\mu}$ as the Arakelov $(1, 1)$-form and the Arakelov-Green function, respectively. A fundamental property of the functions $g_{a,\mu}$ is that they give an inverse to the map $C^\infty \to \mathcal{A}^2$, $f \mapsto (-1/\pi i)\partial\bar{\partial} f$, with

$\mathcal{A}^2$ the sheaf of complex C^∞ 2-forms on X, up to constants. For all f in
$C^\infty(X)$ we have

$$(4.4.4) \qquad f(x) = \int_{y \in X} g(x, y) \frac{-1}{\pi i} (\partial \bar{\partial} f) y + \int_X f \, \mu_X.$$

For a proof of this see [Fal1], p. 393, or Lemme 4 in [Elk].

We note that Stokes's theorem implies $g_{a,\mu}(b) = g_{b,\mu}(a)$ for all a and b
in X. The Arakelov-Green functions determine certain metrics, called admissible metrics, on all line bundles $\mathcal{O}_X(D)$, where D is a divisor on X,
as well as on the holomorphic cotangent bundle Ω_X^1. To start, consider
line bundles of the form $\mathcal{O}_X(a)$ with a a point in X (the general case
with D follows by taking tensor products). Let s be the tautological section of $\mathcal{O}_X(a)$, that is, the constant function 1. We define a smooth Her-
mitian metric $\|\cdot\|_{\mathcal{O}_X(a)}$ on $\mathcal{O}_X(a)$ by putting $\log \|s\|_{\mathcal{O}_X(a)}(b) = g_{a,\mu}(b)$
for any b in X. By property 2 of the Arakelov-Green function, the curva-
ture form $(2\pi i)^{-1} \partial \bar{\partial} \log(\|s\|^2)$ of $\mathcal{O}_X(a)$ is equal to μ. To continue, it is
clear that the functions $g_{a,\mu}$ can be used to put a Hermitian metric on the
line bundle $\mathcal{O}_{X \times X}(\Delta_X)$, where Δ_X is the diagonal on $X \times X$, by putting
$\log \|s\|(a, b) = g_{a,\mu}(b)$ for the tautological section s of $\mathcal{O}_{X \times X}(\Delta_X)$. Restricting to the diagonal we have a canonical adjunction isomorphism be-
tween $\mathcal{O}_{X \times X}(-\Delta_X)|_{\Delta_X}$ and Ω_X^1. We define a Hermitian metric $\|\cdot\|_{\mathrm{Ar}}$ on
Ω_X^1 by insisting that this adjunction isomorphism be an isometry. It is
proved in [Ara] that this gives a smooth Hermitian metric on Ω_X^1, and that
its curvature form is a multiple of μ. From now on we will work with these
metrics on $\mathcal{O}_X(P)$ and Ω_X^1 (as well as on tensor product combinations of
them) and refer to them as Arakelov metrics. Explicitly, for $D = \sum_P n_P P$
a divisor on X, we define $g_{D,\mu} := \sum_P n_P g_{P,\mu}$, and equip $\mathcal{O}_X(D)$ with
the metric $\|\cdot\|$ for which $\log \|1\|(Q) = g_{D,\mu}(Q)$, for all Q away from the
support of D. A metrized line bundle $\mathcal{L}$ in general is called *admissible* if,
up to a constant scaling factor, it is isomorphic to one of the admissible
bundles $\mathcal{O}_X(D)$, or, equivalently, if its curvature form $\mathrm{curv}(\mathcal{L})$ is a multiple
of μ. Note that then necessarily we have $\mathrm{curv}(\mathcal{L}) = (\deg \mathcal{L}) \cdot \mu$ by Stokes's
theorem.

For any admissible line bundle $\mathcal{L}$, Faltings defines a certain metric on the
determinant of cohomology $\lambda(\mathcal{L}) = \det \mathrm{H}^0(X, \mathcal{L}) \otimes \det \mathrm{H}^1(X, \mathcal{L})^\vee$ of the

underlying line bundle. This metric is the unique metric satisfying a set of axioms. We recall these axioms (cf. [Fal1], Theorem 1):

1. any isometric isomorphism $\mathcal{L}_1 \xrightarrow{\sim} \mathcal{L}_2$ of admissible line bundles induces an isometric isomorphism $\lambda(\mathcal{L}_1) \xrightarrow{\sim} \lambda(\mathcal{L}_2)$;

2. if we scale the metric on $\mathcal{L}$ by a factor α, the metric on $\lambda(\mathcal{L})$ is scaled by a factor $\alpha^{\chi(\mathcal{L})}$, where $\chi(\mathcal{L}) = \deg \mathcal{L} - g + 1$ is the Euler-Poincaré characteristic of $\mathcal{L}$;

3. for any divisor D and any point P on X, the exact sequence:

$$0 \to \mathcal{O}_X(D - P) \to \mathcal{O}_X(D) \to P_* P^* \mathcal{O}_X(D) \to 0$$

induces an isometry $\lambda(\mathcal{O}_X(D)) \xrightarrow{\sim} \lambda(\mathcal{O}_X(D - P)) \otimes P^* \mathcal{O}_X(D)$;

4. for $\mathcal{L} = \Omega^1_X$, the metric on $\lambda(\mathcal{L}) \cong \det H^0(X, \Omega^1_X)$ is defined by the Hermitian inner product $(\omega, \eta) \mapsto (i/2) \int_X \omega \wedge \bar{\eta}$ on $H^0(X, \Omega^1_X)$.

In particular, for an admissible line bundle $\mathcal{L}$ of degree $g - 1$, the metric on the determinant of cohomology $\lambda(\mathcal{L})$ is independent of scaling.

It was proved by Faltings that we can relate the metric on the determinant of cohomology to theta functions on the Jacobian of X. Let $\mathbb{H}_g$ be the Siegel upper half-space of complex symmetric g-by-g matrices with positive definite imaginary part. Let τ in $\mathbb{H}_g$ be the period matrix attached to a symplectic basis of $H_1(X, \mathbb{Z})$ and consider the analytic Jacobian $J_\tau(X) = \mathbb{C}^g/(\mathbb{Z}^g + \tau \mathbb{Z}^g)$ attached to τ. On $\mathbb{C}^g$ one has the *Riemann theta function* $\vartheta(z; \tau) = \sum_{n \in \mathbb{Z}^g} \exp(\pi i \, {}^t n \tau n + 2\pi i \, {}^t n z)$, giving rise to a reduced effective divisor Θ_0 and a line bundle $\mathcal{O}(\Theta_0)$ on $J_\tau(X)$. Now consider on the other hand the set $\mathrm{Pic}^{g-1}(X)$ of divisor classes of degree $g-1$ on X. It comes with a canonical subset Θ given by the classes of effective divisors. A fundamental theorem of Abel-Jacobi-Riemann says that there is a canonical bijection $\mathrm{Pic}^{g-1}(X) \xrightarrow{\sim} J_\tau(X)$ mapping Θ onto Θ_0. As a result, we can equip $\mathrm{Pic}^{g-1}(X)$ with the structure of a compact complex manifold, together with a divisor Θ and a line bundle $\mathcal{O}(\Theta)$.

The function ϑ is not well defined on $\mathrm{Pic}^{g-1}(X)$ or $J_\tau(X)$. We can remedy this by putting

(4.4.5) $\qquad \|\vartheta\|(z; \tau) = (\det \Im(\tau))^{1/4} \exp(-\pi \, {}^t y (\Im(\tau))^{-1} y) |\vartheta(z; \tau)|,$

with $y = \Im(z)$. One can check that $\|\vartheta\|$ descends to a function on $J_\tau(X)$. By our identification $\mathrm{Pic}^{g-1}(X) \xrightarrow{\sim} J_\tau(X)$ we obtain $\|\vartheta\|$ as a function on $\mathrm{Pic}^{g-1}(X)$. It can be checked that this function is independent of the choice of τ. Note that $\|\vartheta\|$ gives a canonical way to put a metric on the line bundle $\mathcal{O}(\Theta)$ on $\mathrm{Pic}^{g-1}(X)$. For any line bundle $\mathcal{L}$ of degree $g-1$ there is a canonical isomorphism from $\lambda(\mathcal{L})$ to $\mathcal{O}(-\Theta)[\mathcal{L}]$, the fiber of $\mathcal{O}(-\Theta)$ at the point $[\mathcal{L}]$ in $\mathrm{Pic}^{g-1}(X)$ determined by $\mathcal{L}$. Faltings proves that when we give both sides the metrics discussed above, the norm of this isomorphism is a constant independent of $\mathcal{L}$; he writes it as $\exp(\delta(X)/8)$. In more explicit terms, this means that for any line bundle $\mathcal{L}$ of degree $g-1$ on X with $h^0(\mathcal{L}) = 0$ (and hence $h^1(\mathcal{L}) = 0$) we have

(4.4.6) $\lambda(\mathcal{L}) = \mathbb{C}, \quad \|1\|_{\lambda(\mathcal{L})}^{-1} = \exp(\delta(X)/8) \cdot \|\theta\|([\mathcal{L}]).$

The invariant $\delta(X)$ of X appears in the Noether formula, see below.

We will now turn to intersections on an arithmetic surface. For us, an *arithmetic surface* is a proper, flat morphism $p \colon \mathcal{X} \to B$ with $\mathcal{X}$ a regular scheme, with B the spectrum of the ring of integers O_K in a number field K, and with generic fiber a geometrically connected and smooth curve X/K. We say that $\mathcal{X}$ is of genus g if the generic fiber is of genus g. We will always assume that p is a semistable curve, unless explicitly stated otherwise. After we extend the base field if necessary, any geometrically connected, smooth proper curve X/K of positive genus with K a number field is the generic fiber of a unique semistable arithmetic surface.

An Arakelov divisor on $\mathcal{X}$ is a finite formal integral linear combination of integral closed subschemes of codimension one of $\mathcal{X}$ plus a contribution $\sum_\sigma \alpha_\sigma \cdot F_\sigma$ running over the complex embeddings of K. Here α_σ is a real number, and the symbols F_σ correspond to the compact Riemann surfaces X_σ obtained by base changing X/K to $\mathbb{C}$ via σ. We have an $\mathbb{R}$-valued intersection product $(\cdot, \cdot)$ for such divisors, respecting linear equivalence. When we want to indicate which model $\mathcal{X}$ is used for this intersection product, we will use the notation $(\cdot, \cdot)_{\mathcal{X}}$. The notion of principal divisor is given as follows: let f be a nonzero rational function in $K(X)$, then $(f) = (f)_{\mathrm{fin}} + (f)_{\mathrm{inf}}$ with $(f)_{\mathrm{fin}}$ the usual Weil divisor of f on $\mathcal{X}$, and with $(f)_{\mathrm{inf}} = \sum_\sigma v_\sigma(f) \cdot F_\sigma$ with $v_\sigma(f) = -\int_{X_\sigma} \log |f|_\sigma \mu_\sigma$. For a list of properties of this intersection product we refer to [Ara], [Fal1], or

Chapter II of [Szp].

It is proved in [Ara] that the group of linear equivalence classes of Arakelov divisors is naturally isomorphic to the group $\widehat{\mathrm{Pic}}(\mathcal{X})$ of isometry classes of admissible line bundles on $\mathcal{X}$. By an admissible line bundle on $\mathcal{X}$ we mean the datum of a line bundle $\mathcal{L}$ on $\mathcal{X}$, together with admissible metrics on the restrictions $\mathcal{L}_\sigma$ of $\mathcal{L}$ to the X_σ. In particular we have a canonical admissible line bundle $\omega_{\mathcal{X}/B}$ whose underlying line bundle is the relative dualizing sheaf of p. In many situations it is convenient to treat intersection numbers from the point of view of admissible line bundles.

For example, if $P: B \to \mathcal{X}$ is a section of p, and D is an Arakelov divisor on $\mathcal{X}$, the pullback $P^*\mathcal{O}_{\mathcal{X}}(D)$ is a metrized line bundle on B, and we have

$$(4.4.7) \qquad (D, P) = \deg P^*\mathcal{O}_{\mathcal{X}}(D),$$

where the degree deg of a metrized line bundle is as defined in (4.3.1). As a second example, we mention that by definition of the metric on $\omega_{\mathcal{X}/B}$, we have for each section $P: B \to \mathcal{X}$ of p an adjunction formula,

$$(4.4.8) \qquad (P, P + \omega_{\mathcal{X}/B}) = 0.$$

For an admissible line bundle $\mathcal{L}$ on $\mathcal{X}$, we have the notion of determinant of cohomology on B, in this context denoted by $\det \mathrm{R}p_*\mathcal{L}$ (see Chapter II of [Szp]). By using the description above for its metrization over the complex numbers, we obtain the determinant of cohomology on B as a metrized line bundle. One of its most important features is a metrized Riemann-Roch formula (cf. [Fal1], Theorem 3), also called an *arithmetic Riemann-Roch formula*:

$$(4.4.9) \qquad \deg \det \mathrm{R}p_*\mathcal{L} = \frac{1}{2}(\mathcal{L}, \mathcal{L} \otimes \omega_{\mathcal{X}/B}^{-1}) + \deg \det p_*\omega_{\mathcal{X}/B}$$

for any admissible line bundle $\mathcal{L}$ on $\mathcal{X}$.

The term $\deg \det p_*\omega_{\mathcal{X}/B}$ is also known as the *Faltings height* of X, the definition of which we will now recall. We let J_K be the Jacobian variety of X, and J its Néron model over B. Then we have the locally free O_K-module $\mathrm{Cot}_0(J) := 0^*\Omega^1_{J/O_K}$ of rank g, and hence the invertible O_K-module of rank one:

$$\omega_J := \bigwedge^g \mathrm{Cot}_0(J).$$

For each $\sigma: K \to \mathbb{C}$ we have the scalar product on $\mathbb{C} \otimes_{O_K} \omega_J$ given by

$$\langle \omega, \eta \rangle_\sigma = (i/2)^g (-1)^{g(g-1)/2} \int_{J_\sigma(\mathbb{C})} \omega \, \overline{\eta}.$$

The Faltings height $h_K(X)$ is then defined to be the Arakelov degree of the metrized line bundle ω_J,

(4.4.10) $h_K(X) = \deg(\omega_J),$

and the absolute Faltings height (also called stable Faltings height) $h_{\mathrm{abs}}(X)$ of X is defined as

(4.4.11) $h_{\mathrm{abs}}(X) = [K : \mathbb{Q}]^{-1} \deg(\omega_J).$

We remark that the stable Faltings height of X does not change after base change to larger number fields; that is why it is called stable. Therefore, $h_{\mathrm{abs}}(X)$ can be computed from any model of X over a number field as long as that model has stable reduction over the ring of integers of that number field.

As $\mathcal{X} \to B$ is semistable, a result of Raynaud gives that the connected component of 0 of J is the Picard scheme $\mathrm{Pic}^0_{\mathcal{X}/B}$, whose tangent space at 0 is $R^1 p_* \mathcal{O}_{\mathcal{X}}$. Therefore, $\mathrm{Cot}_0(J)$ is the same as $p_* \omega_{\mathcal{X}/B}$, as locally free O_K-modules. A simple calculation (see Lemme 3.2.1 in Chapter I of [Szp]) shows that, with these scalar products, ω_J and $\det p_* \omega_{\mathcal{X}/B}$ are the same as metrized O_K-modules. Therefore we have

(4.4.12) $h_K(X) = \deg(\omega_J) = \deg \det p_* \omega_{\mathcal{X}/B}.$

One may derive from (4.4.9) the following projection formula: let E be a metrized line bundle on B, and $\mathcal{L}$ an admissible line bundle on $\mathcal{X}$. Then the formula

(4.4.13) $\deg \det \mathrm{R} p_*(\mathcal{L} \otimes p^* E) = \deg \det \mathrm{R} p_* \mathcal{L} + \chi(\mathcal{L}) \cdot \deg E$

holds. Here again $\chi(\mathcal{L})$ is the Euler-Poincaré characteristic of $\mathcal{L}$ on the fibers of p.

Chapter Five

Computing complex zeros of polynomials and power series

by Jean-Marc Couveignes

The purpose of this chapter is twofold. We first want to prove the two Theorems 5.3.1 and 5.4.2 below about the complexity of computing complex roots of polynomials and zeros of power series. The existence of a deterministic polynomial time algorithm for these purposes plays an important role in this book. More important, we want to explain what it means to compute with real or complex data in polynomial time. All the necessary concepts and algorithms already exist, and are provided partly by numerical analysis and partly by algorithmic complexity theory. However, the computational model of numerical analysis is not quite a Turing machine, but rather a real computer with floating point arithmetic. Such a computer makes rounding errors at almost every step in the computation. In this context, it is good enough to estimate the conditioning of the problem and the stability of the algorithm used. Statements about conditioning and stability tend to be local and qualitative. This suffices to identify and overcome most difficulties and to design optimal methods.

Our situation, however, is quite different. We don't really care about efficiency. Being polynomial time is enough for us. On the other hand, we want a rigorous, unconditional, and fully general proof that the algorithms we use are polynomial time and return a result that is correct up to a small error that must be bounded rigorously in any case. For this reason, we shall not use floating point registers: we don't want to worry about the accumulation of rounding errors. Rather, we decompose the computation in big blocks. Inside every block we allow only exact computations (e.g., using integers or rational numbers). We also check that the function computed by every such block is well conditioned, and we make a precise statement for that. Finally, we need to control the accumulation of errors in a chain of big blocks. But this will not be too difficult because, since the blocks are big and efficient enough, the general organization of the algorithm is simple and involves few blocks.

In Section 5.1 we recall basic definitions in computational complexity theory. Section 5.2 deals with the problem of computing square roots. We explain for this simple example what is expected from an algorithm in our context. The more general problem of computing complex roots of polynomials is treated in Section 5.3. Finally, we study in Section 5.4 the problem of finding zeros of a converging power series.

Notation: The symbol Θ in this chapter stands for a positive effective absolute constant. So any statement containing this symbol becomes true if the symbol is replaced in every occurrence by some large enough real number.

Aknowledgments: I thank Pascal Molin for pointing out references [Cart] and [Boa] to me.

5.1 POLYNOMIAL TIME COMPLEXITY CLASSES

In this section we briefly recall classical definitions from computational complexity theory. Since we only need to define the polynomial time complexity classes, we shall not go into the details. We refer the reader to Papadimitriou's book [Pap] for a complete treatment of these matters.

Turing machines are a theoretical model for computers. They are finite automata (they have finitely many inner states), but they can write or read on an infinite tape with a tape head. A Turing machine can be defined by a transition table. For a given inner state and current character read by the head, the transition table provides the next inner state, which character to write on the tape in place of the current one, and how the head should move on the tape (one step left, one step right, or no move at all). See Chapter 2 of [Pap] for a formal definition.

A *decision problem* is a question that must be answered by yes or no: for example, deciding whether an integer is prime. The answer to a *functional problem* is a more general function of the question. For example, factoring an integer is a functional problem. If we want to solve a problem with a Turing machine, we write the *input* on the tape, we run the Turing machine, and we wait until it stops. We then read the *output* on the tape. If the machine always stops and returns the correct answer, we say that it solves the problem in question. The time *complexity* is the number of steps before the Turing machine has solved a given problem. Such a Turing machine is said to be *deterministic* because its behavior only depends on the input. The *size* of the input is the number of bits required to encode it. This is the space used on the tape to write this input. For example, the size of an integer is the number of bits in its binary expansion. A problem is said to be *deterministic polynomial time* if there exists a deterministic Turing machine that solves it in time polynomial in the size of the input. The *class* of all functional problems that can be solved in deterministic polynomial time is denoted **FP** or **FPTIME**. The class of deterministic polynomial time decision problems is denoted **P** or **PTIME**.

There exist other models for complexity theory. For example, one may define *multitape Turing machines*. There also exist *random access machines*. All these models lead to equivalent definitions of the polynomial complexity classes. An *algorithm* is a sequence of elementary operations and instructions. Any algorithm can be turned into a Turing machine, but this is fastidious and rather useless since the conceptual description of the algorithm suffices to decide whether the number of elementary operations performed by the algorithm is polynomial in the size of the input. If this is the case, we say that the algorithm is deterministic polynomial time, and we know that the corresponding problem is in **PTIME** or **FPTIME**.

For example, if we want to multiply two positive integers N_1 and N_2 given by their decimal representations, then the size of the input (N_1, N_2) is the number of digits in N_1 and N_2, and this is

$$\lceil \log_{10}(N_1 + 1) \rceil + \lceil \log_{10}(N_2 + 1) \rceil.$$

The number of elementary operations required by the elementary school algorithm for multiplication is $\Theta \times \lceil \log_{10}(N_1 + 1) \rceil \times \lceil \log_{10}(N_2 + 1) \rceil$. The constant here depends on the (reasonable) definition we have chosen for what an elementary operation is. We don't care about constants anyway.

We say that the elementary school algorithm is deterministic polynomial time. There also exists a deterministic polynomial time algorithm for Euclidean division (e.g., the elementary school one). The extended Euclidean algorithm computes coefficients in Bézout's identity in deterministic polynomial time also. So addition, subtraction, multiplication, and inversion in the ring $\mathbb{Z}/N\mathbb{Z}$ can be performed in time polynomial in $\log N$. The class $a \bmod N$ in $\mathbb{Z}/N\mathbb{Z}$ is represented by its smallest non-negative element. We denote it $a\%N$. This is the remainder in the Euclidean division of a by N.

A very important problem is exponentiation: given $a \bmod N$ with a in $[0, N[$ and an integer $e \geq 1$, compute $a^e \bmod N$.

Computing a^e then reducing modulo N is not a good idea because a^e might be very large. Another option would be to set $a_1 = a$ and compute $a_k = (a_{k-1} \times a)\%N$ for $2 \leq k \leq e$. This requires $e-1$ multiplications and $e - 1$ Euclidean divisions. And we never deal with integers bigger than N^2. The complexity of this method is thus $\Theta \times e \times (\log N)^2$ using elementary school algorithms. It is well known, however, that we can do much better. We write the expansion of e in base 2,

$$e = \sum_{0 \leq k \leq K} \epsilon_k 2^k,$$

and we set $b_0 = a$ and $b_k = b_{k-1}^2 \%N$ for $1 \leq k \leq K$. We then notice that

$$a^e \equiv \prod_{0 \leq k \leq K} b_k^{\epsilon_k} \bmod N.$$

So we can compute $(a^e)\%N$ at the expense of $\Theta \times \log e$ multiplications and Euclidean divisions between integers $\leq N^2$. The total number of elementary operations is thus $\Theta \times \log e \times (\log N)^2$ with this method. So exponentiation in $\mathbb{Z}/N\mathbb{Z}$ lies in **FPTIME**. This is an elementary but decisive result in algorithmic number theory. The algorithm above is called *fast exponentiation* and it makes sense in any group. We shall use it many times and in many different contexts.

A first interesting consequence is that for p an odd prime and a an integer such that $1 \leq a \leq p - 1$, we can compute the Legendre symbol

$$\left(\frac{a}{p} \right) \equiv a^{\frac{p-1}{2}} \bmod p$$

at the expense of $\Theta \times (\log p)^3$ elementary operations. So testing quadratic residues is achieved in polynomial deterministic time. Assume now that we are interested in the following problem:

Given an odd prime integer p, find an integer a such that $1 \le a \le p - 1$ and a is not a square modulo p. $\qquad$ $(\star)$

This looks like a very easy problem because half of the nonzero residues modulo p are not squares. So we may just pick a random integer a between 1 and $p - 1$ and compute the Legendre symbol $\left(\frac{a}{p}\right) = a^{\frac{p-1}{2}}$. If the symbol is -1 we output a. Otherwise we output FAIL. The probability of success is $1/2$ and failing is not a big problem because we can rerun the algorithm: we just pick another random integer a.

This is a typical example of a *randomized Las Vegas* algorithm. The behavior of the algorithm depends on the input, of course, but also on the result of some random choices. One has to flip coins. A nice model for such an algorithm would be a Turing machine that receives besides the input, a long enough (say infinite) one-dimensional array $\mathcal{R}$ consisting of 0s and 1s. Whenever the machine needs to flip a coin, it looks at the next entry in the array $\mathcal{R}$. So the Turing machine does not need to flip coins: we provide enough random data at the beginning. We assume that the running time of the algorithm is bounded from above in terms of the size of the input only (this upper bound should not depend on the random data $\mathcal{R}$). For *each* input, we ask that the probability (on $\mathcal{R}$) that the Turing machine returns the correct answer is $\ge 1/2$. The random data $\mathcal{R}$ takes values in $\{0, 1\}^{\mathbb{N}}$. The measure on this latter set is the limit of the uniform measures on $\{0, 1\}^k$ when k tends to infinity. If the Turing machine fails to return the correct answer, it should return FAIL instead.

We have just proved that finding a nonquadratic residue modulo p can be done in Las Vegas probabilistic polynomial time. At present (June 2010) there is no known algorithm that can be proved to solve this problem in deterministic polynomial time. The class of Las Vegas probabilistic polynomial time decision problems is denoted **ZPP**. We now consider another slightly more difficult problem:

Given an odd prime integer p, find a generating set $(g_i)_{1 \le i \le I}$ for the cyclic group $(\mathbb{Z}/p\mathbb{Z})^*$. $\qquad$ $(\star\star)$

We have a simple probabilistic algorithm for this problem. We compute an integer I such that

$$\log_2(3\log_2(p-1)) \leq I \leq \log_2(3\log_2(p-1)) + 2$$

and we pick I random integers $(a_i)_{1\leq i\leq I}$ in the interval $[1, p-1]$. The a_i are uniformly distributed and pairwise independent. We set $g_i = a_i \bmod p$, and we show that the $(g_i)_{1\leq i\leq I}$ generate the group $(\mathbb{Z}/p\mathbb{Z})^*$ with probability $\geq 2/3$. Indeed, if they don't, they must all lie in a maximal subgroup of $(\mathbb{Z}/p\mathbb{Z})^*$. The maximal subgroups of $(\mathbb{Z}/p\mathbb{Z})^*$ correspond to prime divisors of $p-1$. Let q be such a prime divisor. The probability that the $(g_i)_{1\leq i\leq I}$ all lie in the subgroup of index q is bounded from above by

$$\frac{1}{q^I} \leq \frac{1}{2^I},$$

so the probability that the $(g_i)_{1\leq i\leq I}$ don't generate $(\mathbb{Z}/p\mathbb{Z})^*$ is bounded from above by 2^{-I} times the number of prime divisors of $q-1$. Since the latter is $\leq \log_2(p-1)$, the probability of failure is

$$\leq \frac{\log_2(p-1)}{2^I},$$

and this $\leq 1/3$ by definition of I.

Note that here we have a new kind of probabilistic algorithm: the answer is correct with probability $\geq 2/3$ but when the algorithm fails, it may return a false answer. Such an algorithm (a Turing machine) is called *Monte Carlo* probabilistic. This is weaker than a Las Vegas algorithm. We just proved that problem $(\star\star)$ can be solved in Monte Carlo probabilistic polynomial time. We don't know of any Las Vegas probabilistic polynomial time algorithm for this problem. The class of Monte Carlo probabilistic polynomial time decision problems is denoted **BPP**.

In general, a Monte Carlo algorithm can be turned into a Las Vegas one provided the answer can be checked efficiently, because we then can force the algorithm to admit that it has failed. Note also that if we set

$$I = f + \lceil \log_2(\log_2(p-1)) \rceil,$$

where f is a positive integer; then the probability of failure in the algorithm above is bounded from above by 2^{-f}. So we can make this probability arbitrarily small at almost no cost.

The main purpose of this book is to prove statements about the complexity of computing coefficients of modular forms. For example, Theorem 15.1.1 states that on input a prime integer p, computing the Ramanujan function $\tau(p)$ can be done in deterministic polynomial time in log p. An important intermediate result is that one can compute the two-dimensional Galois representation modulo l associated with the Ramanujan τ-function in time polynomial in l. We shall present two methods for computing such modular representations. Both methods rely on computing approximations. The first method computes complex approximations and leads to a deterministic algorithm. This is explained in Chapter 12 using the main results in this Chapter 5. We also present in Chapter 13 a probabilistic method that relies on computations modulo small auxiliary primes. The main reason why the latter method is probabilistic is that it requires finding generating sets for the Picard group of curves over finite fields. This is a generalization of problem ($\star\star$) and solving it in deterministic polynomial time is out of reach at the moment.

5.2 COMPUTING THE SQUARE ROOT OF A POSITIVE REAL NUMBER

In this section, we consider the following problem:

Given a positive real number a,

compute the positive square root $b = \sqrt{a}$.   ($\star\star\star$)

We need an algorithm that runs in deterministic polynomial time. This raises a few simpleminded but important questions about what should be called an algorithm in this context. In Section 5.2.1 we try to formulate problem ($\star\star\star$) in a more precise way. We explain what is meant by an algorithm in this context, and what properties one would expect from such an algorithm. In Section 5.2.4 we present the classical dichotomy algorithm and check that it has polynomial time complexity. The notions presented in this section are classical and elementary and come from computational complexity theory [Pap] and numerical analysis [Hig, Hen]. The algorithms and methods we present are not original either, and they are far from optimal. We stress that our unique goal here is to prove that a polynomial time algorithm (in a sense that can be made rigorous) exists for some classical computational problem regarding real numbers.

5.2.1 Turing machines and real numbers

We shall use deterministic Turing machines; as defined in Chapter 2 of [Pap], for example. There is something annoying with problem (⋆⋆⋆) however: both the input and the output are real numbers. Both existing computing devices and Turing machines only deal with discrete data. So they can't deal with real numbers. We may imagine a Turing machine or a computer handling registers with real numbers as in [Bl-Sh-Sm]. However, this would not be of great use to us, because we plan to perform computations on real numbers as an intermediate step in the computation of a discrete quantity: our basic idea is to compute a rational number (having some arithmetic significance) from a good enough real approximation of it. In the end, we want a rigorous proof that the discrete information we are interested in can be computed by a deterministic Turing machine. We need to prove that a standard deterministic Turing machine can efficiently and safely compute with real and complex numbers, or at least with approximations of them. One possible approach to this classical problem is interval arithmetic as presented in [Hay]. We shall follow a slightly different track, which is better adapted to our situation. Our goal is to prove that a certain number of more or less elementary calculations on complex numbers can be safely and efficiently performed (in a way that will be made more precise soon) by an ordinary Turing machine. These calculations include root finding of polynomials and power series, computation with divisors on modular curves, and direct and inverse Jacobi problems on these curves. In this section, the problem (⋆⋆⋆) will be used to illustrate a few simple ideas that will be applied more systematically in the sequel. The first question to be addressed concerns the input.

What is the input of problem (⋆⋆⋆) ?

Well, if a classical Turing machine is supposed to solve problem (⋆⋆⋆) it cannot be given the real number a all at one time. That would be too big for it. Instead of that, we assume that the Turing machine is given a black box BOX_a. On input a positive integer m_a, the black box BOX_a returns a decimal fraction $N_a \times 10^{-m_a}$ such that $|a - N_a \times 10^{-m_a}| \leq 10^{-m_a}$. If the black box answers immediately, we will call it an *oracle* for a. A more realistic situation is that the black box answers in polynomial time. This means that on input a positive integer m_a, the black box outputs the expected numerator N_a in time $\leq A_a m_a{}^{d_a}$, where A_a and d_a are positive integers depending on a but not on m_a. We assume that a Turing machine calling to a black box (or an oracle) must take the time to read and copy the integrality of the oracle's answer. For example, a Turing machine with an oracle for π cannot access the 10^{100}th digit without reading the previous ones.

In all the situations we shall be facing, there will be a Turing machine in the black box. However, not every real number can be associated with such a Turing machine: the set of Turing machines is countable and the set of real numbers is not. This is the theoretical reason for introducing black boxes here.

What should be the output of a Turing machine solving problem ($\star\star\star$) ?

Again, we don't expect the Turing machine to provide us with the real $b = \sqrt{a}$ all at once. We would be a bit embarrassed with it anyway. We rather expect the Turing machine, on input a positive integer m_b and a black box for a, to return a decimal fraction $N_b \times 10^{-m_b}$ such that

$$|b - N_b \times 10^{-m_b}| \leq 10^{-m_b}.$$

The square root Turing machine may call the black box for a once or several times. Altogether, the input of the square root Turing machine should consist of a black box BOX_a for a and a positive integer m_b telling it the desired absolute accuracy of the expected result. And the output will be a decimal fraction approximating b.

How do we define the complexity of a square root Turing machine ? What does it mean for such a Turing machine to be polynomial time ?

Assume that we have a Turing machine SQRT that computes square roots. Assume that the input of the square root machine SQRT consists of a black box BOX_a for a and a positive integer m_b (the required absolute accuracy of the result). We look for an upper bound on the number of elementary operations performed by SQRT, as a function of $\log \max(a, 1)$ and m_b. Such a bound will be called a *complexity* estimate for SQRT. Notice that a call to the black box BOX_a will be counted as a single operation.

We assume that there exist two positive integers A_{SQRT} and d_{SQRT} such that the complexity of SQRT is bounded above by a polynomial

$$A_{\text{SQRT}}(m_b + \log \max(a, 1))^{d_{\text{SQRT}}}.$$

Then the number of calls to BOX_a is certainly bounded by this number. And the absolute accuracy required from BOX_a cannot exceed this number either, otherwise the machine SQRT would not even find the time to read the digits provided by BOX_a. So by combining a black box BOX_a for the input and the Turing machine SQRT, we obtain a black box BOX_b for the output b. And if both SQRT and BOX_a have polynomial time complexity, so does the resulting black box BOX_b.

Is problem (★★★) well conditioned?

We have seen that the Turing machine SQRT cannot always access the exact value of the input a. Instead, SQRT is provided with a black box that sends it approximations of a. We want to make sure that a good approximation of $b = \sqrt{a}$ can be deduced from a good approximation of a. Since the function $a \mapsto \sqrt{a}$ is $\frac{1}{2}$-Lipschitz on the interval $[1, \infty]$, we have

(5.2.2) $$|\Delta b| = |\sqrt{a + \Delta a} - \sqrt{a}| \leq \frac{1}{2}|\Delta a|$$

as soon as $a \geq 2$ and $|\Delta a| \leq 1$. So a small perturbation of the input results in a small perturbation of the expected output in that case. One says that the problem is *well conditioned*.

Not every computational problem is well conditioned. For example, computing the rounding function $a \mapsto \lceil a \rfloor$ is not well conditioned if one gets close to $\frac{1}{2}$ because the function is not even continuous there. We shall not need to formalize a definition of conditioning, but we shall check on several occasions that the function we want to evaluate is A-Lipschitz for a reasonable constant A. A weaker condition may suffice in some cases; for example, assume that we want to compute a function $a \mapsto b$ and assume that $-\log \min(1, |\Delta b|)$ is lower bounded by $(-\log \min(1, |\Delta a|))^{\frac{1}{e}}$ for some fixed positive integer e. Then the loss of accuracy is polynomial in some sense: one can obtain m digits of b from Θm^e digits of a.

For example, if we consider the problem of computing the square root of a positive real number a, we notice that the function $a \mapsto \sqrt{a}$ is not Lipschitz on $[0, +\infty[$, but we have

$$(5.2.3) \qquad |\Delta b| = |\sqrt{a + \Delta a} - \sqrt{a}| = \sqrt{a} \times \left| \sqrt{1 + \frac{\Delta a}{a}} - 1 \right|$$

$$\leq \sqrt{a} \times \sqrt{|\Delta a|}$$

whenever $|\Delta a| \leq \min(a, a^2)$. So for small values of a, we lose (no more than) half the absolute accuracy when taking the square root. This is enough for us to say that the problem is well conditioned.

5.2.4 The dichotomy algorithm

Given a real interval $[M_1, M_2]$ and a continuous function

$$f : [M_1, M_2] \to \mathbb{R}$$

such that $f(M_1) f(M_2) < 0$, the dichotomy algorithm finds an approximation of a real zero of f in $[M_1, M_2]$. We use the dichotomy algorithm to compute the positive square root $b = \sqrt{a}$ of a positive decimal number $a = N_a \times 10^{-m_a}$, where $N_a \geq 1$ and $m_a \geq 0$ are integers. So we call $f : [0, \infty[\to [0, \infty[$ the map $x \mapsto x^2 - a$. The algorithm below only handles integers and decimal fractions. Let h be the smallest integer such that $10^h \geq N_a$. If $h - m_a$ is even we set $M_2 = 10^{\frac{h-m_a}{2}}$, otherwise we take $M_2 = 10^{\frac{h-m_a+1}{2}}$. We set $M_1 = M_2/10$. We assume that we are given also a positive integer m_b (the required absolute accuracy of the result).

We use two registers R_1 and R_2 containing decimal fractions. The initial value of R_1 is M_1 and the initial value of R_2 is M_2.

The algorithm then goes as follows:

1. If $f(\frac{R_1+R_2}{2})$ is zero or if $|R_1 - R_2| \leq 10^{-m_b}$, output $\frac{R_1+R_2}{2}$ and stop.

2. If $f(R_1) f(\frac{R_1+R_2}{2}) > 0$, set $R_1 := \frac{R_1+R_2}{2}$ and go to step 1.

3. If $f(R_1) f(\frac{R_1+R_2}{2}) < 0$, set $R_2 := \frac{R_1+R_2}{2}$ and go to step 1.

The algorithm ouputs a decimal fraction $\tilde{b}$ such that $|b - \tilde{b}| \leq 10^{-m_b}$. The loop is not executed more than $\Theta(m_b + h - m_a + 1)$ times. So the denominator of $\tilde{b}$ is bounded above by $10^{\Theta(m_a+m_b+h)}$ and the same holds true for any intermediate result in the course of the algorithm. So the *complexity* of the algorithm is polynomial in m_a, $\log N_a$, and m_b. And so is the size of the output.

We now consider a few questions raised by this algorithm.

What to do if the input a is a real number rather than a decimal one?

In that case, we assume that we are given a black box BOX_a for a and a positive integer m_b (the required absolute accuracy of the result). Our first task is to obtain from BOX_a a positive lower bound for a. So we first ask for a decimal approximation of a within 10^{-10}. If the answer is zero, we ask for a decimal approximation of a within 10^{-20}. If the answer is zero again, we ask for a decimal approximation of a within 10^{-40}. After a few steps, either we obtain a positive lower bound for a or we prove that a is smaller than 10^{-2m_b}. In the latter case, we output 0, which is a good enough approximation for $b = \sqrt{a}$.

The above shows that we can assume that we know an integer s such that $10^{2s+1} \leq a \leq 10^{2s+4}$. Replacing a by $a/10^{2s}$ and b by $b/10^s$, we reduce to the case $s = 0$. So $10 \leq a \leq 10^4$. We ask BOX_a for a decimal approximation $\hat{a}$ of a within 10^{-m_b-1}. We set $b = \sqrt{a}$ and $\hat{b} = \sqrt{\hat{a}}$. Inequality (5.2.2) implies that $|\hat{b} - b| \leq 10^{-m_b-1}$. We send the decimal number $\hat{a}$ to the dichotomy algorithm above and ask it for an approximation of $\hat{b} = \sqrt{\hat{a}}$ within 10^{-m_b-1}.

We obtain a decimal number $\widetilde{b}$ such that $|\widetilde{b} - \hat{b}| \leq 10^{-m_b-1}$. We output $\widetilde{b}$ and check that

$$(5.2.5) \qquad |\widetilde{b} - b| \leq |\widetilde{b} - \hat{b}| + |\hat{b} - b| \leq 2 \times 10^{-m_b-1} \leq 10^{-m_b}.$$

So we have designed an algorithm (a Turing machine) that computes the positive square root of a positive number a in time polynomial in $\log(a+1)$ and the required accuracy.

Is the above algorithm stable?

People in numerical analysis say that an algorithm is *stable* when the value output by the algorithm is not too far from the exact value. As we just proved, the dichotomy algorithm can compute $b = \sqrt{a}$ within 10^{-m_b} in time polynomial in $\log(a+1)$ and $m_b \geq 1$. So it can be said to be stable.

We shall allow ourselves to use this terminology, but we prefer to state and prove clear and accurate complexity estimates like the one above.

It is important to make a distinction between stability and conditioning. A *problem* can be said to be well conditioned. An *algorithm* can be said to be stable. For example, we have proved that the dichotomy algorithm for computing square roots is stable using the fact that that problem itself is well conditioned. This is illustrated by inequality (5.2.5).

5.3 COMPUTING THE COMPLEX ROOTS OF A POLYNOMIAL

In this section, we consider the following problem:

Given a degree $d \geq 2$ monic polynomial with complex coeffi-cients

$$P(x) = x^d + \sum_{0 \leq k \leq d-1} a_k x^k,$$

compute the complex roots of $P(x)$.   ($\star\star\star\star$)

The input of problem ($\star\star\star\star$) consists of an integer $d \geq 2$ and a black box BOX_P for the coefficients of $P(x)$. On input a positive integer m and an index k such that $0 \leq k \leq d - 1$, the black box BOX_P returns a decimal fraction

$(N_1 + N_2 i) \times 10^{-m-1}$ such that $|a_k - (N_1 + N_2 i) \times 10^{-m-1}| \leq 10^{-m}$,

where $i = \sqrt{-1} \in \mathbb{C}$.

A Turing machine ROOTS solving problem ($\star\star\star\star$) should be given also a positive integer m_Z telling it the required accuracy of the result. Let $Z = [z_1] + [z_2] + \cdots + [z_d]$ be the divisor of P. This is the formal sum of complex roots, counted with multiplicities. This is an effective divisor of degree d. On input a positive integer m_Z and a black box for the coefficients of $P(x)$, the machine ROOTS is expected to return an approximation $\hat{Z} = [\hat{z}_1] + [\hat{z}_2] + \cdots + [\hat{z}_d]$ of Z within 10^{-m_Z}. This means that there should exist a permutation of the indices $\tau \in \mathcal{S}_d$ such that $|z_{\tau(k)} - \hat{z}_k| \leq 10^{-m_Z}$ for every $1 \leq k \leq d$. The rest of this section is devoted to proving the following theorem.

5.3.1 Theorem (computing roots of polynomials) *There is a deterministic algorithm that on input a degree d monic polynomial*

$$P(x) = x^d + \sum_{0 \le k \le d-1} a_k x^k$$

in $\mathbb{C}[x]$ and a positive integer m_Z, computes an approximation

$$\hat{Z} = [\hat{z}_1] + [\hat{z}_2] + \cdots + [\hat{z}_d]$$

of the divisor

$$Z = [z_1] + [z_2] + \cdots + [z_d]$$

of P, within 10^{-m_Z}. This means that there exists a permutation of the indices $\tau \in S_d$ such that $|z_{\tau(k)} - \hat{z}_k| \le 10^{-m_Z}$ for every $1 \le k \le d$. The running time is polynomial in d, $\log A$ and the required accuracy $m_Z \ge 1$. Here A is the smallest integer bigger than the absolute value of all coefficients in $P(x)$.

In Section 5.3.2 we prove that the problem is well conditioned in some sense. In Section 5.3.7 we recall the principles of Weyl's quadtree algorithm. We recall in Section 5.3.10 that there exists an exclusion function that is sharp enough and easy to compute. This finishes the proof of Theorem 5.3.1.

5.3.2 Conditioning

Our first concern is to check that the problem of finding roots is well conditioned in some sense. We first need to define *clusterings* of roots. Let ϵ be a positive real number. An ϵ-clustering for the roots of $P(x)$ consists of a positive integer K and a pair (c_k, n_k) for every $1 \le k \le K$ such that the following conditions are satisfied:

- c_k is a complex number and n_k is a positive integer for $1 \le k \le K$,

- if $k_1 \ne k_2$ then $|c_{k_1} - c_{k_2}| > 2\epsilon$,

- there are n_k roots of $P(x)$, counting multiplicities, in the open disk $D(c_k, \epsilon/2)$ of center c_k and radius $\epsilon/2$,

- there are n_k roots of $P(x)$ in the open disk $D(c_k, \epsilon)$, counting multiplicities,

- $\sum_{1 \le k \le K} n_k = d$.

So we want to squeeze the complex roots of $P(x)$ into small disks that are distant enough from each other. Note that there may not exist an ϵ-clustering for every ϵ. Problems may occur when the distance between two roots of P is close to ϵ. However, for every positive ϵ, there exist an ϵ' such that

$$2^{-d^2} \epsilon \le \epsilon' \le \epsilon$$

and an ϵ'-clustering for $P(x)$. Indeed, we consider the interval

$$S = [-d^2 \log 2 + \log \epsilon, \log \epsilon],$$

and for every pair (z, z') of distinct roots of $P(x)$ we remove the interval

$$[\log |z - z'| - \log 2, \log |z - z'| + \log 2]$$

to S. The resulting set T is not empty because there are at most $d(d-1)/2$ pairs of distinct roots. Any ϵ' such that $\log \epsilon'$ belongs to T is fine.

Now, call $Z = [z_1] + [z_2] + \cdots + [z_d]$ the divisor of P. We call A the smallest integer bigger than the absolute values of the coefficients of $P(x)$. Let $\Delta(x)$ be a polynomial of degree $\le d - 1$. Let δ be the maximum of the absolute values of the coefficients of $\Delta(x)$. We want to compare the roots of P and the roots of $P + \Delta$. The absolute values of the roots of P are $\le dA$. We fix an $\epsilon \le 1$. We know that there exists an ϵ'-clustering $(c_k, n_k)_{1 \le k \le K}$ of the roots of P for some $2^{-d^2} \epsilon \le \epsilon' \le \epsilon$. Let z be a complex number such that $|z - c_k| = \epsilon'$ for some $1 \le k \le K$. The absolute value of $P(z)$ is lower bounded by $(\epsilon')^d 2^{-d}$. On the other hand, the absolute value of $\Delta(z)$ is upper bounded by $d\delta(dA + 2\epsilon')^{d-1}$. So if

(5.3.3) $$\delta < \left(\frac{\epsilon'}{2(dA + 2)} \right)^d$$

we deduce from Rouché's theorem that $P + \Delta$ has n_k roots inside $D(c_k, \epsilon')$. As a consequence, the roots $\hat{z}_1, \hat{z}_2, \ldots, \hat{z}_d$ of $P + \Delta$ can be indexed in such a way that $|z_j - \hat{z}_j| \le 2\epsilon' \le 2\epsilon$ for every $1 \le j \le d$. Roughly speaking, the meaning of inequality (5.3.3) is that when passing from coefficients to roots, the accuracy is divided by (no significantly more than) the degree d of the polynomial P. We thus have proved the following lemma.

5.3.4 Lemma (conditioning of the roots) *There exists a positive constant Θ such that the following is true. Let $d \geq 1$ be an integer and let $P(x)$ be a degree d monic polynomial with complex coefficients. Let A be the smallest integer bigger than the absolute values of the coefficients of P. Let $\Delta(x)$ be a degree $d - 1$ polynomial with coefficients bounded above by δ in absolute value. Let ϵ be the unique positive real such that*

$$(5.3.5) \qquad\qquad \log \delta = d \log \epsilon - d \log A - \Theta d^3.$$

Assume that $\epsilon \leq 1$. Let $Z = [z_1] + [z_2] + \cdots + [z_d]$ be the divisor of P and let $\hat{Z} = [\hat{z}_1] + [\hat{z}_2] + \cdots + [\hat{z}_d]$ be the divisor of $P(x) + \Delta(x)$. There exists a permutation of the indices $\tau \in \mathcal{S}_d$ such that $|z_{\tau(k)} - \hat{z}_k| \leq \epsilon$ for every $1 \leq k \leq d$.

This lemma tells us that if we are looking for an approximation within 10^{-m_Z} of the divisor $Z = [z_1] + [z_2] + \cdots + [z_d]$ of a monic polynomial $P(x) \in \mathbb{C}[x]$ given by a blackbox BOX_P, we may replace P by a good enough approximation of it having, for instance, decimal coefficients. Indeed, we first compute the smallest integer A bigger than the absolute values of all coefficients of $P(x)$. We assume that m_Z is positive, and we set $\epsilon = 10^{-m_Z - 1}$ and let δ be the real number given by equation (5.3.5). Let m_P be an integer such that $10^{-m_P} \leq \delta/d \leq 10^{-m_P + 2}$. We call the black box BOX_P and obtain for every $0 \leq k \leq d - 1$ a decimal fraction $\hat{a}_k$ with denominator $10^{m_P + 1}$ such that $|\hat{a}_k - a_k| \leq 10^{-m_P} \leq \delta/d$. Let $\hat{P} = x^d + \sum_{0 \leq k \leq d-1} \hat{a}_k x^k$. We compute the discriminant of $\hat{P}$. If it is zero, we replace $\hat{P}$ by $\hat{P} + 10^{-m_P}$ (we increase the constant term by 10^{-m_P}). If the discriminant is zero again, we add 10^{-m_P} to the constant term again. We go on like that until the discriminant of $\hat{P}$ is nonzero. This process stops after d steps at most (seen as a polynomial in the indeterminate a_0, the discriminant has degree $d - 1$, so it cannot cancel d times). In the end we obtain a monic polynomial $\hat{P}$ with decimal coefficients $\hat{a}_k$ such that $|\hat{a}_k - a_k| \leq \delta$ and the discriminant of $\hat{P}$ is nonzero. If $\hat{Z} = [\hat{z}_1] + [\hat{z}_2] + \cdots + [\hat{z}_d]$ is the divisor of $\hat{P}$, there exists a permutation of the indices $\tau \in \mathcal{S}_d$ such that $|z_{\tau(k)} - \hat{z}_k| \leq \epsilon$ for every $1 \leq k \leq d$. The coefficients of $\hat{P}$ are decimal fractions with denominator 10^{m_P+1}, where m_P is an integer such that

$$(5.3.6) \qquad m_P \le \Theta d(m_Z + d^2 + \log A).$$

In order to approximate the roots of P within 10^{-m_Z}, it suffices to approximate the roots of $\hat{P}$ within 10^{-m_Z-1}. So in the sequel we shall assume that we are given a monic polynomial with coefficients in $\mathbb{Z}[i, \frac{1}{10}]$ having no multiple root.

5.3.7 Weyl's quadtree algorithm

We now describe a simpleminded variant of the celebrated Weyl's quadtree algorithm to compute complex roots of polynomials. Let

$$P(x) = x^d + \sum_{0 \le k \le d-1} a_k x^k$$

be a degree $d \ge 1$ monic polynomial with coefficients in $\mathbb{Z}[i, \frac{1}{10}]$. So every coefficient a_k is a fraction $N_k \times 10^{-m_P-1}$ with $N_k = N_{1,k} + N_{2,k}i$ and $N_{1,k}, N_{2,k}$ in $\mathbb{Z}$. We assume that a_0 is not zero. Let g be the smallest positive integer such that the distance between any two distinct roots of P is at least 10^{-g}. For every complex number z we denote by $r(z)$ the distance between z and the closest root of P. We denote by $R(z)$ the distance between z and the furthest root of P. Computing $r(z)$ seems difficult unless one already knows the roots of P. However, we assume that we can compute for every z in $\mathbb{Z}[i, \frac{1}{10}]$ a decimal $\rho(z)$ such that $\rho(z) \le r(z) \le 1.01 \times \rho(z)$. Such a ρ is called an *exclusion function*. We shall give in Section 5.3.10 an example of such an exclusion function that can be computed efficiently.

We first construct a square $\mathcal{Q}$ in the complex plane $\mathbb{C}$ that is large enough to contain all roots of P. We take for $\mathcal{Q}$ a square with center the origin and side length

$$s = 2dA,$$

where A is the smallest integer bigger than the absolute values of all coefficients in P.

Now we divide $\mathcal{Q}$ into four squares of side length $s/2$: the top left square $\mathcal{Q}_1$, the top right square $\mathcal{Q}_2$, the bottom left square $\mathcal{Q}_3$, and the bottom right square $\mathcal{Q}_4$. For each $1 \leq k \leq 4$ we evaluate the exclusion function ρ at the center c_k of $\mathcal{Q}_k$. If $\rho(c_k)$ is bigger than the half-diagonal $s/(2\sqrt{2})$ of $\mathcal{Q}_k$, then we know that there is no root of P in $\mathcal{Q}_k$. So we erase this square.

Next we consider all those squares that have not been erased and divide each of them into four smaller squares with side length $s/4$. We evaluate the exclusion function at the center of every such square. If the value of the exclusion function if bigger than the half-diagonal $s/(4\sqrt{2})$ of the square in question, we erase it.

We go on like that, dividing all remaining squares into four smaller ones at each step. The number of remaining squares is never bigger than $4d$. The reason is that a given root of P cannot compromise more than 4 squares at a time (such a situation would occur if the root in question were very close to the intersection of four contiguous squares).

After n steps, the side length of the remaining squares is $s/2^n$. If $s/2^n$ is much smaller than the minimum distance between two roots of P, then there remains exactly d groups of contiguous squares, and they each contain a single root of P. So the number of steps is bounded by a constant times

$$(5.3.8) \qquad\qquad \log(dA) + m_Z + g$$

where A is the smallest integer bigger than the absolute values of all coefficients in P, m_Z is the required accuracy of the result, and 10^{-g} is a lower bound for the distance between any two roots of P. We illustrate this process in Figure 5.1. The two roots are represented by two bullets.

Note that if the discriminant of P is nonzero, its absolute value is at least $10^{-(m_P+1)(2d-1)}$, so the distance between any two distinct roots is at least $10^{-(m_P+1)(2d-1)}(2dA)^{-d(d-1)}$. We deduce

$$g \leq \Theta d(m_P + d^2 + d \log A).$$

Combining this with the estimates in equations (5.3.8) and (5.3.6), we deduce that the number of steps in Weyl's quadtree algorithm is

$$(5.3.9) \qquad\qquad \leq \Theta d^2 (\log(A) + d^2 + m_Z),$$

where d is the degree of the polynomial, A the smallest integer bigger than the coefficients, and m_Z the required accuracy for the roots of $P(x)$.

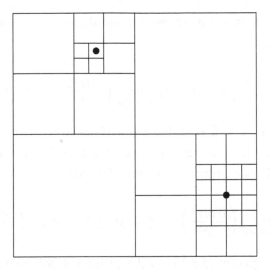

Figure 5.1 Weyl's quadtree method

To finish the proof of Theorem 5.3.1 it remains to prove that there exists an exclusion function ρ that can be evaluated quickly enough. This is done in Section 5.3.10.

5.3.10 Buckholtz inequalities

In this section we recall useful inequalities due to Buckholtz [Buc1, Buc2] and we explain how to deduce a nice exclusion function following [Pan]. Let $P(x) = x^d + \sum_{0 \le k \le d-1} a_k x^k$ be a degree $d \ge 2$ monic polynomial with complex coefficients. Assume that a_0 is not zero. Let

$$Z = [z_1] + [z_2] + \cdots + [z_d]$$

be the divisor of P. For every integer v (positive or negative) we call

$$v_v = z_1^v + \cdots + z_d^v$$

the vth power sum. Buckholtz has shown the inequality:

$$(5.3.11) \qquad \frac{1}{5} \max_{1 \le v \le d} |z_v| \le \max_{1 \le v \le d} \left(\frac{|v_v|}{d} \right)^{\frac{1}{v}} \le \max_{1 \le v \le d} |z_v|.$$

One can easily deduce a more general and sharper statement. If $M \geq 1$ is an integer, then

$$(5.3.12) \quad 5^{-\frac{1}{M}} \max_{1 \leq v \leq d} |z_v| \leq \max_{1 \leq v \leq d} \left(\frac{|v_M v|}{d} \right)^{\frac{1}{Mv}} \leq \max_{1 \leq v \leq d} |z_v|.$$

Recall that for any $z \in \mathbb{C}$ we call $r(z)$ the distance between z and the closest root of P and $R(z)$ the distance between z and the furthest root of P. From equation (5.3.12) we deduce an estimate for $R(0)$,

$$(5.3.13) \quad \max_{1 \leq v \leq d} \left(\frac{|v_M v|}{d} \right)^{\frac{1}{Mv}} \leq R(0) \leq 5^{\frac{1}{M}} \max_{1 \leq v \leq d} \left(\frac{|v_M v|}{d} \right)^{\frac{1}{Mv}}.$$

If we apply inequality (5.3.13) to the reciprocal polynomial of $P(x)$ we obtain

$$(5.3.14) \quad \frac{5^{-\frac{1}{M}}}{\max_{1 \leq v \leq d} \left(\frac{|v_{-M}v|}{d} \right)^{\frac{1}{Mv}}} \leq r(0) \leq \frac{1}{\max_{1 \leq v \leq d} \left(\frac{|v_{-M}v|}{d} \right)^{\frac{1}{Mv}}}.$$

The power sum v_v can be computed using Newton formulas. Assume that $P(x)$ has coefficients $a_k = N_k \times 10^{-m_P - 1}$ with $N_k = N_{k,1} + N_{k,2}i$, and $N_{k,1}, N_{k,2}$ are in $\mathbb{Z}$. Let A be the smallest integer bigger than the absolute values of the coefficients a_k. Then computing v_v takes time polynomial in v, m_P, and $\log A$.

Extracting the vth power of a decimal fraction $N \times 10^{-m}$ can be done in time polynomial in $v, \log N, m$, and the required accuracy, using a dichotomy algorithm as in Section 5.2. If $M \geq 200$ then the approximation factor $5^{\frac{1}{N}}$ is smaller than 1.01. So we obtain a very sharp estimate for $r(0)$.

For any $z \in \mathbb{C}$ we apply inequality (5.3.14) to the polynomial $P(x + z)$ and obtain a good approximation of $r(z)$. Assume that $P(x)$ has coefficients $a_k = N_k \times 10^{-m_P - 1}$, with $N_k = N_{k,1} + N_{k,2}i$. Assume that $z = (z_1 + z_2 i) \times 10^{-m_z}$ and z_1, z_2 are in $\mathbb{Z}$. Let A be the smallest integer bigger than the absolute values $|a_k|$. Then, using equation (5.3.14) for $N = 200$ and the change of variable $x \mapsto x + z$, we can compute a decimal number $\rho(z)$ such that

$$\rho(z) \leq r(z) \leq 1.01 \times \rho(z),$$

which takes time polynomial in d, m_P, $\log A$, m_z, and

$$\log \max(|z_1|, |z_2|, 1).$$

In the course of Weyl's quadtree algorithm, m_z is the number of steps, which is bounded in (5.3.9). Both $|z_1|$ and $|z_2|$ are bounded by $dA10^{m_z}$. An upper bound for m_P is given in (5.3.6). So the calculation of any value of the exclusion function that is required in the course of Weyl's algorithm takes polynomial time in d, $\log A$, and the required accuracy m_z. This finishes the proof of Theorem 5.3.1.

5.4 COMPUTING THE ZEROS OF A POWER SERIES

Given a power series $f(x) = f_0 + \sum_{k\geq 1} f_k x^k \in \mathbb{C}[[x]]$ with positive radius of convergence R, we may want to compute the zeros of f inside the open disk $D(0, R)$. However, the number of such zeros may very well be infinite. So we should restrict to a smaller disk $D(0, r)$ with $0 < r < R$. Then it makes sense to wonder how many zeros there are in the disk $D(0, r)$ and try to compute approximations of these zeros. However, the problem of counting zeros in such a disk is not a well-conditioned problem, because if f has a zero z of absolute value exactly r, then an infinitesimal perturbation of f may push this zero inside or outside $D(0, r)$. So we should allow the algorithm to choose an r' that is very close to r and compute the number of zeros inside $D(0, r')$. Then it makes sense to ask for approximations of these zeros. The input of a Turing machine ZERO computing zeros of power series would consist of a black box BOX_f for the coefficients of $f(x)$. On input an integer $K \geq 0$ and an integer $m \geq 1$, the black box BOX_f returns for every $0 \leq k \leq K$ a decimal fraction $(N_{1,k} + N_{2,k}i) \times 10^{-m-1}$ such that $|f_k - (N_{1,k} + N_{2,k}i) \times 10^{-m-1}| \leq 10^{-m}$. Note that the Turing machine ZERO is not allowed to ask about one coefficient individually. In particular, it needs at least time Θk to receive any information about coefficient f_k from the black box BOX_f. Unfortunately this is not enough to compute an approximation of the zeros, because there may be a huge coefficient f_k very far away in the development of f. If we want ZERO to be able to compute zeros without knowing all coefficients of the power series, we should at least provide it with an upper bound for the coefficients in $f(x)$. We introduce the following definition.

5.4.1 Definition (type of a power series) Let $A \geq 1$ be a real and $n \geq 1$ an integer. A power series $f(x) = f_0 + \sum_{k \geq 1} f_k x^k$ is said to be of type (A, n) if for every $k \geq 0$ we have

$$|f_k| \leq A(k+1)^n.$$

The radius of convergence of a power series of type (A, n) is at least 1. Elementary results about series of type (A, n) are collected in Section 5.4.3. We shall assume that the Turing machine ZERO is given two integers $a \geq 1$ and $n \geq 1$ such that the power series $f(x)$ is of type $(\exp(a), n)$.

There is still one difficulty to overcome. For any finite set $S \subset \mathbb{C}$ we may consider the polynomial $P_S(x) = \prod_{z \in S}(x - z)$. If we divide $P_S(x)$ by its L^∞ norm we obtain a polynomial $Q_S(x)$ having all its coefficient bounded by 1 in absolute value. We may regard Q_S as a power series of type $(1, 1)$. Since S can be arbitrarily large, we deduce that we cannot bound the complexity of finding zeros of a power series in a given disk $D(0, r)$ just in terms of the type (A, n) of the series. However, we guess that a power series of type $(1, 1)$, having too many zeros inside a small compact set contained in its disk of convergence, must be small everywhere on this compact set. Also we shall assume that the Turing machine ZERO is given a lower bound for the maximum of f on the closed disk $\bar{D}(0, r)$. More precisely, we assume that $1/2 \leq r < 1$ and provide the Turing machine with a positive integer μ such that $|f(z)| \geq \exp(-\mu)$ for *at least one z* in $\bar{D}(0, 1/2)$. In this section, we shall prove the existence of an algorithm that computes approximations of zeros of such an f.

5.4.2 Theorem (approximating the zeros of a power series) *There is a deterministic algorithm that on input an integer $m \geq 1$, an integer $a \geq 1$, an integer $n \geq 1$, an integer $\mu \geq 1$, a rational number $r = 1 - 1/o$ where $o \geq 2$ is an integer, and a power series*

$$f(x) = f_0 + \sum_{k \geq 1} f_k x^k \in \mathbb{C}[[x]]$$

of type $(\exp(a), n)$, such that $|f(z)| \geq \exp(-\mu)$ for at least one z in the closed disk $\bar{D}(0, 1/2)$ returns

- a rational number r' such that $|r - r'| \le 10^{-m}$,

- the number J of zeros of $f(x)$ in the closed disk $\bar{D}(0, r')$, or equivalently the degree J of the divisor $Z = [z_1] + [z_2] + \cdots + [z_J]$ of f restricted to the closed disk $\bar{D}(0, r')$,

- assuming $J \ge 1$, for every $1 \le j \le J$, a decimal

$$\hat{z}_j = (M_{1,j} + M_{2,j} i) \times 10^{-m-1},$$

where $M_{1,j}$ and $M_{2,j}$ are integers, such that

$$\hat{Z} = [\hat{z}_1] + [\hat{z}_2] + \cdots + [\hat{z}_J]$$

approximates the divisor $Z = [z_1] + [z_2] + \cdots + [z_J]$ within 10^{-m}. More precisely, there exists a permutation of the indices $\tau \in \mathcal{S}_J$ such that $|z_{\tau(j)} - \hat{z}_j| \le 10^{-m}$ for every $1 \le j \le J$.

The algorithm runs in time polynomial in $a, n, \mu, o = (1 - r)^{-1}$, and m.

Here is the main idea in the proof of this theorem. For every positive integer u we write $f(x) = P_u(x) + R_u(x)$, where $P_u(x) = \sum_{0 \le k \le u-1} f_k x^k$ is the principal part and $R_u(x) = \sum_{k \ge u} f_k x^k$ is the remainder term of order u. We expect that if u is large enough, the roots of $P_u(x)$ in $D(0, r)$ sharply approximate the zeros of $f(x)$ in $D(0, r)$. It would then suffice to apply the algorithm and theorem in Section 5.3. So our main concern will be to prove that the problem of finding zeros of a power series is well conditioned in some sense: a small perturbation does not affect the zeros too much. We cannot use Lemma 5.3.4 about the conditioning of roots of polynomials because both the statement and the proof involve the degree of the polynomial. We shall use Jensen's and Borel-Carathéodory's formulas for analytic functions. Both formulas are consequences of Green's integration formula.

In the next Section 5.4.3 we recall simpleminded inequalities relating coefficients and values of power series in one variable. Section 5.4.8 proves a simple effective analytic continuation theorem. We explain in Section 5.4.13 how Borel-Carathéodory's inequality implies that an analytic function cannot vary too much inside a zero free disk. We recall Jensen's formula in

Section 5.4.17 and derive upper bounds for the number of zeros of analytic functions. In Section 5.4.21 we prove that the problem of finding zeros of power series is well conditioned. Theorem 5.4.2 follows using Theorem 5.3.1.

5.4.3 Upper bounds

Let f be a function of the complex variable z, which is analytic in the open disk $D(c, R)$ with center c and radius R. For $0 \leq r < R$, we denote by $M(f, c, r)$ (resp. $m(f, c, r)$) the maximum (resp. minimum) absolute value of $f(z)$ for $|z - c| = r$. The maximum modulus principle implies that $M(f, c, r)$ is a nondecreasing function of r.

The so called Cauchy's estimate is an upper bound for the coefficients in the expansion of f at c. Let $f(c + (z - c)) = \sum_{k \geq 0} f_k \times (z - c)^k$ be the expansion of f at c. This is a series with positive radius of convergence $\geq R$. If $0 < r < R$, then

$$(5.4.4) \qquad |f_k| = \left| \frac{1}{2i\pi} \int_{|z-c|=r} \frac{f(z)}{(z-c)^{k+1}} dz \right| \leq \frac{M(f, c, r)}{r^k}.$$

Conversely, an upper bound on the coefficients f_k results in an upper bound on the values of f. Assume that $c = 0$. Recall Definition 5.4.1 of the *type* of a power series in one variable, and let $f(x) = \sum_{k \geq 0} f_k x^k$ be a power series of type (A, n) with $A \geq 1$ a real number and $n \geq 1$ an integer. Such a series defines an analytic function in the open disk $D(0, 1)$. Let $u \geq 0$ be an integer. We write $f(x) = P_u(x) + R_u(x)$, where $P_u(x) = \sum_{0 \leq k \leq u-1} f_k x^k$ is the principal part and $R_u(x) = \sum_{k \geq u} f_k x^k$ is the remainder term of order u. We first want to bound $|R_u(z)|$ for $z \in D(0, 1)$.

$$\begin{aligned} |R_u(z)| = \left| \sum_{k \geq u} f_k z^k \right| &\leq \sum_{k \geq u} |f_k||z|^k \\ &\leq A \sum_{k \geq u} (k+1)^n |z|^k \\ &\leq A|z|^u \sum_{k \geq 0} (k+u+1)^n |z|^k. \end{aligned}$$

So

(5.4.5)

$$|R_u(z)| \le A|z|^u (u+1)^n \sum_{k \ge 0} (k+1)^n |z|^k$$

$$\le A|z|^u (u+1)^n \frac{n!}{(1-|z|)^{n+1}}.$$

We set

$$\kappa = \frac{n! A|z|^{\frac{u}{2}}}{(1-|z|)^{n+1}},$$

and we show that $|R_u(z)| \le \kappa$ provided $u \ge 1$ and $u \ge \frac{4n^2}{(\log|z|)^2}$. Indeed, this is evident if $z = 0$. Otherwise, assuming $u \ge \frac{4n^2}{(\log|z|)^2}$ we have $n \log(u+1) \le n\sqrt{u} \le \frac{u|\log|z||}{2}$ so

$$u \log|z| + n \log(u+1) \le \frac{u \log|z|}{2}.$$

Using (5.4.5) we deduce

$$\log|R_u(z)| \le \log A + \frac{u \log|z|}{2} + \log \frac{n!}{(1-|z|)^{n+1}} = \log \kappa.$$

We thus have proved the following lemma.

5.4.6 Lemma (bounding the remainder of a power series) *Let $f(x)$ be a power series of type (A, n), where $A \ge 1$ is a real number and $n \ge 1$ is an integer. Let z be a complex number such that $|z| < 1$. Let $u \ge 0$ be an integer and let $R_u(x) = \sum_{k \ge u} f_k x^k$ be the remainder of order u of $f(x)$. We have*

$$|R_u(z)| \le A|z|^u (u+1)^n \frac{n!}{(1-|z|)^{n+1}},$$

and if $u \ge 1$ and $u \ge \frac{4n^2}{(\log|z|)^2}$, then

(5.4.7)

$$|R_u(z)| \le \frac{n! A|z|^{\frac{u}{2}}}{(1-|z|)^{n+1}}.$$

Note that if we set $u = 0$ in (5.4.5) we obtain an upper bound for f itself,

$$|f(z)| \le \frac{n! A}{(1-|z|)^{n+1}}$$

so

$$M(f, 0, r) \leq \frac{n!A}{(1-r)^{n+1}},$$

provided $0 \leq r < 1$.

5.4.8 A lower bound for the maximum

Let $f(x) = f_0 + \sum_{k \geq 1} f_k x^k \in \mathbb{C}[[x]]$ be a nonzero power series with positive radius of convergence R. For every r such that $0 \leq r < R$ we write $M(r)$ for $M(f, 0, r)$. This is an increasing function of r.

Assume now that $f(x)$ is of type (A, n) for $A \geq 1$ a real number and $n \geq 1$ an integer. Assume further that $0 < r < 1/2$. We want to show that if $M(r)$ is small enough, then $M(1/2)$ must be small too. More precisely, we look for an upper bound for $M(1/2)$ in terms of $M(r)$.

We choose an integer $u \geq 1$ and write $f(x) = P_u(x) + R_u(x)$, where $P_u(x) = \sum_{0 \leq k \leq u-1} f_k x^k$ is the principal part and $R_u(x)$ is the remainder term of order u. Let z be a complex number with absolute value $1/2$. We bound $P_u(z)$ and $R_u(z)$ separately.

On the one hand, using Cauchy's estimate (5.4.4) we obtain

$$(5.4.9) \qquad |P_u(z)| \leq \sum_{0 \leq k \leq u-1} \left| f_k z^k \right| \leq M(r) \sum_{0 \leq k \leq u-1} \left(\frac{1}{2r} \right)^k$$

$$\leq u M(r) \left(\frac{1}{2r} \right)^{u-1}.$$

On the other hand, Lemma 5.4.6 implies

$$(5.4.10) \qquad |R_u(z)| \leq A 2^{-u} (u+1)^n n! 2^{n+1}.$$

Let K be an absolute positive constant (to be chosen later). We set

$$u = \left\lfloor \frac{\log M(r)}{K \log r} \right\rfloor.$$

and we assume that

$$\frac{\log M(r)}{K \log r} \geq K(n^2 + \log A).$$

If K has been chosen large enough, then equation (5.4.10) implies that

$$|R_u(z)| \leq M(r)^{\frac{0.693}{K|\log r|}}.$$

As for the principal part, we deduce from equation (5.4.9) that

$$|P_u(z)| \leq M(r)^{0.999},$$

assuming K has been chose large enough. We deduce that

$$M(1/2) \leq M(r)^{\frac{1}{2K|\log r|}},$$

provided K has been chosen large enough. We deduce the following weak effective form of the analytic continuation theorem.

5.4.11 Lemma (effective analytic continuation) *There exists an absolute effective constant Θ such that the following holds true. Let $f(x)$ be a power series of type (A, n), where $A \geq 1$ is a real number and $n \geq 1$ is an integer. Let r be a real number such that $0 < r < 1/2$. If*

$$-\log M(r) \geq \Theta \times (n^2 + \log A) \times |\log r|,$$

then

$$M(1/2) \leq M(r)^{\frac{1}{\Theta|\log r|}}.$$

The lemma below states a weaker but simpler form of the above inequality.

5.4.12 Lemma (lower bound for the maximum) *There exists an absolute effective constant Θ such that the following holds true. Let $f(x)$ be a power series of type (A, n), where $A \geq 1$ is a real number and $n \geq 1$ is an integer. Let r be a real number such that $0 < r < 1/2$. Set*

$$\mu = \log^-(M(1/2)) = \max(0, -\log M(1/2)).$$

Then

$$-\log M(r) \leq \Theta \times (\mu + n^2 + \log A) \times |\log r|.$$

5.4.13 Zero-free analytic functions

If f is an analytic function in a disk, and has no zero there, we can bound it from below using Borel-Carathéodory's inequality.

5.4.14 Theorem (Borel-Carathéodory) *Let $f(z)$ be an analytic function in the open disk with center 0 and positive radius R. For every r such that $0 \leq r < R$, let*

$$Q(r) = \max_{|z|=r} \Re f(z)$$

be the maximum of the real part of f on the circle with center 0 and radius r. Let r and r' be real numbers such that $0 < r < r' < R$. Then for $|z| = r$ we have

$$|f(z)| \leq |f(0)| + \frac{2r}{r'-r}\left(Q(r') - \Re f(0)\right).$$

This is Theorem 1.3.1 in Boas's book [Boa]. We deduce the following estimate for the minimum of a zero-free analytic function.

5.4.15 Lemma (lower bound for the minimum I) *Let $r < r' < R$ be three positive real numbers, and let $f(z)$ be an analytic function in the open disk with center 0 and radius R. Assume that f has no zero in $D(0,R)$. The minimum $m(r) = m(f,0,r)$ of $|f(z)|$ for $|z| = r$ verifies*

$$\log m(r) \geq \frac{r'+r}{r'-r}\log|f(0)| - \frac{2r}{r'-r}\log M(r').$$

This lemma is proved applying Borel-Carathéodory's inequality to the logarithm of $f(z)/f(0)$. In case $f(0)$ is too small, we may look for a bigger value of f in the neighborhood of 0. So we choose some real r'' such that

$$0 \leq r'' < (r'-r)/2,$$

and we set $r_1 = r + r''$ and $r_2 = r' - r''$. Let a be a complex number with absolute value r'' and such that $f(a)$ has maximum possible absolute value $M(r'')$. We check that the closed disk $\bar{D}(a, r_2)$ is contained in $\bar{D}(0, r')$. And the closed disk $\bar{D}(a, r_1)$ contains the closed disk $\bar{D}(0, r)$. Lemma 5.4.15 implies that $\log|f(z)|$ is

$$\geq \frac{r_2 + r_1}{r_2 - r_1}\log|f(a)| - \frac{2r_1}{r_2 - r_1}\log M(r')$$

for any z on the circle with center a and radius r_1. So

$$\log m(r) \geq \frac{r' + r}{r' - r - 2r''} \log M(r'') - \frac{2(r + r'')}{r' - r - 2r''} \log M(r').$$

We deduce the following lemma.

5.4.16 Lemma (lower bound for the minimum II) *Let* $r < r' < R$ *be three positive real numbers, and let* $f(z)$ *be an analytic function in the open disk with center* 0 *and radius* R. *Assume that* f *has no zero in* $D(0, R)$. *Let* r'' *be a real such that* $0 \leq r'' < (r'-r)/2$. *The minimum* $m(r) = m(f, 0, r)$ *of* $|f(z)|$ *for* $|z| = r$ *verifies*

$$\log m(r) \geq \frac{r' + r}{r' - r - 2r''} \log M(r'') - \frac{2(r + r'')}{r' - r - 2r''} \log M(r').$$

5.4.17 Counting zeros in disks

Let $f(z)$ be an analytic function in the open disk $D(0, R)$ for some $R > 0$. Let v be the valuation of f at 0. So $f(z) = f_v z^v + O(z^{v+1})$ for some nonzero f_v. For every real number s with $0 \leq s < R$ we write $n(s)$ for the number of zeros of f having absolute value $\leq s$, counting multiplicities. Let r be a real number such that $0 \leq r < R$. We recall Jensen's formula [Cart] Section 1.34, in that situation:

$$\int_0^r \frac{n(s) - n(0)}{s} ds + n(0) \log r = \frac{1}{2\pi} \int_0^{2\pi} \log |f(r \exp(i\phi))| \, d\phi - \log |f_v|.$$

Assume now that $v = 0$. We have $M(0) = |f(0)|$, and we can use Jensen's formula to count the zeros of f. Indeed, let r' be a real number such that $r < r' < R$. Since $n(s)$ is an increasing function of s we have

$$n(r) \log \frac{r'}{r} = n(r) \int_r^{r'} \frac{ds}{s} \leq \int_r^{r'} \frac{n(s) ds}{s} \leq \log M(r') - \log |f(0)|.$$

We deduce the following lemma.

5.4.18 Lemma (upper bound for the number of zeros I) *Let* $f(z)$ *be analytic in the open disk* $D(0, R)$ *for some positive* R. *Assume that* $f(0) \neq 0$. *Let* r *and* r' *be two real numbers such that* $0 < r < r' < R$. *Let* $n(r)$ *be the*

number of zeros of f having absolute value $\leq r$, counting multiplicities.
Then

$$n(r) \leq \frac{\log(M(r')/M(0))}{\log(r'/r)}.$$

In case $f(0)$ is zero or too small for our purpose, we may look for a bigger
value of f in the neighborhood of 0. So we choose some real r'' such that
$0 \leq r'' < \frac{r'-r}{2}$, and we set $r_1 = r + r''$ and $r_2 = r' - r''$. Let a be a
complex number with absolute value r'' and such that $f(a)$ has maximum
possible absolute value $M(r'')$. We check that the closed disk $\bar{D}(0, r)$ is
contained in $\bar{D}(a, r_1)$, and the closed disk $\bar{D}(a, r_2)$ is contained in $\bar{D}(0, r')$.
Lemma 5.4.18 implies that the number of zeros in the closed disk $\bar{D}(a, r_1)$
is $\leq \log(M(r')/M(r''))/\log(r_2/r_1)$. We deduce the following lemma.

5.4.19 Lemma (upper bound for the number of zeros II) *Let f be analytic in the open disk $D(0, R)$ for some positive R. Assume that f is
nonzero. Let r, r', r'' be three real numbers such that $0 < r < r' < R$
and $r'' < (r' - r)/2$. Let $n(r)$ be the number of zeros of f having absolute
value $\leq r$, counting multiplicities. Then*

$$n(r) \leq \frac{\log(M(r')/M(r''))}{\log((r' - r'')/(r + r''))}.$$

Now let $f(x) = \sum_{k \geq 0} f_k x^k$ be a nonzero power series of type (A, n),
with $A \geq 1$ a real number and $n \geq 1$ an integer. Set $R = 1, r' = (r + 1)/2$,
$r'' = (r' - r)/3$ in Lemma 5.4.19. We use Lemma 5.4.6 to bound $\log M(r')$
from above. We use Lemma 5.4.12 to bound $-\log M(r'')$ from above. We
obtain the following lemma.

5.4.20 Lemma (upper bound for the number of zeros III) *There exists
an absolute effective constant Θ such that the following holds true. Let
$f(x)$ be a nonzero power series of type (A, n) with $A \geq 1$ a real number
and $n \geq 1$ an integer. Let $\mu = \max(0, -\log M(1/2))$, where $M(1/2)$ is
the maximum absolute value of $f(z)$ for $|z| = 1/2$. Let r be a real number
such that $0 < r < 1$. Then*

$$n(r) \leq \Theta \times (n^2 + \log A + \mu) \times (1 + |\log(1 - r)|) \times (1 - r)^{-1}.$$

5.4.21 Conditioning of the zeros of a series

Let $f(x) = \sum_{k \geq 0} f_k x^k$ be a nonzero power series of type (A, n), where $A \geq 1$ is a real number and $n \geq 1$ is an integer. Let

$$\mu = \max(0, -\log M(f, 1/2)),$$

where $M(f, 1/2)$ is the maximum absolute value of $f(z)$ for $|z| = 1/2$. Let r_1 be a real number such that $3/4 \leq r_1 < 1$. We want to prove that the zeros of f in the disk $\bar{D}(0, r_1)$ are well conditioned in the sense that they are not too much affected by a small perturbation of f.

Set $r_2 = (r_1 + 1)/2$ and $r_3 = (r_2 + 1)/2$. We encourage the reader to make a drawing. Let ϵ_1 be a real number such that $0 < \epsilon_1 \leq (1 - r_1)/3$. So

$$3/4 \leq r_1 < r_1 + \epsilon_1 < r_2 < r_3 < 1.$$

According to Lemma 5.4.20 the number $n(r_3)$ of zeros of f with absolute value $\leq r_3$ is at most $\Theta \times (n^2 + \log A + \mu) \times (1 - r_1)^{-1} \times |\log(1 - r_1)|$.

So there exist three real numbers r_4, r_5, and r_6 such that

$$r_1 \leq r_4 < r_5 < r_6 \leq r_1 + \epsilon_1$$

and

$$|r_5 - r_4| = |r_6 - r_5| \geq \frac{\epsilon_1 \times (1 - r_1)}{\Theta \times (n^2 + \log A + \mu) \times |\log(1 - r_1)|},$$

and f has no zero with absolute value in the interval $[r_4, r_6]$.

Similarly, there exist three real numbers r_7, r_8 and r_9 such that

$$r_2 \leq r_7 < r_8 < r_9 \leq r_3$$

and

$$|r_8 - r_7| = |r_9 - r_8| \geq \frac{(1 - r_1)^2}{\Theta \times (n^2 + \log A + \mu) \times |\log(1 - r_1)|},$$

and f has no zero with absolute value in the interval $[r_7, r_9]$.

Let $P(x)$ be the monic polynomial whose roots are the zeros of f in the closed disk $\bar{D}(0, r_3)$. Set $F(z) = f(z)/P(z)$. This is an analytic function in the disk $D(0, 1)$. It has no zero in the disk $\bar{D}(0, r_3)$. We want to bound $F(z)$ from below when $|z| = r_5$. We set $R = r_3$, $r = r_5$, $r' = r_8$, and

$r'' = (r' - r)/3$ in Lemma 5.4.16. We check that $r' - r \geq (1 - r_1)/6$, so $r' - r - 2r'' \geq (1 - r_1)/18$.

The maximum $M(F, r'')$ is at least $M(f, r'')$, so

$$-\log(M(F, r'')) \leq \Theta \times (\mu + n^2 + \log A) \times |\log(1 - r_1)|$$

using Lemma 5.4.12. Applying Lemma 5.4.6 to f we find that

$$\log M(f, r') \leq \Theta \times (n^2 + \log A + n|\log(1 - r_1)|).$$

So $\log M(F, r')$ is

$$\leq \Theta \times (n^2 + \log A + \mu) \times (1 - r_1)^{-1} \times \log^2(1 - r_1) \times \log(1 + n + \mu + \log A).$$

The conclusion in Lemma 5.4.16 then implies that $-\log m(F, r_5)$ is

$$\leq \Theta \times (n^2 + \log A + \mu) \times (1 - r_1)^{-2} \times \log^2(1 - r_1) \times \log(1 + n + \mu + \log A).$$

Let d be the degree of $P(x)$. This is the number $n(r_3)$ of zeros of f in the closed disk $\bar{D}(0, r_3)$, counting multiplicities. Set $\epsilon_2 = (r_5 - r_4)/2$. We know from Section 5.3.2 that there exist a positive ϵ_3 and an ϵ_3-clustering $(c_k, n_k)_{1 \leq k \leq K}$ of the roots of $P(x)$ such that $2^{-d^2}\epsilon_2 \leq \epsilon_3 \leq \epsilon_2$. Every closed disk $\bar{D}(c_k, \epsilon_3)$ is either contained in $D(0, r_5)$ or in the interior of its complementary set. Let T be the complementary set in $\bar{D}(0, r_5)$ of the union of all $D(c_k, \epsilon_3)$. Then

$$T = \bar{D}(0, r_5) - \cup_{1 \leq k \leq K} \bar{D}(c_k, \epsilon_3).$$

The boundary ∂T of T consists of $K' + 1$ circles, where K' is the number of disks $\bar{D}(c_k, \epsilon_3)$ that are contained in $D(0, r_5)$. The distance ϵ_4 between ∂T and the set of roots of $P(x)$ is at least $\epsilon_3/2$. So $\epsilon_4 \geq 2^{-d^2-1}\epsilon_2$, and

$$-\log \epsilon_4 \leq \Theta \times (d^2 + |\log \epsilon_1| + |\log(1 - r_1)| + \log(1 + n + \mu + \log A)).$$

Let δ be the minimum of the absolute value of $f(z)$ for z in T. Then $\delta \geq m(F, r_5) \times \epsilon_4^d$ and

$$-\log \delta \leq \Theta \times |\log \epsilon_1| \times (n^2 + \log A + \mu) \times (1 - r_1)^{-1} \times |\log(1 - r_1)|$$
$$+ \Theta \times (n^2 + \log A + \mu)^3 \times (1 - r_1)^{-3} \times |\log(1 - r_1)|^3.$$

Let u be a positive integer and write $f(x) = P_u(x) + R_u(x)$, where P_u is the principal part and $R_u(x) = \sum_{k \geq u} f_k x^k$ is the remainder term of order u. If the absolute value of $R_u(z)$ is $< \delta$ for every z in $D(0, r_6) \supset T$, then P_u has no zero in T and for every $\bar{D}(c_k, \epsilon_3)$ that is contained in $\bar{D}(0, r_5)$, the number of roots of f and P_u in $\bar{D}(c_k, \epsilon_3)$, counting multiplicities, is the same. So f and P_u have the same number of roots inside $\bar{D}(0, r_5)$, and the roots of $P_u(x)$ approximate the roots of f within $2\epsilon_3 \leq \epsilon_1$.

According to Lemma 5.4.6 the absolute value of $R_u(z)$ is $< \delta$ for every z in $D(0, r_6)$, provided

$$u \geq \Theta \times |\log \epsilon_1| \times (n^2 + \log A + \mu) \times (1 - r_1)^{-2} \times |\log(1 - r_1)|$$
$$+ \Theta \times (n^2 + \log A + \mu)^3 \times (1 - r_1)^{-4} \times |\log(1 - r_1)|^3.$$

This proves the following lemma.

5.4.22 Lemma (conditioning of the zeros of a power series) *There is an absolute effective constant Θ such that the following holds true. Let $f(x)$ be a power series of type (A, n) with $A \geq 1$ a real number and $n \geq 1$ an integer. Let $\mu = \max(0, -\log M(1/2))$, where $M(1/2)$ is the maximum absolute value of $f(z)$ for $|z| = 1/2$. Let r be a real number such that $3/4 \leq r < 1$. Let ϵ be a positive real number.*

Then there exists a real number r' such that

1. *$r \leq r' < r + \epsilon$;*

2. *for every holomorphic function Δ in $D(0, r + \epsilon)$ such that $-\log|\Delta|$ is lower bounded by*

$$\Theta \times |\log \epsilon| \times (n^2 + \log A + \mu) \times (1 - r)^{-1} \times |\log(1 - r)|$$
$$+ \Theta \times (n^2 + \log A + \mu)^3 \times (1 - r)^{-3} \times |\log(1 - r)|^3,$$

the number of zeros of f and $f + \Delta$ in $\bar{D}(0, r')$, counting multiplicities, is the same, and the zeros of $f + \Delta$ approximate the zeros of f within ϵ;

3. *in particular, for every integer* u *such that*

$$u \geq \Theta \times |\log \epsilon| \times (n^2 + \log A + \mu) \times (1 - r)^{-2} \times |\log(1 - r)|$$
$$+ \Theta \times (n^2 + \log A + \mu)^3 \times (1 - r)^{-4} \times |\log(1 - r)|^3$$

the number of zeros of f *and* $P_u(x) = \sum_{0 \leq k \leq u-1} f_k x^k$ *in* $\bar{D}(0, r')$, *counting multiplicities, is the same, and the zeros of* $P_u(x)$ *approximate the zeros of* f *within* ϵ.

Now Theorem 5.4.2 follows from Lemma 5.4.22 and Theorem 5.3.1.

Chapter Six

Computations with modular forms and Galois representations

by Johan Bosman

In this chapter we will discuss several aspects of the practical side of computating with modular forms and Galois representations. We start by discussing computations with modular forms and from there work toward the computation of polynomials that give the Galois representations associated with modular forms. Throughout this chapter, we will denote the space of cusp forms of weight k, group $\Gamma_1(N)$, and character ε by $S_k(N, \varepsilon)$.

6.1 MODULAR SYMBOLS

Modular symbols provide a way of doing symbolic calculations with modular forms, as well as the homology of modular curves. In this section our intention is to give the reader an idea of what is going on rather than a complete and detailed account. For more details and further reading on the subject of modular symbols, the reader might take a look at [Man1], [Sho], and [Mer]. A computational approach to the material can be found in [Ste1] and [Ste2].

6.1.1 Definitions

Let A be the free Abelian group on the symbols $\{\alpha, \beta\}$ with $\alpha, \beta \in \mathbb{P}^1(\mathbb{Q})$. Consider the subgroup $I \subset A$ generated by all elements of the forms

$$\{\alpha, \beta\} + \{\beta, \gamma\} + \{\gamma, \alpha\}, \quad \{\alpha, \beta\} + \{\beta, \alpha\}, \quad \text{and} \quad \{\alpha, \alpha\}.$$

We define the group

$$\mathbb{M}_2 := (A/I)/\text{torsion}$$

as the quotient of A/I by its torsion subgroup. By a slight abuse of notation, we will denote the class of $\{\alpha, \beta\}$ in this quotient also by $\{\alpha, \beta\}$. We have an action $\mathrm{GL}_2^+(\mathbb{Q})$ on $\mathbb{M}_2$ by

$$\gamma\{\alpha, \beta\} := \{\gamma\alpha, \gamma\beta\},$$

where γ acts on $\mathbb{P}^1(\mathbb{Q})$ by fractional linear transformations.

For $k \geq 2$, we consider also the Abelian group $\mathbb{Z}[x, y]_{k-2} \subset \mathbb{Z}[x, y]$ of homogeneous polynomials of degree $k - 2$, and we let matrices in $\mathrm{GL}_2^+(\mathbb{Q})$ with integer coefficients act on it on the left by

$$\begin{pmatrix} a & b \\ c & d \end{pmatrix} P(x, y) := P(dx - by, -cx + ay).$$

We define

$$\mathbb{M}_k := \mathbb{Z}[x, y]_{k-2} \otimes \mathbb{M}_2,$$

and we equip $\mathbb{M}_k$ with the componentwise action of integral matrices in $\mathrm{GL}_2^+(\mathbb{Q})$ (that is $\gamma(P \otimes \alpha) = \gamma(P) \otimes \gamma(\alpha)$).

6.1.2 Definition Let $k \geq 2$ be an integer. Let $\Gamma \subset \mathrm{SL}_2(\mathbb{Z})$ be a subgroup of finite index, and let $I \subset \mathbb{M}_k$ be the subgroup generated by all elements of the form $\gamma x - x$ with $\gamma \in \Gamma$ and $x \in \mathbb{M}_k$. Then we define the space of *modular symbols* of weight k for Γ to be the quotient of $\mathbb{M}_k/I$ by its torsion subgroup and denote this space by $\mathbb{M}_k(\Gamma)$:

$$\mathbb{M}_k(\Gamma) := (\mathbb{M}_k/I)/\text{torsion}.$$

In the special case $\Gamma = \Gamma_1(N)$, which we will mostly be interested in, $\mathbb{M}_k(\Gamma)$ is called the space of modular symbols of weight k and level N. The class of $\{\alpha, \beta\}$ in $\mathbb{M}_k(\Gamma)$ will be denoted by $\{\alpha, \beta\}_\Gamma$ or, if no confusion exists, by $\{\alpha, \beta\}$.

The group $\Gamma_0(N)$ acts naturally on $\mathbb{M}_k(\Gamma_1(N))$ and induces an action of $(\mathbb{Z}/N\mathbb{Z})^\times$ on $\mathbb{M}_k(\Gamma_1(N))$. We denote this action by the diamond symbol $\langle d \rangle$. The operator $\langle d \rangle$ on $\mathbb{M}_k(\Gamma_1(N))$ is called a *diamond operator*. This leads to the notion of modular symbols with character.

6.1.3 Definition Let $\varepsilon: (\mathbb{Z}/N\mathbb{Z})^{\times} \to \mathbb{C}^{\times}$ be a Dirichlet character. Denote by $\mathbb{Z}[\varepsilon] \subset \mathbb{C}$ the subring generated by all values of ε. Let I be the $\mathbb{Z}[\varepsilon]$-submodule of $\mathbb{M}_k(\Gamma_1(N)) \otimes \mathbb{Z}[\varepsilon]$ generated by all elements of the form $\langle d \rangle x - \varepsilon(d)x$ with $d \in (\mathbb{Z}/N\mathbb{Z})^{\times}$ and $x \in \mathbb{M}_k(\Gamma_1(N))$. Then we define the space $\mathbb{M}_k(N, \varepsilon)$ of modular symbols of weight k, level N and character ε as the $\mathbb{Z}[\varepsilon]$-module

$$\mathbb{M}_k(N, \varepsilon) := \left(\mathbb{M}_k(\Gamma_1(N)) \otimes \mathbb{Z}[\varepsilon]/I\right)/\text{torsion}.$$

We denote the elements of $\mathbb{M}_k(N, \varepsilon)$ by $\{\alpha, \beta\}_{N,\varepsilon}$ or simply $\{\alpha, \beta\}$. If ε is trivial, then we have $\mathbb{M}_k(N, \varepsilon) \cong \mathbb{M}_k(\Gamma_0(N))$.

Let $\mathbb{B}_2$ be the free Abelian group on the symbols $\{\alpha\}$ with $\alpha \in \mathbb{P}^1(\mathbb{Q})$, equipped with action of $\mathrm{SL}_2(\mathbb{Z})$ by $\gamma\{\alpha\} = \{\gamma\alpha\}$, and define $\mathbb{B}_k$ as the tensor product $\mathbb{Z}[x, y]_{k-2} \otimes \mathbb{B}_2$ with componentwise $\mathrm{SL}_2(\mathbb{Z})$-action. Elements of $\mathbb{B}_k$ are called *boundary modular symbols* . For a subgroup $\Gamma < \mathrm{SL}_2(\mathbb{Z})$ of finite index, we define $\mathbb{B}_k(\Gamma)$ as

$$\mathbb{B}_k(\Gamma) := (\mathbb{B}_k/I)/\text{torsion},$$

where I is the subgroup of $\mathbb{B}_k$ generated by all elements $\gamma x - x$ with $\gamma \in \Gamma$ and $x \in \mathbb{B}_k$. We define

$$\mathbb{B}_k(N, \varepsilon) := ((\mathbb{B}_k(\Gamma_1(N)) \otimes \mathbb{Z}[\varepsilon])/I)/\text{torsion},$$

where I is the $\mathbb{Z}[\varepsilon]$-submodule of $\mathbb{B}_k(\Gamma_1(N)) \otimes \mathbb{Z}[\varepsilon]$ generated by the elements $\gamma x - \varepsilon(\gamma)x$ with $\gamma \in \Gamma_0(N)$.

We have *boundary homomorphisms*

$$\delta: \mathbb{M}_k(\Gamma) \to \mathbb{B}_k(\Gamma) \quad \text{and} \quad \delta: \mathbb{M}_k(N, \varepsilon) \to \mathbb{B}_k(N, \varepsilon),$$

defined by

$$\delta(P \otimes \{\alpha, \beta\}) = P \otimes \{\beta\} - P \otimes \{\alpha\}.$$

The spaces of *cuspidal modular symbols* , denoted by $\mathbb{S}_k(\Gamma)$ and $\mathbb{S}_k(N, \varepsilon)$, respectively, are defined as the kernel of δ.

6.1.4 Properties

One can interpret the symbol $\{\alpha, \beta\}$ as a smooth path in $\mathbb{H} \cup \mathbb{P}^1(\mathbb{Q})$ from the cusp α to the cusp β, lying in $\mathbb{H}$ except for the end points α and β. It can be shown that this interpretation induces an isomorphism

$$\mathbb{M}_2(\Gamma) \cong H_1(X_\Gamma, \text{Cusps}, \mathbb{Z}).$$

Here the homology is taken of the topological pair $(X_1(N), \text{Cusps})$. We also get an isomorphism

$$\mathbb{S}_2(\Gamma) \cong H_1(X_\Gamma, \mathbb{Z}).$$

So we immediately see that there is a perfect pairing

$$(\mathbb{S}_2(\Gamma(N)) \otimes \mathbb{C}) \times \left(S_2(\Gamma(N)) \oplus \overline{S}_2(\Gamma(N))\right) \to \mathbb{C}$$

defined by

$$(\{\alpha, \beta\}, f \oplus g) \mapsto \int_\alpha^\beta \left(f \frac{dq}{q} + g \frac{d\overline{q}}{\overline{q}}\right).$$

More generally, there is a pairing

(6.1.5) $$\mathbb{M}_k(\Gamma_1(N)) \times \left(S_k(\Gamma_1(N)) \oplus \overline{S}_k(\Gamma_1(N))\right) \to \mathbb{C}$$

defined by

$$(P \otimes \{\alpha, \beta\}, f \oplus g) \mapsto 2\pi i \int_\alpha^\beta \left(f(z)P(z, 1)dz - g(z)P(\overline{z}, 1)d\overline{z}\right),$$

which becomes perfect if we restrict the left factor to $S_k(\Gamma(N))$ and then tensor to $S_k(\Gamma(N)) \otimes \mathbb{C}$. This pairing induces a pairing

$$(\mathbb{M}_k(N, \varepsilon)) \times \left(S_k(N, \varepsilon) \oplus \overline{S}_k(N, \varepsilon)\right) \to \mathbb{C},$$

which becomes perfect when the left factor is restricted and then tensored to $S_k(N, \varepsilon) \otimes_{\mathbb{Z}[\varepsilon]} \mathbb{C}$. From now on we will denote all these pairings with the notation

$$(x, f) \mapsto \langle x, f \rangle.$$

6.1.6 The star involution

On the spaces $\mathbb{M}_k(\Gamma_1(N))$ and $\mathbb{M}_k(N, \varepsilon)$ we have an involution ι^*:

$$\iota^*(P(x, y) \otimes \{\alpha, \beta\}) := -P(x, -y) \otimes \{-\alpha, -\beta\},$$

called the *star involution*. This involution preserves cuspidal subspaces. We define the subspaces $\mathbb{S}_k(\Gamma_1(N))^+$ and $\mathbb{S}_k(\Gamma_1(N))^-$ of $\mathbb{S}_k(\Gamma_1(N))$, where ι^* acts as $+1$ and -1, respectively, and we use similar definitions for $\mathbb{S}_k(N, \varepsilon)^{\pm}$. It can be shown that the pairing (6.1.5) induces perfect pairings:

$$(\mathbb{S}_k(\Gamma_1(N))^+ \otimes \mathbb{C}) \times S_k(\Gamma_1(N)) \to \mathbb{C}$$

and:

$$(\mathbb{S}_k(\Gamma_1(N))^- \otimes \mathbb{C}) \times \overline{S}_k(\Gamma_1(N)) \to \mathbb{C}$$

and similarly for the spaces with character. This allows us to work sometimes in modular symbols spaces of half the dimension of the full cuspidal space.

6.1.7 Hecke operators

Let $k \geq 2$ and $N \geq 1$ be given. Then for $\gamma \in \mathrm{GL}_2^+(\mathbb{Q}) \cap \mathrm{M}_2(\mathbb{Z})$ we define an operator T_γ on $\mathbb{M}_k(\Gamma_1(N))$ by letting $\gamma_1, \ldots, \gamma_r$ be double coset representatives for $\Gamma_1(N) \setminus \Gamma_1(N) \gamma \Gamma_1(N)$ and putting

$$(6.1.8) \qquad T_\gamma(x) := \sum_{i=1}^{r} \gamma_i x \quad \text{for } x \in \mathbb{M}_k(\Gamma_1(N)).$$

It follows from [Sho], Theorem 4.3, that this operator is well defined. For a prime number p we put $T_p = T_\gamma$ for $\gamma = \begin{pmatrix} 1 & 0 \\ 0 & p \end{pmatrix}$, and for positive integers n we define T_n by means of the formal identity (2.2.12). The operators T_n are called Hecke operators.

The Hecke operators preserve the subspace $\mathbb{S}_k(\Gamma_1(N))$ and induce an action on the spaces $\mathbb{M}_k(N, \varepsilon)$ and $\mathbb{S}_k(N, \varepsilon)$. Furthermore, from [Sho], Theorem 4.3, one can conclude that the diamond and Hecke operators are self-adjoint with respect to the pairings defined in the previous subsection: one has

$$(6.1.9) \qquad \langle Tx, f \rangle = \langle x, Tf \rangle$$

for any modular symbol x, cusp form f, and diamond or Hecke operator T for which this relation is well defined. Here, for anti-holomorphic cusp forms we define the Hecke action by $T\overline{f} = \overline{Tf}$. Also, the Hecke operators commute with the star involution ι^*.

In conclusion, we have seen how we can write cusp forms spaces as the dual of modular symbols spaces. The computation of Hecke operators on these modular symbols spaces would enable us to compute q-expansions of cusp forms: q-coefficients of newforms can be computed once we can compute the eigenvalues of Hecke operators. But because of (6.1.9) this reduces to the computation of the eigenvalues of Hecke operators on modular symbols spaces. In computations one often works with the spaces $\mathbb{S}_k(N, \varepsilon)^+ \otimes_{\mathbb{Z}[\varepsilon]} \mathbb{Q}(\varepsilon)$ because these have smaller dimension than $\mathbb{S}_k(\Gamma_1(N)) \otimes \mathbb{Q}$. Since we also know how all cusp forms arise from newforms of possibly lower level (see (2.2.18)), this allows us to compute the q-expansions of a basis for the spaces $S_k(\Gamma_1(N))$ and $S_k(N, \varepsilon)$. For precise details on how these computations work, please read [Ste2], Chapter 9.

6.1.10 Manin symbols

If we want to do symbolic calculations with modular symbols, then the above definitions are not quite applicable, since the groups of which we take quotients are not finitely generated. The *Manin symbols* enable us to give finite presentations for the spaces of modular symbols.

First we need some definitions and lemmas. For a positive integer N we define a set

$$E_N := \left\{ (c, d) \in (\mathbb{Z}/N\mathbb{Z})^2 : \gcd(N, c, d) = 1 \right\}.$$

Define the following equivalence relation on E_N:

$$(c, d) \sim (c', d') \overset{\text{def}}{\Longleftrightarrow} \text{ there is an } a \in (\mathbb{Z}/N\mathbb{Z})^\times \text{ with } (c, d) = (ac', ad')$$

and the denote the quotient by P_N:

$$(6.1.11) \qquad\qquad\qquad P_N := E_N / \sim.$$

The following lemma is easily verified.

6.1.12 Lemma *Let N be a positive integer. Then the maps*

$$\Gamma_1(N) \setminus \mathrm{SL}_2(\mathbb{Z}) \to E_N : \overline{\begin{pmatrix} a & b \\ c & d \end{pmatrix}} \mapsto (\overline{c}, \overline{d}) \quad \text{and}$$

$$\Gamma_0(N) \setminus \mathrm{SL}_2(\mathbb{Z}) \to P_N : \overline{\begin{pmatrix} a & b \\ c & d \end{pmatrix}} \mapsto \overline{(c, d)}$$

are well defined and bijective.

This lemma enables us to write down an explicit set of coset representatives for the orbit spaces $\Gamma_1(N) \setminus \mathrm{SL}_2(\mathbb{Z})$ and $\Gamma_0(N) \setminus \mathrm{SL}_2(\mathbb{Z})$. The following lemma provides us a first step in reducing the set of generators for the spaces of modular symbols.

6.1.13 Lemma *Each space $\mathbb{M}_2(\Gamma_1(N))$ or $\mathbb{M}_2(N, \varepsilon)$ is generated by the symbols $\{a/c, b/d\}$ with $a, b, c, d \in \mathbb{Z}$ and $ad - bc = 1$, where in this notation a fraction with denominator zero denotes the cusp at infinity.*

Calculating the continued fraction expansion at each cusp in $\mathbb{Q}$ gives us immediately an algorithm to write a given element of $\mathbb{M}_2$ in terms of the generators in the lemma. Furthermore, note that

$$\left\{ \frac{a}{c}, \frac{b}{d} \right\} = \begin{pmatrix} a & b \\ c & d \end{pmatrix} \{\infty, 0\},$$

so that we can write each element of $\mathbb{M}_2$ as a $\mathbb{Z}$-linear combination of $\gamma\{\infty, 0\}$ with γ in $\mathrm{SL}_2(\mathbb{Z})$.

Let's consider the space $\mathbb{M}_2(\Gamma_1(N))$. As we saw, it is generated by the elements $\gamma\{\infty, 0\}$, where γ runs through $\mathrm{SL}_2(\mathbb{Z})$. Now, two matrices γ define the same element this way if they are in the same coset of the quotient $\Gamma_1(N) \setminus \mathrm{SL}_2(\mathbb{Z})$. According to Lemma 6.1.12 such a coset can be uniquely identified with a pair $(c, d) \in (\mathbb{Z}/N\mathbb{Z})^2$. The corresponding element in $\mathbb{M}_2(\Gamma_1(N))$ is also denoted by (c, d). This element (c, d) is called a *Manin symbol*. Clearly, there are only a finite number of Manin symbols, so we now know a finite set of generators for $\mathbb{M}_2(\Gamma_1(N))$.

For arbitrary k we define the Manin symbols in $\mathbb{M}_k(\Gamma_1(N))$ as the symbols of the form $P \otimes (c, d)$ where P is a monomial in $\mathbb{Z}[x, y]_{k-2}$ and (c, d) a Manin symbol in $\mathbb{M}_2(\Gamma_1(N))$. In this case as well there are finitely many Manin symbols and they generate the whole space.

In the modular symbols spaces with a given character ε we have, for all $\gamma \in \Gamma_0(N)$, that $\gamma(\alpha) = \varepsilon(\alpha)$. Now for each element of P_N we choose according to Lemma 6.1.12 a corresponding element $\gamma \in \mathrm{SL}_2(\mathbb{Z})$ and hence an element in $\mathbb{M}_2(N, \varepsilon)$, which we call again a Manin symbol. Note that this Manin symbol depends on the choice of γ, but because of the relation $\gamma(x) = \varepsilon(x)$ these chosen Manin symbols always form a finite set of generators for $\mathbb{M}_2(N, \varepsilon)$ as a $\mathbb{Z}[\varepsilon]$-module. Likewise, $\mathbb{M}_k(N, \varepsilon)$ is generated by elements $P \otimes (c, d)$ with P a monomial in $\mathbb{Z}[x, y]_{k-2}$ and (c, d) a Manin symbol in $\mathbb{M}_2(N, \varepsilon)$.

If we want to do symbolic calculations, then besides generators we also need to know the relations between the Manin symbols. For $\mathbb{M}_k(\Gamma_1(N))$ one can do the following.

6.1.14 Proposition *Let N be a positive integer and let A be the free Abelian group on the Manin symbols of the space $\mathbb{M}_k(\Gamma_1(N))$. Let $I \subset A$ be the subgroup generated by the following elements:*

$$P(x, y) \otimes (c, d) + P(-y, x) \otimes (-d, -c),$$
$$P(x, y) \otimes (c, d) + P(-y, x - y) \otimes (-d, -c - d)$$
$$+ P(-x + y, -x) \otimes (-c - d, -c),$$
$$P(x, y) \otimes (c, d) - P(-x, -y) \otimes (c, d),$$

where $P(x, y) \otimes (c, d)$ runs through all Manin symbols. Then $\mathbb{M}_k(\Gamma_1(N))$ is naturally isomorphic to the quotient of A/I by its torsion subgroup.

For the modular symbols spaces $\mathbb{M}_k(N, \varepsilon)$ we have a similar proposition.

6.1.15 Proposition *Let N and ε be given. Let A be the free $\mathbb{Z}[\varepsilon]$-module on the Manin symbols of $\mathbb{M}_k(N, \varepsilon)$. Let $I \subset A$ be the submodule generated by the elements given in Proposition 6.1.14 plus for each $n \in (\mathbb{Z}/N\mathbb{Z})^\times$ the elements*

$$P(x, y) \otimes \overline{(nc, nd)} - \varepsilon(n) P(x, y) \otimes \overline{(c, d)}.$$

Then $\mathbb{M}_k(N, \varepsilon)$ is naturally isomorphic to the quotient of A/I by its torsion submodule.

These presentations enable us to perform symbolic calculations very efficiently.

A remark on the computation of Hecke operators is in order here. The formula (6.1.8) does not express the Hecke action on Manin symbols in terms of Manin symbols. However, one can use other formulas to compute Hecke operators. The following theorem, due to Merel, allows us to express Hecke operators more directly in terms of Manin symbols:

6.1.16 Theorem (see [Mer], Theorem 2) *On the spaces* $\mathbb{M}_k(\Gamma_1(N))$ *and* $\mathbb{M}_k(N, \varepsilon)$ *the Hecke operator* T_n *satisfies the following relation:*

$$T_n(P(x, y) \otimes (u, v)) = \sideset{}{'}\sum_{\substack{a>b\geq 0 \\ d>c\geq 0 \\ ad-bc=n}} P(ax+by, cx+dy) \otimes (au+cv, bu+dv),$$

where the prime in the notation for the sum means that the terms such that $\gcd(N, au + cv, bu + dv) \neq 1$ *have to be omitted.*

One would also like to express $\mathbb{S}_k(\Gamma_1(N))$ and $\mathbb{S}_k(N, \varepsilon)$ in terms of the Manin symbols. The following proposition will help us.

6.1.17 Proposition (see [Mer], Proposition 4) *Let N be in $\mathbb{Z}_{\geq 1}$ and let k be in $\mathbb{Z}_{\geq 2}$. Define an equivalence relation $\sim$ on the $\mathbb{Q}$-vector space $\mathbb{Q}[\Gamma_1(N) \setminus \mathbb{Q}^2]$ by*

$$[\overline{\lambda x}] \sim \text{sign}(\lambda)^k [\overline{x}] \quad \text{for } \lambda \in \mathbb{Q}^{\times} \text{ and } x \in \mathbb{Q}^2.$$

Then the map

$$\mu: \mathbb{B}_k(\Gamma_1(N)) \to \mathbb{Q}[\Gamma_1(N) \setminus \mathbb{Q}^2]/\sim$$

given by

$$\mu: P \otimes \left\{\frac{a}{b}\right\} \mapsto P(a, b)\left[\overline{\binom{a}{b}}\right] \quad (a, b \text{ coprime integers})$$

is well defined and injective.

The vector space $\mathbb{Q}[\Gamma_1(N)\setminus\mathbb{Q}^2]/\sim$ is finite dimensional. The above proposition shows that $\mathbb{S}_k(\Gamma_1(N))$ is the kernel of $\mu\delta$, which is a map that can be computed in terms of Manin symbols. The computation of $\mathbb{S}_k(N, \varepsilon)$ can be done in a similar way, see [Ste2], Section 8.4.

6.2 INTERMEZZO: ATKIN-LEHNER OPERATORS

In the rest of this chapter, we will be using Atkin-Lehner operators on $S_k(\Gamma_1(N))$ from time to time. This section provides a brief treatment of the properties we need. The main reference for this material is [At-Li].

Let Q be a positive divisor of N such that $\gcd(Q, N/Q) = 1$. Let $w_Q \in \mathrm{GL}_2^+(\mathbb{Q})$ be any matrix of the form

$$(6.2.1) \qquad\qquad w_Q = \begin{pmatrix} Qa & b \\ Nc & Qd \end{pmatrix}$$

with $a, b, c, d \in \mathbb{Z}$ and $\det(w_Q) = Q$. The assumption $\gcd(Q, N/Q) = 1$ ensures that such a w_Q exists. A straightforward verification shows that $f|_k w_Q$ is in $S_k(\Gamma_1(N))$. Now, given Q, this $f|_k w_Q$ still depends on the choice of a, b, c, d. However, we can use a normalization in our choice of a, b, c, d which will ensure that $f|_k w_Q$ only depends on Q. Be aware of the fact that different authors use different normalizations here. The one we will be using is

$$(6.2.2) \qquad\qquad a \equiv 1 \bmod N/Q, \quad b \equiv 1 \bmod Q,$$

the normalization used in [At-Li]. We define

$$
\begin{aligned}
(6.2.3) \quad W_Q(f) &:= Q^{1-k/2} f|_k w_Q \\
&= \frac{Q^{k/2}}{(Ncz + Qd)^k} f\left(\frac{Qaz + b}{Ncz + Qd} \right),
\end{aligned}
$$

which is now independent of the choice of w_Q, and call W_Q an *Atkin-Lehner operator*. In particular we have

$$W_N(f) = \frac{1}{N^{k/2} z^k} f\left(\frac{-1}{Nz} \right).$$

An unfortunate thing about these Atkin-Lehner operators is that they do not preserve the spaces $S_k(N, \varepsilon)$. But we can say something about it. Let $\varepsilon \colon (\mathbb{Z}/N\mathbb{Z})^\times \to \mathbb{C}^\times$ be a character and suppose that f is in $S_k(N, \varepsilon)$. By the Chinese remainder theorem, one can write ε in a unique way as $\varepsilon = \varepsilon_Q \varepsilon_{N/Q}$, such that ε_Q is a character on $(\mathbb{Z}/Q\mathbb{Z})^\times$ and $\varepsilon_{N/Q}$ is a character on $(\mathbb{Z}/(N/Q)\mathbb{Z})^\times$. It is a fact that

$$W_Q(f) \in S_k(N, \overline{\varepsilon}_Q \varepsilon_{N/Q}).$$

Also, there is a relation between the q-expansions of f and $W_Q(f)$.

6.2.4 Theorem *Let f be a newform in $S_k(N, \varepsilon)$. Take Q dividing N with $\gcd(Q, N/Q) = 1$. Then*

$$W_Q(f) = \lambda_Q(f)g$$

with $\lambda_Q(f)$ an algebraic number in $\mathbb{C}$ of absolute value 1 and g a newform in $S_k(N, \bar{\varepsilon}_Q \varepsilon_{N/Q})$. Suppose now that n is a positive integer and write $n = n_1 n_2$, where n_1 consists only of prime factors dividing Q and n_2 consists only of prime factors not dividing Q. Then we have

$$a_n(g) = \varepsilon_{N/Q}(n_1)\bar{\varepsilon}_Q(n_2)\overline{a_{n_1}(f)}a_{n_2}(f).$$

The number $\lambda_Q(f)$ is called a *pseudo-eigenvalue* for the Atkin-Lehner operator. In some cases there exists a closed expression for it. In the notation of the following theorem, $g(\chi)$ denotes the Gauss sum of a Dirichlet character χ of conductor $N(\chi)$:

$$(6.2.5) \qquad g(\chi) := \sum_{v \in (\mathbb{Z}/N(\chi)\mathbb{Z})^\times} \chi(v) \exp\left(\frac{2\pi i v}{N(\chi)}\right).$$

6.2.6 Theorem *Let $f \in S_k(N, \varepsilon)$ be a newform and suppose q is a prime that divides N exactly once. Then we have*

$$\lambda_q(f) = \begin{cases} g(\varepsilon_q)q^{-k/2}\overline{a_q(f)} & \text{if } \varepsilon_q \text{ is nontrivial,} \\ -q^{1-k/2}\overline{a_q(f)} & \text{if } \varepsilon_q \text{ is trivial.} \end{cases}$$

6.2.7 Theorem (see [Asa], Theorem 2) *Let $f \in S_k(N, \varepsilon)$ be a newform with N square-free. For $Q \mid N$ we have:*

$$\lambda_Q(f) = \varepsilon(Qd - \frac{N}{Q}a) \prod_{q|Q} \varepsilon(Q/q)\lambda_q(f).$$

Here, a and d are defined by (6.2.1). Moreover, this identity holds without any normalization assumptions on the entries of w_Q, as long as we define $\lambda_q(f)$ by the formula given in Theorem 6.2.6.

6.3 BASIC NUMERICAL EVALUATIONS

In this section we will describe how to perform basic numerical evaluations, such as the evaluation of a cusp form at a point in $\mathbb{H}$ and the evaluation of an integral of a cusp form between to points in $\mathbb{H} \cup \mathbb{P}^1(\mathbb{Q})$. Again, the focus will be on performing actual computations.

6.3.1 Period integrals: the direct method

In this subsection we will stick to the case $k = 2$, referring to [Ste2], Chapter 10 for a more general approach (see also [Cre], Section 2.10 for a treatment of $\Gamma_0(N)$). So fix a positive integer N and an $f \in S_2(\Gamma_1(N))$. Our goal is to efficiently evaluate the integral pairing $\langle x, f \rangle$ for $x \in \mathbb{S}_2(\Gamma_1(N))$.

Let us indicate why it suffices to look at newforms f. Because of (2.2.18), it suffices to look at $f = \alpha_d(f')$ with $f' \in S_k(\Gamma_1(M))$ a newform for some $M \mid N$ and $d \mid N/M$. By [Sho], Theorem 4.3, we have

$$\langle x, f \rangle = \langle x, \alpha_d(f') \rangle = d^{1-k} \left\langle \begin{pmatrix} d & 0 \\ 0 & 1 \end{pmatrix} x, f' \right\rangle,$$

so that computing period integrals for f reduces to computing period integrals of the newform f'.

Let us now make the important remark that for each $z \in \mathbb{H}$ we can numerically compute $\int_\infty^z f \, dq/q$ by formally integrating the q-expansion of f:

$$(6.3.2) \qquad \int_\infty^z f \frac{dq}{q} = \sum_{n \geq 1} \frac{a_n(f)}{n} q^n, \qquad \text{where } q = \exp(2\pi i z).$$

The radius of convergence of this series is 1 and the coefficients are small (that is, estimated by $\tilde{O}(n^{(k-3)/2})$). So if $\Im z \gg 0$ then we have $|q| \ll 1$ and the series converges rapidly. To be more concrete, for $\Im z > M$ we have $|q^n| < \exp(-2\pi M n)$, so if we want to compute $\int_\infty^z f \, dq/q$ to a precision of p decimals, we need to compute about $\frac{p \log 10}{2\pi M} \approx 0.37 \frac{p}{M}$ terms of the series.

To compute a period integral we remark that for any $\gamma \in \Gamma_1(N)$ and any $z \in \mathbb{H} \cup \mathbb{P}^1(\mathbb{Q})$ any continuous, piecewise smooth path δ in $\mathbb{H} \cup \mathbb{P}^1(\mathbb{Q})$ from z to γz, the homology class of δ pushed forward to $X_1(N)(\mathbb{C})$ depends only on γ [Man1], Proposition 1.4. Let us denote this homology class by

$$\{\infty, \gamma\infty\} \in S_2(\Gamma_1(N)) \cong H_1(X_1(N)(\mathbb{C}), \mathbb{Z})$$

and remark that all elements of $H_1(X_1(N)(\mathbb{C}), \mathbb{Z})$ can be written in this way. As we also have $S_2(\Gamma_1(N)) \cong H^0(X_1(N)_\mathbb{C}, \Omega^1)$, this means we can calculate $\int_{\{\infty, \gamma\infty\}} f \frac{dq}{q}$ by choosing a smart path in $\mathbb{H} \cup \mathbb{P}^1(\mathbb{Q})$:

$$\int_\infty^{\gamma\infty} f \frac{dq}{q} = \int_z^{\gamma z} f \frac{dq}{q} = \int_\infty^{\gamma z} f \frac{dq}{q} - \int_\infty^z f \frac{dq}{q}.$$

If we write $\gamma = \begin{pmatrix} a & b \\ c & d \end{pmatrix}$ then a good choice for z is

$$z = -\frac{d}{c} + \frac{i}{|c|}.$$

In this case we have $\Im z = \Im\gamma z = 1/|c|$ so in view of (6.3.2), to compute the integral to a precision of p decimals we need about $\frac{pc\log 10}{2\pi} \approx 0.37pc$ terms of the series.

Another thing we can use is the Hecke compatibility from (6.1.9). Put

$$W_f := \left(S_2(\Gamma_1(N))/I_f S_2(\Gamma_1(N))\right) \otimes \mathbb{Q},$$

where I_f is the Hecke ideal belonging to f, the kernel of the map $\mathbb{T} \to \mathbb{C}$ that sends T_n to a_n for all n (here, as usual, $\mathbb{T}$ denotes the Hecke algebra attached to $S_2(\Gamma_1(N))$). The space W_f has the structure of a vector space over $(\mathbb{T}/I_f) \otimes \mathbb{Q} \cong K_f$ of dimension 2. This means that computing any period integral of f, we only need to precompute 2 period integrals. So one tries to find a K_f-basis of W_f consisting of elements $\{\infty, \gamma\infty\}$, where $\gamma \in \Gamma_1(N)$ has a very small c-entry. In practice it turns out that we do not need to search very far.

6.3.3 Period integrals: the twisted method

In this subsection we have the same set-up as in the previous subsection. There is another way of computing period integrals for $f \in S_2(\Gamma_1(N))$ that sometimes beats the method described in the previous subsection. The method described in this subsection is similar to [Cre], Section 2.11, and makes use of winding elements and twists.

The *winding element* of $\mathbb{M}_2(\Gamma_1(N))$ is simply defined as the element $\{\infty, 0\}$ (some authors define it as $\{0, \infty\}$, a matter of sign convention). Integration over this element is easy because we can break up the path in a very neat way:

$$\int_\infty^0 f \frac{dq}{q} = \int_\infty^{i/\sqrt{N}} f \frac{dq}{q} + \int_{i/\sqrt{N}}^0 f \frac{dq}{q}$$

$$= \int_\infty^{i/\sqrt{N}} f \frac{dq}{q} + \int_{i/\sqrt{N}}^\infty W_N(f) \frac{dq}{q}$$

$$= \int_\infty^{i/\sqrt{N}} (f - W_N(f)) \frac{dq}{q}.$$

Now, choose an odd prime ℓ not dividing N and a primitive Dirichlet character $\chi: \mathbb{Z} \to \mathbb{C}$ of conductor ℓ. If $f \in S_k(\Gamma_1(N))$ is a newform, then $f \otimes \chi$ is a newform in $S_k(\Gamma_1(N\ell^2))$, where

$$f \otimes \chi = \sum_{n \geq 1} a_n(f) \chi(n) q^n.$$

The following formula to express χ as a linear combination of additive characters is well known:

$$\chi(n) = \frac{g(\chi)}{\ell} \sum_{v=1}^{\ell-1} \overline{\chi}(-v) \exp\left(\frac{2\pi i v n}{\ell}\right),$$

where $g(\chi)$ is the Gauss sum of χ (see (6.2.5)). It follows now immediately that

$$f \otimes \chi = \frac{g(\chi)}{\ell} \sum_{v=1}^{\ell-1} \chi(-v) f\left(z + \frac{v}{\ell}\right) = \frac{g(\chi)}{\ell} \sum_{v=1}^{\ell-1} \chi(-v) f \left| \begin{pmatrix} \ell & v \\ 0 & \ell \end{pmatrix} \right. .$$

For $f \in S_2(\Gamma_1(N))$ we now get the following useful formula for free:

$$(6.3.4) \qquad \langle \{\infty, 0\}, f \otimes \chi \rangle = \frac{g(\chi)}{\ell} \left\langle \sum_{v=0}^{l-1} \chi(-v) \left\{\infty, \frac{v}{\ell}\right\}, f \right\rangle.$$

The element $\sum_{v=0}^{l-1} \chi(-v) \{\infty, \frac{v}{\ell}\}$ of $\mathbb{M}_k(\Gamma_1(N)) \otimes \mathbb{Z}[\chi]$ or of some other modular symbols space where it is well defined is called a *twisted winding element* or, more precisely the χ-*twisted winding element*. Because of formula (6.3.4), we can calculate the pairings of newforms in $S_2(\Gamma_1(N))$ with twisted winding elements quite efficiently as well.

We can describe the action of the Atkin-Lehner operator $W_{N\ell^2}$ on $f \otimes \chi$:

$$W_{N\ell^2}(f \otimes \chi) = \frac{g(\chi)}{g(\overline{\chi})}\varepsilon(\ell)\chi(-N)\lambda_N(f)\tilde{f} \otimes \overline{\chi},$$

where $\tilde{f} = \sum_{n \geq 1} \overline{a_n(f)}q^n$ (see, for example, [At-Li], Section 3). So in particular we have the following integral formula for a newform f in the space $S_2(N, \varepsilon)$:

(6.3.5)

$$\int_\infty^0 f \otimes \chi \frac{dq}{q} = \int_\infty^{i/(\ell\sqrt{N})} (f \otimes \chi - W_{N\ell^2}(f \otimes \chi))\frac{dq}{q}$$

$$= \int_\infty^{i/(\ell\sqrt{N})} \left(f \otimes \chi - \frac{g(\chi)}{g(\overline{\chi})}\chi(-N)\varepsilon(\ell)\lambda_N(f)\tilde{f} \otimes \overline{\chi} \right) \frac{dq}{q}.$$

So to calculate

$$\left\langle \sum_{\nu=0}^{l-1} \chi(-\nu)\left\{\infty, \frac{\nu}{\ell}\right\}, f \right\rangle$$

we need to evaluate the series (6.3.2) at z in $\mathbb{H}$ with $\Im z = 1/(\ell\sqrt{N})$, which means that for a precision of p decimals we need to sum about $\frac{p\ell\sqrt{N}\log 10}{2\pi} \approx 0.37p\ell\sqrt{N}$ terms of the series. In the spirit of the previous subsection, we try several ℓ and χ, as well as the untwisted winding element $\{\infty, 0\}$, until we can make a K_f-basis for W_f. It follows from [Shi2], Theorems 1 and 3, that we can always find such a basis. Also here, it turns out that in practice we do not need to search very far. The method that requires the least amount of of q-expansion terms is preferred.

6.3.6 Computation of q-expansions at various cusps

The upper half-plane $\mathbb{H}$ is covered by neighborhoods of the cusps. If we want to evaluate a cusp form $f \in S_k(\Gamma_1(N))$ or an integral of a cusp form at a point in such a neighborhood, then it is useful to be able to calculate the q-expansion of f at the corresponding cusp. We shall mean by this the following. A cusp a/c can be written as $\gamma\infty$ with $\gamma = \begin{pmatrix} a & b \\ c & d \end{pmatrix} \in SL_2(\mathbb{Z})$.

Then a q-expansion of f at a/c is simply the q-expansion of $f|_k\gamma$. This notation is abusive, since it depends on the choice of γ. The q-expansion will be an element of the power series ring $\mathbb{C}[[q^{1/w}]]$, where w is the width of the cusp a/c and $q^{1/w} = \exp(2\pi i z/w)$.

If the level N is square-free this can be done symbolically. For general N it is not known how to do this, but we shall give some attempts that at least give numerical computations of q-expansions. We use that we can compute the q-expansions of newforms in $S_k(\Gamma_1(N))$ at ∞ using modular symbols methods.

The case of square-free N

The method we present here is due to [Asa]. Let N be square-free, and let $f \in S_k(\Gamma_1(N))$ be a newform of character ε. The main reason for being able to compute q-expansions at all cusps in this case is because the group generated by $\Gamma_0(N)$ and all w_Q (see (6.2.1)) acts transitively on the cusps, something that is not true when N is not square-free.

So let $\gamma = \left(\begin{smallmatrix} a & b \\ c & d \end{smallmatrix}\right) \in \mathrm{SL}_2(\mathbb{Z})$ be given. Put

$$c' = \frac{c}{\gcd(N,c)} \quad \text{and} \quad Q = \frac{N}{\gcd(N,c)}.$$

Let $r \in \mathbb{Z}$ be such that $d \equiv cr \bmod Q$ and define $b', d' \in \mathbb{Z}$ by

$$Qd' = d - cr \quad \text{and} \quad b' = b - ar.$$

Then we have

$$\begin{pmatrix} a & b \\ c & d \end{pmatrix} = \begin{pmatrix} Qa & b' \\ Nc' & Qd' \end{pmatrix} \begin{pmatrix} Q^{-1} & rQ^{-1} \\ 0 & 1 \end{pmatrix}.$$

Theorems 6.2.4 and 6.2.7 tell us how $\left(\begin{smallmatrix} Qa & b' \\ Nc' & Qd' \end{smallmatrix}\right)$ acts on q-expansions. The action of $\left(\begin{smallmatrix} Q^{-1} & rQ^{-1} \\ 0 & 1 \end{smallmatrix}\right)$ on q-expansions is simply

$$\sum_{n\geq 1} a_n q^n \mapsto Q^{1-k} \sum_{n\geq 1} a_n \zeta_Q^{rn} q^{n/Q} \quad \text{with} \quad \zeta_Q = \exp\left(\frac{2\pi i}{Q}\right).$$

This shows how the q-expansion of $f|_k\gamma$ can be derived from that of f.

Let us now explain how to do it for oldforms as well. By induction and (2.2.18) we may suppose that $f = \alpha_p(f')$, where $p \mid N$ is prime and

f' is in $S_k(\Gamma_1(N/p))$, and that we know how to compute the q-expansions of f' at all the cusps. Let $\gamma = \begin{pmatrix} a & b \\ c & d \end{pmatrix}$ be given. Then we have

$$f|_k \gamma = p^{1-k} f'|_k \begin{pmatrix} p & 0 \\ 0 & 1 \end{pmatrix} \gamma = p^{1-k} f'|_k \begin{pmatrix} pa & pb \\ c & d \end{pmatrix}.$$

We will now distinguish two cases: $p \mid c$ and $p \nmid c$. If $p \mid c$ then we have a decomposition

$$\begin{pmatrix} pa & pb \\ c & d \end{pmatrix} = \begin{pmatrix} a & pb \\ c/p & d \end{pmatrix} \begin{pmatrix} p & 0 \\ 0 & 1 \end{pmatrix},$$

and we know how both matrices on the right-hand side act on q-expansions. If $p \nmid c$, choose b', d' with $pad' - b'c = 1$. Then we have

$$\begin{pmatrix} pa & pb \\ c & d \end{pmatrix} = \begin{pmatrix} pa & b' \\ c & d' \end{pmatrix} \beta$$

with $\beta \in \mathrm{GL}_2^+(\mathbb{Q})$ upper triangular, so also in this case we know how both matrices on the right-hand side act on q-expansions.

The general case

In a discussion with Peter Bruin, the author figured out an attempt to drop the assumption that N be square-free and compute q-expansions of cusp forms numerically in this case. The idea is to generalize the W_Q operators from Section 6.2.

So let N be given. Let Q be a divisor of N, and put $R = \gcd(Q, N/Q)$. Let w_Q be any matrix of the form

$$w_Q = \begin{pmatrix} RQa & b \\ RNc & Qd \end{pmatrix} \quad \text{with } a, b, c, d \in \mathbb{Z},$$

such that $\det w_Q = QR^2$ (the conditions guarantee us that such matrices do exist). One can then verify

$$\Gamma_1(NR^2) < w_Q^{-1} \Gamma_1(N) w_Q,$$

so that slashing with w_Q defines a linear map

$$S_k(\Gamma_1(N)) \oplus \overline{S}_k(\Gamma_1(N)) \xrightarrow{\ |w_Q\ } S_k(\Gamma_1(NR^2)) \oplus \overline{S}_k(\Gamma_1(NR^2)),$$

which is injective since the slash operator defines a group action on the space
of all functions $\mathbb{H} \to \mathbb{C}$.

On the other hand, w_Q defines an operation on $\mathbb{M}_k$ that can be shown to
induce a linear map

$$w_Q \colon \mathbb{S}_k(\Gamma_1(NR^2)) \otimes \mathbb{Q} \to \mathbb{S}_k(\Gamma_1(N)) \otimes \mathbb{Q}$$

satisfying the following compatibility with respect to the integration pairing
between modular symbols and cusp forms (see [Sho], Theorem 4.3):

$$(6.3.7) \qquad\qquad \langle w_Q x, f \rangle = \langle x, f|_k w_Q \rangle.$$

Let $(x_1, \ldots, x_r)$ be a basis of $\mathbb{S}_k(\Gamma_1(N)) \otimes \mathbb{Q}$ and $(y_1, \ldots, y_s)$ be a basis of
$\mathbb{S}_k(\Gamma_1(NR^2)) \otimes \mathbb{Q}$. Then we can write a matrix A in terms of these bases
that describes the map w_Q, since we can express any symbol $P \otimes \{\alpha, \beta\}$ in
terms of Manin symbols. The matrix A^t then defines the action of w_Q in
terms of the bases of the cusp forms spaces that are dual to $(x_1, \ldots x_r)$ and
$(y_1, \ldots, y_s)$.

Now, let $(f_1, \ldots, f_r)$ be a basis of $S_k(\Gamma_1(N)) \oplus \overline{S}_k(\Gamma_1(N))$ and let
$(g_1, \ldots, g_s)$ be a basis of $S_k(\Gamma_1(NR^2)) \oplus \overline{S}_k(\Gamma_1(NR^2))$ (for instance we
could take bases consisting of eigenforms for the Hecke operators away
from N). Define matrices:

$$B := \big(\langle x_i, f_j \rangle \big)_{i,j} \quad \text{and} \quad C := \big(\langle y_i, g_j \rangle \big)_{i,j}.$$

These can be computed numerically as the entries are period integrals. Then
the matrix $C^{-1} A^t B$ describes the map $\cdot|_k w_Q$ with respect to the bases
$(f_1, \ldots, f_r)$ and $(g_1, \ldots, g_s)$. Hence if we can invert C efficiently, then we
can numerically compute the q-expansion of $f|_k w_Q$ with $f \in S_k(\Gamma_1(N))$.

Let now a matrix $\gamma = \begin{pmatrix} a & b \\ c & d \end{pmatrix} \in SL_2(\mathbb{Z})$ be given. Put:

$$c' := \gcd(N, c) \quad \text{and} \quad Q := N/c'.$$

Because of $\gcd(c/c', Q) = 1$ we can find $\alpha \in (\mathbb{Z}/Q\mathbb{Z})^\times$ with $\alpha c/c' \equiv 1$
mod Q. If we lift α to $(\mathbb{Z}/N\mathbb{Z})^\times$, then we have $\alpha c \equiv c'$ mod N. Let now
$d' \in \mathbb{Z}$ be a lift of $\alpha d \in (\mathbb{Z}/N\mathbb{Z})^\times$. Because αc and αd together generate
$\mathbb{Z}/N\mathbb{Z}$ we have $\gcd(c', d') = 1$, and so we can find $a', b' \in \mathbb{Z}$ that satisfy

$a'd' - b'c' = 1$. According to Lemma 6.1.12, we have

$$\gamma = \gamma_0 \begin{pmatrix} a' & b' \\ c' & d' \end{pmatrix} \quad \text{with } \gamma_0 \in \Gamma_0(N).$$

Put $R = \gcd(c', Q)$. Then we have $\gcd(NR, Q^2 Ra') = QR \gcd(c', Qa')$ and hence $\gcd(NR, Q^2 Ra') = QR^2$, so there exist $b'', d'' \in \mathbb{Z}$, with

$$w_Q := \begin{pmatrix} QRa' & b'' \\ NR & Qd'' \end{pmatrix}$$

having determinant QR^2. One can now verify that we have $\begin{pmatrix} a' & b' \\ c' & d' \end{pmatrix} = w_Q \beta$ with $\beta \in \mathrm{GL}_2^+(\mathbb{Q})$ upper triangular. So in the decomposition

$$\gamma = \gamma_0 w_Q \beta$$

we can compute the slash action of all three matrices on the right-hand side in terms of q-expansions, hence also of γ.

In conclusion we see that in this method we have to increase the level and go to $S_k(\Gamma_1(NR^2))$ for the square divisors R^2 of N to compute q-expansions of cusp forms in $S_k(\Gamma_1(N))$ at arbitrary cusps.

6.3.8 Numerical evaluation of cusp forms

For $f \in S_k(\Gamma_1(N))$ and a point $P \in \mathbb{H}$ we wish to compute $f(P)$ to a high numerical precision. Before we do this let us say some words on how P should be represented. The transformation property of modular forms implies that representing P as $x + iy$ with $x, y \in \mathbb{R}$ is not a good idea, as this would be numerically very unstable when P is close to the real line. Instead, we represent P as

(6.3.9) $P = \gamma z \quad \text{with } \gamma \in \mathrm{SL}_2(\mathbb{Z}), \ z = x + iy, \ x \ll \infty, \text{ and } y \gg 0.$

For instance, we could demand that z be in the standard fundamental domain:

$$\mathcal{F} := \{z \in \mathbb{H} : |\Re z| \leq 1/2 \text{ and } |z| \geq 1\}$$

for $\mathrm{SL}_2(\mathbb{Z})$ acting on $\mathbb{H}$, although this is not strictly necessary.

So let $P = \gamma z$ be given, with $\gamma = \begin{pmatrix} a & b \\ c & d \end{pmatrix} \in \mathrm{SL}_2(\mathbb{Z})$ and $\Im z > M$, say. Let $w = w(\gamma)$ be the width of the cusp $\gamma\infty$ with respect to $\Gamma_1(N)$. To compute $f(P)$ we make use of a q-expansion of f at $\gamma\infty$,

$$f(P) = (cz + d)^k (f|_k\gamma)(z) = (cz + d)^k \sum_{n \geq 1} a_n q^{n/w}.$$

The radius of convergence is 1 and the coefficients are small (estimated by $\tilde{O}(n^{(k-1)/2})$). So to compute $f(P)$ to a precision of p decimals we need about $\frac{pw \log 10}{2\pi M} \approx 0.37 \frac{pw}{M}$ terms of the q-expansion of $f|_k\gamma$.

Of course, we have some freedom in choosing γ and z to write down P. We want to find γ such that $P = \gamma z$ with $\Im z / w(\gamma)$ as large as possible. In general, one can always write $P = \gamma z$ with $z \in \mathcal{F}$, so one obtains

$$(6.3.10) \qquad\qquad \max_{\gamma \in \mathrm{SL}_2(\mathbb{Z})} \frac{\Im \gamma^{-1} P}{w(\gamma)} \geq \frac{\sqrt{3}}{2N}.$$

We see that in order to calculate $f(P)$ to a precision of p decimals it suffices to use about $\frac{pN \log 10}{\sqrt{3}\pi} \approx 0.42pN$ terms of the q-expansions at each cusp. Although for many points P there is a better way of writing it as γz in this respect than taking $z \in \mathcal{F}$, it seems hard to improve the bound $\frac{\sqrt{3}}{2N}$ in general.

We wish to adjust the representation $P = \gamma z$ sometimes to another one $P = \gamma' z'$, where $\gamma' \in \mathrm{SL}_2(\mathbb{Z})$, for instance, because during our calculations $\Re z$ has become too large or $\Im z$ has become too small (but still within reasonable bounds). We can make $\Re z$ smaller by putting $z' := z - n$ for appropriate $n \in \mathbb{Z}$ and putting $\gamma' := \gamma \begin{pmatrix} 1 & n \\ 0 & 1 \end{pmatrix}$. Making $\Im z$ larger is rather easy as well. We want to find $\gamma'' = \begin{pmatrix} a & b \\ c & d \end{pmatrix} \in \mathrm{SL}_2(\mathbb{Z})$ such that

$$\Im \gamma'' z = \frac{\Im z}{|cz + d|^2}$$

is large. This simply means that we have to find a small vector $cz + d$ in the lattice $\mathbb{Z}z + \mathbb{Z}$, something that can be done easily if $\Re z \ll \infty$ and $\Im z \gg 0$. If c and d are not coprime, we can divide both by their greatest common divisor to obtain a smaller vector. The matrix γ'' can now be completed, and we put $z' := \gamma'' z$ and $\gamma' := (\gamma'')^{-1}$.

6.3.11 Numerical evaluation of integrals of cusp forms

In this subsection we will describe for $f \in S_2(\Gamma_1(N))$ and $P \in \mathbb{H}$ how to evaluate the integral $\int_\infty^P f dq/q$. As in the previous subsection, we assume P to be given by means of (6.3.9). The path of integration will be broken into two parts: we go from ∞ to a cusp α near P, and then from α to P.

Integrals over paths between cusps

The pairing (6.1.5) gives a map

$$\Theta \colon \mathbb{M}_2(\Gamma_1(N)) \to \mathrm{Hom}_\mathbb{C}\left(S_2(\Gamma_1(N)), \mathbb{C}\right),$$

which is injective when restricted to $\mathbb{S}_2(\Gamma_1(N))$. The image of Θ is a lattice of full rank; hence the induced map

$$\mathbb{S}_2(\Gamma_1(N)) \otimes \mathbb{R} \to \mathrm{Hom}_\mathbb{C}\left(S_2(\Gamma_1(N)), \mathbb{C}\right)$$

is an isomorphism. In particular we obtain a map

$$\Phi \colon \mathbb{M}_2(\Gamma_1(N)) \to \mathbb{S}_2(\Gamma_1(N)) \otimes \mathbb{R},$$

which is an interesting map to compute if we want to calculate integrals of cusp forms along paths between cusps. The map Φ is called a *period mapping*.

The Manin-Drinfel'd theorem (see [Man1], Corollary 3.6, and [Dri], Theorem 1), tells us that $\mathrm{im}(\Phi) \subset \mathbb{S}_2(\Gamma_1(N)) \otimes \mathbb{Q}$. This is equivalent to saying that each degree 0 divisor of $X_1(N)$ which is supported on cusps defines a torsion point of $J_1(N)$. The proof given in [Dri] already indicates how to compute Φ with symbolic methods. Let p be a prime that is 1 mod N. Then the operator $p + 1 - T_p$ on $\mathbb{M}_2(\Gamma_1(N))$ has its image in $\mathbb{S}_2(\Gamma_1(N))$. The same operator is invertible on $\mathbb{S}_2(\Gamma_1(N)) \otimes \mathbb{Q}$. So we simply have

$$\Phi = (p + 1 - T_p)^{-1}(p + 1 - T_p),$$

where the second $p+1-T_p$ denotes the map $\mathbb{M}_2(\Gamma_1(N)) \to \mathbb{S}_2(\Gamma_1(N))$ and the first $p+1-T_p$ denotes the invertible operator on $\mathbb{S}_2(\Gamma_1(N)) \otimes \mathbb{Q}$. For other methods to compute Φ, see [Ste2], Section 10.6. So we can express the integral of $f dq/q$ between any two cusps α and β in terms of period integrals, which we have already seen how to compute:

$$\int_\alpha^\beta f \frac{dq}{q} = \langle \Phi(\{\alpha, \beta\}), f \rangle.$$

Integrals over general paths

We can pretty much imitate the previous subsection. Write $P \in \mathbb{H}$ as $P = \gamma z$ with $\gamma \in SL_2(\mathbb{Z})$ such that $\Im z / w(\gamma \infty)$ is as large as possible. Then we have

(6.3.12)
$$\int_\infty^P f \frac{dq}{q} = \int_\infty^{\gamma\infty} f \frac{dq}{q} + \int_{\gamma\infty}^{\gamma z} f \frac{dq}{q}$$
$$= \int_\infty^{\gamma\infty} f \frac{dq}{q} + \int_\infty^z (f|_2\gamma) \frac{dq}{q}.$$

The integral $\int_\infty^{\gamma\infty} f \frac{dq}{q}$ is over a path between two cusps, so we can compute it by the above discussion, and the integral $\int_\infty^z (f|_2\gamma) \frac{dq}{q}$ can be computed using the q-expansion of $f|_2\gamma$,

$$\int_\infty^z (f|_2\gamma) \frac{dq}{q} = w \sum_{n \geq 1} \frac{a_n}{n} q^{n/w},$$

where $w = w(\gamma)$, $q^{1/w} = \exp(2\pi i z / w)$ and $f|_2\gamma = \sum a_n q^{n/w}$. Because of (6.3.10), computing about $\frac{pN \log 10}{\sqrt{3}\pi} \approx 0.42 pN$ terms of the series should suffice to compute $\int_\infty^P f \frac{dq}{q}$ for any $P \in \mathbb{H}$.

   Note also that we can use formula (6.3.12) to compute the pseudo-eigen-value $\lambda_Q(f)$ by plugging in $\gamma = w_Q$ and a z, for which both $\Im z$ and $\Im w_Q z$ are high and $\int_\infty^z W_q(f) dq/q$ is not too close to zero.

6.4 NUMERICAL CALCULATIONS AND GALOIS REPRESENTATIONS

Let f be a newform (of some level and weight), and let $\lambda \mid \ell$ be a prime of its coefficient field. From Section 2.5 we know that a residual Galois representation $\bar{\rho} = \bar{\rho}_{f,\lambda}$ is attached to the pair (f, λ). The fixed field K_λ of $\ker(\bar{\rho})$ in $\overline{\mathbb{Q}}$ is a number field. The results from Chapter 14 point out that we know that computing $\bar{\rho}$ essentially boils down to computing a polynomial that has K_λ as splitting field. In this section we describe how numerical calculations can be used to compute such a polynomial. We will follow ideas from Chapter 3.

Theorem 2.5.7 shows that we can reduce this problem to the case of a form of weight 2 in most interesting cases. Hence we will assume that f is a newform in $S_2(\Gamma_1(N))$. Assume that the representation $\overline{\rho}_{f,\lambda}$ is absolutely irreducible, and let $\mathbb{T}$ be the Hecke algebra acting on $J_1(N)$. There is a subspace V_λ of $J_1(N)(\overline{\mathbb{Q}})[\ell]$ on which both $\mathbb{T}$ and $\mathrm{Gal}(\overline{\mathbb{Q}}/\mathbb{Q})$ act, such that the action of $\mathrm{Gal}(\overline{\mathbb{Q}}/\mathbb{Q})$ defines $\overline{\rho}_{f,\lambda}$.

6.4.1 Approximation of torsion points

The Jacobian $J_1(N)_{\mathbb{C}}$ can be described as follows. Pick a basis $f_1, \ldots, f_g$ of $S_2(\Gamma_1(N))$. Put

$$\Lambda := \left\{ \int_\gamma (f_1, \ldots, f_g) \frac{dq}{q} : [\gamma] \in \mathrm{H}_1(X_1(N)(\mathbb{C}), \mathbb{Z}) \right\} \subset \mathbb{C}^g.$$

This is a lattice in $\mathbb{C}^g$ of full rank. By the Abel-Jacobi theorem we have an isomorphism:

$$J_1(N)(\mathbb{C}) \xrightarrow{\sim} \mathbb{C}^g/\Lambda,$$

$$\left[\sum_i ([Q_i] - [R_i]) \right] \mapsto \sum_i \int_{R_i}^{Q_i} (f_1, \ldots, f_g) \frac{dq}{q}.$$

Let now a divisor $\sum_{i=1}^g [R_i]$ on $X_1(N)$ be given. Identifying $J_1(N)(\mathbb{C})$ with $\mathbb{C}^g/\Lambda$ in this way, we get a birational morphism

$$\phi: \mathrm{Sym}^g X_1(N)(\mathbb{C}) \to \mathbb{C}^g/\Lambda,$$

$$(Q_1, \ldots, Q_g) \mapsto \sum_{i=1}^g \int_{R_i}^{Q_i} (f_1, \ldots, f_g) \frac{dq}{q}.$$

The homology group $\mathrm{H}_1(X_1(N)(\mathbb{C}), \mathbb{Z})$ is canonically isomorphic to the modular symbols space $\mathbb{S}_2(\Gamma_1(N))$. The period lattice Λ can thus be computed numerically using the methods from Sections 6.3.1 and 6.3.3. Since we can compute the action of $\mathbb{T}$ on $\mathbb{S}_2(\Gamma_1(N)) \cong \Lambda$, we can write down the points in $\frac{1}{\ell}\Lambda/\Lambda \subset \mathbb{C}^g/\Lambda$ that correspond to the points of V_λ. The aim is now to compute the divisors on $X_1(N)_{\mathbb{C}}$ that map to these points along ϕ. In our computations, we assume without proof that V_λ lies beneath the good locus of ϕ, that is, the map $X_1(N)^g \to \mathbb{C}^g/\Lambda$ induced by ϕ is étale above V_λ.

We start calculating with small precision. Let P in $V_\lambda(\mathbb{C}) \subset \mathbb{C}^g/\Lambda$ be nonzero. First we try out a lot of random points $Q = (Q_1, \ldots, Q_g)$ in $X_1(N)(\mathbb{C})^g$. Here, each Q_i will be written in the form $Q_i = \gamma_i w_i$, with γ_i in a set of representatives for $\Gamma_1(N) \backslash \mathrm{SL}_2(\mathbb{Z})$ and $w_i \in \mathcal{F}$. We can compute $\phi(Q)$ using methods from Section 6.3.11. We work with the point Q for which $\phi(Q)$ is closest to P. If we in fact already know some points Q with $\phi(Q)$ approximately equal to a point in $V_\lambda(\mathbb{C})$, then we could also take one of those points as a starting point Q.

The next thing to do is adjust Q so that $\phi(Q)$ comes closer to P. For this, we will make use of the Newton-Raphson approximation method. Let $\phi' : \mathbb{H}^g \to \mathbb{C}^g/\Lambda$ be the function defined by

$$\phi'(z_1, \ldots, z_g) = \phi(\gamma_1 z_1, \ldots, \gamma_g z_g).$$

We observe that for a small vector $h = (h_1, \ldots, h_g) \in \mathbb{C}^g$ we have

$$\phi'(w_1 + h_1, \ldots, w_g + h_g) = \phi(Q) + hD + O(\|h\|^2),$$

with

$$D = \begin{pmatrix} \frac{\partial \phi'_1}{\partial z_1} & \cdots & \frac{\partial \phi'_g}{\partial z_1} \\ \vdots & \ddots & \vdots \\ \frac{\partial \phi'_1}{\partial z_g} & \cdots & \frac{\partial \phi'_g}{\partial z_g} \end{pmatrix} \Bigg|_{(w_1, \ldots, w_g)} .$$

From the definition of ϕ we can immediately deduce

$$\frac{\partial \phi'_i}{\partial z_j}(w_1, \ldots, w_g) = 2\pi i \cdot (f_i |_2 \gamma_j)(w_j),$$

where we apologize for the ambiguous i. We can thus compute the matrix D using the methods of Section 6.3.8. Now choose a small vector $v = (v_1, \ldots, v_g) \in \mathbb{C}^g$ such that $\phi(Q) + v$ is closer to P than $\phi(Q)$ is. For example, v can be chosen among all vectors of a bounded length so that $\phi(Q) + v$ is closest to P. If we write:

$$h = vD^{-1},$$

then we expect that $\phi'(w_1+h_1, \ldots, w_g+h_g)$ is almost equal to $\phi(Q)+v$. If this is not the case, then we try the same thing with a smaller v. It could be that this still fails, for instance because we are too close to the bad locus of the map ϕ. In that case, we start with a new random point Q.

We repeat the above adjustments until we are (almost) as close as we can get considering our calculation precision. It might happen that the w_i become too wild, that is, $|\Re w_i|$ becomes too large or $\Im w_i$ becomes too small. If this is the case, we adjust the way we write Q_i as $\gamma_i w_i$ using the method described in Section 6.3.8. We can always replace the γ_i then by a small matrix in the same coset of $\Gamma_1(N) \setminus SL_2(\mathbb{Z})$.

Once we have for each $P \in V_\lambda - \{0\}$ a point Q such that $\phi(Q)$ is approximately equal to P, we can start increasing the precision. We double our calculation precision and repeat the above adjustments ($\phi(Q)+v$ will in this case be equal to P). We repeat this a few times until we have very good approximations.

6.4.2 Computation of polynomials

Now, we will choose a function in $h \in \mathbb{Q}(X_1(N))$ and evaluate it at the components of the points in $\phi^{-1}(V_\lambda)$. With the discussion of Chapter 3 in mind, we want h to take values of small height. Since h multiplies heights of points roughly by $\deg(h)$, we want to find a function of small degree. Take any k and a basis $h_1, \ldots, h_n$ of $S_k(\Gamma_1(N))$ such that the q-expansions of the h_i lie in $\mathbb{Z}[[q]]$ and such that the exponents of the first nonzero terms of these q-expansions form a strictly increasing sequence. We propose to use $h = W_N(h_{n-1})/W_N(h_n)$ as a function to use (assuming $n \geq 2$). Remember from Section 2.2 that $S_k(\Gamma_1(N))$ is the space of global sections of the line bundle $\mathcal{L} = \omega^{\otimes k}(-\text{Cusps})$ on $X_1(N)$, base changed to $\mathbb{C}$. Be aware of the fact that the cusp ∞ of $X_1(N)$ is not defined over $\mathbb{Q}$, but the cusp 0 is. Since we demand the q-expansions to have rational coefficients, the sections $W_N(h_1), \ldots, W_N(h_n)$, with W_N an Atkin-Lehner operator, are defined over $\mathbb{Q}$ and they have increasing order at 0. One can now verify that for $h = W_N(h_{n-1})/W_N(h_n)$ we have

$$\deg(h) \leq \deg(\mathcal{L}) - v_\infty(h_{n-1}) \leq \deg(\mathcal{L}) - \dim H^0(\mathcal{L}) + 2 \leq g + 1.$$

For $k = 2$ and $g \geq 2$ we have $\mathcal{L} \cong \Omega^1(X_1(N))$ and we get g as an upper bound for $\deg(h)$. Using methods from Section 6.3.8, we can evaluate h numerically. The author is not aware of a sophisticated method for finding a function $h \in \mathbb{Q}(X_1(N))$ of minimal degree in general; this minimal degree is called the *gonality* of the curve $X_1(N)$. Published results on these matters seem to be limited to $X_0(N)$ or to concern only *lower* bounds for the gonality of modular curves; see, for example, [Abr], [Bak], Chapter 3, or [Poo1].

Now put, for $P \in V_\lambda(\mathbb{C}) - \{0\}$,

$$\alpha_P = \sum_{i=1}^{g} h(Q_i), \quad \text{where } \phi(Q_1, \ldots, Q_g) = P.$$

We work out the product in

$$P_\lambda(x) := \prod_{P \in V_\lambda(\mathbb{C}) - \{0\}} (x - \alpha_P) = \sum_{k=0}^{n} a_k x^k, \quad \text{where } n = \deg P_\lambda.$$

The coefficients a_k are rational numbers that we have computed numerically. Since the height of P_λ is expected to be not too large, the denominators of the a_k should have a relative small common denominator. The LLL algorithm can be used to compute integers $p_0, \ldots, p_{n-1}, q$ such that $|p_k - a_k q|$ is small for all k; see [Le-Le-Lo], Proposition 1.39. If the sequence (a_k) is arbitrary, then we'll be able to find p_k and q such that $|p_k - a_k q|$ is roughly of order $q^{-1/n}$ for each k, but not much better than that. So if it happens that we find p_k and q with $|p_k - a_k q|$ much smaller than $q^{-1/n}$ for all k, then we guess that a_k is equal to p_k/q. If we cannot find such p_k and q then we double the precision and repeat the calculations above.

Heuristically, the calculation precision that is needed to find the true value of the a_k is about $(1 + 1/n) \cdot \text{height}(P_\lambda)/\log(10)$ decimals. Another way of finding rational approximations of the a_k is by approximating them using continued fractions. For this method, the precision needed to find the true value of the a_k would be about $2 \cdot \text{height}(P_\lambda)/\log(10)$ decimals.

Since the degree of P_λ will be quite large, we won't be able to do many further calculations with it. In particular it may be hard to verify whether all the guesses we made were indeed correct. Instead, we will look at the following variant. If $\mathfrak{m}$ is the Hecke ideal of f mod λ, then V_λ is a vector space over $\mathbb{T}/\mathfrak{m}$. The representation $\overline{\rho}_{f,\lambda}$ induces an action $\tilde{\rho}_\lambda$ of $\mathrm{Gal}(\overline{\mathbb{Q}}/\mathbb{Q})$ on the set $\mathbb{P}(V_\lambda)$ of lines in V_λ. We can attach a polynomial $\tilde{P}_\lambda$ to this projectivized representation $\tilde{\rho}_\lambda$, analogous to the way this was done for $\overline{\rho}$. This polynomial will have a smaller degree than P_λ. We put

$$\tilde{P}_\lambda(x) = \prod_{L \in \mathbb{P}(V_\ell)} \left(x - \sum_{P \in L - \{0\}} \alpha_P\right) = \sum_{k=0}^{m} b_k x^k, \quad \text{where } m = \deg \tilde{P}_\lambda.$$

As above, if the calculation precision is sufficient we can use lattice reduction algorithms to compute the exact values of the b_k.

6.4.3 Reduction of polynomials

Although the polynomial $\tilde{P}_\lambda$ will not have a very huge height, its height is still too large to do any useful computations with it. The first step in making a polynomial of smaller height defining the same number field is computing the maximal order of that number field. Let q be the common denominator of the coefficients and put $p_k = b_k q$. Consider the polynomial

$$Q(x) = q \cdot \tilde{P}_\lambda(x) = q x^m + p_{m-1} x^{m-1} + \cdots + p_0.$$

We make ourselves confident that we correctly computed $Q(x)$ (although we won't prove anything at this point yet). For instance, we verify that $Q(x)$ is irreducible and that its discriminant has the prime factors of $N\ell$ in it. We can also compute for several primes p not dividing $\mathrm{Disc}(Q(x))$ the decomposition type of $Q(x)$ modulo p, and verify that it could be equal to the cycle type of $\tilde{\rho}(\mathrm{Frob}_p)$. If not, we again double the precision and repeat the above calculations.

Let now α be a root of $\tilde{P}_\lambda(x)$ and write down the order:

$$\mathcal{O} := \mathbb{Z} + \sum_{k=1}^{m-1} \left(\mathbb{Z} \cdot \sum_{j=0}^{k-1} a_{m-j} \alpha^{k-j}\right),$$

which is an order that is closer to the maximal order than $\mathbb{Z}[q\alpha]$ (see [Len2], Section 2.10). Being confident in the correctness of $Q(x)$, we know where the number field K defined by it ramifies and thus we can compute its maximal order (see [Bu-Le], Section 6, and Theorems 1.1 and 1.4). Having done this, we embed $\mathcal{O}_K$ as a lattice into $\mathbb{C}^m$ in the usual way and we use the LLL algorithm to compute a basis of small vectors in $\mathcal{O}_K$. We can then search for an element of small length in $\mathcal{O}_K$ that generates K over $\mathbb{Q}$. Its defining polynomial $\tilde{P}'_\lambda$ will have small coefficients. See also [Co-Di].

In the computation of the polynomials P_λ and $\tilde{P}'_\lambda$ we have made several guesses and assumptions that we cannot prove to be correct. In Chapter 7, we work out in special cases how we can use established parts of Serre's conjecture to prove for polynomials of the style $\tilde{P}'_\lambda$ that they indeed belong to the modular Galois representations that we claim they belong to. See [Bos1] for another example of this. In the unlikely case that such tests fail, we can of course make adjustments like choosing another function h or another divisor to construct ϕ.

6.4.4 Further refinements

The Jacobian $J_1(N)$ has large dimension (for N prime this dimension is $(N-5)(N-7)/24$. It could be that our newform f is an element of $S_2(\Gamma)$ with $\Gamma_1(N) \lneqq \Gamma < \Gamma_0(N)$. In that case we work with the curve X_Γ, which is given its $\mathbb{Q}$-structure by defining it as a quotient of $X_1(N)$. The Jacobian J_Γ of X_Γ is isogenous to an Abelian subvariety of $J_1(N)$ that contains V_λ, so this works perfectly well.

In the case $\Gamma = \Gamma_0(N)$ we can sometimes go a step farther. The operator W_N on $X_0(N)$, sending z to $-1/Nz$, is defined over $\mathbb{Q}$. If f is invariant under W_N, one can work with the curve $X_0^+(N) := X_0(N)/\langle W_N \rangle$. Its Jacobian $J_0^+(N)$ is isogenous to an Abelian subvariety of $J_1(N)$ that contains V_λ, so it also works here. Some words on the computation of the homology of $X_0^+(N)$ are in order. The action of W_N on $X_0(N)$ induces an action on $H_1(X_0(N)(\mathbb{C}), \mathbb{Z})$ and on $H_1(X_0(N)(\mathbb{C}), \text{Cusps}, \mathbb{Z})$. Since paths between cusps on $X_0^+(N)(\mathbb{C})$ lift to paths between cusps on $X_0(N)(\mathbb{C})$ we have a surjection

$$H_1(X_0(N), \text{Cusps}, \mathbb{Z}) \twoheadrightarrow H_1(X_0^+(N)(\mathbb{C}), \text{Cusps}, \mathbb{Z}).$$

The kernel of this map consists of the elements $[\gamma]$ in $H_1(X_0(N), \text{Cusps}, \mathbb{Z})$ satisfying $W_N([\gamma]) = -[\gamma]$. So modular symbols methods allow us to compute the $\mathbb{Z}$-module $H_1(X_0^+(N)(\mathbb{C}), \text{Cusps}, \mathbb{Z})$ as a quotient of $\mathbb{M}_2(\Gamma_0(N))$.

Let $\mathbb{B}_2^+(\Gamma_0(N))$ be the free Abelian group on the cusps of $X_0^+(N)(\mathbb{C})$ and define

$$\delta: H_1(X_0^+(N)(\mathbb{C}), \text{Cusps}, \mathbb{Z}) \to \mathbb{B}_2^+(\Gamma_0(N)), \quad \{\alpha, \beta\} \mapsto \{\beta\} - \{\alpha\}.$$

Then $H_1(X_0^+(N)(\mathbb{C})) = \ker(\delta)$.

Chapter Seven

Polynomials for projective representations of level one forms

by Johan Bosman

7.1 INTRODUCTION

In this chapter we explicitly compute mod-ℓ Galois representations attached to modular forms. To be precise, we look at cases with $\ell \leq 23$, and the modular forms considered will be cusp forms of level 1 and weight up to 22. We present the result in terms of polynomials associated with the projectivized representations. As an application, we will improve a known result on Lehmer's nonvanishing conjecture for Ramanujan's tau function (see [Leh], p. 429).

To fix a notation, for any $k \in \mathbb{Z}$ satisfying $\dim S_k(\mathrm{SL}_2(\mathbb{Z})) = 1$ we will denote the unique normalized cusp form in $S_k(\mathrm{SL}_2(\mathbb{Z}))$ by Δ_k. We will denote the coefficients of the q-expansion of Δ_k by $\tau_k(n)$:

$$\Delta_k(z) = \sum_{n \geq 1} \tau_k(n)q^n \in S_k(\mathrm{SL}_2(\mathbb{Z})).$$

From $\dim S_k(\mathrm{SL}_2(\mathbb{Z})) = 1$ it follows that the numbers $\tau_k(n)$ are integers. For every Δ_k and every prime ℓ there is a continuous representation

$$\overline{\rho}_{\Delta_k,\ell} : \mathrm{Gal}(\overline{\mathbb{Q}}/\mathbb{Q}) \to \mathrm{GL}_2(\mathbb{F}_\ell)$$

such that for every prime number $p \neq \ell$ the characteristic polynomial of $\overline{\rho}_{\Delta_k,\ell}(\mathrm{Frob}_p)$ is congruent to $X^2 - \tau_k(p)X + p^{k-1}$ mod ℓ. For a summary on the exceptional representations $\overline{\rho}_{\Delta_k,\ell}$ and the corresponding congruences for $\tau_k(n)$, see [Swi].

7.1.1 Notational conventions

Throughout this chapter, for every field K we will fix an algebraic closure $\overline{K}$, and all algebraic extension fields of K will be regarded as subfields of $\overline{K}$. Furthermore, for each prime number p we will fix an embedding $\overline{\mathbb{Q}} \hookrightarrow \overline{\mathbb{Q}}_p$ and hence an embedding $\mathrm{Gal}(\overline{\mathbb{Q}}_p/\mathbb{Q}_p) \hookrightarrow \mathrm{Gal}(\overline{\mathbb{Q}}/\mathbb{Q})$, whose image we call D_p. We will use I_p to denote the inertia subgroup of $\mathrm{Gal}(\overline{\mathbb{Q}}_p/\mathbb{Q}_p)$.

All representations (either linear or projective) in this chapter will be *continuous*. For any field K, a linear representation $\rho: G \to \mathrm{GL}_n(K)$ defines a projective representation $\tilde{\rho}: G \to \mathrm{PGL}_n(K)$ via the canonical map $\mathrm{GL}_n(K) \to \mathrm{PGL}_n(K)$. We say that $\tilde{\rho}: G \to \mathrm{PGL}_n(K)$ is *irreducible* if the induced action of G on $\mathbb{P}^{n-1}(K)$ fixes no proper subspace. So for $n = 2$ this means that every point of $\mathbb{P}^1(K)$ has its stabilizer subgroup not equal to G.

7.1.2 Statement of results

7.1.3 Theorem *For every pair (k, ℓ) occurring in the table in Section 7.5, let the polynomial $P_{k,\ell}$ be defined as in that same table. Then the splitting field of each $P_{k,\ell}$ is the fixed field of $\mathrm{Ker}(\tilde{\rho}_{\Delta_k,\ell})$ and has Galois group $\mathrm{PGL}_2(\mathbb{F}_\ell)$. Furthermore, if $\alpha \in \overline{\mathbb{Q}}$ is a root of $P_{k,\ell}$ then the subgroup of $\mathrm{Gal}(\overline{\mathbb{Q}}/\mathbb{Q})$ fixing α corresponds via $\tilde{\rho}_{\Delta_k,\ell}$ to a subgroup of $\mathrm{PGL}_2(\mathbb{F}_\ell)$ fixing a point of $\mathbb{P}^1(\mathbb{F}_\ell)$.*

For completeness we also include the pairs (k, ℓ) for which $\rho_{k,\ell}$ is isomorphic to the action of $\mathrm{Gal}(\overline{\mathbb{Q}}/\mathbb{Q})$ on the ℓ-torsion of an elliptic curve. These are the pairs in the table in Section 7.5 with $\ell = k - 1$, as there the representation is the ℓ-torsion of $J_0(\ell)$, which happens to be an elliptic curve for $\ell \in \{11, 17, 19\}$. A simple calculation with division polynomials ([Lan5], Chapter II) can be used to treat these cases. In the general case, one has to work in the more complicated Jacobian variety $J_1(\ell)$, which has dimension 12 for $\ell = 23$, for instance.

We can apply Theorem 7.1.3 to verify the following result.

7.1.4 Corollary *The nonvanishing of* $\tau(n)$ *holds for all* n *such that*

$$n < 22798241520242687999 \approx 2 \cdot 10^{19}.$$

In [Jo-Ke], the nonvanishing of $\tau(n)$ was verified for all n such that

$$n < 22689242781695999 \approx 2 \cdot 10^{16}.$$

To compute the polynomials, the author used the approach described in Section 6.4. After the initial computations some of the polynomials had coefficients of almost 2000 digits, so reduction techniques were absolutely necessary. The used algorithms do not give a proved output, so we have to concentrate on the verification. We will show how to verify the correctness of the polynomials in Section 7.3 after setting up some preliminaries about Galois representations in Section 7.2. In Section 7.4 we will point out how to use Theorem 7.1.3 in a calculation that verifies Corollary 7.1.4. All the calculations were perfomed using MAGMA (see [Magma]).

7.2 GALOIS REPRESENTATIONS

This section states some results on Galois representations that we will need in the proof of Theorem 7.1.3.

7.2.1 Liftings of projective representations

Let G be a topological group, K a topological field, and $\tilde{\rho}: G \to \mathrm{PGL}_n(K)$ a projective representation. Let L be an extension field of K. By a *lifting* of $\tilde{\rho}$ over L we shall mean a representation $\rho: G \to \mathrm{GL}_n(L)$ that makes the following diagram commute:

$$
\begin{array}{ccc}
G_\rho & \xrightarrow{\;\tilde{\rho}\;} & \mathrm{PGL}_n(K) \\
\downarrow & & \uparrow \\
\mathrm{GL}_n(L) & \longrightarrow\!\!\!\!\!\to & \mathrm{PGL}_n(L),
\end{array}
$$

where the maps on the bottom and the right are the canonical ones. If the field L is not specified, then by a lifting of $\tilde{\rho}$ we shall mean a lifting over $\overline{K}$.

An important theorem of Tate arises in the context of liftings. For the proof we refer to [Ser6], Section 6. Note that in the reference representations over $\mathbb{C}$ are considered, but the proof works for representations over arbitrary algebraically closed fields.

7.2.2 Theorem (Tate) *Let $\tilde{\rho}: \mathrm{Gal}(\overline{\mathbb{Q}}/\mathbb{Q}) \to \mathrm{PGL}_n(K)$ be a projective representation of $\mathrm{Gal}(\overline{\mathbb{Q}}/\mathbb{Q})$ over a field K. Then for each prime number p there exists a lifting $\rho'_p: D_p \to \mathrm{GL}_n(\overline{K})$ of $\tilde{\rho}|_{D_p}$. If these liftings ρ'_p have been chosen so that all but finitely many of them are unramified, then there is a unique lifting $\rho: \mathrm{Gal}(\overline{\mathbb{Q}}/\mathbb{Q}) \to \mathrm{GL}_n(\overline{K})$ such that for all primes p we have*

$$\rho|_{I_p} = \rho'_p|_{I_p}.$$

7.2.3 Lemma *Let p be a prime number and let K be a field. Suppose that we are given a projective representation $\tilde{\rho}_p: \mathrm{Gal}(\overline{\mathbb{Q}}_p/\mathbb{Q}_p) \to \mathrm{PGL}_n(K)$ that is unramified. Then there exists a lifting $\rho_p: \mathrm{Gal}(\overline{\mathbb{Q}}_p/\mathbb{Q}_p) \to \mathrm{GL}_n(\overline{K})$ of $\tilde{\rho}_p$ that is unramified as well.*

Proof Since $\tilde{\rho}$ is an unramified representation of $\mathrm{Gal}(\overline{\mathbb{Q}}_p/\mathbb{Q}_p)$, it factors through $\mathrm{Gal}(\overline{\mathbb{F}}_p/\mathbb{F}_p) \cong \hat{\mathbb{Z}}$ and is determined whenever we know the image of $\mathrm{Frob}_p \in \mathrm{Gal}(\overline{\mathbb{F}}_p/\mathbb{F}_p)$. By continuity, this image is an element of $\mathrm{PGL}_n(K)$ of finite order, say of order m. If we take any lift F of $\tilde{\rho}(\mathrm{Frob}_p)$ to $\mathrm{GL}_n(K)$, then we have $F^m = a$ for some $a \in K^\times$. So $F' := \alpha^{-1}F$, where $\alpha \in \overline{K}$ is any mth root of a, has order m in $\mathrm{GL}_n(\overline{K})$. Hence the homomorphism $\mathrm{Gal}(\overline{\mathbb{Q}}_p/\mathbb{Q}_p) \to \mathrm{GL}_n(\overline{K})$ obtained by the composition

$$\mathrm{Gal}(\overline{\mathbb{Q}}_p/\mathbb{Q}_p) \twoheadrightarrow \mathrm{Gal}(\overline{\mathbb{F}}_p/\mathbb{F}_p) \xrightarrow{\sim} \hat{\mathbb{Z}} \twoheadrightarrow \mathbb{Z}/m\mathbb{Z} \xrightarrow{1 \mapsto F'} \mathrm{GL}_n(\overline{K})$$

lifts $\tilde{\rho}$ and is continuous as well as unramified. $\qquad\square$

7.2.4 Serre invariants and Serre's conjecture

Let ℓ be a prime. A Galois representation $\rho: \mathrm{Gal}(\overline{\mathbb{Q}}/\mathbb{Q}) \to \mathrm{GL}_2(\overline{\mathbb{F}}_\ell)$ has a *level* $N(\rho)$ and a *weight* $k(\rho)$. The definitions were introduced by Serre (see [Ser9], Sections 1.2, and 2). Later on, Edixhoven found an improved definition for the weight; see [Edi1], Section 4. The definitions agree in the cases of our interest, but in the general formulation of Theorem 7.2.7 later on, Edixhoven's definition applies. The level $N(\rho)$ is defined as the prime-to-ℓ part of the Artin conductor of ρ and equals 1 if ρ is unramified outside ℓ. The weight is defined in terms of the local representation $\rho|_{D_\ell}$; its definition is rather lenghty so we will not write it out here. When we need results about the weight we will just state them. Let us for now mention that one can consider the weights of the twists $\rho \otimes \chi$ of a representation $\rho \colon \mathrm{Gal}(\overline{\mathbb{Q}}/\mathbb{Q}) \to GL_2(\overline{\mathbb{F}}_\ell)$ by a character $\chi \colon \mathrm{Gal}(\overline{\mathbb{Q}}/\mathbb{Q}) \to \overline{\mathbb{F}}_\ell^\times$. If we choose χ so that $k(\rho \otimes \chi)$ is minimal, then $1 \le k(\rho \otimes \chi) \le \ell + 1$, and we can in fact choose our χ to be a power of the mod ℓ cyclotomic character.

Serre conjectured ([Ser9], Conjecture 3.2.4) that if ρ is irreducible and odd, then ρ belongs to a modular form of level $N(\rho)$ and weight $k(\rho)$. A proof of this conjecture in the case $N(\rho) = 1$ has been published by Khare, building on ideas of himself and Wintenberger:

7.2.5 Theorem (Khare and Wintenberger, [Kha], Theorem 1.1) *Let ℓ be a prime number and let $\rho \colon \mathrm{Gal}(\overline{\mathbb{Q}}/\mathbb{Q}) \to GL_2(\overline{\mathbb{F}}_\ell)$ be an odd irreducible representation of level $N(\rho) = 1$. Then there exist a modular form f of level 1 and weight $k(\rho)$ which is a normalized eigenform and a prime $\lambda \mid \ell$ of K_f such that ρ and $\overline{\rho}_{f,\lambda}$ become isomorphic after a suitable embedding of $\mathbb{F}_\lambda$ into $\overline{\mathbb{F}}_\ell$.*

7.2.6 Weights and discriminants

If a representation $\rho \colon \mathrm{Gal}(\overline{\mathbb{Q}}/\mathbb{Q}) \to GL_2(\overline{\mathbb{F}}_\ell)$ is wildly ramified at ℓ, it is possible to relate the weight to discriminants of certain number fields. In this subsection we will present a theorem of Moon and Taguchi ([Mo-Ta], Theorem 3), on this matter and derive some results from it that are of use to us.

7.2.7 Theorem (Moon and Taguchi) *Consider a wildly ramified representation from $\rho \colon \mathrm{Gal}(\overline{\mathbb{Q}}_\ell/\mathbb{Q}_\ell) \to GL_2(\overline{\mathbb{F}}_\ell)$. Let $\alpha \in \mathbb{Z}$ be such that $k(\rho \otimes \chi_\ell^{-\alpha})$ is minimal, where $\chi_\ell \colon \mathrm{Gal}(\overline{\mathbb{Q}}_\ell/\mathbb{Q}_\ell) \to \mathbb{F}_\ell^\times$ is the mod ℓ cyclotomic character. Put $\tilde{k} = k(\rho \otimes \chi_\ell^{-\alpha})$, put $d = \gcd(\alpha, \tilde{k} - 1, \ell - 1)$ and define m in $\mathbb{Z}$ by letting ℓ^m be the wild ramification degree of $K := \overline{\mathbb{Q}}_\ell^{\mathrm{Ker}(\rho)}$ over $\mathbb{Q}_\ell$. Then we have*

$$
v_\ell(\mathcal{D}_{K/\mathbb{Q}_\ell}) = \begin{cases} 1 + \dfrac{\tilde{k}-1}{\ell-1} - \dfrac{\tilde{k}-1+d}{(\ell-1)\ell^m} & \text{if } 2 \le \tilde{k} \le \ell, \\[2mm] 2 + \dfrac{1}{(\ell-1)\ell} - \dfrac{2}{(\ell-1)\ell^m} & \text{if } \tilde{k} = \ell + 1, \end{cases}
$$

where $\mathcal{D}_{K/\mathbb{Q}_\ell}$ denotes the different of K over $\mathbb{Q}_\ell$ and v_ℓ is normalized by $v_\ell(\ell) = 1$.

We can simplify this formula to one that is useful in our case. In the following corollaries v_ℓ denotes a valuation at a prime above ℓ that is normalized by $v_\ell(\ell) = 1$.

7.2.8 Corollary *Let $\tilde{\rho}\colon \mathrm{Gal}(\overline{\mathbb{Q}}/\mathbb{Q}) \to \mathrm{PGL}_2(\mathbb{F}_\ell)$ be an irreducible projective representation that is wildly ramified at ℓ. Take a point in $\mathbb{P}^1(\mathbb{F}_\ell)$, let $H \subset \mathrm{PGL}_2(\mathbb{F}_\ell)$ be its stabilizer subgroup and let K be the number field defined as*

$$K = \overline{\mathbb{Q}}^{\tilde{\rho}^{-1}(H)}.$$

Then the ℓ-primary part of $\mathrm{Disc}(K/\mathbb{Q})$ is related to the minimal weight k of the liftings of $\tilde{\rho}$ by the formula

$$v_\ell(\mathrm{Disc}(K/\mathbb{Q})) = k + \ell - 2.$$

Proof Let ρ be a lifting of $\tilde{\rho}$ of minimal weight. Since ρ is wildly ramified, after a suitable conjugation in $\mathrm{GL}_2(\overline{\mathbb{F}}_\ell)$ we may assume

$$(7.2.9) \qquad\qquad \rho|_{I_\ell} = \begin{pmatrix} \chi_\ell^{k-1} & * \\ 0 & 1 \end{pmatrix},$$

where $\chi_\ell\colon I_\ell \to \mathbb{F}_\ell^\times$ denotes the mod ℓ cyclotomic character; this follows from the definition of weight. The canonical map $\mathrm{GL}_2(\overline{\mathbb{F}}_\ell) \to \mathrm{PGL}_2(\overline{\mathbb{F}}_\ell)$ is injective on the subgroup $\begin{pmatrix} * & * \\ 0 & 1 \end{pmatrix}$, so the subfields of $\overline{\mathbb{Q}}_\ell$ cut out by $\rho|_{I_\ell}$ and $\tilde{\rho}|_{I_\ell}$ are equal, call them K_2. Also, let $K_1 \subset K_2$ be the fixed field of the diagonal matrices in $\mathrm{Im}\,\rho|_{I_\ell}$. We see from (7.2.9) that in the notation of Theorem 7.2.7 we can put $\alpha = 0$, $m = 1$, and $d = \gcd(\ell - 1, k - 1)$. So we have the following diagram of field extensions:

$$
\begin{array}{c}
& K_2 \\
\chi_\ell^{k-1} \nearrow & \Big| \begin{pmatrix} \chi_\ell^{k-1} & * \\ 0 & 1 \end{pmatrix}, \\
K_1 & \\
\deg = \ell \searrow & \Big| \\
& \mathbb{Q}_\ell^{\mathrm{un}}
\end{array}
$$

The extension K_2/K_1 is tamely ramified of degree $(\ell - 1)/d$, hence we have

$$v_\ell(\mathcal{D}_{K_2/K_1}) = \frac{(\ell - 1)/d - 1}{(\ell - 1)\ell/d} = \frac{\ell - 1 - d}{(\ell - 1)\ell}.$$

Consulting Theorem 7.2.7 for the case $2 \le k \le \ell$ now yields

$$v_\ell(\mathcal{D}_{K_1/\mathbb{Q}_\ell^{un}}) = v_\ell(\mathcal{D}_{K_2/\mathbb{Q}_\ell^{un}}) - v_\ell(\mathcal{D}_{K_2/K_1})$$
$$= 1 + \frac{k - 1}{\ell - 1} - \frac{k - 1 + d}{(\ell - 1)\ell} - \frac{\ell - 1 - d}{(\ell - 1)\ell} = \frac{k + \ell - 2}{\ell},$$

and also in the case $k = \ell + 1$ we get

$$v_\ell(\mathcal{D}_{K_1/\mathbb{Q}_\ell^{un}}) = 2 + \frac{1}{(\ell - 1)\ell} - \frac{2}{(\ell - 1)\ell} - \frac{\ell - 2}{(\ell - 1)\ell} = \frac{k + \ell - 2}{\ell}.$$

Let L be the number field $\overline{\mathbb{Q}}^{\mathrm{Ker}(\tilde\rho)}$. From the irreducibility of $\tilde\rho$ and the fact that $\mathrm{Im}\,\tilde\rho$ has an element of order ℓ it follows that the induced action of $\mathrm{Gal}(\overline{\mathbb{Q}}/\mathbb{Q})$ on $\mathbb{P}^1(\mathbb{F}_\ell)$ is transitive and hence that L is the normal closure of K in $\overline{\mathbb{Q}}$. This in particular implies that $K/\mathbb{Q}$ is wildly ramified. Now from $[K : \mathbb{Q}] = \ell + 1$ it follows that there are two primes in K above ℓ: one is unramified and the other has inertia degree 1 and ramification degree ℓ. From the considerations above it now follows that any ramification subgroup of $\mathrm{Gal}(L/\mathbb{Q})$ at ℓ is isomorphic to a subgroup of $\left(\begin{smallmatrix} * & * \\ 0 & 1 \end{smallmatrix}\right) \subset \mathrm{GL}_2(\overline{\mathbb{F}}_\ell)$ of order $(\ell - 1)\ell/d$ with $d \mid \ell - 1$. Up to conjugacy, the only subgroup of index ℓ is the subgroup of diagonal matrices. Hence K_1 and $K_{\lambda_2}^{un}$ are isomorphic field extensions of $\mathbb{Q}_\ell^{un}$, from which

$$v_\ell(\mathrm{Disc}(K/\mathbb{Q})) = v_\ell(\mathrm{Disc}(K_1/\mathbb{Q}_\ell^{un})) = \ell \cdot v_\ell(\mathcal{D}_{K_1/\mathbb{Q}_\ell^{un}}) = k + \ell - 2$$

follows. $\square$

7.2.10 Corollary *Let $\tilde\rho \colon \mathrm{Gal}(\overline{\mathbb{Q}}/\mathbb{Q}) \to \mathrm{PGL}_2(\mathbb{F}_\ell)$ be an irreducible projective representation, and let ρ be a lifting of $\tilde\rho$ of minimal weight. Let K be the number field belonging to a point of $\mathbb{P}^1(\mathbb{F}_\ell)$, as in the notation of Corollary 7.2.8. If $k \ge 3$ is such that*

$$v_\ell(\mathrm{Disc}(K/\mathbb{Q})) = k + \ell - 2,$$

then we have $k(\rho) = k$.

Proof From $v_\ell(\mathrm{Disc}(K/\mathbb{Q})) = k + \ell - 2 \geq \ell + 1$ it follows that $\tilde{\rho}$ is wildly ramified at ℓ, so we can apply Corollary 7.2.8. $\qquad\square$

7.3 PROOF OF THE THEOREM

To prove Theorem 7.1.3 we need to do several verifications. We will derive representations from the polynomials $P_{k,\ell}$ and verify that they satisfy the conditions of Theorem 7.2.5. Then we know there are modular forms attached to them that have the right level and weight, and uniqueness follows easily.

First we we will verify that the polynomials $P_{k,\ell}$ from the table in Section 7.5 have the right Galois group. The algorithm described in [Ge-Kl], Algorithm 6.1, can be used perfectly to do this verification; proving that $A_{\ell+1} \not< \mathrm{Gal}(P_{k,\ell})$ is the most time-consuming part of the calculation here. It turns out that in all cases we have

$$(7.3.1) \qquad\qquad \mathrm{Gal}(P_{k,\ell}) \cong \mathrm{PGL}_2(\mathbb{F}_\ell).$$

That the action of $\mathrm{Gal}(P_{k,\ell})$ on the roots of $P_{k,\ell}$ is compatible with the action of $\mathrm{PGL}_2(\mathbb{F}_\ell)$ follows from the following lemma.

7.3.2 Lemma *Let ℓ be a prime and let G be a subgroup of $\mathrm{PGL}_2(\mathbb{F}_\ell)$ of index $\ell+1$. Then G is the stabilizer subgroup of a point in $\mathbb{P}^1(\mathbb{F}_\ell)$. In particular, any transitive permutation representation of $\mathrm{PGL}_2(\mathbb{F}_\ell)$ of degree $\ell+1$ is isomorphic to the standard action on $\mathbb{P}^1(\mathbb{F}_\ell)$.*

Proof This follows from [Suz], Proof of Theorem 6.25. $\qquad\square$

So now we have shown that the second assertion in Theorem 7.1.3 follows from the first one.

Next we will verify that we can obtain representations from this that have the right Serre invariants. Let us first note that every automorphism of the group $\mathrm{PGL}_2(\mathbb{F}_\ell)$ is an inner automorphism. This implies that for every $P_{k,\ell}$, any two isomorphisms as in (7.3.1) define isomorphic representations $\mathrm{Gal}(\overline{\mathbb{Q}}/\mathbb{Q}) \to \mathrm{PGL}_2(\mathbb{F}_\ell)$ via composition with the canonical map $\mathrm{Gal}(\overline{\mathbb{Q}}/\mathbb{Q}) \twoheadrightarrow \mathrm{Gal}(P_{k,\ell})$. In other words, every $P_{k,\ell}$ gives a projective representation $\tilde{\rho} \colon \mathrm{Gal}(\overline{\mathbb{Q}}/\mathbb{Q}) \to \mathrm{PGL}_2(\mathbb{F}_\ell)$ that is defined up to isomorphism.

Now, for each (k, ℓ) in the table in Section 7.5, the polynomial $P_{k,\ell}$ is irreducible and hence defines a number field

$$K_{k,\ell} := \mathbb{Q}[x]/(P_{k,\ell}),$$

whose ring of integers we will denote by $\mathcal{O}_{k,\ell}$. It is possible to compute $\mathcal{O}_{k,\ell}$ using the algorithm from [Bu-Le], Section 6 (see also [Bu-Le], Theorems 1.1, 1.4), since we know what kind of ramification behavior to expect. In all cases it turns out that we have

$$\text{Disc}(K_{k,\ell}/\mathbb{Q}) = (-1)^{(\ell-1)/2}\ell^{k+\ell-2}.$$

We see that for each (k, ℓ) the representation $\tilde{\rho}_{k,\ell}$ is unramified outside ℓ. From Lemma 7.2.3 it follows that for each $p \neq \ell$, the representation $\tilde{\rho}_{k,\ell}|_{\text{Gal}(\overline{\mathbb{Q}}_p/\mathbb{Q}_p)}$ has an unramified lifting. Above we saw that via $\tilde{\rho}_{k,\ell}$ the action of $\text{Gal}(\overline{\mathbb{Q}}/\mathbb{Q})$ on the set of roots of $P_{k,\ell}$ is compatible with the action of $\text{PGL}_2(\mathbb{F}_\ell)$ on $\mathbb{P}^1(\mathbb{F}_\ell)$, hence we can apply Corollary 7.2.10 to show that the minimal weight of a lifting of $\tilde{\rho}_{k,\ell}$ equals k. Theorem 7.2.2 now shows that every $\tilde{\rho}_{k,\ell}$ has a lifting $\rho_{k,\ell}$ that has level 1 and weight k. From $\text{Im}\,\tilde{\rho}_{k,\ell} = \text{PGL}_2(\mathbb{F}_\ell)$ it follows that each $\rho_{k,\ell}$ is absolutely irreducible.

To apply Theorem 7.2.5 we should still verify that $\rho_{k,\ell}$ is odd in each case. I thank Robin de Jong for pointing out that this is immediate: Since the weight of $\rho_{k,\ell}$ is k, we have $\det \rho_{k,\ell}|_{I_\ell} = \chi_\ell^{k-1}|_{I_\ell}$, where χ_ℓ is the mod ℓ cyclotomic character. Now, $\det \rho_{k,\ell}|_{D_p}$ is unramified for $p \neq \ell$ and hence $\det \rho_{k,\ell}$ must equal χ_ℓ^{k-1} on all of $\text{Gal}(\overline{\mathbb{Q}}/\mathbb{Q})$ (apply Theorem 7.2.2 with $n = 1$, for instance). But then, since k is even, we have that $\det \rho_{k,\ell}$ evaluated at a complex conjugation equals -1, hence $\rho_{k,\ell}$ is odd.

Now that we have verified all the conditions of Theorem 7.2.5 we remark as a final step that all spaces of modular forms $S_k(\text{SL}_2(\mathbb{Z}))$ involved here are one-dimensional. So the modularity of each $\rho_{k,\ell}$ implies immediately the isomorphism $\rho_{k,\ell} \cong \overline{\rho}_{\Delta_k,\ell}$, hence also $\tilde{\rho}_{k,\ell} \cong \tilde{\rho}_{\Delta_k,\ell}$, which completes the proof of Theorem 7.1.3.

7.4 PROOF OF THE COROLLARY

If τ vanishes somewhere, then the smallest positive integer n for which $\tau(n)$ is zero is a prime. This was observed by Lehmer ([Leh], Theorem 2) and can also be seen using the following argument. Suppose n is the smallest positive integer with $\tau(n) = 0$. From the multiplicative property of τ it follows that n is a power of a prime p. If $\tau(p) \neq 0$, then from $|\tau(p)| < p^6$ and the recursion for $\tau(p^r)$ it follows that $v_p(\tau(p^r)) = r \cdot v_p(\tau(p))$ for all r, so $\tau(p^r)$ would never be zero.

Using results on the exceptional representations for $\tau(p)$, Serre pointed out in [Ser8], Section 3.3 that if p is a prime number with $\tau(p) = 0$ then p can be written as

$$p = hM - 1$$

with

$$M = 2^{14} 3^7 5^3 691 = 3094972416000,$$
$$\left(\frac{h+1}{23} \right) = 1, \quad \text{and} \quad h \equiv 0, 30 \text{ or } 48 \bmod 49.$$

In fact p is of this form if and only if $\tau(p) \equiv 0 \bmod 23 \cdot 49 \cdot M$ holds. Knowing this, we will do a computer search on these primes p and verify whether $\tau(p) \equiv 0 \bmod \ell$ for $\ell \in \{11, 13, 17, 19\}$. To do that we will use the following lemma.

7.4.1 Lemma *Let K be a field of characteristic not equal to 2. Then the following conditions on $M \in \mathrm{GL}_2(K)$ are equivalent:*

(1) $\mathrm{tr}\, M = 0$.

(2) For the action of M on $\mathbb{P}^1(K)$, there are 0 or 2 orbits of length 1 and all other orbits have length 2.

(3) The action of M on $\mathbb{P}^1(K)$ has an orbit of length 2.

Proof We begin with verifying $(1) \Rightarrow (2)$. Suppose $\mathrm{tr}\, M = 0$. Matrices of trace 0 in $\mathrm{GL}_2(K)$ have distinct eigenvalues in $\overline{K}$ because $\mathrm{char}(K) \neq 2$. It follows that two such matrices are conjugate if and only if their characteristic polynomials coincide. Hence M and $M' := \begin{pmatrix} 0 & 1 \\ -\det M & 0 \end{pmatrix}$ are conjugate, so without loss of generality we assume $M = M'$. Since M^2 is a scalar matrix, all the orbits of M on $\mathbb{P}^1(K)$ have length 1 or 2. If there are at least 3 orbits of length 1, then K^2 itself is an eigenspace of M, hence M is scalar, which is not the case. If there is exactly one orbit of length 1 then M has a nonscalar Jordan block in its Jordan decomposition, which contradicts the fact that the eigenvalues are distinct.

The implication $(2) \Rightarrow (3)$ is trivial, so that leaves proving $(3) \Rightarrow (1)$. Suppose that M has an orbit of length 2 in $\mathbb{P}^1(K)$. After a suitable conjugation, we may assume that this orbit is $\{[\binom{1}{0}], [\binom{0}{1}]\}$. But this means that $M \sim \left(\begin{smallmatrix} 0 & a \\ b & 0 \end{smallmatrix}\right)$ for certain $a, b \in K$, hence $\operatorname{tr} M = 0$. $\square$

In view of the above lemma it follows from Theorem 7.1.3 that for ℓ in the set $\{11, 13, 17, 19\}$ and $p \neq \ell$ we have $\tau(p) \equiv 0 \bmod \ell$ if and only if the prime p decomposes in the number field $\mathbb{Q}[x]/(P_{12,\ell})$ as a product of primes of degree 1 and 2, with degree 2 occurring at least once. For $p \nmid \operatorname{Disc}(P_{12,\ell})$, which is a property that all primes p satisfying Serre's criteria possess, we can verify this condition by checking whether $P_{12,\ell}$ has an irreducible factor of degree 2 over $\mathbb{F}_p$. This can be easily done by verifying:

$$\overline{x}^{p^2} = \overline{x} \quad \text{and} \quad \overline{x}^p \neq \overline{x} \quad \text{in} \quad \mathbb{F}_p[x]/(\overline{P}_{12,\ell}).$$

A computer search shows that the first few primes satisfying Serre's criteria as well as $\tau(p) \equiv 0 \bmod 11 \cdot 13 \cdot 17 \cdot 19$ are

$$22798241520242687999, \; 60707199950936063999,$$
$$93433753964906495999, \; 102797608484376575999.$$

Remark. The unpublished paper [Jo-Ke] in which Jordan and Kelly obtained the previous bound for the verification of Lehmer's conjecture seems to be unfindable. Kevin Buzzard asked me what method they could have used. If we weaken the above search to using only the prime $\ell = 11$, we obtain the same bound as Jordan and Kelly did. So our speculation is that they searched for primes p satisfying Serre's criteria as well as $\tau(p) \equiv 0 \bmod 11$. This congruence can be verified using an elliptic curve computation, as was already remarked in Subsection 7.1.2.

7.5 THE TABLE OF POLYNOMIALS

In this section we present the table of polynomials that is referred to throughout this chapter.

Polynomials Belonging to Projective Modular Representations

(k, ℓ)	$P_{k,\ell}$
$(12, 11)$	$x^{12} - 4x^{11} + 55x^9 - 165x^8 + 264x^7 - 341x^6 + 330x^5$ $- 165x^4 - 55x^3 + 99x^2 - 41x - 111$
$(12, 13)$	$x^{14} + 7x^{13} + 26x^{12} + 78x^{11} + 169x^{10} + 52x^9 - 702x^8$ $- 1248x^7 + 494x^6 + 2561x^5 + 312x^4 - 2223x^3$ $+ 169x^2 + 506x - 215$
$(12, 17)$	$x^{18} - 9x^{17} + 51x^{16} - 170x^{15} + 374x^{14} - 578x^{13}$ $+ 493x^{12} - 901x^{11} + 578x^{10} - 51x^9 + 986x^8 + 1105x^7$ $+ 476x^6 + 510x^5 + 119x^4 + 68x^3 + 306x^2 + 273x + 76$
$(12, 19)$	$x^{20} - 7x^{19} + 76x^{17} - 38x^{16} - 380x^{15} + 114x^{14}$ $+ 1121x^{13} - 798x^{12} - 1425x^{11} + 6517x^{10} + 152x^9$ $- 19266x^8 - 11096x^7 + 16340x^6 + 37240x^5 + 30020x^4$ $- 17841x^3 - 47443x^2 - 31323x - 8055$
$(16, 17)$	$x^{18} - 2x^{17} - 17x^{15} + 204x^{14} - 1904x^{13} + 3655x^{12}$ $+ 5950x^{11} - 3672x^{10} - 38794x^9 + 19465x^8 + 95982x^7$ $- 280041x^6 - 206074x^5 + 455804x^4 + 946288x^3$ $- 1315239x^2 + 606768x - 378241$
$(16, 19)$	$x^{20} + x^{19} + 57x^{18} + 38x^{17} + 950x^{16} + 4389x^{15}$ $+ 20444x^{14} + 84018x^{13} + 130359x^{12} - 4902x^{11}$ $- 93252x^{10} + 75848x^9 - 1041219x^8 - 1219781x^7$ $+ 3225611x^6 + 1074203x^5 - 3129300x^4 - 2826364x^3$ $+ 2406692x^2 + 6555150x - 5271039$

Continued on next page

Table continued from previous page

(k, ℓ)	$P_{k,\ell}$
$(16, 23)$	$x^{24} + 9x^{23} + 46x^{22} + 115x^{21} - 138x^{20} - 1886x^{19}$ $+ 1058x^{18} + 59639x^{17} + 255599x^{16} + 308798x^{15}$ $- 1208328x^{14} - 6156732x^{13} - 10740931x^{12}$ $+ 2669403x^{11} + 52203054x^{10} + 106722024x^{9}$ $+ 60172945x^{8} - 158103380x^{7} - 397878081x^{6}$ $- 357303183x^{5} + 41851168x^{4} + 438371490x^{3}$ $+ 484510019x^{2} + 252536071x + 55431347$
$(18, 17)$	$x^{18} - 7x^{17} + 17x^{16} + 17x^{15} - 935x^{14} + 799x^{13}$ $+ 9231x^{12} - 41463x^{11} + 192780x^{10} + 291686x^{9}$ $- 390014x^{8} + 6132223x^{7} - 3955645x^{6} + 2916112x^{5}$ $+ 45030739x^{4} - 94452714x^{3} + 184016925x^{2}$ $- 141466230x + 113422599$
$(18, 19)$	$x^{20} + 10x^{19} + 57x^{18} + 228x^{17} - 361x^{16} - 3420x^{15}$ $+ 23446x^{14} + 88749x^{13} - 333526x^{12} - 1138233x^{11}$ $+ 1629212x^{10} + 13416014x^{9} + 7667184x^{8}$ $- 208954438x^{7} + 95548948x^{6} + 593881632x^{5}$ $- 1508120801x^{4} - 1823516526x^{3} + 2205335301x^{2}$ $+ 1251488657x - 8632629109$
$(18, 23)$	$x^{24} + 23x^{22} - 69x^{21} - 345x^{20} - 483x^{19} - 6739x^{18}$ $+ 18262x^{17} + 96715x^{16} - 349853x^{15} + 2196684x^{14}$ $- 7507476x^{13} + 59547x^{12} + 57434887x^{11}$ $- 194471417x^{10} + 545807411x^{9} + 596464566x^{8}$ $- 9923877597x^{7} + 33911401963x^{6} - 92316759105x^{5}$ $+ 157585411007x^{4} - 171471034142x^{3}$ $+ 237109280887x^{2} - 93742087853x + 97228856961$

Continued on next page

Table continued from previous page

(k, ℓ)	$P_{k,\ell}$
$(20, 19)$	$x^{20} - 5x^{19} + 76x^{18} - 247x^{17} + 1197x^{16} - 8474x^{15}$ $+ 15561x^{14} - 112347x^{13} + 325793x^{12} - 787322x^{11}$ $+ 3851661x^{10} - 5756183x^9 + 20865344x^8$ $- 48001353x^7 + 45895165x^6 - 245996344x^5$ $+ 8889264x^4 - 588303992x^3 - 54940704x^2$ $- 538817408x + 31141888$
$(20, 23)$	$x^{24} - x^{23} - 23x^{22} - 184x^{21} - 667x^{20} - 5543x^{19}$ $- 22448x^{18} + 96508x^{17} + 1855180x^{16}$ $+ 13281488x^{15} + 66851616x^{14} + 282546237x^{13}$ $+ 1087723107x^{12} + 3479009049x^{11} + 8319918708x^{10}$ $+ 8576048755x^9 - 19169464149x^8 - 111605931055x^7$ $- 227855922888x^6 - 193255204370x^5$ $+ 176888550627x^4 + 1139040818642x^3$ $+ 1055509532423x^2 + 1500432519809x$ $+ 314072259618$
$(22, 23)$	$x^{24} - 2x^{23} + 115x^{22} + 23x^{21} + 1909x^{20} + 22218x^{19}$ $+ 9223x^{18} + 121141x^{17} + 1837654x^{16} - 800032x^{15}$ $+ 9856374x^{14} + 52362168x^{13} - 32040725x^{12}$ $+ 279370098x^{11} + 1464085056x^{10} + 1129229689x^9$ $+ 3299556862x^8 + 14586202192x^7 + 29414918270x^6$ $+ 45332850431x^5 - 6437110763x^4 - 111429920358x^3$ $- 12449542097x^2 + 93960798341x - 31890957224$

Chapter Eight

Description of $X_1(5l)$

by Bas Edixhoven

8.1 CONSTRUCTION OF A SUITABLE CUSPIDAL DIVISOR ON $X_1(5l)$

In this section we put ourselves in the situation of Theorem 2.5.13: l is a prime number, k is an integer such that $2 < k \leq l+1$, and f is a surjective ring morphism $\mathbb{T}(1,k) \to \mathbb{F}$ a with $\mathbb{F}$ a finite field of characteristic l, such that the associated Galois representation $\rho\colon \mathrm{Gal}(\overline{\mathbb{Q}}/\mathbb{Q}) \to \mathrm{GL}_2(\mathbb{F})$ is absolutely irreducible. We let V denote the two-dimensional $\mathbb{F}$-vector space in $J_1(l)(\overline{\mathbb{Q}})[l]$ that realizes ρ.

As explained in Chapter 3, we would like to have an effective divisor D_0 on $X_1(l)_{\mathbb{Q}}$ of degree the genus of $X_1(l)$, such that for all nonzero x in the submodule V of $J_1(l)(\overline{\mathbb{Q}})$ we have $h^0(\mathcal{L}_x(D_0)) = 1$. It would be nice to have a cuspidal divisor (that is, a divisor supported on the cusps) with this property. The first complication is that the cusps are not all rational over $\mathbb{Q}$: half of them have the maximal real subfield of $\mathbb{Q}(\zeta_l)$ as field of definition. Moreover, even working with all the cusps, we have not succeeded in finding a cuspidal divisor D_0 with the desired properties. On the other hand, below we will give explicitly a cuspidal divisor D_0 on the curve $X_1(5l)_{\mathbb{Q}(\zeta_l)}$ with the property that $h^0(\mathcal{L}_x(D_0)) = 1$ for each x in $J_1(5l)(\overline{\mathbb{Q}})$ that specializes to 0 at some place of $\overline{\mathbb{Q}}$ over l. In particular, D_0 has the required property for V embedded in $J_1(5l)(\overline{\mathbb{Q}})$ in an arbitrary way, provided that the image of ρ contains $\mathrm{SL}_2(\mathbb{F})$ (this will be shown in Section 8.2). We have chosen to work with $X_1(5l)$, but the same method will work for modular curves corresponding to some level structure if the prime to l part of the level structure is fine, and of genus zero.

For the rest of this section, our assumptions are the following. We let
l be a prime number, not equal to 5, and $X := X_1(5l)_{\mathbb{Q}(\zeta_l)}$ over $\mathbb{Q}(\zeta_l)$.
The genus of X is $(l-2)^2$. References for facts about X that we use can be
found in [Gro], and also in [Edi1]; they are derived from results in [De-Ra]
and [Ka-Ma].

The curve $X_0(5)_{\mathbb{Q}}$ has 2 cusps, both $\mathbb{Q}$-rational, called 0 and ∞ (after
the points of $\mathbb{P}^1(\mathbb{Q})$ from which they come). The cusp ∞ has as moduli
interpretation the degenerate elliptic curve (generalized elliptic curve in the
terminology of [De-Ra]): the 1-gon, equipped with the unique subgroup
of order 5 of $\mathbb{G}_m$. The cusp 0 corresponds to the 5-gon, equipped with a
subgroup of order 5 that meets all 5 components. The group $\mathbb{F}_5^\times$ acts (via
diamond operators) on $X_1(5)$, with quotient $X_0(5)$; the subgroup $\{\langle\pm1\rangle\}$
acts trivially, and the quotient $\mathbb{F}_5^\times/\{\langle\pm1\rangle\}$ by this subgroup acts faithfully.
The inverse images of 0 and ∞ both consist of two cusps. Those over 0
are $\mathbb{Q}$-rational (the subgroup of order 5 of the 5-gon is the constant group-
scheme $\mathbb{Z}/5\mathbb{Z}$), whereas those over ∞ are conjugated over $\mathbb{Q}(\sqrt{5})$. We fix
one $\mathbb{Q}$-rational cusp c of $X_1(5)$.

The group $\mathbb{F}_l^\times$ acts faithfully, and in fact even freely, on X. The set of
cusps of X over the cusp c of $X_1(5)$ form two $\mathbb{F}_l^\times$-orbits, corresponding to
the type of degenerate elliptic curve that they correspond to: 5-gon or $5l$-
gon. The orbit corresponding to the 5-gon consists of points rational over
$\mathbb{Q}(\zeta_l)$, all conjugates of each other. The orbit corresponding to the $5l$-gon
consists of $\mathbb{Q}$-points.

We let J denote the Jacobian of X. What we want is an effective divisor
D_0 of degree g on X (with g the genus of X), supported on the cusps over c,
such that for all x in $J(\overline{\mathbb{Q}})$ that specialize to 0 at some place of $\overline{\mathbb{Q}}$ over l we
have $h^0(X_{\overline{\mathbb{Q}}}, \mathcal{L}_x(D_0)) = 1$. For the notion of specialization we use Néron
models; the reader is referred to [Bo-Lu-Ra] for this notion. For x in $J(K)$
with $K \subset \overline{\mathbb{Q}}$ a finite extension of $\mathbb{Q}(\zeta_l)$, and λ a place of $\overline{\mathbb{Q}}$ over l, we say
that x specializes to 0 at λ if x, viewed as element of $J_{O_K}(O_K)$, with J_{O_K}
the Néron model of J over O_K, specializes to 0 at the place of K given
by λ. For $K \subset K'$ a finite extension we have $J(K) \subset J(K')$, hence we can
also view x as element of $J(K')$. The notion of x specializing to zero is the
same for K and K', because $J_{\mathbb{Z}[\zeta_l]}$ is semistable at l.

The moduli interpretation of X gives a semistable model $X_{\mathbb{Z}[\zeta_l,1/5]}$ over

$\mathbb{Z}[\zeta_l, 1/5]$, described in [Gro], for example. A result of Raynaud identifies the connected component of the Néron model J_{O_K} with the connected component of the Picard scheme of X_{O_K} (see Section 9.5 of [Bo-Lu-Ra]). This means that for x in $J(K)$ specializing to 0 at λ, the line bundle $\mathcal{L}_x$ on X_K associated with x can be extended uniquely over the local ring $O_{K,\lambda}$ to a line bundle $\mathcal{L}_x$ on $X_{O_{K,\lambda}}$ such that the restriction $\overline{\mathcal{L}_x}$ of $\mathcal{L}_x$ to the special fiber $X_{\mathbb{F}_\lambda}$ is trivial. The divisor D_0 on X_K extends, by taking the Zariski closure, to an effective Cartier divisor on $X_{O_{K,\lambda}}$.

We now note that $h^0(X_K, \mathcal{L}_x(D_0)) \geq 1$, by Riemann-Roch, and that $h^0(X_K, \mathcal{L}_x(D_0)) \leq h^0(X_{\mathbb{F}_\lambda}, \overline{\mathcal{L}_x}(\overline{D_0}))$ by semicontinuity of cohomology of coherent sheaves. As $\overline{\mathcal{L}_x}(\overline{D_0})) = \mathcal{O}(\overline{D_0})$, it now suffices to take D_0 such that $h^0(X_{\overline{\mathbb{F}}_l}, \mathcal{O}(\overline{D_0})) = 1$. We do this by looking at the geometry of $X_{\overline{\mathbb{F}}_l}$. As $\mathbb{Z}[\zeta_l]$ has a unique morphism to $\overline{\mathbb{F}}_l$, the curve $X_{\overline{\mathbb{F}}_l}$ does not depend on K. The scheme of cusps of $X_{\mathbb{Z}_l[\zeta_l]}$ is finite étale over $\mathbb{Z}_l[\zeta_l]$, hence the cusps lying over c specialize injectively to $X_{\overline{\mathbb{F}}_l}$.

The curve $X_{\overline{\mathbb{F}}_l}$ is the union of two irreducible components, X_1 and X_2, say, both isomorphic to the Igusa curve of level l over $X_1(5)$ over $\overline{\mathbb{F}}_l$, that meet transversally in the set Σ of supersingular points. We will take $\overline{D_0} = D_1 + D_2$, with D_1 on X_1 and D_2 on X_2; note that the cusps are disjoint from Σ, so $\overline{D_0}$ lies in the smooth locus of $X_{\overline{\mathbb{F}}_l}$.

In order to simplify the notation, we let X and D_0 denote $X_{\overline{\mathbb{F}}_l}$ and $\overline{D_0}$, from now on, in this section. We let Ω_X be the dualizing sheaf on X (see Section 8 of [Gro], or [Ma-Ri]): it is the invertible $\mathcal{O}_X$-module obtained by gluing $\Omega^1_{X_1}(\Sigma)$ and $\Omega^1_{X_2}(\Sigma)$ along Σ via the residue maps at the points of Σ on X_1 and minus the residues maps at the points of Σ on X_2. By Riemann-Roch, what we want is that $h^1(X, \mathcal{O}(D_0)) = 0$, and hence, by Serre duality, that $h^0(X, \Omega_X(-D_0)) = 0$. In other words, an element of $H^0(X, \Omega_X)$ that vanishes on D_0 must be zero. Restriction to X_1 gives a short exact sequence:

$$0 \to H^0(X_2, \Omega^1_{X_2/\overline{\mathbb{F}}_l}(-D_2)) \to H^0(X, \Omega_X(-D_0))$$
$$\to H^0(X_1, \Omega^1_{X_1/\overline{\mathbb{F}}_l}(\Sigma - D_1)) \to 0.$$

Hence, it suffices to take D_1 such that $H^0(X_1, \Omega^1_{X_1/\overline{\mathbb{F}}_l}(\Sigma - D_1)) = 0$ and D_2 such that $H^0(X_2, \Omega^1_{X_2/\overline{\mathbb{F}}_l}(-D_2)) = 0$. Let us now first see of which degrees d_1 and d_2 we want to take D_1 and D_2. Let g_1 and g_2 be the genera of X_1 and X_2 (note: they are equal). Then we have that $g = g_1 + g_2 + \#\Sigma - 1$, and $g = \deg D_0 = d_1 + d_2$. It can be shown in several ways that the degree of the sheaf $\underline{\omega}$ on $X_1(5)_{\mathbb{F}_l}$ is one, either by explicit computation, using the equations of Proposition 8.2.8, or by the following argument. The curve $X_1(5)_{\mathbb{F}_l}$ over $\mathbb{F}_l$ has genus zero. The Kodaira-Spencer isomorphism on $X_1(5)_{\mathbb{F}_l}$, from $\underline{\omega}^{\otimes 2} \to \Omega^1(\text{Cusps})$, plus the fact that the divisor of cusps has degree 4, gives that the degree of $\underline{\omega}^{\otimes 2}$ is 2. Therefore, the Hasse invariant, being a global section of $\underline{\omega}^{\otimes l - 1}$, has exactly $l - 1$ zeros on $X_1(5)$ over $\overline{\mathbb{F}}_l$ and therefore we have

(8.1.1) $\#\Sigma = l - 1$.

Applying Hurwitz's formula to the covering $X_1 \to X_1(5)_{\overline{\mathbb{F}}_l}$, which is totally ramified over Σ and unramified outside it, gives

$$2g_1 - 2 = -2(l - 1) + (l - 1)(l - 2),$$

and hence

(8.1.2) $g_1 = \dfrac{1}{2}(l - 2)(l - 3).$

This implies that we want to take

(8.1.3)
$$d_1 = g_1 + \#\Sigma - 1 = \frac{1}{2}(l - 1)(l - 2),$$
$$d_2 = g_2 = \frac{1}{2}(l - 2)(l - 3).$$

Now we use equations to compute with. We choose a coordinate z on $X_1(5)_{\overline{\mathbb{F}}_l}$, that is, an isomorphism from $X_1(5)_{\overline{\mathbb{F}}_l}$ to $\mathbb{P}^1_{\overline{\mathbb{F}}_l}$, such that $z(\Sigma)$ does not contain 0 or ∞ and such that $z^{-1}0$ is the (rational) cusp 0 of $X_1(5)_{\overline{\mathbb{F}}_l}$. Let f be the monic polynomial in z whose zeros are the elements of Σ, each with multiplicity one. Then X_1 and X_2 are both isomorphic to the cover of $X_1(5)_{\overline{\mathbb{F}}_l}$ given by the equation $y^{l-1} = f$, by the following argument. The complete local rings of X_1 at the points of Σ, with their $\mathbb{F}_l^\times$-actions, are all isomorphic to each other because, by a theorem of Serre and Tate, these can all be described in terms of the deformation theory of one l-divisible group over $\overline{\mathbb{F}}_l$; see [Ka-Ma], Sections 5.2, 5.3. There is a general theory of cyclic possibly ramified covers such as $X_1 \to X_1(5)_{\overline{\mathbb{F}}_l}$, based on the decomposition of $\mathcal{O}_{X_1}$ as $\mathcal{O}$-module on $X_1(5)_{\overline{\mathbb{F}}_l}$ for the $\mathbb{F}_l^\times$-action. It shows that the cover $X_1 \to X_1(5)_{\overline{\mathbb{F}}_l}$ is the cover of $(l-1)$th roots of the global section 1 of the invertible sheaf $\mathcal{O}(\Sigma)$ on $X_1(5)_{\overline{\mathbb{F}}_l}$, in an invertible $\mathcal{O}$-module $\mathcal{L}$ with a given isomorphism $\mathcal{O}(\Sigma) \to \mathcal{L}^{\otimes(l-1)}$, where the $\mathbb{F}_l^\times$-action may have been changed by an automorphism of $\mathbb{F}_l^\times$. As $X_1(5)_{\overline{\mathbb{F}}_l}$ has genus zero, we can take $\mathcal{L}$ to be $\mathcal{O}_{X_1}(z^{-1}\infty)$. In fact, Section 12.8 of [Ka-Ma] shows that X_1 is obtained from $X_1(5)_{\overline{\mathbb{F}}_l}$ by extracting the $(l-1)$th root of the Hasse invariant, in $\underline{\omega}$.

We compute a basis of $\mathrm{H}^0(X_1, \Omega^1_{X_1/\overline{\mathbb{F}}_l}(\Sigma))$. On X_1 we have

$$(8.1.4) \qquad\qquad -y^{l-2}dy = f'dz.$$

Hence $(dz)/y^{l-2} = -(dy)/f'$ is a generating section of $\Omega^1_{X_1/\overline{\mathbb{F}}_l}$ on the affine part given by our equation. Hence $(dz)/y^{l-1}$ is generating section of $\Omega^1_{X_1/\overline{\mathbb{F}}_l}(\Sigma)$ on the affine part, and it is $\mathbb{F}_l^\times$-invariant. At each point of X_1 over the point where z has its pole, both z and y have a simple pole, and $(dz)/y^{l-1}$ has order $-2 + l - 1 = l - 3$. So we have a basis:

$$(8.1.5) \qquad \mathrm{H}^0(X_1, \Omega^1_{X_1/\overline{\mathbb{F}}_l}(\Sigma)) = \bigoplus_{i+j\le l-3} \overline{\mathbb{F}}_l z^i y^j \cdot (dz)/y^{l-1}.$$

Note that this agrees with the fact that $d_1 = \frac{1}{2}(l - 1)(l - 2)$.

We can now say how to choose D_1. At each of the $l-1$ points where z has a zero we must give a multiplicity. In the coordinate system given by z and y, these points are the ones of the form $(0, b)$, with $b \in \overline{\mathbb{F}}_l^\times$ satisfying $b^{l-1} = f(0)$. Here is how we choose D_1: just distribute the multiplicities $(0, 1, \ldots, l - 2)$ over these points. Then any linear combination of our basis elements that vanishes on D_1 is zero, shown as follows. At all points of D_1, z has a simple zero. Let ω be an element of $\mathrm{H}^0(X_1, \Omega^1_{X_1/\overline{\mathbb{F}}_l}(\Sigma))$, with coordinates $\lambda_{i,j}$ in the basis (8.1.5). Assume that ω vanishes on D_1 (taking multiplicities into account). As there are $l-2$ points in D_1 with multiplicity > 0, the polynomial $\sum_j \lambda_{0,j} y^j$, being of degree $\le l-3$, must be zero. As there are $l-3$ points in D_1 with multiplicity > 1, the polynomial $\sum_j \lambda_{1,j} zy^j$, being of degree $\le l-4$, must be zero. And so on.

Now we do D_2. A basis is the following:

$$(8.1.6) \qquad H^0(X_2, \Omega^1_{X_2/\overline{\mathbb{F}}_l}) = \bigoplus_{i+j \leq l-4} \overline{\mathbb{F}}_l z^i y^j \cdot (dz)/y^{l-2}.$$

We note that this agrees with $g_2 = (l-3)(l-2)/2$. So, for D_2, just distribute the multiplicities $(0, 0, 1, \ldots, l-3)$ over the points where z has a zero. The same argument as the one we used for D_1 shows that any ω in $H^0(X_2, \Omega^1_{X_2/\overline{\mathbb{F}}_l})$ that vanishes on D_2 is zero.

As the action of $\mathbb{F}_5^\times$ permutes the two $\mathbb{Q}$-rational cusps of $X_1(5)$, our arguments above work for both of them, and we have proved the following theorem.

8.1.7 Theorem *Let l be a prime number not equal to 5. Let c be one of the two $\mathbb{Q}$-rational cusps of $X_1(5)$. Then the cusps of $X_1(5l)$ over c are $\mathbb{Q}(\zeta_l)$-rational and consist of two $\mathbb{F}_l^\times$-orbits, on which $\mathbb{F}_l^\times$ acts freely. Let D_1 be a divisor on $X_1(5l)_{\mathbb{Q}(\zeta_l)}$ obtained by distributing the multiplicities $(0, 1, \ldots, l-2)$ over one of these two orbits. Let D_2 be the divisor obtained by distributing the multiplicities $(0, 0, 1, \ldots, l-3)$ over the other orbit. Then $D_0 := D_1 + D_2$ has degree equal to the genus of $X_1(5l)$ and has the property that for any $\overline{\mathbb{Q}}$-point x of the Jacobian of $X_1(5l)$ that specializes to 0 at some place over l we have $h^0(X_1(5l)_{\overline{\mathbb{Q}}}, \mathcal{L}_x(D_0)) = 1$.*

8.2 THE EXACT SETUP FOR THE LEVEL ONE CASE

In Chapter 3 we described our strategy for computing the residual Galois representations attached to a fixed newform. That strategy depends on properties of divisors D_0 and functions f to be chosen, on modular curves of varying level. These D_0 and f must satisfy a number of conditions. In general we do not know how to choose divisors D_0 for which we can prove, without a computer computation, that they have the required property. This is the main reason we will now restrict ourselves to the case of modular forms of level one.

The aim of this section is to describe exactly our strategy for computing the residual representations V in the situation of Theorem 2.5.13: Here l is a prime number, k is an integer such that $2 < k \leq l+1$, and f is a surjective ring morphism $\mathbb{T}(1, k) \to \mathbb{F}$ a with $\mathbb{F}$ a finite field of characteristic l, such that the associated Galois representation $\rho\colon \mathrm{Gal}(\overline{\mathbb{Q}}/\mathbb{Q}) \to GL_2(\mathbb{F})$ is absolutely irreducible, under the extra hypothesis that the image of ρ contains $SL_2(\mathbb{F})$. By Theorem 2.5.20, this hypothesis holds when ρ is irreducible and $l \geq 6k - 5$. We let V denote the two-dimensional $\mathbb{F}$-vector space in $J_1(l)(\overline{\mathbb{Q}})[l]$ that realizes ρ.

Theorem 8.1.7 gives us a divisor D_0 on $X_1(5l)_{\mathbb{Q}(\zeta_l)}$ that we want to use. Therefore, we want to embed V into $J_1(5l)(\overline{\mathbb{Q}})[l]$.

Let $\pi\colon X_1(5l) \to X_1(l)$ be the standard map (that is, the one that forgets the 5-part of the level structure, denoted $B_{5l,l,1}$ in Section 2.2). Then the degree of π is $5^2 - 1 = 24$, which is prime to l. This implies that $\pi_*\pi^*$ is multiplication by 24 on $J_1(l)$, and that π^* is injective on $J_1(l)(\overline{\mathbb{Q}})[l]$. We have a projector

$$(8.2.1) \quad \frac{1}{24}\pi^*\pi_* \colon J_1(5l)(\overline{\mathbb{Q}})[l] \twoheadrightarrow \pi^*J_1(l)(\overline{\mathbb{Q}})[l] \subset J_1(5l)(\overline{\mathbb{Q}})[l].$$

We embed V in $J_1(5l)(\overline{\mathbb{Q}})[l]$ via its embedding into $J_1(l)(\overline{\mathbb{Q}})[l]$, followed by π^*.

8.2.2 Proposition *Let l be a prime number, k be an integer in $\{3, \ldots, l+1\}$, and let f be a surjective ring morphism $\mathbb{T}(1, k) \to \mathbb{F}$ with $\mathbb{F}$ a finite field of characteristic l, such that the image of the associated Galois representation $\rho\colon \mathrm{Gal}(\overline{\mathbb{Q}}/\mathbb{Q}) \to GL_2(\mathbb{F})$ contains $SL_2(\mathbb{F})$. We let V denote the pullback as above of the two-dimensional $\mathbb{F}$-vector space in $J_1(l)(\overline{\mathbb{Q}})[l]$ that realizes ρ. Let D_0 be a divisor on $X_1(5l)_{\mathbb{Q}(\zeta_l)}$ as given in Theorem 8.1.7. Then, for every x in V_l, we have $h^0(X_1(5l)_{\overline{\mathbb{Q}}}, \mathcal{L}_x(D_0)) = 1$.*

Proof In view of Theorem 8.1.7, it suffices to show that for each x in V there is a place of $\overline{\mathbb{Q}}$ over l at which x specializes to 0. (The notion of specialization is explained in Section 8.1.) Let $J_{\mathbb{Z}[\zeta_l]}$ denote the Néron model of $J := J_1(5l)$ over $\mathbb{Z}[\zeta_l]$. Then V is the group of $\overline{\mathbb{Q}}$-points of an $\mathbb{F}$-vector space scheme $\mathcal{V}_{\mathbb{Q}(\zeta_l)}$ in $J_{\mathbb{Q}(\zeta_l)}$. Let $\mathcal{V}$ be the Zariski closure of $\mathcal{V}_{\mathbb{Q}(\zeta_l)}$ in $J_{\mathbb{Z}[\zeta_l]}$. Then it is shown in Section 12 of [Gro] and in Section 6 of [Edi1] that $\mathcal{V}_{\mathbb{Z}_l[\zeta_l]}$ is finite locally free over $\mathbb{Z}_l[\zeta_l]$, and that the dimension as $\mathbb{F}$-vector space scheme of the local part of $\mathcal{V}_{\mathbb{Z}_l[\zeta_l]}$ is 1 if $f(T_l) \neq 0$ and 2 if $f(T_l) = 0$. We note that it does not matter if we take Zariski closure in $J_1(l)$ or in $J_1(5l)$, as π^* gives a closed immersion of the l-torsion of $J_1(l)$ over $\mathbb{Z}_l$ into that of $J_1(5l)$.

This means that at each place of $\overline{\mathbb{Q}}$ over l there is a nonzero x in V that specializes to 0. Under our assumptions, the image of $\text{Gal}(\overline{\mathbb{Q}}/\mathbb{Q}(\zeta_l))$ acting on V is $\text{SL}(V)$. Hence $\text{Gal}(\overline{\mathbb{Q}}/\mathbb{Q}(\zeta_l))$ acts transitively on $V - \{0\}$. Hence for each x in $V - \{0\}$ there is at least one place of $\overline{\mathbb{Q}}$ over l where x specializes to 0. $\qquad\square$

The fact that our divisor D_0 lives on $X_1(5l)_{\mathbb{Q}(\zeta_l)}$, and not on $X_1(5l)_{\mathbb{Q}}$, forces us to work over $\mathbb{Q}(\zeta_l)$, and not over $\mathbb{Q}$, as in Chapter 3.

We let X_l denote $X_1(5l)_{\mathbb{Q}}$ and g_l its genus, and we let $A_{l,\mathbb{Q}(\zeta_l)}$ denote the $\mathbb{Q}(\zeta_l)$-algebra that corresponds to the $\text{Gal}(\overline{\mathbb{Q}}/\mathbb{Q}(\zeta_l))$-set V. In order to explain the notation $A_{l,\mathbb{Q}(\zeta_l)}$, we note that this $\mathbb{Q}(\zeta_l)$-algebra is obtained from the $\mathbb{Q}$-algebra A_l (that corresponds to the $\text{Gal}(\overline{\mathbb{Q}}/\mathbb{Q})$-set V) by extension of scalars.

Proposition 8.2.2 gives that for each x in V there is a unique effective divisor $D_x = \sum_{i=1}^{g_l} Q_{x,i}$ of degree g_l on $X_{l,\overline{\mathbb{Q}}}$ such that $x = [D_x - D_0]$ in $J_l(\overline{\mathbb{Q}})$. Note that, for $x = 0$, it is indeed true that $D_x = D_0$, hence the two notations are consistent. The following observation will make the exposition in Chapter 12 somewhat easier. As for each x in V specializes to 0 at some place of $\overline{\mathbb{Q}}$ above l, the divisor D_x specializes to the cuspidal divisor D_0 at such a place, and hence none of all $Q_{x,i}$ can be a CM-point. In particular,

(8.2.3)   for all x and i: $j(Q_{x,i}) \notin \{0, 1728\}$.

The uniqueness of D_x implies that

(8.2.4) $D_{gx} = gD_x$,   for all x in V and g in $\text{Gal}(\overline{\mathbb{Q}}/\mathbb{Q}(\zeta_l))$.

We write each D_x as

$$(8.2.5) \qquad\qquad\qquad D_x = D_x^{\text{fin}} + D_x^{\text{cusp}},$$

where D_x^{cusp} is supported on the cusps of $X_{l,\overline{\mathbb{Q}}}$ and D_x^{fin} is disjoint from the cusps. The next lemma shows that D_x^{fin} determines x; its proof uses only that ρ is absolutely irreducible, not that its image contains $\text{SL}_2(\mathbb{F})$.

8.2.6 Lemma *In this situation, the map from V to the set of effective divisors on $X_{l,\overline{\mathbb{Q}}}$ that sends x to D_x^{fin} is injective.*

Proof Suppose that it is not. We take x_1 and x_2 in V, distinct, such that $D^{\text{fin}}_{x_1} = D^{\text{fin}}_{x_2}$. Then the element $x_1 - x_2$ in V is nonzero and is represented by the cuspidal divisor $D_{x_1} - D_{x_2}$. The cusps of X_l are rational over $\mathbb{Q}(\zeta_{5l})$. Hence $x_1 - x_2$ gives an injection $\mathbb{F} \to V$ of representations of $\text{Gal}(\overline{\mathbb{Q}}/\mathbb{Q}(\zeta_{5l}))$, where $\mathbb{F}$ has trivial action. But that gives, by adjunction of induction and restriction, a nonzero map from the regular representation of $\text{Gal}(\mathbb{Q}(\zeta_{5l})/\mathbb{Q})$ over $\mathbb{F}$ to V, necessarily surjective because V is irreducible. But then the image of $\rho\colon \text{Gal}(\overline{\mathbb{Q}}/\mathbb{Q}) \to \text{GL}_2(\mathbb{F})$ is Abelian. As ρ is odd, the two eigenspaces in V of any complex conjugation then decompose V, in contradiction with the irreducibility of V. $\qquad\square$

As the cusps of $X_{l,\overline{\mathbb{Q}}}$ form a $\text{Gal}(\overline{\mathbb{Q}}/\mathbb{Q})$-stable subset of $X_{l,\overline{\mathbb{Q}}}$ we have

(8.2.7)   for all g in $\text{Gal}(\overline{\mathbb{Q}}/\mathbb{Q}(\zeta_l))$: $D^{\text{fin}}_{gx} = gD^{\text{fin}}_x, \quad D^{\text{cusp}}_{gx} = gD^{\text{cusp}}_x.$

Hence the map that sends x in V to D^{fin}_x is $\text{Gal}(\overline{\mathbb{Q}}/\mathbb{Q}(\zeta_l))$-equivariant.

We will now produce a suitable function $f_l\colon X_l \to \mathbb{P}^1_{\mathbb{Q}}$, in order to push the set of $\{D^{\text{fin}}_x \mid x \in V\}$ injectively and $\text{Gal}(\overline{\mathbb{Q}}/\mathbb{Q}(\zeta_l))$-equivariantly to the set $\{f_{l,*}D^{\text{fin}}_x \mid x \in V\}$ of divisors on $\mathbb{A}^1_{\mathbb{Q}}$.

We start by giving an explicit description of the curve $Y_1(5)$ over $\mathbb{Z}[1/5]$. In order to do that, we determine a universal triple $(E/S, P)$, where E/S is an elliptic curve over an arbitrary scheme, and P in $E(S)$ is everywhere of order 5, that is, for every $\text{Spec}(A) \to S$ with A nonzero, the image of P in $E(A)$ has order 5. The base of this triple is the open part $Y_1(5)'$ of the model $Y_1(5)$ over $\mathbb{Z}$ (constructed in Chapter 8 of [Ka-Ma]) where P has order 5 (that is, $Y_1(5)'$ is the complement of the irreducible component of $Y_1(5)_{\mathbb{F}_5}$ where the point P generates the kernel of Frobenius). The equation of this universal triple can also be found on p. 7 of Tom Fisher's thesis, see [Fis].

8.2.8 Proposition *Let E/S be an elliptic curve, and $P \in E(S)$ a point that is everywhere of order 5. Then $(E/S, P)$ arises via a unique base change from the following triple:*

$$\begin{cases} E : y^2 + (b+1)xy + by = x^3 + bx^2, \\ Y_1(5)' = \text{Spec}(\mathbb{Z}[b, 1/\text{discr}(E)]), \quad \text{discr}(E) = -b^5(b^2 + 11b - 1), \\ P = (0,0). \end{cases}$$

The j-invariant of E is given by:

$$j(E) = -(b^4 + 12b^3 + 14b^2 - 12b + 1)^3 / b^5 (b^2 + 11b - 1).$$

Proof Our proof is modeled on Section 2.2 of [Ka-Ma]; basic properties of Weierstrass equations for elliptic curves are used without being mentioned.

Let $(E/S, P)$ be given, with P everywhere of order 5. Choose a parameter t at 0, up to order 2, that is, a trivialization of $\omega_{E/S}$. Note: we are working locally on S, here; in the end, as we will succeed in making things unique, our construction will be global. Note: t is unique up to $t' = ut$, with $u \in R^\times$ ($S = \mathrm{Spec}(R)$ now).

Choose x a global function on $E - 0(S)$ such that $x = t^{-2} + \cdots$. Then x is unique up to $x' = x + a$, $a \in R$. Make x unique by demanding that $x(P) = 0$ (this is all right because 0 and P are disjoint).

Choose $y = t^{-3} + \cdots$ regular on $E - 0(S)$. Then y is unique up to $y' = y + ax + b$. Make y unique by demanding that $y(P) = 0$ and that the tangent of E at P is the line given by the equation $y = 0$. Indeed, use b (uniquely) to get $y(P) = 0$, then note that the tangent at P is nowhere the line given by $x = 0$ because P is nowhere annihilated by 2.

The equation for E is of the form

$$y^2 + a_1 xy + a_3 y = x^3 + a_2 x^2,$$

because the coefficients usually called a_4 and a_6 are zero. We also see that a_3 is a unit because E is smooth at $(0, 0)$. The coefficient a_2 is a unit, because P is nowhere annihilated by 3.

Now we try to get rid of u (the ambiguity in the choice of t). If $t' = u^{-1}t$, then $a'_i = t^i a_i$, hence we can make t unique by demanding that $a_2 = a_3$. We do that, and then we have the following equations.

The elliptic curve E and the point P are given by

$$y^2 + axy + by = x^3 + bx^2, \quad P = (0, 0).$$

We have $5 \cdot P \equiv 5 \cdot 0$, hence there is a unique f on $E - 0(S)$ of the form

$$f = xy + \alpha y + \beta x^2 + \gamma x + \delta$$

such that the divisor of f is $5 \cdot P - 5 \cdot 0$. As x and y have order one and two, respectively, at P, we have $\gamma = \delta = 0$. The function f with divisor

$5 \cdot P - 5 \cdot 0$ is given by:

$$f = xy + \alpha y + \beta x^2.$$

Here we know that b, α and β are in $R^\times$, because $v_P(x) = 1$ and $v_P(y) = 2$ everywhere on S. Now we have to compute what it means that $v_P(f) = 5$. This means that the intersection multiplicity of the two curves E and $V(f)$ at $(0, 0)$ is 5. A systematic way to compute that is to do successive blow-ups; that works nicely, but a much faster way is to take suitable linear combinations of the equations for E and f directly. One finds the equations

$$\beta = -\alpha, \quad a = \alpha^{-1}b + 1, \quad \alpha = 1.$$

$\square$

The reason we give such a detailed description of $Y_1(5)$ is that it gives us functions on all the $Y_1(5l)$, at least over $\mathbb{Z}[1/5l]$, as stated in the following proposition.

8.2.9 Proposition *Let $l \neq 5$ be prime. Let $E/Y_1(5)'$ be the elliptic curve given in Proposition 8.2.8. Then $Y_1(5l)$ and $E[l] - \{0\}$ agree over $\mathbb{Z}[1/5l]$: for S a $\mathbb{Z}[1/5l]$-scheme and Q in $(E[l] - \{0\})(S)$ we get, by pullback from $Y_1(5)'$, an elliptic curve over S with an S-valued point P_5 that is everywhere of order 5, and an S-valued point P_l that is everywhere of order l. In particular, the functions b, x, and y on $E[l] - \{0\}$ give functions b_l, x_l, and y_l on $Y_1(5l)$ over $\mathbb{Z}[1/5l]$ that generate its coordinate ring.*

Proof This is standard. The construction above gives a morphism, over $\mathbb{Z}[1/5l]$, from $E[l] - \{0\}$ to $Y_1(5l)$. Conversely, an elliptic curve over S with such points P_5 and P_l gives a point of $E[l] - \{0\}$ by the universality of $Y_1(5)'$ and the fact that $E[l]$ is finite étale over $Y_1(5)'$ away from characteristic l. The second statement follows from the fact that $E[l] - \{0\}$ is a closed subscheme of the affine scheme $E - \{0\}$. $\square$

The functions b_l, x_l, and y_l on $Y_1(5l)$ over $\mathbb{Z}[1/5l]$ have the following moduli interpretations. For any $\mathbb{Z}[1/5l]$-algebra A, and any Q in $Y_1(5l)(A)$, a point corresponding to a triple (E, P_5, P_l) with E an elliptic curve over A, P_5 in $E(A)$ a point that is everywhere of order 5 and P_l in $E(A)$ a point that is everywhere of order l, there is a unique element $b_l(Q)$ in A such that $(E/A, P_5)$ is uniquely isomorphic to the pair given by

$$y^2 + (b_l(Q) + 1)xy + b_l(Q)y = x^3 + b_l(Q)x^2, \quad P_5 = (0, 0).$$

Then, in these coordinates, we have

$$P_l = (x_l(Q), y_l(Q)).$$

Similarly, we define regular functions x_l' and y_l' on $Y_1(5l)_{\mathbb{Z}[1/5l]}$ by the condition that, in the coordinates above, we have:

$$l^{-1} P_5 + P_l = (x_l'(Q), y_l'(Q)),$$

where $l^{-1} P_5$ is the unique point R in $E[5](A)$ with $lR = P_5$.

We note that the pair of functions (b_l, x_l') embeds $Y_1(5l)_{\mathbb{Z}[1/5l]}$ in the affine plane $\mathbb{A}^2_{\mathbb{Z}[1/5l]}$. Indeed, assume that Q and Q' are in $Y_1(5l)(A)$, corresponding to (E, P_5, P_l) and (E', P_5', P_l'), with $b_l(Q) = b_l(Q')$ and $x_l'(Q) = x_l'(Q')$. Then, by Proposition 8.2.8, (E, P_5) is uniquely isomorphic to (E', P_5'), and so we simply consider them to be equal. Then, $l^{-1} P_5 + P_l$ and $l^{-1} P_5 + P_l'$ have the same x-coordinate. Hence, locally on $\mathrm{Spec}(A)$, $l^{-1} P_5 + P_l = \pm(l^{-1} P_5 + P_l')$. Multiplying by l we see that the sign cannot be a minus.

Using the functions b_l and x_l', we can now say how we will choose the function f_l. We return to the situation right after Lemma 8.2.6.

For $x \in V$, let d_x be the degree of D_x^{fin}, and write D_x as a sum of points in $X_l(\overline{\mathbb{Q}})$ as follows:

$$D_x = \sum_{i=1}^{g} Q_{x,i}, \quad \text{with} \quad D_x^{\mathrm{fin}} = \sum_{i=1}^{d_x} Q_{x,i}, \quad D_x^{\mathrm{cusp}} = \sum_{i=d_x+1}^{g} Q_{x,i}.$$

We note that $d_0 = 0$, as D_0 is a cuspidal divisor, and that for all nonzero x in V the d_x are equal, as they are permuted transitively by $\mathrm{Gal}(\overline{\mathbb{Q}}/\mathbb{Q}(\zeta_l))$.

The set S of points in $\mathbb{A}^2(\overline{\mathbb{Q}})$ consisting of the $(b_l(Q_{x,i}), x_l'(Q_{x,i}))$, with x in V and i in $\{1, \ldots, d_x\}$ has at most $g \cdot (\#\mathbb{F})^2$ elements. We want to project S injectively into $\mathbb{A}^1(\overline{\mathbb{Q}})$ with a map of the form $(a, b) \mapsto a + nb$ for a suitable integer n. As there are at most $g^2 \cdot (\#\mathbb{F})^4$ pairs of distinct elements in S, at most that number of integers n is excluded. Hence there exists an integer n with $0 \leq n \leq g^2 \cdot (\#\mathbb{F})^4$ such that the function $f_l := b_l + n x_l'$ has the required property that the $f_{l,*} D_x^{\text{fin}}$, for $x \in V$, are all distinct.

Let f_l be such a function. For each x in V, $f_{l,*} D_x^{\text{fin}}$ gives us a polynomial $P_{D_0,f_l,x}$ with coefficients in $\overline{\mathbb{Q}}$ given by

$$P_{D_0,f_l,x}(t) = \prod_{i=1}^{d_x} (t - f_l(Q_{x,i})) \quad \text{in } \overline{\mathbb{Q}}[t].$$

Conversely, each $P_{D_0,f_l,x}$ gives us the divisor $f_{l,*} D_x^{\text{fin}}$ by taking the roots, with multiplicity. Therefore, the map that sends x to $P_{D_0,f_l,x}$ is injective, and $\text{Gal}(\overline{\mathbb{Q}}/\mathbb{Q}(\zeta_l))$-equivariant.

The next step is to "encode" each $P_{D_0,f_l,x}$ in a single element of $\overline{\mathbb{Q}}$, respecting the action of $\text{Gal}(\overline{\mathbb{Q}}/\mathbb{Q}(\zeta_l))$. We do this by evaluating at a suitable integer m, that is, by sending $P_{D_0,f_l,x}$ to $P_{D_0,f_l,x}(m)$. For a given m, this map is injective if and only if for any distinct x_1 and x_2 in V, m is not a root of the difference of P_{D_0,f_l,x_1} and P_{D_0,f_l,x_2}. Each of these differences has at most g roots, and as there are less than $(\#\mathbb{F})^4$ such differences, there are at most $g \cdot (\#\mathbb{F})^4$ integers to avoid. So there is a suitable m with $0 \leq m \leq g \cdot (\#\mathbb{F})^4$. Composing our maps, we obtain a generator for the $\mathbb{Q}(\zeta_l)$-algebra $A_{l,\mathbb{Q}(\zeta_l)}$ associated with V:

$$a_{D_0,f_l,m} \colon V \to \overline{\mathbb{Q}}, \quad x \mapsto P_{D_0,f_l,x}(m).$$

We let

(8.2.10) $$P_{D_0,f_l,m} := \prod_{x \in V} (T - a_{D_0,f_l,m}(x)) \quad \text{in } \mathbb{Q}(\zeta_l)[T]$$

be the minimal polynomial over $\mathbb{Q}(\zeta_l)$ of $a_{D_0,f_l,m}$.

Chapter Nine

Applying Arakelov theory

by Bas Edixhoven and Robin de Jong

In this chapter we start applying Arakelov theory in order to derive a bound for the height of the coefficients of the polynomials $P_{D_0,f_l,m}$ as in (8.2.10). We proceed in a few steps. The first step, taken in Section 9.1, is to relate the height of the $b_l(Q_{x,i})$ as in Section 8.2 to intersection numbers on X_l. The second step, taken in Section 9.2, is to get some control on the difference of the divisors D_0 and D_x as in (3.4). Certain intersection numbers concerning this difference are bounded in Theorem 9.2.5, in terms of a number of invariants in the Arakelov theory on modular curves X_l. These invariants will then be bounded in terms of l in Sections 11.1, 11.2, and 11.3. Finally, in Section 11.7, the height of the coefficients of the $P_{D_0,f_l,m}$ will be bounded. In this chapter, we do our best to formulate the most important results, Theorem 9.1.3, Theorem 9.2.1, and Theorem 9.2.5, in the context of curves over number fields, that is, outside the context of modular curves.

9.1 RELATING HEIGHTS TO INTERSECTION NUMBERS

We pick up the notation at the end of Section 8.2, so we have a modular curve $X_l = X_1(5l)_{\mathbb{Q}}$ with $l > 5$, nonconstant morphisms b_l and $x_l' \colon X_l \to \mathbb{P}^1_{\mathbb{Q}}$, and certain divisors $D_x^{\text{fin}} = \sum_{i=1}^{d_x} Q_{x,i}$ on $X_{l,\overline{\mathbb{Q}}}$ that have support outside the cusps. It is our objective in this section to link the absolute height $h(b_l(Q_{x,i}))$ of the algebraic numbers $b_l(Q_{x,i})$ to certain quantities coming from Arakelov intersection theory. The height of $x_l'(Q_{x,i})$ will be bounded in terms of $h(b_l(Q_{x,i}))$ in Section 11.5. The final estimates for these heights, depending only on l, will be done in Section 11.7.

9.1.1 Theorem *Let x be in V, and let i be in $\{1, \dots, d_x\}$. Let K be a number field containing $\mathbb{Q}(\zeta_{5l})$ and such that $Q_{x,i}$ is defined over K. Let $\mathcal{X}$ be the minimal regular model of X_l over K. Then we have*

$$h(b_l(Q_{x,i})) \leq \frac{1}{[K : \mathbb{Q}]} \left((Q_{x,i}, b_l^* \infty)_{\mathcal{X}} + l^2 \sum_\sigma \sup_{X_\sigma} g_\sigma \right.$$
$$\left. + \frac{1}{2} \sum_\sigma \int_{X_\sigma} \log(|b_l|^2 + 1) \mu_{X_\sigma} \right) + \frac{1}{2} \log 2 .$$

Here σ runs through the embeddings of K into $\mathbb{C}$ and g_σ is the Arakelov-Green function on $X_{l,\sigma}$.

In the next chapters we shall derive bounds that are polynomial in l for all terms in the above estimate.

Theorem 9.1.1 will be derived from Theorem 9.1.3 below, which states a fairly general result. We start with a definition. Let K be a number field and consider $\mathbb{P}^1_{O_K}$. Let ∞ denote the O_K-point $(1 : 0)$ of $\mathbb{P}^1_{O_K}$. For any section P in $\mathbb{P}^1_{O_K}(O_K)$ we define by $(P, \infty)_{\mathbb{P}^1}$ the degree (see (4.3.1)) of $P^* O_{\mathbb{P}^1}(\infty)$, where $O_{\mathbb{P}^1}(\infty)$ has the Fubini-Study metric, that is, the metric $\| \cdot \|_{\mathbb{P}^1}$ given by

$$(9.1.2) \qquad \|1\|_{\mathbb{P}^1}(x_0 : x_1) := \frac{|x_1|}{(|x_0|^2 + |x_1|^2)^{1/2}}$$

over $\mathbb{P}^1_{\mathbb{C}}$. Here 1 is the tautological section of $O_{\mathbb{P}^1}(\infty)$.

9.1.3 Theorem *Let X be a geometrically irreducible, smooth, and complete curve of positive genus over a number field K, and let $\mathcal{X}$ be a proper semistable model of X over the ring of integers O_K of K. Suppose that we have a nonconstant morphism $f : X \to \mathbb{P}^1_K$ and a K-rational point Q of X with $f(Q) \neq \infty$. Write the divisor $\operatorname{div}(f)$ of f as $\operatorname{div}(f)_+ - \operatorname{div}(f)_-$, with $\operatorname{div}(f)_+$ and $\operatorname{div}(f)_-$ effective and without common irreducible components. Assume that the Zariski closure of $\operatorname{Supp}(\operatorname{div}(f)_+) \cup \operatorname{Supp}(\operatorname{div}(f)_-)$ in $\mathcal{X}$ is étale over O_K. For any closed point s of $\operatorname{Spec}(O_K)$, denote by $m_s(f)$ the supremum of the multiplicities of $\operatorname{div}(f)_-$ on $\mathcal{X}$ along the irre-*

ducible components of the fiber at s of $\mathcal{X}$. Then we have the inequality

$$(f(Q), \infty)_{\mathbb{P}^1} \leq (Q, f^*\infty)_{\mathcal{X}} + \deg f \sum_\sigma \sup_{X_\sigma} g_\sigma$$

$$+ \frac{1}{2} \sum_\sigma \int_{X_\sigma} \log(|f|^2 + 1)\mu_{X_\sigma} + \sum_s m_s(f) \log \#k(s).$$

Here the first sum runs over the embeddings of K into $\mathbb{C}$, and the last sum runs over the closed points of $\mathrm{Spec}(O_K)$.

Proof Note that the locus of indeterminacy of f on $\mathcal{X}$ consists of finitely many closed points. This implies that there exists a blow-up $\tilde{\mathcal{X}} \to \mathcal{X}$ of $\mathcal{X}$ such that f extends to a regular map $f \colon \tilde{\mathcal{X}} \to \mathbb{P}^1_{O_K}$. For any such $\tilde{\mathcal{X}}$ we have by construction

$$(f(Q), \infty)_{\mathbb{P}^1} = \deg f(Q)^*(O_{\mathbb{P}^1}(\infty), \|\cdot\|_{\mathbb{P}^1})$$
$$= \deg Q^* f^*(O_{\mathbb{P}^1}(\infty), \|\cdot\|_{\mathbb{P}^1})$$
$$= \deg Q^*(O_{\tilde{\mathcal{X}}}(f^*\overline{\infty}), \|\cdot\|_{\mathbb{P}^1}),$$

where we write $\overline{\infty}$ to emphasize that $f^*\overline{\infty}$ is the inverse image under f of $\infty(\mathrm{Spec}(O_K))$, and not the Zariski closure $\overline{f^*\infty}$ of the inverse image under f of $\infty(\mathrm{Spec}(K))$. If we let $\|\cdot\|_X$ denote the canonical Arakelov metric on $O_X(f^*\infty)$ then we can write

$$\deg Q^*(O_{\tilde{\mathcal{X}}}(f^*\overline{\infty}), \|\cdot\|_{\mathbb{P}^1}) = \deg Q^*(O_{\tilde{\mathcal{X}}}(f^*\overline{\infty}), \|\cdot\|_X \cdot \frac{\|\cdot\|_{\mathbb{P}^1}}{\|\cdot\|_X})$$

$$= \deg Q^*(O_{\tilde{\mathcal{X}}}(f^*\overline{\infty}), \|\cdot\|_X)$$
$$\quad - \sum_\sigma \log((\frac{\|\cdot\|_{\mathbb{P}^1}}{\|\cdot\|_X})(Q_\sigma))$$

$$= (Q, f^*\overline{\infty})_{\tilde{\mathcal{X}}} + \sum_\sigma \log((\frac{\|\cdot\|_X}{\|\cdot\|_{\mathbb{P}^1}})(Q_\sigma)).$$

A bound for $\log((\|\cdot\|_X / \|\cdot\|_{\mathbb{P}^1})(Q_\sigma))$ follows by testing on the tautological section 1, giving

$$\log \|1\|_X(Q_\sigma) - \log \|1\|_{\mathbb{P}^1}(Q_\sigma) = g_\sigma(f^*\infty, Q_\sigma) + \frac{1}{2} \log(|f(Q_\sigma)|^2 + 1).$$

Applying Proposition 9.1.4 below at this point shows that

$$g_\sigma \left(f^* \infty, Q_\sigma \right) + \frac{1}{2} \log \left(|f(Q_\sigma)|^2 + 1 \right)$$

$$\leq (\deg f) \sup_{X_\sigma} g_\sigma + \frac{1}{2} \int_{X_\sigma} \log(|f|^2 + 1) \, \mu_{X_\sigma} \, .$$

This accounts for the second and third terms in the bound of the theorem. We are finished once we prove that $(Q, f^* \overline{\infty})_{\tilde{\mathcal{X}}} - (Q, \overline{f^* \infty})_{\mathcal{X}}$ is bounded by $\sum_s m_s(f) \log \#k(s)$ for a particular choice of $\tilde{\mathcal{X}}$. (The usual projection formula shows that in fact $(Q, f^* \overline{\infty})_{\tilde{\mathcal{X}}}$ is independent of the choice of $\tilde{\mathcal{X}}$.) On any $\tilde{\mathcal{X}}$ we write $f^* \overline{\infty}$ as a sum $f^* \overline{\infty} = (f^* \overline{\infty})_{\text{hor}} + (f^* \overline{\infty})_{\text{vert}}$ of a horizontal and a vertical part. Note that $(f^* \overline{\infty})_{\text{hor}} = \overline{f^* \infty}$, with the Zariski closure now taken in $\tilde{\mathcal{X}}$. Since the local intersection multiplicities of Q and $\overline{f^* \infty}$ do not go up when passing from $\mathcal{X}$ to $\tilde{\mathcal{X}}$, we have $(Q, (f^* \overline{\infty})_{\text{hor}})_{\tilde{\mathcal{X}}} = (Q, \overline{f^* \infty})_{\tilde{\mathcal{X}}} \leq (Q, \overline{f^* \infty})_{\mathcal{X}}$, and thus we are reduced to proving that, for a particular choice of $\tilde{\mathcal{X}}$, $(Q, (f^* \overline{\infty})_{\text{vert}})_{\tilde{\mathcal{X}}}$ is bounded from above by $\sum_s m_s(f) \log \#k(s)$.

We exhibit a specific blow-up, and we calculate which multiplicities f acquires along the irreducible components of the vertical fibers of this blow-up. Note that the locus of indeterminacy of f on $\mathcal{X}$ consists precisely of the closed points of $\mathcal{X}$ where an irreducible component C from the zero divisor $\text{div}(f)_+$ of f on $\mathcal{X}$ and an irreducible component C' from its polar divisor $\text{div}(f)_-$ meet. Now since by assumption the Zariski closure of $\text{Supp}(\text{div}(f)_+) \cup \text{Supp}(\text{div}(f)_-)$ in $\mathcal{X}$ is étale over O_K, this can only happen when at least one of C, C' is vertical. In points where this happens we have to perform a sequence of successive blowings-up until a component arises with multiplicity 0 for f, so that the components with positive multiplicities and the components with negative multiplicities are separated from each other.

We begin by observing that it will cause no harm if we pass to a finite extension $K \to K'$. Indeed, both the left-hand side and the right-hand side of the inequality that we wish to prove get multiplied by $[K' : K]$ if we do this. Here is why. For the terms $(f(Q), \infty)_{\mathbb{P}^1}$ and $(Q, f^* \infty)$ the scaling by a factor $[K' : K]$ follows from general properties of the Arakelov intersection product (cf. [Fal1], p. 404, for example). (Note that it is understood that over K', intersection products are taken on the minimal resolution of the pullback of the model $\mathcal{X}$.) That the term $\sum_\sigma \sup_{X_\sigma} g_\sigma$ scales by a factor $[K' : K]$ is obvious. Finally, fix a closed point s of $\mathrm{Spec}(O_K)$ and let s' be any closed point of $\mathrm{Spec}(O_{K'})$ above it. Denoting by $e_{s'}$ the ramification index of s' over s and by $f_{s'}$ the degree of the residue field extension of s' over s, we see that for any s' above s, the integer $m_s(f)$ gets multiplied by $e_{s'}$, and the number $\log \#k(s)$ gets multiplied by $f_{s'}$. Using that $\sum_{s'} e_{s'} f_{s'} = [K' : K]$, the sum running over the closed points s' above s, we see finally that also the term $\sum_s m_s(f) \log \#k(s)$ gets multiplied by $[K' : K]$.

Starting with $\mathcal{X}$ over O_K, we first do the following. Let x be a closed point on $\mathcal{X}$ that is the intersection of a vertical component C and a horizontal component C' having nonzero multiplicities m and m' for f that have different signs. After blowing up in x, we obtain an exceptional divisor E whose multiplicity for f is $m+m'$. We have two distinguished points on E, one lying on the strict transform of C, and one lying on the strict transform of C'. At exactly one of them there is a sign change for the multiplicities, or $m + m' = 0$. If a sign change happens at the double point lying on the strict transform of C', then we repeat the process. If a sign change happens at the double point lying on the strict transform of C or if $m + m' = 0$, we stop, and continue with a new point x', if available.

We end up with a blow-up $\mathcal{X}' \to \mathcal{X}$ such that an intersection of two irreducible components C, C' that have different signs in $\mathrm{div}(f)$ on $\mathcal{X}'$ only occurs for C, C' both vertical. For this, we did not yet need to extend the ground field K. In order to continue, we note the following. Suppose that we have a closed point x on the model $\mathcal{X}'$ of X over O_K which is a double point of a vertical fiber, and two irreducible components C, C' of that vertical fiber pass through x, having nonzero multiplicities m and m' for f that differ in sign. Assume that $K \to K'$ is a Galois extension that ramifies over the image of x in $\mathrm{Spec}(O_K)$, with a ramification index e that is a multiple of $m - m'$. In passing to the minimal resolution $\tilde{\mathcal{X}}$ of $\mathcal{X}'_{O_{K'}}$, the point x gets replaced by a chain of $e-1$ projective lines of self-intersection -2. The multiplicities of f along these components change in e steps from em to em', so that the steps are $m - m'$ and a multiplicity 0 will appear somewhere, because $m - m'$ is a divisor of em.

Thus we see how we can reach our goal. Take a Galois extension $K \to K'$

that ramifies as specified above over the images in $\mathrm{Spec}(O_K)$ of the double points where components meet with a different sign for f. (This is always possible.) By our remarks above, it suffices to prove the inequality over K'. By construction, the morphism f extends over the model $\tilde{X}$ that arises in this way. Moreover, it follows from the construction that for s' a closed point of $\mathrm{Spec}(O_{K'})$ and s its image in $\mathrm{Spec}(O_K)$ we have $m_{s'}(f) \le e_{s'}m_s(f)$. Hence the sum of the local intersection numbers $(Q, (f^*\overline{\infty})_{\mathrm{vert}})_{s'}$ for all s' over s is bounded from above by $[K':K]m_s(f)\log\#k(s)$. This is what we needed to prove. $\qquad\square$

9.1.4 Proposition *Let* $f: X \to \mathbb{P}^1$ *be a finite morphism of Riemann surfaces with X connected and of positive genus. Consider on $X - f^{-1}\{\infty\}$ the function*

$$h(x) = g(f^*\infty, x) + \frac{1}{2}\log(|f(x)|^2 + 1).$$

Then h extends uniquely to an element of $C^\infty(X)$, also denoted h. For all $x \in X$ we have

$$h(x) \le (\sup_X g) \deg f + \frac{1}{2}\int_X \log(|f|^2 + 1)\,\mu_X.$$

Proof Let us first show that h extends to a C^∞ function on X. In fact, as the beginning of the proof of Theorem 9.1.3 indicates, h is the logarithm of the function $x \mapsto \|\cdot\|_X(x)/\|\cdot\|_{\mathbb{P}^1}(x)$ that gives the quotient of two metrics on $O_X(f^*\infty)$: the Arakelov metric $\|\cdot\|_X$ and the pullback of the Fubini-Study metric $\|\cdot\|_{\mathbb{P}^1}$ (see 9.1.2). But then h is in $C^\infty(X)$. We have, by (4.4.4),

$$h(x) = \int_{y \in X} -g(x, y)\frac{1}{\pi i}(\partial\bar{\partial}h)y + \int_X h\,\mu_X.$$

For s any local holomorphic generator of $O_X(f^*\infty)$ we have

$$\frac{1}{\pi i}\partial\bar{\partial}h = \frac{1}{2\pi i}\partial\bar{\partial}\log(\|s\|_X^2) - \frac{1}{2\pi i}\partial\bar{\partial}\log(\|s\|_{\mathbb{P}^1}^2) = (\deg f)\mu_X - f^*\mu_{\mathbb{P}^1},$$

where $\mu_{\mathbb{P}^1}$ is the curvature form of the Fubini-Study metric on $O_{\mathbb{P}^1}(\infty)$. Substituting this in the previous equality, and using that for all x in X we have $\int_{y \in X} g(x, y)\mu_X(y) = 0$, we get

$$h(x) = \int_{y \in X} g(x, y)(f^*\mu_{\mathbb{P}^1})y + \int_X \frac{1}{2}\log(|f|^2 + 1)\,\mu_X.$$

As $\mu_{\mathbb{P}^1}$ is defined as the curvature form of the Fubini-Study metric on $O_{\mathbb{P}^1}(\infty)$, we have $\int_{\mathbb{P}^1} \mu_{\mathbb{P}^1} = \deg(O_{\mathbb{P}^1}(\infty)) = 1$. As the metric is invariant under the transitive action of SU_2, $\mu_{\mathbb{P}^1}$ is everywhere positive. In fact, one can compute that $\mu_{\mathbb{P}^1} = (i/2\pi)\, dz\, d\bar{z}/(1 + |z|^2)^2$. We end up with

$$h(x) \leq (\sup_X g) \deg f + \frac{1}{2} \int_X \log(|f|^2 + 1)\mu_X$$

as required. $\square$

Proof [Proof of Theorem 9.1.1] In order to simplify our notation, we drop the subscript l in b_l. For P in $\mathbb{P}^1(O_K)$ we put

$$h'(P) = (P, \infty)/[K : \mathbb{Q}].$$

Then we have $h(P) \leq h'(P) + \frac{1}{2}\log 2$, as for all x in $\mathbb{C}^2$ we have

$$|x_1|^2 + |x_2|^2 \leq 2\max\{|x_1|, |x_2|\}^2.$$

In order to bound $h'(b(Q_{x,i}))$ from above we want to apply Theorem 9.1.3. It follows from the definition of the morphism b that both the zero divisor $\text{div}(b)_+$ and the polar divisor $\text{div}(b)_-$ of b on X_l have as their support only K-rational closed points, namely, cusps. In particular, we never have $b(Q_{x,i}) = \infty$, by construction of D_x''. We have also seen that the Zariski closure in $\mathcal{X}$ of $\text{Supp}(\text{Cusps})$ is étale over O_K (as O_K-valued points the cusps are disjoint), and hence the same holds for the Zariski closure in $\mathcal{X}$ of $\text{Supp}(\text{div}(b)_+) \cup \text{Supp}(\text{div}(b)_-)$. Theorem 9.1.3 now gives us that

$$[K : \mathbb{Q}] \cdot h'(b(Q_{x,i})) \leq (Q_{x,i}, b^*\infty)_{\mathcal{X}} + \deg b \sum_\sigma \sup_{X_\sigma} g_\sigma$$

$$+ \frac{1}{2}\sum_\sigma \int_{X_\sigma} \log(|b|^2 + 1)\mu_{X_\sigma}$$

$$+ \sum_s m_s(b) \log \#k(s),$$

with $m_s(b)$ the supremum of the multiplicities of $\text{div}(b)_-$ on $\mathcal{X}$ along the irreducible components of the fibers of $\mathcal{X}$ at s. We are done if we can prove that $\deg b$ is at most l^2, and that $m_s(b) = 0$ for all s. The definition of b shows directly that its degree is $l^2 - 1$ (it is the degree of the natural morphism from $X_1(5l)$ to $X_1(5)$). Let us now show that $m_s(b) = 0$ for all s. For this we evidently need information on the divisor $\text{div}(b)$ on $\mathcal{X}$.

We start with working over $\mathbb{Q}(\zeta_{5l})$. From the discussion in Section 8.1 we recall that there is a fine moduli scheme $Y_1(5l)_{\mathbb{Z}[\zeta_{5l}]}$ over $\mathbb{Z}[\zeta_{5l}]$ of elliptic curves with balanced level structure (terminology from [Ka-Ma]). Let $E(5l) \to Y_1(5l)_{\mathbb{Z}[\zeta_{5l}]}$ be the universal elliptic curve, and let P_5, P_l be the tautological points of order 5 and l. From Proposition 8.2.8 we recall the elliptic curve $E \to Y_1(5)'$ with $Y_1(5)' = \text{Spec}(\mathbb{Z}[b, 1/\text{discr}(E)])$. The elliptic curve $E(5l) \to Y_1(5l)_{\mathbb{Z}[\zeta_{5l},1/5]}$ arises from $E \to Y_1(5)'$ by a unique base change $Y_1(5l)_{\mathbb{Z}[\zeta_{5l},1/5]} \to Y_1(5)'$. This gives the regular function b on $Y_1(5l)_{\mathbb{Z}[\zeta_{5l},1/5]}$. As b is invertible on $Y_1(5)'$ it is invertible on $Y_1(5l)_{\mathbb{Z}[\zeta_{5l},1/5]}$. We conclude that $\text{div}(b)$ on $X_1(5l)_{\mathbb{Z}[\zeta_{5l}]}$ is a certain linear combination of the irreducible components of the closed subschemes Cusps and $X_1(5l)_{\mathbb{F}_5[\zeta_l]}$. In order to find this linear combination, we examine the multiplicities of b along the irreducible components that we have isolated.

We start with the multiplicities along the irreducible components of the divisor Cusps. It is sufficient to study the situation over $\mathbb{C}$, and here we can make a beginning by looking at $\text{div}(b)$ on $X_1(5)_{\mathbb{C}}$. From the equations in Proposition 8.2.8 we obtain that over the cusp 0 of $X_0(5)_{\mathbb{C}}$ lie two cusps, say c_1 and c_2, with c_1, say, corresponding to the 5-gon with the tautological point of order 5 being on a component adjacent to the connected component of 0, and the other, c_2, corresponding to the 5-gon with the tautological point of order 5 being on a component that is not adjacent to the connected component of 0. We have $\text{div}(b) = \pm(c_1 - c_2)$ on $X_1(5)_{\mathbb{C}}$; we could compute the exact sign, but that is not important for us. The divisor of b on $X_1(5l)_{\mathbb{C}}$ is obtained by pulling back its divisor on $X_1(5)_{\mathbb{C}}$ via the forgetful map $X_1(5l)_{\mathbb{C}} \to X_1(5)_{\mathbb{C}}$. Hence, pulling back the divisor $c_1 - c_2$ we get plus or minus the divisor of b on $X_1(5l)_{\mathbb{C}}$; the multiplicities are just the ramification indices above the cusps c_1 and c_2. Since these are in $\{1, l\}$, we obtain that the multiplicities of b along the irreducible components of Cusps are just 1 or l in absolute value.

Next we calculate the multiplicities of b along the irreducible components of $X_1(5l)_{\mathbb{F}_5[\zeta_l]}$. The structure of a connected component of the scheme $X_1(5l)_{\mathbb{F}_5[\zeta_l]}$ is as follows: it consists of two irreducible components, one on which P_5 has order 1, and one having an open part where P_5 has order 5. These two irreducible components intersect (transversally) in the supersingular points.

We denote by Γ the union of the irreducible components over $\mathbb{F}_5$ on which P_5 has order 1. The construction of the scheme $Y_1(5)'$ immediately gives us a forgetful map $X_1(5l)_{\mathbb{Z}[\zeta_{5l}]} - \Gamma - \mathrm{Supp}(\mathrm{Cusps}) \to Y_1(5)'$. Since b is invertible on $Y_1(5)'$, the same holds for b along the irreducible components of $X_1(5l)_{\mathbb{F}_5[\zeta_l]}$, except possibly for the irreducible components in Γ. But the multiplicity of b along such an irreducible component is then also zero, as can be seen by the following argument. Let $C \cup C'$ be a connected component of $X_1(5l)_{\mathbb{F}_5[\zeta_l]}$, with the irreducible component C corresponding to P_5 having order 1. All the horizontal components of $\mathrm{div}(b)$ on $X_1(5l)_{\mathbb{Z}[\zeta_{5l}]}$ specialize to C'. We know that b has multiplicity 0 along C' and hence it restricts to a nontrivial rational function, also denoted b, on C'. The degree of b on C' is zero, or equivalently $m(C', C) + (C', \mathrm{div}(b)_{\mathrm{hor}}) = 0$, where m is the multiplicity of b along C. Now, since $(C', \mathrm{div}(b)_{\mathrm{hor}})$ is zero and (C', C) isn't, we get $m = 0$.

All in all, we conclude that the absolute values of the multiplicities of the irreducible components in $\mathrm{div}(b)$ on $X_1(5l)_{\mathbb{Z}[\zeta_{5l}]}$ are bounded by a constant times l, and that all multiplicities along irreducible components of fibers over closed points s of $\mathrm{Spec}(\mathbb{Z}[\zeta_{5l}])$ are zero. In particular, for all closed points s we have $m_s(b) = 0$. This implies in fact that the rational function b on $X_1(5l)_{\mathbb{Z}[\zeta_{5l}]}$ extends to a morphism to $\mathbb{P}^1_{\mathbb{Z}}$. As this is a useful fact, we record it in a Proposition. This completes the proof of Theorem 9.1.1. $\square$

9.1.5 Proposition *Let $l > 5$ be prime a prime number. The rational function b_l on $X_1(5l)_{\mathbb{Z}[\zeta_{5l}]}$ from Proposition 8.2.9 extends to a morphism to $\mathbb{P}^1_{\mathbb{Z}}$.*

9.2 CONTROLLING $D_x - D_0$

In this subsection, the hypotheses are as follows (unless stated otherwise). We let K be a number field, O_K its ring of integers, $B := \mathrm{Spec}(O_K)$, $p: \mathcal{X} \to B$ a regular, split semistable curve over B whose generic fiber $X \to \mathrm{Spec}\, K$ is geometrically irreducible and of genus $g \geq 1$. We let D be the closure in $\mathcal{X}$ of an effective divisor of degree g (also denoted D) on X.

We let x be a K-rational torsion point of the Jacobian of X, that is, a torsion element of $\mathrm{Pic}(X)$, which has the property that there is a unique effective divisor D_x on X such that $x = [D_x - D]$. Finally, we let $P \colon B \to \mathcal{X}$ be a section of p, that is, an element of $\mathcal{X}(B)$.

We denote by $\Phi_{x,P}$ the unique finite vertical *fractional* divisor Φ (that is, with rational coefficients that are not necessarily integral) on $\mathcal{X}$ such that $(D_x - D - \Phi, C) = 0$ for all irreducible components C of fibers of p, and such that $P(B)$ is disjoint from the support of Φ. It is not difficult to see that a Φ satisfying the first condition exists and that it is unique up to adding multiples of fibers of p (the intersection pairing restricted to the divisors with support in a fiber is negative semidefinite); see Lemme 6.14.1 of [Mor2]. The second condition removes the ambiguity of adding multiples of fibers.

We denote by δ_s the number of singular points in the geometric fiber at a closed point s of B.

9.2.1 Theorem *The O_B-module $\mathrm{R}^1 p_* O_{\mathcal{X}}(D_x)$ is a torsion module on B, and we have:*

$$(D_x, P) + \log \#\mathrm{R}^1 p_* O_{\mathcal{X}}(D_x) + \frac{1}{8}(\omega_{\mathcal{X}/B}, \omega_{\mathcal{X}/B}) + \frac{1}{8}\sum_s \delta_s \log \#k(s)$$

$$= (D, P) - \frac{1}{2}(D + \Phi_{x,P}, D + \Phi_{x,P} - \omega_{\mathcal{X}/B}) + \frac{1}{2}\deg \det p_*\omega_{\mathcal{X}/B}$$

$$+ \sum_\sigma \int_{X_\sigma} \log \|\vartheta\| (D_x^\sigma - Q) \cdot \mu_\sigma(Q) + \frac{g}{2}[K : \mathbb{Q}] \log(2\pi).$$

Here s runs over the closed points of B, and σ runs through the complex embeddings of K.

We derive Theorem 9.2.1 from three lemmas. For the moment we work in $\mathbb{Q} \otimes_{\mathbb{Z}} \widehat{\mathrm{Pic}}(\mathcal{X})$.

9.2.2 Lemma *The admissible line bundles*

$$O_{\mathcal{X}}(D_x - D) \otimes p^* P^* O_{\mathcal{X}}(D_x - D)^{\vee} \quad \text{and} \quad O_{\mathcal{X}}(\Phi_{x,P})$$

are numerically equivalent. That is, for any admissible line bundle F on $\mathcal{X}$ we have

$$(O_{\mathcal{X}}(D_x - D) \otimes p^* P^* O_{\mathcal{X}}(D_x - D)^{\vee}, F) = (O_{\mathcal{X}}(\Phi_{x,P}), F).$$

Proof In this proof we just write Φ for $\Phi_{x,P}$. We denote the first line bundle in the lemma by Ψ. Since $D_x - D$ is torsion, there is a positive integer N such that $\Psi^{\otimes N}$ is trivial on the generic fiber as a classical line bundle (that is, without taking the metrics into account). We have a canonical isomorphism $P^*\Psi \xrightarrow{\sim} O_B$ on B. Combining, we find that $\Psi^{\otimes N}$ has a rational section s with $\mathrm{div}_{\mathcal{X}}(s)$ vertical and $P^*s \mapsto 1$. The latter condition implies that P intersects to zero with $\mathrm{div}_{\mathcal{X}}(s)$ for the Arakelov intersection product. On the other hand, as $p^*P^*O_{\mathcal{X}}(D_x - D)^{\vee}$ is trivial on the fibers of p over finite places of B, we have $(N(D_x - D) - \mathrm{div}_{\mathcal{X}}(s), C) = 0$ for all irreducible components C of fibers of p. Hence in fact $\Phi = \frac{1}{N}\mathrm{div}_{\mathcal{X}}(s)$. To prove the lemma, it suffices now to prove that $\Psi^{\otimes N} \xrightarrow{\sim} O_{\mathcal{X}}(\mathrm{div}_{\mathcal{X}}(s))$ given by $s \mapsto 1$, with 1 the tautological section, is an isometry. Because of admissibility, it suffices to check that this is so when restricted to P; but here we get the canonical isomorphism $P^*\Psi \xrightarrow{\sim} O_B$. This is indeed an isometry by the definition of Ψ. □

9.2.3 Lemma *Let X be a compact Riemann surface of genus $g \geq 1$. Let D be an effective divisor on X of degree g satisfying $h^0(D) = 1$. Then the determinant of cohomology $\lambda(O_X(D))$ is identified with $H^0(X, O_X(D))$. Further, the formula*

$$\log \|1\| + \frac{\delta(X)}{8} + \int_X \log \|\vartheta\|(D - Q) \cdot \mu_X(Q) = 0$$

holds for the length (with respect to Faltings' metrization of the determinant of cohomology) of the tautological section 1 of $H^0(X, O_X(D))$.

Proof Since $h^0(D) = 1$, $H^0(X, O_X(D)) = \mathbb{C}$. Therefore, the set of points Q on X such that $h^0(D - Q) > 0$ is the support of D. Let Q be a point outside the support of D. Then $h^0(D - Q) = 0$. According to the axioms for the metrization of the determinant of cohomology, the exact sequence

$$0 \to O_X(D - Q) \to O_X(D) \to Q_*Q^*O_X(D) \to 0$$

gives rise to an isometry

$$\lambda(O_X(D)) \xrightarrow{\sim} \lambda(O_X(D - Q)) \otimes Q^*O_X(D)$$
$$\cong O(-\Theta)[O_X(D - Q)] \otimes Q^*O_X(D).$$

Taking the norm on left and right of a tautological section we obtain, using (4.4.6),

$$\|1\| = \exp(-\delta(X)/8) \cdot \|\vartheta\| (D - Q)^{-1} \cdot G(D, Q),$$

where $\log G(D, Q) = g_{D,\mu_X}(Q)$. Taking logarithms and then integrating against $\mu_X(Q)$ gives the result. $\square$

9.2.4 Lemma *(Noether formula) We have*

$$12 \deg \det p_* \omega_{X/B} = (\omega_{X/B}, \omega_{X/B}) + \sum_s \delta_s \log \#k(s)$$
$$+ \sum_\sigma \delta(X_\sigma) - 4g[K : \mathbb{Q}] \log(2\pi),$$

the first sum running over the closed points of B, the second sum running over the complex embeddings of K.

Proof See [Fal1] and [Mor3]. $\square$

Proof [Proof of Theorem 9.2.1] We first show that $R^1 p_* O_X(D_x)$ is a torsion module. As it is a coherent O_B-module, it suffices to show that it is zero on the generic point of B, that is, that $H^1(X, O_X(D_x))$ is zero. By Riemann-Roch, we have $h^0(D_x) - h^1(D_x) = 1$. By definition of D_x, we have $O_X(D_x) \cong \mathcal{L}_x(D)$. And by construction of D, we have $h^0(\mathcal{L}_x(D)) = 1$. This shows that $h^1(D_x) = 0$.

Let us now prove the identity in Theorem 9.2.1. We start by noting that, by (4.4.7)

$$(D_x - D, P) = \deg P^* O_X(D_x - D).$$

By Lemma 9.2.2, $O_X(D_x) \otimes p^* P^* O_X(D_x - D)^\vee$ and $O_X(D + \Phi_{x,P})$ are numerically equivalent. The Riemann-Roch theorem then gives

$$\deg \det R p_* (O_X(D_x) \otimes p^* P^* O_X(D_x - D)^\vee)$$
$$= \frac{1}{2}(D + \Phi_{x,P}, D + \Phi_{x,P} - \omega_{X/B}) + \deg \det p_* \omega_{X/B}.$$

By the projection formula for the determinant of cohomology we can write the left-hand side as

$$\deg \det \mathrm{R}p_*(O_\mathcal{X}(D_x) \otimes p^* P^* O_\mathcal{X}(D_x - D)^\vee)$$
$$= \deg \det \mathrm{R}p_* O_\mathcal{X}(D_x) - \deg P^* O_\mathcal{X}(D_x - D).$$

Since $p_* O_\mathcal{X}(D_x)$ is canonically trivialized by the function 1, we have

$$\deg \det \mathrm{R}p_* O_\mathcal{X}(D_x) = -\sum_\sigma \log \|1\|_\sigma - \log \#\mathrm{R}^1 p_* O_\mathcal{X}(D_x),$$

where for each complex embedding σ, the norm $\|1\|_\sigma$ is the length of the tautological section 1 of $\lambda(O_{X_\sigma}(D_x)) = H^0(O_{X_\sigma}(D_x))$. By Lemma 9.2.3 we can then write

$$\deg \det \mathrm{R}p_* O_\mathcal{X}(D_x) = \sum_\sigma \int_{X_\sigma} \log \|\vartheta\|_\sigma (D_x^\sigma - Q) \cdot \mu_\sigma(Q)$$
$$+ \sum_\sigma \delta(X_\sigma)/8 - \log \#\mathrm{R}^1 p_* O_\mathcal{X}(D_x).$$

Combining everything gives

$$(D_x - D, P) = -\frac{1}{2}(D + \Phi_{x,P}, D + \Phi_{x,P} - \omega_{\mathcal{X}/B}) - \deg \det p_* \omega_{\mathcal{X}/B}$$
$$+ \sum_\sigma \int_{X_\sigma} \log \|\vartheta\| (D_x^\sigma - Q) \cdot \mu_\sigma(Q)$$
$$+ \sum_\sigma \delta(X_\sigma)/8 - \log \#\mathrm{R}^1 p_* O_\mathcal{X}(D_x).$$

We obtain the required formula upon eliminating $\sum_\sigma \delta(X_\sigma)/8$ with the Noether formula, Lemma 9.2.4. $\square$

9.2.5 Theorem *We have an upper bound*

$$(D_x, P) + \log \#\mathrm{R}^1 p_* O_\mathcal{X}(D_x) \leq -\frac{1}{2}(D, D - \omega_{\mathcal{X}/B})$$
$$+ 2g^2 \sum_{s \in B} \delta_s \log \#k(s)$$
$$+ \sum_\sigma \log \|\vartheta\|_{\sigma,\sup} + \frac{g}{2}[K : \mathbb{Q}] \log(2\pi)$$
$$+ \frac{1}{2} \deg \det p_* \omega_{\mathcal{X}/B} + (D, P),$$

where s runs through the closed points of B, and where the supremum norm
$\|\vartheta\|_{\sigma,\mathrm{sup}}$ *is taken over* $\mathrm{Pic}^{g-1}(X_\sigma)$.

The required upper bound follows directly from Theorem 9.2.1 by using
Lemma 9.2.6 below and the fact that $(\omega_{X/B}, \omega_{X/B}) \geq 0$ (see Theorem 5
of [Fal1]).

9.2.6 Lemma *We have an upper bound*

$$-\frac{1}{2}(D + \Phi_{x,P}, D + \Phi_{x,P} - \omega_{X/B})$$

$$\leq -\frac{1}{2}(D, D - \omega_{X/B}) + 2g^2 \sum_{s \in B} \delta_s \log \#k(s),$$

with s running through the closed points of B.

Proof In this proof we just write Φ for $\Phi_{x,P}$. By the definition of Φ, we
have $(D_x - D - \Phi, \Phi) = 0$, or, in other words, $(\Phi, \Phi) = (D_x - D, \Phi)$.
Using this we can write

$$-\frac{1}{2}(D+\Phi, D+\Phi-\omega_{X/B}) = -\frac{1}{2}(D, D-\omega_{X/B}) + \frac{1}{2}(\Phi, \omega_{X/B}-D-D_x).$$

We write $\Phi = \sum_C \Phi(C) \cdot C$, and for any finite fiber F_s of p we put
$A_s := \sup_C |\Phi(C)|$ with C running through the irreducible components
of F_s. Since $\omega_{X/B}, D$ and D_x intersect any irreducible component C with
non-negative multiplicity, we find

$$\frac{1}{2}(\Phi, \omega_{X/B} - D - D_x) \leq \frac{1}{2}\left(\sum_s A_s F_s, \omega_{X/B} + D + D_x\right)$$

$$\leq 2g \sum_s A_s \log \#k(s).$$

We are going to prove that $A_s \leq g\delta_s$, and then we are done. So let s be a
finite place of B. Let S_0 be the set of irreducible components of F_s, and let
S_1 be the set of double points on F_s. Let Γ_s be the dual graph of F_s (thus, the
set of vertices of Γ_s corresponds to S_0, the set of edges corresponds to S_1,
and the graph is defined by the incidence relations). Choose an orientation
on Γ_s. This gives rise to the usual source and target maps s and $t: S_1 \to S_0$.
Consider the boundary and coboundary maps $d_* = t_* - s_*: \mathbb{Q}^{S_1} \to \mathbb{Q}^{S_0}$

and $d^* = t^* - s^* : \mathbb{Q}^{S_0} \to \mathbb{Q}^{S_1}$. Then $d_* d^* : \mathbb{Q}^{S_0} \to \mathbb{Q}^{S_0}$ is given by minus the intersection matrix of F_s. In particular, the map $d_* d^*$ sends Φ to the map $u : C \mapsto -(\Phi, C) = (D - D_x, C)$. The kernel of $d_* d^*$ consists exactly of the constant functions, and the image consists of the orthogonal complement of the constant functions. Now consider the graph Γ_s as an electric circuit, where each edge has a resistance of 1 ohm. By Ohm's law and by spelling out the maps d_* and d^* we see that if we let at each vertex C a current of $u(C)$ amperes enter the circuit, subject to the condition that $\sum_C u(C) = 0$, the potentials $\varphi(C)$ at each vertex C will be given, up to addition of a constant function, by a solution of the equation $d_* d^* \varphi = u$. Hence Φ is the potential corresponding to the current $C \mapsto (D - D_x, C)$, normalized by the condition that $\Phi(C_P) = 0$ with C_P the component that P specializes to. We must bound the $|\Phi(C)|$ for C varying over S_0. The worst case that may happen is that Γ_s is a chain, with D' and D specializing entirely to the beginning and end point, respectively. In this case, the biggest potential difference is $g \cdot (\#S_0 - 1)$ in absolute value, so that we arrive at $|\Phi(C)| \le g \cdot (\#S_0 - 1)$. Now note that Γ_s is connected and that X/K has split semistable reduction. This gives $\#S_0 - 1 \le \delta_s$ and hence $|\Phi(C)| \le g\delta_s$, as required. $\qquad\qquad\square$

Chapter Ten

An upper bound for Green functions on Riemann surfaces

by Franz Merkl

We begin with explaining the setup and the results of this chapter. Let X be a compact Riemann surface, endowed with a 2-form $\mu \geq 0$ that fulfills $\int_X \mu = 1$. Let $*$ denote rotation by $90°$ in the cotangential spaces (with respect to the holomorphic structure); in a coordinate $z = x + iy$ this means $*dx = dy$, $*dy = -dx$ and, equivalently, $*dz = -i\,dz$, $*d\bar{z} = i\,d\bar{z}$. In particular, the Laplace operator on real C^∞ functions on X can be written as $d*d = 2i\,\partial\bar{\partial}$.

For $a, b \in X$, let $g_{a,b} \colon X - \{a, b\} \to \mathbb{R}$ be the (unique) solution on X (in the sense of distributions) of the differential equation

$$d*dg_{a,b} = \delta_a - \delta_b \quad \text{on } X$$

with the normalizing condition

$$\int_{X-\{a,b\}} g_{a,b}\,\mu = 0.$$

Note that $g_{a,a} = 0$. The distributional differential equation for $g_{a,b}$ is equivalent to the following two more elementary conditions. First, $g_{a,b}$ is a real-valued harmonic function on $X - \{a, b\}$. Second, it has logarithmic singularities near a and near b of the following type: for any local coordinate z near a, the function $P \mapsto g_{a,b}(P) - (2\pi)^{-1} \log |z(P) - z(a)|$ extends to a harmonic function in a neighborhood of a, and for any local coordinate w near b, the function $P \mapsto g_{a,b}(P) + (2\pi)^{-1} \log |w(P) - w(b)|$ extends to a harmonic function in a neighborhood of b. The existence of such a function

$g_{a,b}$ is shown in the theorem in II.4.3 on p. 49 in [Fa-Kr]. However, for a close to b, the proof of Lemma 10.4 below also shows the existence of $g_{a,b}$ as a byproduct. The difference of any two solutions of the differential equation for $g_{a,b}$ extends to a global harmonic function on X and thus is a constant. Hence, the normalizing condition for $g_{a,b}$ determines the function $g_{a,b}$ uniquely.

Now, for $x \in X - \{a\}$, let

$$g_{a,\mu}(x) := \int_{b \in X - \{x\}} g_{a,b}(x)\, \mu(b).$$

Then we have

$$d * dg_{a,\mu} = \delta_a - \mu$$

in the sense of distributions, and

$$\int_{X - \{a\}} g_{a,\mu}\, \mu = 0.$$

We consider an atlas of X consisting of n local coordinates

$$z^{(j)} : U^{(j)} \to \mathbb{C}, \quad j = 1, \dots, n,$$

such that each range $z^{(j)}[U^{(j)}]$ contains the closed unit disk. For any radius $0 < r \le 1$ and $j \in \{1, \dots, n\}$, we define the disk

$$U_r^{(j)} = \{P \in U^{(j)} \mid |z^{(j)}(P)| < r\}.$$

We fix a radius $0 < r_1 < 1$ once and for all. Our aim is to prove the following result.

10.1 Theorem *Assume that the $U_{r_1}^{(j)}$ with j in $\{1, \dots, n\}$ cover X. Next, assume that c_1 is a positive real number such that for all j in $\{1, \dots, n\}$ we have*

$$\mu \le c_1 |dz^{(j)} \wedge d\bar{z}^{(j)}| \quad \text{on } U_1^{(j)}.$$

Finally, assume that for all j and k in $\{1, \dots, n\}$,

$$\sup_{U_1^{(j)} \cap U_1^{(k)}} \left| \frac{dz^{(j)}}{dz^{(k)}} \right| \le M$$

*holds with some constant $M \geq 1$. Then for some positive constants c_7, c_9,
c_{10}, and c_{11}, depending only on r_1, we have, for all a in X,*

$$(10.2) \qquad g_{a,\mu} \leq n(c_{10} + c_1 c_{11} + c_7 \log M) + \frac{\log 2}{2\pi}$$

and, for all j such that $a \in U_{r_1}^{(j)}$,

$$(10.3) \qquad \lim_{x \to a} \left| g_{a,\mu}(x) - \frac{1}{2\pi} \log |z^{(j)}(x) - z^{(j)}(a)| \right|$$
$$\leq n(c_{10} + c_1 c_{11} + c_7 \log M) + \frac{\log M}{2\pi} + c_9.$$

We start by considering just one coordinate $z = z^{(j)}$ for a fixed j. To simplify the notation at this point, we drop the superscript "(j)" in $U = U^{(j)}$,
$z = z^{(j)}$, and so on. We fix three radii $0 < r_1 < r_2 < r_3 < 1$ once
and for all. The radii r_2 and r_3 should depend only on r_1, for example,
$r_2 = (2r_1 + 1)/3$; $r_3 = (r_1 + 2)/3$ is an admissible choice. Furthermore, we fix a partition of unity: let $\chi: X \to [0, 1]$ be a C^∞ function which
is compactly supported in the interior of U_1 with $\chi = 1$ on $\overline{U_{r_2}}$, and set
$\chi^c = 1 - \chi$. More specifically, we take $\chi = \tilde{\chi}(|z|)$ on U_1 with a smooth
function $\tilde{\chi}: \mathbb{R} \to [0, 1]$ such that $\tilde{\chi}(r) = 0$ for $r \geq 1 - \epsilon$ with some $\epsilon > 0$,
and $\tilde{\chi}(r) = 1$ for $r \leq r_2$. The shape function $\tilde{\chi}$ may be taken independently
of X and the choice of the coordinate z, only depending on r_2.

We shall use the 2-norm of a (real-valued) 1-form ω over a measurable
set $Y \subseteq X$ defined by

$$\|\omega\|_Y := \left(\int_Y \omega \wedge *\omega \right)^{1/2} = \left(2i \int_Y \omega_{1,0} \wedge \omega_{0,1} \right)^{1/2},$$

where $\omega = \omega_{1,0} + \omega_{0,1}$ is the decomposition of ω in its components in
$T_{(1,0)}X$ and $T_{(0,1)}X$. In the case $Y = X$, we just write $\|\omega\|$ for $\|\omega\|_X$.

Given a and b in U_{r_1}, we define the following function, having logarithmic singularities in a and b:

$$f_{a,b} := \frac{1}{2\pi} \log \left| \frac{(z - z(a))(\overline{z(a)}z - 1)}{(z - z(b))(\overline{z(b)}z - 1)} \right| \quad \text{on } U_1 - \{a, b\}.$$

Note that the singularities at $1/\overline{z(a)}$ and $1/\overline{z(b)}$ do not lie within the unit
disk. We note that

$$d*df_{a,b} = \delta_a - \delta_b$$

holds on U_1 in the sense of distributions, and that $f_{a,b}$ fulfills Neumann boundary conditions on ∂U_1. One can see this as follows. The meromorphic function on U_1 given by

$$P \mapsto \frac{(z(P) - z(a))(\overline{z(a)}z(P) - 1)}{(z(P) - z(b))(\overline{z(b)}z(P) - 1)} = \frac{(z(P) - z(a))(\frac{1}{z(P)} - \overline{z(a)})}{(z(P) - z(b))(\frac{1}{z(P)} - \overline{z(b)})}$$

takes positive real values on ∂U_1. Let Log denote the principal branch of the logarithm. The function

$$q_{a,b}(P) := \frac{1}{2\pi} \operatorname{Log} \frac{(z(P) - z(a))(\overline{z(a)}z(P) - 1)}{(z(P) - z(b))(\overline{z(b)}z(P) - 1)}$$

is defined and holomorphic for P in a neighborhood of ∂U_1, and it takes real values for $P \in \partial U_1$. As a consequence, the directional derivative of the imaginary part $\Im q_{a,b}$ tangential to ∂U_1 vanishes on ∂U_1. Using holomorphy, this implies that the directional derivative of the real part $\Re q_{a,b}$ in normal direction to ∂U_1 vanishes also on ∂U_1. Using $\Re q_{a,b}(P) = f_{a,b}(P)$ for P in a neighborhood of U_1, this proves the claimed Neumann boundary conditions for $f_{a,b}$.

Finally, for $a \in U_{r_1}$ and $P \in U_1 - \{a\}$, we set

$$l_a(P) := \frac{1}{2\pi} \chi(P) \log |z(P) - z(a)|,$$

extended by 0 to $X - \{a\}$.

Our first step in the proof of Theorem 10.1 is the following key lemma.

10.4 Lemma *For a and b in U_{r_1}, the supremum $\sup_X |g_{a,b} - l_a + l_b|$ is bounded by a constant $c_2 = c_4 + c_1 c_5$, with c_4, c_5 depending only on r_1.*

10.5 Remark Note that $g_{a,b} - l_a + l_b$ has removable singularities at a and b, since the logarithmic singularities cancel. The constant c_2 is uniform in the choice of the Riemann surface X, and uniform in the choice of $a, b \in U_{r_1}$. The choice of the coordinate z influences c_2 only via the dependence of c_1 on the choice of z. The radii r_2, r_3 and the shape function $\tilde{\chi}$ are viewed as r_1-dependent parameters; this is why we need not emphasize in the lemma that c_2 also depends on these quantities.

Proof   (of Lemma 10.4) We define the 2-form

$$u_{a,b} := d*d(\chi^c f_{a,b}) \quad \text{on } U_1 - \overline{U_{r_1}}$$

and extend it by 0 to the whole surface X. Note that $u_{a,b}$ is supported in $\overline{U_1} - U_{r_2}$, since χ^c varies only there, and since $f_{a,b}$ is harmonic. Consider the following variational principle on square integrable 1-forms ω. We want to minimize $\|\omega\|^2$ with the constraint

$$d*\omega = u_{a,b}$$

in the sense of distributions. Writing the constraint with test functions, we see that the minimization problem is taken over the following closed affine linear subspace of $L^2(X, T^*X)$:

$$V = \{\omega \in L^2(X, T^*X) : -\int_X dg \wedge *\omega = \int_X g u_{a,b} \text{ for all } g \in C^\infty(X)\}.$$

The space V is nonempty, since $\tilde{\omega}_{a,b} \in V$ holds for the following 1-form:

$$\tilde{\omega}_{a,b} = \begin{cases} d(\chi^c f_{a,b}) & \text{on } U_1 - U_{r_1}, \\ 0 & \text{otherwise.} \end{cases}$$

Indeed, using Stokes's theorem, we have:

$$-\int_X dg \wedge *\tilde{\omega}_{a,b} = -\int_{U_1} dg \wedge *\tilde{\omega}_{a,b}$$

$$= -\int_{\partial U_1} g *\tilde{\omega}_{a,b} + \int_{U_1} g \, d*\tilde{\omega}_{a,b}.$$

The first summand in the last expression vanishes by the Neumann boundary conditions of $\chi^c f_{a,b} = f_{a,b}$ on ∂U_1, and the second summand equals

$$\int_{U_1} g \, d*\tilde{\omega}_{a,b} = \int_X g u_{a,b}$$

by the definition of $u_{a,b}$.

Our minimization problem has a unique solution $\omega_{a,b} \in V$. It fulfills:

$$(10.6) \qquad \int_X \omega_{a,b} \wedge \sigma = 0 \quad \text{for all closed } C^\infty \text{ 1-forms } \sigma.$$

Indeed, if $d\sigma = 0$, then $\omega_{a,b} + t * \sigma \in V$ holds for all $t \in \mathbb{R}$, since $\omega_{a,b} \in V$ and $d*(*\sigma) = -d\sigma = 0$. Thus

$$0 = \frac{d}{dt}\|\omega_{a,b} + t * \sigma\|^2\bigg|_{t=0} = -2\int_X \omega_{a,b} \wedge \sigma.$$

In particular,

$$\int_X \omega_{a,b} \wedge dg = 0$$

for all $g \in C^\infty(X)$, that is, $d\omega_{a,b} = 0$ in the sense of distributions. Since $d*\omega_{a,b} = u_{a,b}$ and $d\omega_{a,b} = 0$, we get that $\omega_{a,b}$ is smooth. This follows from (hypo-)elliptic regularity, as treated in Corollary 4.1.2 on p. 101 in [Hor]. Precisely speaking, this corollary treats only the case of a single partial differential equation. However, as is mentioned at the end of Section 4.0 on p. 97 in [Hor], the extension of the result to systems of partial differential equations with as many equations as unknowns, as needed here, follows trivially. Then equation (10.6) implies that $\omega_{a,b}$ is exact:

$$\omega_{a,b} = d\tilde{g}_{a,b}$$

for some $\tilde{g}_{a,b} \in C^\infty(X)$; see, for example, [For], Corollary 19.13. We normalize $\tilde{g}_{a,b}$ such that

(10.7) $$\int_X \tilde{g}_{a,b}\mu = 0,$$

to make it uniquely determined.

We set

$$h_{a,b} := \tilde{g}_{a,b} + \chi f_{a,b} - \int_X \chi f_{a,b}\mu.$$

We are now going to prove that $d*dh_{a,b} = \delta_a - \delta_b$. We claim that $d*d(\chi f_{a,b}) = -u_{a,b} + \delta_a - \delta_b$ holds. We prove this equality separately on the three sets $X - \mathrm{Supp}\chi$, $X - \mathrm{Supp}\chi^c$, and $U_1 - \overline{U}_{r_1}$, which cover X. The claimed equality holds on $X - \mathrm{Supp}\chi$, because both sides vanish there. It holds also on $X - \mathrm{Supp}\chi^c$, because there $u_{a,b} = 0$ and $d*d(\chi f_{a,b}) = d*df_{a,b} = \delta_a - \delta_b$ are valid. Finally, on $U_1 - \overline{U}_{r_1}$, the function $f_{a,b}$ is harmonic, hence $d*d(\chi f_{a,b}) = -d*d(\chi^c f_{a,b}) = -u_{a,b}$ on this annulus, which contains neither a nor b. So the claim that $d*d(\chi f_{a,b})$ equals $-u_{a,b} + \delta_a - \delta_b$ holds in all cases.

Combining this with the fact $d*d\tilde{g}_{a,b} = u_{a,b}$, we conclude

$$d*dh_{a,b} = d*d\tilde{g}_{a,b} + d*d(\chi f_{a,b}) = u_{a,b} - u_{a,b} + \delta_a - \delta_b,$$

and thus

$$d*dh_{a,b} = \delta_a - \delta_b.$$

Furthermore, using the normalization $\int_X \mu = 1$ and $\int_X \tilde{g}_{a,b}\mu = 0$, we observe

$$\int_X h_{a,b}\mu = 0.$$

Because $g_{a,b}$ is uniquely characterized by its properties $d*dg_{a,b} = \delta_a - \delta_b$ and $\int_X g_{a,b}\mu = 0$, we conclude $g_{a,b} = h_{a,b}$. Thus, we have shown

$$g_{a,b} = \tilde{g}_{a,b} + \chi f_{a,b} - \int_X \chi f_{a,b}\mu.$$

The function

$$g_{a,b}^{(1)} = \tilde{g}_{a,b} + \chi f_{a,b}$$

is harmonic on $X - \{a, b\}$, and

$$g_{a,b}^{(2)} = \tilde{g}_{a,b} - \chi^c f_{a,b}$$

is harmonic on U_1; in particular, both functions are harmonic on the annulus $A := U_1 - \overline{U}_{r_2}$. Now for every harmonic function g on A, we have a bound

$$\max_{\partial U_{r_3}} g - \min_{\partial U_{r_3}} g \leq c_3 \|dg\|_A$$

with some positive constant c_3 depending only on r_2 and r_3; note that the circle ∂U_{r_3} is relatively compact in the annulus A. We bound $\|dg_{a,b}^{(2)}\|_A$ from above:

$$\|dg_{a,b}^{(2)}\|_A \leq \|d\tilde{g}_{a,b}\|_A + \|d(\chi^c f_{a,b})\|_A.$$

We estimate the first summand as follows, using that $\omega_{a,b} = d\tilde{g}_{a,b}$ solves the above variational problem:

$$\|d\tilde{g}_{a,b}\|_A \leq \|d\tilde{g}_{a,b}\| = \|\omega_{a,b}\| \leq \|\tilde{\omega}_{a,b}\| = \|\tilde{\omega}_{a,b}\|_A = \|d(\chi^c f_{a,b})\|_A;$$

we used that $\tilde{\omega}_{a,b}$ is supported in A. Thus we have

$$\|dg_{a,b}^{(2)}\| \leq 2\|d(\chi^c f_{a,b})\|_A,$$

which is bounded by a constant, uniformly in a and b in U_{r_1}.

This also allows us to estimate $g_{a,b}^{(1)}$: on A, we know $g_{a,b}^{(1)} = g_{a,b}^{(2)} + f_{a,b}$, hence

$$\|dg_{a,b}^{(1)}\|_A \leq \|dg_{a,b}^{(2)}\|_A + \|df_{a,b}\|_A \leq 2\|d(\chi^c f_{a,b})\|_A + \|df_{a,b}\|_A.$$

Both summands on the right-hand side are bounded by constants, depending only on r_1 and r_2, but uniformly in a and b in U_{r_1}. To summarize, we have shown that

$$\max_{\partial U_{r_3}} g_{a,b}^{(j)} - \min_{\partial U_{r_3}} g_{a,b}^{(j)}$$

$(j = 1, 2)$ are uniformly bounded by a constant depending only on r_1, r_2, and r_3. However, $g_{a,b}^{(1)}$ is harmonic on $X - U_{r_3}$, and $g_{a,b}^{(2)}$ is harmonic on $\overline{U}_{r_3}$, which both have the same boundary ∂U_{r_3}. Thus, by the maximum principle,

$$\max_{X-U_{r_3}} g_{a,b}^{(1)} - \min_{X-U_{r_3}} g_{a,b}^{(1)} = \max_{\partial U_{r_3}} g_{a,b}^{(1)} - \min_{\partial U_{r_3}} g_{a,b}^{(1)}$$

and

$$\max_{\overline{U}_{r_3}} g_{a,b}^{(2)} - \min_{\overline{U}_{r_3}} g_{a,b}^{(2)} = \max_{\partial U_{r_3}} g_{a,b}^{(2)} - \min_{\partial U_{r_3}} g_{a,b}^{(2)}.$$

Moreover, $\max_{X-U_{r_3}} |\chi f_{a,b}|$ and $\max_{\overline{U}_{r_3}} |\chi^c f_{a,b}|$ are bounded, uniformly in a and b in U_{r_1}, by a constant depending only on r_1 and r_3. Using $\tilde{g}_{a,b} = g_{a,b}^{(1)} - \chi f_{a,b}$ on $X - U_{r_3}$ and $\tilde{g}_{a,b} = g_{a,b}^{(2)} + \chi^c f_{a,b}$ on $\overline{U}_{r_3}$, we conclude that $\max_X \tilde{g}_{a,b} - \min_X \tilde{g}_{a,b}$ is bounded on $X = (X - U_{r_3}) \cup \overline{U}_{r_3}$ by a constant c_6 depending only on the radii r_1, r_2, and r_3. Using the normalization condition (10.7), we know that

$$\max_X \tilde{g}_{a,b} \geq 0 \geq \min_X \tilde{g}_{a,b}$$

holds; thus

$$\max_X |\tilde{g}_{a,b}| \leq \max_X \tilde{g}_{a,b} - \min_X \tilde{g}_{a,b}$$

is also bounded by the same constant.

From this we get a bound for

$$g_{a,b} - \chi f_{a,b} = \tilde{g}_{a,b} - \int_X \chi f_{a,b}\mu.$$

Indeed, we estimate

$$\left| \int_X \chi f_{a,b}\mu \right| \leq \int_{U_1} |f_{a,b}|\mu \leq c_1 \int_{U_1} |f_{a,b}\, dz \wedge d\bar{z}|,$$

which is uniformly bounded for $a, b \in U_{r_1}$ by a constant $c_1 c_5$ with c_5 depending only on r_1; note that the logarithmic singularities are integrable. Combining the bounds for $\max_X |\tilde{g}_{a,b}|$ and $\left| \int_X \chi f_{a,b}\mu \right|$, we conclude that $\sup_X |g_{a,b} - \chi f_{a,b}|$ is bounded by a constant $c_6 + c_1 c_5$ with c_6, c_5 depending on r_1. Since

$$\sup_X |\chi f_{a,b} - l_a + l_b| = \frac{1}{2\pi} \sup_{U_1} \left| \chi \log \left| \frac{\overline{z(a)}z - 1}{\overline{z(b)}z - 1} \right| \right|$$

is bounded, uniformly in $a, b \in U_{r_1}$ and X, the key lemma follows (with c_4 being the sum of c_6 and the uniform upper bound last mentioned).   □

Proof (of Theorem 10.1) Since we now work with varying coordinates, we include again the superscript coordinate index (j) in the coordinate $z^{(j)}$, its domain $U^{(j)}$, and also in $U_r^{(j)}$, $\chi^{(j)}$, and $l_a^{(j)}$.

10.8 Lemma *Consider two coordinates $z^{(j)}$ and $z^{(k)}$, where k and j are in $\{1, \ldots, n\}$. Assume that x is in $U_{r_1}^{(j)} \cap U_{r_1}^{(k)}$ and that y is in $U_{r_2}^{(j)}$ with $|z^{(j)}(y) - z^{(j)}(x)| < (r_2 - r_1)/M$. Then y is in $U_{r_2}^{(k)}$.*

Proof The intersection $U_{r_1}^{(j)} \cap U_{r_1}^{(k)}$ is an open neighborhood of x. Assume that there exists $y \in \overline{U}_{r_2}^{(j)}$ with $|z^{(j)}(y) - z^{(j)}(x)| < (r_2 - r_1)/M$ and $y \notin U_{r_2}^{(k)}$. Then there is also such a point y with minimal distance $|z^{(j)}(y) - z^{(j)}(x)|$ from x, since $\overline{U}_{r_2}^{(j)} - U_{r_2}^{(k)}$ is compact. For this point y, we conclude $y \in \partial U_{r_2}^{(k)} \subseteq \overline{U}_{r_2}^{(k)}$, and the straight line from x to y in the $z^{(j)}$-coordinate is contained in $\overline{U}_{r_2}^{(j)} \cap \overline{U}_{r_2}^{(k)}$. By the mean value theorem, we conclude $|z^{(k)}(y) - z^{(k)}(x)| \leq M|z^{(j)}(y) - z^{(j)}(x)| < r_2 - r_1$, hence $|z^{(k)}(y)| < r_2$, since $|z^{(k)}(x)| \leq r_1$. This contradicts $y \in \partial U_{r_2}^{(k)}$.   □

We choose a smooth partition of unity $\phi^{(j)} : X \to [0, 1]$, $j = 1, \ldots, n$, such that $\phi^{(j)}$ is supported in $U_{r_1}^{(j)}$. For $a \in X$, we set

$$h_a := \sum_j \phi^{(j)}(a) l_a^{(j)}.$$

10.9 Lemma Let $a \in U_{r_1}^{(k)}$, $y \in X$, $y \neq a$. Then we have

$$l_a^{(k)}(y) \leq \frac{\log 2}{2\pi}.$$

Proof This follows immediately from the definition of the function $l_a^{(k)}$, since $|z^{(k)}(y) - z^{(k)}(a)| \leq 2$ whenever $y \in \mathrm{Supp}(\chi^{(k)})$. $\qquad\square$

10.10 Lemma For all $a, b \in X$ we have the inequality

$$\sup_X |g_{a,b} - h_a + h_b| \leq n(c_{10} + c_1 c_5 + c_7 \log M)$$

with constants c_{10}, c_5, and c_7 depending only on r_1.

Proof We first show for $a \in U_{r_1}^{(k)} \cap U_{r_1}^{(j)}$ that

$$\sup_X |l_a^{(k)} - l_a^{(j)}| \leq \frac{1}{2\pi}[\log M + |\log(r_2 - r_1)| + \log 2].$$

To prove this, let $y \in X$. We distinguish 3 cases in order to prove that $l_a^{(k)}(y) - l_a^{(j)}(y)$ is bounded from above by the right-hand side.

Case 1:

$y \in U_1^{(j)}$ with $|z^{(j)}(y) - z^{(j)}(a)| < (r_2 - r_1)/M$. In particular, we have $|z^{(j)}(y)| < |z^{(j)}(a)| + (r_2 - r_1)/M \leq r_2$ (recall that $M \geq 1$), hence $a, y \in U_{r_2}^{(j)}$. Consequently, the straight line $[a, y]^{(j)}$ from a to y in the $z^{(j)}$-coordinate is contained in $U_{r_2}^{(j)}$. Then Lemma 10.8 implies that $[a, y]^{(j)} \subseteq U_{r_2}^{(k)}$. Using $\chi^{(j)}(y) = \chi^{(k)}(y) = 1$, we conclude by the mean value theorem that

$$l_a^{(k)}(y) - l_a^{(j)}(y) = \frac{1}{2\pi} \log \left| \frac{z^{(k)}(y) - z^{(k)}(a)}{z^{(j)}(y) - z^{(j)}(a)} \right| \leq \frac{\log M}{2\pi},$$

which is bounded by the right-hand side.

Case 2:

$y \notin U_1^{(j)}$. Then $l_a^{(j)}(y) = 0$, and we conclude, using Lemma 10.9, that

$$l_a^{(k)}(y) - l_a^{(j)}(y) = l_a^{(k)}(y) \le \frac{\log 2}{2\pi}.$$

Case 3:

$y \in U_1^{(j)}$ and $|z^{(j)}(y) - z^{(j)}(a)| \ge (r_2 - r_1)/M$; thus

$$l_a^{(k)}(y) - l_a^{(j)}(y) \le \frac{\log 2}{2\pi} - l_a^{(j)}(y) \le \frac{1}{2\pi}(\log 2 - \chi^{(j)}(y)\log[(r_2 - r_1)/M]),$$

which is also bounded by the right-hand side.

The upper bound for $l_a^{(j)}(y) - l_a^{(k)}(y)$ in our claim is obtained by exchanging j and k. Thus the claim is proved.

We conclude:

(10.11)
$$\begin{aligned}
|h_a - l_a^{(j)}| &\le \sum_k \phi^{(k)}(a)|l_a^{(k)} - l_a^{(j)}| \\
&\le \frac{1}{2\pi}(\log M + |\log(r_2 - r_1)| + \log 2).
\end{aligned}$$

Combining this with Lemma 10.4, we conclude for $a, b \in U_{r_1}^{(j)}$,

$$\begin{aligned}
|g_{a,b} - h_a + h_b| &\le |g_{a,b} - l_a^{(j)} + l_b^{(j)}| + |h_a - l_a^{(j)}| + |h_b - l_b^{(j)}| \\
&\le c_{10} + c_1 c_5 + c_7 \log M,
\end{aligned}$$

with some constants c_{10}, c_5, c_7 depending only on r_1 (a possible choice is $c_7 = (\log M)/\pi$ and $c_{10} = (|\log(r_2 - r_1)| + \log 2)/\pi + c_4$).

Finally, for general $a, b \in X$, we choose a finite sequence of points $a = a_0, a_1, \ldots, a_m = b$ in X and indices $j_1, \ldots, j_m$ with $m \le n$ and $a_{i-1}, a_i \in U_{r_1}^{(j_i)}$ for all $i = 1, \ldots, m$. Using

$$g_{a,b} = \sum_{i=1}^m g_{a_{i-1}, a_i},$$

we get by estimating

$$|g_{a,b} - h_a + h_b| \le \sum_{i=1}^m |g_{a_{i-1}, a_i} - h_{a_{i-1}} + h_{a_i}| \le n(c_{10} + c_1 c_5 + c_7 \log M),$$

the claim of the lemma. $\square$

We define

$$h_\mu(x) := \int_{b \in X} h_b(x)\, \mu(b) \qquad (x \in X).$$

10.12 Lemma *We have*

$$\sup_X |h_\mu| \le nc_1c_8,$$

with some universal constant c_8. Furthermore, we have

$$\sup_{\substack{b,x \in X \\ b \ne x}} h_b(x) \le \frac{\log 2}{2\pi}.$$

Proof We observe first that for all $w \in \mathbb{C}$ with $|w| \le 1$, the integral

$$\frac{1}{2\pi} \int_{|z| \le 1} |\log|z - w||\, |dz \wedge d\bar{z}|$$

is bounded from above by a universal constant c_8. We conclude that for all $x \in X$ we have

$$\int_{b \in U_{r_1}^{(j)}} |l_b^{(j)}(x)||\phi^{(j)}(b)\, \mu(b) \le c_1 \int_{b \in U_{r_1}^{(j)}} |l_b^{(j)}(x)|\, |dz^{(j)} \wedge \overline{dz^{(j)}}|$$

$$\le c_1c_8.$$

Let $x \in X$. We get the first estimate:

$$|h_\mu(x)| \le \sum_{j=1}^{n} \int_{U_1^{(j)}} |l_b^{(j)}(x)||\phi^{(j)}(b)\, \mu(b) \le nc_1c_8.$$

Finally, the second estimate follows from Lemma 10.9:

$$h_b = \sum_{j=1}^{n} \phi^{(j)}l_b^{(j)} \le \frac{\log 2}{2\pi},$$

as required. $\square$

10.13 Proposition *For some positive constants c_{10}, c_7, and c_{11} that depend only on r_1 we have, uniformly in a and $x \ne a$ on X,*

$$|g_{a,\mu}(x) - h_a(x)| \le n(c_{10} + c_1c_{11} + c_7 \log M).$$

Proof Indeed, averaging Lemma 10.10 over b with respect to μ, we obtain

$$\sup_X |g_{a,\mu} - h_a + h_\mu| \leq n(c_{10} + c_1 c_5 + c_7 \log M).$$

By Lemma 10.12, one has $|h_\mu| \leq n c_1 c_8$. Combining gives what we want (we can take $c_{11} = c_5 + c_8$). □

10.14 Proposition *Let c_{10}, c_7, and c_{11} be as in Proposition 10.13, and let a be in X. Then $\lim_{x \to a} |g_{a,\mu}(x) - h_a(x)|$ exists, and we have*

$$\lim_{x \to a} |g_{a,\mu}(x) - h_a(x)| \leq n(c_{10} + c_1 c_{11} + c_7 \log M).$$

Proof The functions $g_{a,\mu}$ and h_a have the same logarithmic singularity at a; hence the limit exists. The estimate is then a consequence of Proposition 10.13. □

We can now finish the proof of Theorem 10.1. We have seen in (10.11) that:

$$|h_a - l_a^{(j)}| \leq \frac{1}{2\pi} (\log M + |\log(r_2 - r_1)| + \log 2).$$

Combining this with Proposition 10.14 and using the definition of $l_a^{(j)}$ gives the second estimate of the theorem. As to the first estimate, using

$$g_{a,\mu} \leq h_a + |g_{a,\mu} - h_a|$$

we obtain it by applying the upper bound for h_a in Lemma 10.12 and the upper bound for $|g_{a,\mu} - h_a|$ in Proposition 10.13. This ends the proof of Theorem 10.1. □

Chapter Eleven

Bounds for Arakelov invariants of modular curves

by Bas Edixhoven and Robin de Jong

In this chapter, we give bounds for all quantities on the right-hand side in the inequality in Theorems 9.1.1 and 9.2.5, in the context of the modular curves $X_1(5l)$ with $l > 5$ prime, using the upper bounds for Green functions from the previous chapter. The final estimates are given in the last section.

11.1 BOUNDING THE HEIGHT OF $X_1(pl)$

As before, for $l > 5$ prime, we let X_l be the modular curve $X_1(5l)$, over a suitable base that will be clear from the notation. We let g_l denote the genus of X_l; we have $g_l > 1$. A model $X_{l,\mathbb{Z}}$ is given by [Ka-Ma], as well as a model $X_{l,\mathbb{Z}[\zeta_{5l}]}$ that is semistable; see Chapter 8. The aim of this section is to prove a suitable bound for the stable Faltings height of X_l; see (4.4.10). We will in fact give such a bound for the modular curves $X_1(pl)$ with p and l distinct primes. Before we get to that, we prove some intermediate results that will also be important in the next section.

11.1.1 Lemma *Let $N \geq 1$ be an integer, and let*

$$\mathcal{B}_2(N) := \coprod_{M|N} \coprod_{d|(N/M)} B_{N,M,d}^* \, S_2(\Gamma_1(M))^{\text{new}}$$

be the basis of $S_2(\Gamma_1(N))$ obtained from newforms of levels dividing N as explained in (2.2.18). Let $f = \sum_{n\geq 1} a_n(f)q^n$ be an element of $\mathcal{B}_2(N)$. Then we have for all $n \geq 1$

$$|a_n(f)| \leq 2n.$$

Proof As $a_n(B^*_{N,M,d} f) = a_{n/d}(f)$ (see 2.2.17), it suffices to treat the case that f is a newform of some level M dividing N. We use the Weil bounds on the $a_p(f)$ for all primes p. We recall from Section 1.8 of [De-Se] that we have an equality of formal Dirichlet series:

$$\sum_{n\geq 1} a_n(f)n^{-s} = \prod_{p|M}(1-a_p(f)p^{-s})^{-1} \prod_{p\nmid M}(1-\alpha_p p^{-s})^{-1}(1-\beta_p p^{-s})^{-1}$$

with the following properties. For $p \nmid M$ we have $|\alpha_p| = |\beta_p| = \sqrt{p}$. For $p|M$ we have

$$\begin{cases} a_p(f) = 0 & \text{if } p^2|M, \\ a_p(f) = 0 & \text{if } \varepsilon_f \text{ factors through } (\mathbb{Z}/(M/p)\mathbb{Z})^\times, \\ |a_p(f)| = p^{1/2} & \text{if } \varepsilon_f \text{ does not factor through } (\mathbb{Z}/(M/p)\mathbb{Z})^\times, \\ |a_p(f)| = 1 & \text{if } p^2|M \text{ and } \varepsilon_f \text{ factors through } (\mathbb{Z}/(M/p)\mathbb{Z})^\times. \end{cases}$$

Using that

$$(1 - a_p(f)p^{-s})^{-1} = \sum_{k\geq 0} a_p(f)^k p^{-sk},$$

and that

$$(1 - \alpha_p p^{-s})^{-1}(1 - \beta_p p^{-s})^{-1} = \left(\sum_{k\geq 0} \alpha_p^k p^{-sk}\right)\left(\sum_{k\geq 0} \beta_p^k p^{-sk}\right)$$

we find that for arbitrary n we have $|a_n(f)| \leq \sigma_{0,M}(n)\sqrt{n}$, where $\sigma_{0,M}(n)$ is the number of positive divisors of n that are prime to M, and a simple estimate leads to $|a_n(f)| \leq 2n$. $\square$

The following lemma states a very well-known lower bound for the Petersson norm of a normalized cusp form.

11.1.2 Lemma Let $N \geq 1$ and let $\omega = f dq/q$ be the holomorphic 1-form on $X_1(N)(\mathbb{C})$ attached to a cusp form $f = \sum_n a_n(f)q^n$ in $S_2(\Gamma_1(N))$ with $a_1(f) = 1$. Then we have

$$\|\omega\|^2 = \frac{i}{2}\int_{X_1(N)} \omega \wedge \overline{\omega} \geq \pi e^{-4\pi}.$$

Proof We have $\omega = \sum_{n \geq 1} a_n(f) q^n \, dq/q$ in the coordinate $q = e^{2\pi i z}$, where z is the standard coordinate on the upper half-plane $\mathbb{H}$. If we let x and y be the real and imaginary parts of z we have

$$\frac{i}{2} \omega \wedge \overline{\omega} = 4\pi^2 |f|^2 \, dx \, dy \, .$$

Let F be the region in $\mathbb{H}$ given by the conditions $|x| < 1/2$ and $y > 1$. Then

$$\|\omega\|^2 \geq \int_F 4\pi^2 |f(z)|^2 dx dy$$

$$= 4\pi^2 \sum_{m,n \geq 1} a_m(f) \overline{a_n(f)} \int_{-1/2}^{1/2} e^{2\pi i (m-n)x} \int_1^{\infty} e^{-2\pi (m+n)y} dy$$

$$= 4\pi^2 \sum_{n \geq 1} |a_n(f)|^2 e^{-4\pi n}/4\pi n \, .$$

From the first term (note that $a_1(f) = 1$) we obtain $\|\omega\|^2 \geq \pi e^{-4\pi}$. $\quad\square$

We now specialize to a slightly less special case than our curves X_l: the curves $X_1(pl)$ with p and l two distinct prime numbers. We call an Atkin-Lehner basis for $\Omega^1(X_1(pl))$ any basis of $\Omega^1(X_1(pl))$ given by an ordering of the set $B_2(pl)$. We start by describing, in a notation slightly different from the one used in (2.2.15), the degeneracy maps that are used for the definition of $B_2(pl)$. This time, we call them source and target maps:

$$\begin{cases} s_l : X_1(pl) \to X_1(p), & (E, P, L) \mapsto (E, P), \\ t_l : X_1(pl) \to X_1(p), & (E, P, L) \mapsto (E/\langle L \rangle, P), \\ s_p : X_1(pl) \to X_1(l), & (E, P, L) \mapsto (E, L), \\ t_p : X_1(pl) \to X_1(l), & (E, P, L) \mapsto (E/\langle P \rangle, L), \end{cases}$$

where (E, P, L) denotes an elliptic curve E with a point P of order p and a point L of order l. Note that s_l and t_l have degree $l^2 - 1$, and that s_p and t_p have degree $p^2 - 1$. For any integer $M \geq 1$ we denote by $\Omega^1(X_1(M))^{\text{new}}$ the set of holomorphic 1-forms in $\Omega^1(X_1(M))$ of the form $f dq/q$ with f in $S_2(X_1(M))^{\text{new}}$. Our next goal is to get information on the Gram matrix of an Atkin-Lehner basis of $\Omega^1(X_1(pl))$. As described above, the contribution to $\Omega^1(X_1(pl))$ of each f in $S_2(\Gamma_1(pl))^{\text{new}}$ is the subspace $\mathbb{C} f dq/q$. The contribution of an f in $S_2(\Gamma_1(p))^{\text{new}}$ is the two-dimensional space generated by $s_l^* f dq/q$ and $t_l^* f dq/q$, and, of course, each f in $S_2(\Gamma_1(l))^{\text{new}}$ contributes the two-dimensional space generated by $s_p^* f dq/q$ and $t_p^* f dq/q$.

11.1.3 Lemma *For f in $S_2(\Gamma_1(l))^{\text{new}}$ and $\omega = f dq/q$ we have*

$$\langle s_p^* \omega, s_p^* \omega \rangle = (p^2 - 1)\|\omega\|^2 ,$$
$$\langle t_p^* \omega, t_p^* \omega \rangle = (p^2 - 1)\|\omega\|^2 ,$$
$$\langle s_p^* \omega, t_p^* \omega \rangle = (p - 1)\overline{a_p(f)}\|\omega\|^2 .$$

We have similar equalities with p and l switched.

Proof The first two equalities are clear. As to the latter, note first that

$$\langle s_p^* \omega, t_p^* \omega \rangle = \frac{i}{2} \int_{X_1(pl)} s_p^* \omega \wedge \overline{t_p^* \omega} = \frac{i}{2} \int_{X_1(l)} s_{p,*}(s_p^* \omega \wedge \overline{t_p^* \omega})$$
$$= \frac{i}{2} \int_{X_1(l)} \omega \wedge \overline{s_{p,*} t_p^* \omega} .$$

Next note that $s_p \colon X_1(pl) \to X_1(l)$ and $t_p \colon X_1(pl) \to X_1(l)$ factor through the forgetful map $X_1(pl) \to X_1(l; p)$, where the latter curve corresponds to the moduli problem (E, P, G) with P of order l and G a subgroup of order p. This forgetful map has degree $p-1$, and the correspondence on $X_1(l)$ induced by $X_1(l; p)$ is the standard Hecke correspondence T_p. We find that $s_{p,*} t_p^* \omega = (p-1) T_p^* \omega$. By the standard relation between eigenvalues and q-coefficients we have $T_p^* \omega = a_p(f)\omega$, so finally

$$\langle s_p^* \omega, t_p^* \omega \rangle = \frac{i}{2} \int_{X_1(l)} \omega \wedge \overline{(p - 1) T_p^* \omega} = (p - 1)\overline{a_p(f)}\|\omega\|^2$$

as required.   □

11.1.4 Corollary *Let p and l be two distinct primes. The structure of the Gram matrix $(\langle \omega_i, \omega_j \rangle)_{i,j}$ of holomorphic 1-forms attached to an Atkin-Lehner basis for $\Omega^1(X_1(pl))$ is as follows. Subspaces attached to distinct elements of the union of $S_2(\Gamma_1(pl))^{\text{new}}$, $S_2(\Gamma_1(l))^{\text{new}}$, and $S_2(\Gamma_1(p))^{\text{new}}$ are orthogonal to each other, hence the Gram matrix decomposes into blocks corresponding to these subspaces. The contribution of an element f in*

$S_2(\Gamma_1(pl))^{\text{new}}$ is the 1-by-1 block $\|fdq/q\|^2$. The contribution of an element f in $S_2(\Gamma_1(l))^{\text{new}}$ is the 2-by-2 block

$$(p-1)\|fdq/q\|^2 \begin{pmatrix} p+1 & \overline{a_p(f)} \\ a_p(f) & p+1 \end{pmatrix},$$

where the norm $\|fdq/q\|^2$ is taken on $X_1(l)$). The contribution of an element f in $S_2(\Gamma_1(p))^{\text{new}}$ is the 2-by-2 block

$$(l-1)\|fdq/q\|^2 \begin{pmatrix} l+1 & \overline{a_l(f)} \\ a_l(f) & l+1 \end{pmatrix},$$

where the norm $\|fdq/q\|^2$ is taken on $X_1(l)$.

11.1.5 Corollary *The determinant of the Gram matrix of the holomorphic 1-forms attached to an Atkin-Lehner basis for $\Omega^1(X_1(pl))$ is bounded below by $(\pi e^{-4\pi})^g$.*

Proof By the Weil-Ramanujan-Deligne bounds (or, in this case, the Weil bounds, as the weight of the modular forms here is two), the determinant of a 2-by-2 block as in Corollary 11.1.4 is bounded below by $\|fdq/q\|^4$. We obtain our corollary by invoking Lemma 11.1.2. $\qquad\square$

11.1.6 Corollary *The Arakelov $(1,1)$-form μ on $X_1(pl)$ is given by:*

$$\mu = \frac{i}{2g} \sum_\omega \frac{\omega \wedge \overline{\omega}}{\|\omega\|^2} + \frac{i}{2g} \sum_\omega \left(\frac{(p+1)s_p^*(\omega \wedge \overline{\omega}) + (p+1)t_p^*(\omega \wedge \overline{\omega})}{(p-1)\|\omega\|^2((p+1)^2 - |a_p(f_\omega)|^2)} \right.$$
$$\left. - \frac{a_p(f_\omega)s_p^*\omega \wedge \overline{t_p^*\omega} + \overline{a_p(f_\omega)}t_p^*\omega \wedge \overline{s_p^*\omega}}{(p-1)\|\omega\|^2((p+1)^2 - |a_p(f_\omega)|^2)} \right)$$
$$+ \frac{i}{2g} \sum_\omega \left(\frac{(l+1)s_l^*\omega \wedge \overline{s_l^*\omega} + (l+1)t_l^*\omega \wedge \overline{t_l^*\omega}}{(l-1)\|\omega\|^2((l+1)^2 - |a_l(f_\omega)|^2)} \right.$$
$$\left. - \frac{a_l(f_\omega)s_l^*\omega \wedge \overline{t_l^*\omega} + \overline{a_l(f_\omega)}t_l^*\omega \wedge \overline{s_l^*\omega}}{(l-1)\|\omega\|^2((l+1)^2 - |a_l(f_\omega)|^2)} \right),$$

with the first sum running over $\Omega^1(X_1(pl))^{\text{new}}$, the second sum running over $\Omega^1(X_1(l))^{\text{new}}$, the third sum running over $\Omega^1(X_1(p))^{\text{new}}$, and f_ω defined by $\omega = f_\omega dq/q$.

Proof Consider first an arbitrary compact Riemann surface, X, say, and let $\omega = (\omega_1, \ldots, \omega_g)$ be an arbitrary basis of $\Omega^1(X)$. Let $a := \langle \omega, \omega \rangle$ be the g-by-g matrix given by $a_{i,j} = \langle \omega_i, \omega_j \rangle$. Note that $\bar{a} = a^t$. Let $b = a^{t,-1}$, the inverse of the transpose of a. Then we claim that the Arakelov $(1, 1)$-form on X can be written as:

$$\mu = \frac{i}{2g} \sum_{i,j} b_{i,j} \omega_i \wedge \overline{\omega_j}.$$

To see this, note that for ω an orthonormal basis this is the correct expression, and that changing to $\omega' = \omega \cdot g$ with any invertible g does not change μ, as one may directly calculate.

In our case, the basis $(\omega_1, \ldots, \omega_g)$ that we take is an Atkin-Lehner basis. Using Corollary 11.1.4 one obtains the expression we gave. $\square$

We remark that Abbes and Ullmo in [Ab-Ul] have determined the Arakelov $(1, 1)$-form on $X_0(n)$ for all square-free $n \geq 1$ such that $X_0(n)$ has genus at least one. It should not be hard to generalize their result to $X_1(n)$ for square-free n.

Now we arrive at the main result of this section. We recall that the Faltings height of a curve, and its stable or absolute version, have been briefly described in (4.4.11).

11.1.7 Theorem *For the stable Faltings height of $X_1(pl)$, for distinct prime numbers p and l, one has:*

$$h_{\mathrm{abs}}(X_1(pl)) = O((pl)^2 \log(pl)).$$

Proof This proof is an adaptation of an argument in Section 5 of [Co-Ed], where the case $X_0(p)$ with p prime was treated. We may and do assume that $X_1(pl)$ has genus at least one.

We start with a general observation. For X_K a curve over a number field, and $K \to L$ a finite extension, we claim that

$$[L : \mathbb{Q}]^{-1} h_L(X_L) \leq [K : \mathbb{Q}]^{-1} h_K(X_K).$$

This inequality simply results from the fact that for the Néron models of the Jacobians the identity morphism on the generic fibers extends to a morphism:

$$(J_{O_K})_{O_L} \longrightarrow J_{O_L}.$$

For n a positive integer, we let $X_\mu(n)_\mathbb{Q}$ denote the modular curve corresponding to elliptic curves with an embedding of μ_n. The reason for considering this variant of $X_1(n)$ is that the cusp ∞ of $X_\mu(n)$ is $\mathbb{Q}$-rational. Of course, $X_1(n)_\mathbb{Q}$ and $X_\mu(n)_\mathbb{Q}$ become isomorphic over $\mathbb{Q}(\zeta_n)$, and therefore we have, for all n

$$h_{\mathrm{abs}}(X_1(n)_\mathbb{Q}) = h_{\mathrm{abs}}(X_\mu(n)_\mathbb{Q}).$$

For more details about these curves $X_\mu(n)$ we refer to Sections 9.3 and 12.3 of [Di-Im].

The general observation above gives

$$h_{\mathrm{abs}}(X_\mu(n)_\mathbb{Q}) \le h_\mathbb{Q}(X_\mu(n)_\mathbb{Q}).$$

Because of this, it suffices to establish the bound of the theorem for the $h_\mathbb{Q}(X_\mu(pl)_\mathbb{Q})$.

Let p and l be given, and let X be the model over $\mathbb{Z}$ of $X_\mu(pl)\mathbb{Q}$ obtained by normalization of the j-line $\mathbb{P}^1_\mathbb{Z}$ in the function field of $X_\mu(pl)\mathbb{Q}$. As X is proper over $\mathbb{Z}$, the $\mathbb{Q}$-rational point ∞ extends to an element ∞ in $X(\mathbb{Z})$, which is known to lie in the open part X^{sm} of X where the structure morphism to $\mathrm{Spec}(\mathbb{Z})$ is smooth; see [Di-Im]. In terms of the Tate curve over $\mathbb{Z}((q)) = \mathbb{Z}((1/j))$, the cusp ∞ is the immersion of μ_n, over $\mathbb{Z}$, in the n-torsion of the Tate curve (see Sections 8.6–8.11 of [Ka-Ma]).

We let J be the Néron model over $\mathbb{Z}$ of the Jacobian of the curve $X_\mathbb{Q}$. Then, by the defining property, the embedding of $X_\mathbb{Q}$ into $J_\mathbb{Q}$ that sends ∞ to 0 extends to a morphism from X^{sm} to J. This morphism induces via pullback of differential forms a morphism from $\mathrm{Cot}_0(J)$ to $S(\mathbb{Z})$, the sub-$\mathbb{Z}$-module of $\Omega^1(X_\mathbb{Q})$ of forms whose q-expansion at ∞ has coefficients in $\mathbb{Z}$ (see around (2.4.9)). As $\mathrm{Cot}_0(J)$ and $S(\mathbb{Z})$ are both $\mathbb{Z}$-structures on $\Omega^1(X(\mathbb{C}))$, we have (see around (4.4.10)):

$$h_\mathbb{Q}(X_\mathbb{Q}) = \deg\left(\bigwedge^g 0^* \mathrm{Cot}_0(J) \right)$$

$$= -\log \mathrm{Vol}((\mathbb{R} \otimes \mathrm{Cot}_0(J))/\mathrm{Cot}_0(J))$$

$$\le -\log \mathrm{Vol}(\mathbb{R} \otimes S(\mathbb{Z})/S(\mathbb{Z})),$$

where the volume form on $\mathbb{R} \otimes \mathrm{Cot}_0(J)$ comes from integration over $J(\mathbb{C})$, and that on $\mathbb{R} \otimes S(\mathbb{Z})$ from integration over $X(\mathbb{C})$.

Let $\mathbb{T} \subset \mathrm{End}(J)$ be the Hecke algebra generated by all T_i, $i \geq 1$, and the $\langle a \rangle$, a in $(\mathbb{Z}/pl\mathbb{Z})^\times$. We have a perfect pairing (see (2.4.9)):

$$\mathbb{T} \times S(\mathbb{Z}) \to \mathbb{Z}, \quad (t, \omega) \mapsto a_1(t\omega).$$

Using the duality we can write:

$$-\log \mathrm{Vol}(\mathbb{R} \otimes S(\mathbb{Z})/S(\mathbb{Z})) = \log \mathrm{Vol}(\mathbb{R} \otimes \mathbb{T}/\mathbb{T})$$

where the volume form on $\mathbb{R} \otimes \mathbb{T}$ is dual to the one on $\mathbb{R} \otimes S(\mathbb{Z})$. Now consider an Atkin-Lehner basis $(\omega_1, \ldots, \omega_g)$ of $\Omega^1(X)_\mathbb{C}$. Let Vol' denote the volume with respect to the volume form on $\mathbb{R} \otimes \mathbb{T}$ induced by the one on $\mathbb{C} \otimes S(\mathbb{Z})$ for which the basis $(\omega_1, \ldots, \omega_g)$ is an orthonormal basis. Then we have

$$\log \mathrm{Vol}(\mathbb{R} \otimes \mathbb{T}/\mathbb{T}) = \log \mathrm{Vol}'(\mathbb{R} \otimes \mathbb{T}/\mathbb{T}) - \frac{1}{2} \log \det(\langle \omega, \omega \rangle),$$

where $\langle \omega, \omega \rangle$ is the matrix whose (i, j)-coefficient is $\langle \omega_i, \omega_j \rangle$. By Corollary 11.1.5 we have:

$$-\log \det(\langle \omega, \omega \rangle) \leq g(4\pi - \log \pi) = O((pl)^2).$$

It remains to bound $\log \mathrm{Vol}'(\mathbb{R} \otimes \mathbb{T}/\mathbb{T})$. Let Γ be the set of integers $i \geq 1$ such that there exists an ω in $\Omega^1(X(\mathbb{C}))$ with a zero of exact order $i - 1$ at ∞. Then Γ is the set of integers $i \geq 1$ such that $h^0(X(\mathbb{C}), \Omega^1(-i\infty))$ is strictly less (and hence exactly one less) than $h^0(X(\mathbb{C}), \Omega^1((-i + 1)\infty))$. As $h^0(X(\mathbb{C}), \Omega^1) = g$, and $h^0(X(\mathbb{C}), \Omega^1(-(2g - 1)\infty)) = 0$, there are exactly g such integers, and we can write $\Gamma = \{i_1, \ldots, i_g\}$ with

$$1 = i_1 < i_2 < \ldots < i_g \leq 2g - 1.$$

Under the pairing between $\mathbb{T}$ and $S(\mathbb{Z})$, each Hecke operator T_i is sent to the element $\omega \mapsto a_i(\omega)$ of the dual of $S(\mathbb{Z})$, where the $a_i(\omega)$ are given by the q-expansion

$$\omega = \sum_{i \geq 1} a_i(\omega) q^i \, (dq/q) = \sum_{i \geq 1} a_i(\omega) q^{i-1} \, dq.$$

It follows that the elements $T_{i_1}, \ldots, T_{i_g}$ of the free $\mathbb{Z}$-submodule $\mathbb{T}$ are linearly independent. Hence $\mathbb{T}'$, the submodule of $\mathbb{T}$ generated by these T_{i_j}, has finite index. We thus find

$$\log \mathrm{Vol}(\mathbb{R} \otimes \mathbb{T}/\mathbb{T}) \leq \log \mathrm{Vol}'(\mathbb{R} \otimes \mathbb{T}'/\mathbb{T}') - \frac{1}{2} \log \det(\langle \omega, \omega \rangle) .$$

Now we have $g = r_1 + 2r_2$, where r_1 is the number of elements of our basis $(\omega_1, \ldots, \omega_g)$ of $\mathbb{C} \otimes \mathbb{T}^\vee$ that are fixed by the complex conjugation. We let

$$\phi : \mathbb{R} \otimes \mathbb{T} \longrightarrow \mathbb{R}^{r_1} \times \mathbb{C}^{r_2} \times \mathbb{C}^{r_2} \longrightarrow \mathbb{R}^{r_1} \times \mathbb{C}^{r_2}$$

be the map obtained from our basis (each ω_i gives $t \mapsto a_1(t\omega_i)$), composed with the projection. We view $\mathbb{R}^{r_1} \times \mathbb{C}^{r_2}$ as $\mathbb{R}^g$ by decomposing each factor $\mathbb{C}$ as $\mathbb{R} \oplus \mathbb{R}i$. Then we have

$$\mathrm{Vol}'(\mathbb{R} \otimes \mathbb{T}/\mathbb{T}') = 2^{r_2} |\det(\phi(T_{i_1}), \ldots, \phi(T_{i_g}))| .$$

By construction, each $\phi(T_{i_j})_k$ is the real or imaginary part of some $a_{i_j}(\omega_l)$. Hence by Lemma 11.1.1 we have

$$|\phi(T_{i_j})_k| \leq 2i_j .$$

We obtain

$$|\det(\phi(T_{i_1}), \ldots, \phi(T_{i_g}))| \leq \prod_{j=1}^{g} (2i_j \sqrt{g}) \leq (4g^2)^g .$$

Hence, finally,

$$\log \mathrm{Vol}'(\mathbb{R} \otimes \mathbb{T}/\mathbb{T}) \leq r_2 \log 2 + g(\log 4 + 2 \log g) .$$

Noting that $r_2 \leq g/2$ and that $g = O((pl)^2)$ completes our proof. $\square$

11.2 BOUNDING THE THETA FUNCTION ON $\mathrm{Pic}^{g-1}(X_1(pl))$

The aim of this section is to give a bound for the supnorm of the theta function that occurs in Theorem 9.2.5.

11.2.1 Theorem *For* $X = X_1(pl)$, *with* p *and* l *distinct primes for which the genus of* $X_1(pl)$ *is at least one, we have* $\log \|\vartheta\|_{\sup} = O((pl)^6)$.

We start with two lemmas, which are possibly of independent interest.

11.2.2 Lemma *Let $X = V/\Lambda$ be a principally polarized complex Abelian variety and let $H: V \times V \to \mathbb{C}$ be its Riemann form. Let $\lambda_1, \ldots, \lambda_{2g}$ be the successive minima of the lattice Λ, with norm defined by $\|x\|^2 = H(x, x)$. Let $(e_1, \ldots, e_{2g})$ be a symplectic basis of Λ, that is, a basis with respect to which the matrix of the symplectic form $\Im(H)$ is, in g by g block form, equal to $\left(\begin{smallmatrix} 0 & 1 \\ -1 & 0 \end{smallmatrix}\right)$. Then one has*

$$\sqrt{\det \Im(\tau)} \leq \frac{(2g)!}{2^g} \frac{V_{2g}}{V_g} \lambda_{g+1} \cdots \lambda_{2g},$$

where τ is the period matrix in $\mathbb{H}_g$ corresponding to $(e_1, \ldots, e_{2g})$. Here V_n denotes the volume of the unit ball in $\mathbb{R}^n$ with its standard Euclidean inner product.

Proof We consider the lattice $M = \mathbb{Z} \cdot e_1 \oplus \cdots \oplus \mathbb{Z} \cdot e_g$ in the real subvector space $W = \mathbb{R} \cdot e_1 \oplus \cdots \oplus \mathbb{R} \cdot e_g$ of V. Denote by $\mu_1, \ldots, \mu_g$ the successive minima of M, where the norm is given by restricting H to W (note that M is isotropic for the symplectic form, so that H takes real values on W). We have $(H(e_i, e_j))_{i,j} = (\Im(\tau))^{-1}$, so that the volume (with respect to the inner product on W given by H) of W/M is equal to $(\det \Im(\tau))^{-1/2}$, and hence by Minkowski's second fundamental inequality,

$$\mu_1 \cdots \mu_g \leq 2^g \frac{(\det \Im(\tau))^{-1/2}}{V_g}.$$

On the other hand we have

$$\mu_1 \cdots \mu_g \geq \lambda_1 \cdots \lambda_g = (\lambda_1 \cdots \lambda_{2g}) \cdot (\lambda_{g+1} \cdots \lambda_{2g})^{-1},$$

and since the volume of V/Λ is 1 we obtain by Minkowski's first fundamental inequality:

$$\lambda_1 \cdots \lambda_{2g} \geq \frac{2^{2g}}{(2g)!} \frac{1}{V_{2g}}.$$

Combining we find a lower bound:

$$\mu_1 \cdots \mu_g \geq \frac{2^{2g}}{(2g)!} \frac{1}{V_{2g}} \cdot (\lambda_{g+1} \cdots \lambda_{2g})^{-1}.$$

Combining this with the upper bound for $\mu_1 \cdots \mu_g$ we obtain the required formula. $\qquad\qquad\qquad\qquad\qquad\qquad\qquad\qquad\qquad\qquad\qquad\qquad$ $\square$

11.2.3 Lemma *Let $N \geq 3$ be an integer. The group $\Gamma_1(N)$ is generated by its elements whose entries are bounded from above in absolute value by $N^6/4$.*

Proof We first note the following: let G be a group, and let $S \subset G$ be a set of generators. Let X be a transitive G-set and let x be in X. For each y in X, let g_y be an element of G such that $g_y x = y$; we demand that $g_x = 1$. Then the $g_{sy}^{-1} s g_y$, for s in S and y in X, form a system of generators for the stabilizer G_x of x. To see this, first replace S by $S \cup S^{-1}$. Let g be in G_x. Write $g = s_n \cdots s_1$ with s_i in S. Then we can write

$$g = s_n \cdots s_1 = g_{s_n y_n}^{-1} s_n g_{y_n} \cdots g_{s_1 y_1}^{-1} s_1 g_{y_1} \quad \text{with } y_i = (s_{i-1} \cdots s_1)x.$$

The equality holds because $s_i y_i = y_{i+1}$ and $g_{y_1} = 1$ and $g_{s_n y_n} = 1$. Now we apply this to our case. We take $G = \mathrm{SL}_2(\mathbb{Z})$, and take X to be the subset of $(\mathbb{Z}/N\mathbb{Z})^2$ consisting of the elements of order N. This is a transitive G-set. Let $x = (1,0)$; then G_x is identified with $\Gamma_1(N)$. Let S be the set consisting of $\left(\begin{smallmatrix} 1 & 1 \\ 0 & 1 \end{smallmatrix}\right)$ and $\left(\begin{smallmatrix} 1 & 0 \\ 1 & 1 \end{smallmatrix}\right)$ and their inverses. Then S generates G. We now apply the previous argument to find generators of $\Gamma_1(N)$. So let $y = (\overline{a}, \overline{b})$ be in X. Then, thinking of $\mathbb{Z}/N\mathbb{Z}$ as the product of its local rings, we see that there is a u in $\mathbb{Z}$ with $|u| \leq N/2$ and $\overline{a + bu}$ in $(\mathbb{Z}/N\mathbb{Z})^\times$. Put $a_1 := a + bu$ and $b_1 := b$. Then

$$\begin{pmatrix} a_1 \\ b_1 \end{pmatrix} = \begin{pmatrix} 1 & u \\ 0 & 1 \end{pmatrix} \begin{pmatrix} a \\ b \end{pmatrix}.$$

Next, there is a v in $\mathbb{Z}$ with $|v| \leq N/2$ and $b_1 + a_1 v = 1 \bmod N$, that is,

$$\begin{pmatrix} a_1 \\ 1 \end{pmatrix} = \begin{pmatrix} 1 & 0 \\ v & 1 \end{pmatrix} \begin{pmatrix} a_1 \\ b_1 \end{pmatrix} \bmod N.$$

Finally, let w be in $\mathbb{Z}$ with $|w| \leq N/2$ and with image $\overline{a_1}$ in $\mathbb{Z}/N\mathbb{Z}$. Then we have

$$\begin{pmatrix} a \\ b \end{pmatrix} = \begin{pmatrix} 1 & -u \\ 0 & 1 \end{pmatrix} \begin{pmatrix} 1 & 0 \\ -v & 1 \end{pmatrix} \begin{pmatrix} 1 & w \\ 0 & 1 \end{pmatrix} \begin{pmatrix} 0 & -1 \\ 1 & 0 \end{pmatrix} \begin{pmatrix} 1 \\ 0 \end{pmatrix} \bmod N.$$

Writing out, we have

$$\begin{pmatrix} 1 & -u \\ 0 & 1 \end{pmatrix}\begin{pmatrix} 1 & 0 \\ -v & 1 \end{pmatrix}\begin{pmatrix} 1 & w \\ 0 & 1 \end{pmatrix}\begin{pmatrix} 0 & -1 \\ 1 & 0 \end{pmatrix} = \begin{pmatrix} w - u(1 - vw) & -1 - uv \\ 1 - vw & v \end{pmatrix},$$

so that we can put

$$g_y = g_{(a,b)} = \begin{pmatrix} w - u(1 - vw) & -1 - uv \\ 1 - vw & v \end{pmatrix}.$$

The absolute values of the coefficients of $g_{(a,b)}$ are smaller than $N^3/4$, if $N \geq 3$, and the lemma follows. $\square$

Proof [Proof of Theorem 11.2.1] Recall that $\|\vartheta\|(z; \tau)$ is given by

$$\|\vartheta\|(z; \tau) = (\det \Im(\tau))^{1/4} \exp(-\pi\, {}^t y (\Im(\tau))^{-1} y)|\vartheta(z; \tau)|,$$

where $y = \Im(z)$ and where τ is a period matrix in the Siegel upper half-plane $\mathbb{H}_g$ corresponding to X. We first deal with the factor $\det \Im(\tau)$; for this we invoke Lemma 11.2.2. Choose once more an Atkin-Lehner basis $(\omega_1, \ldots, \omega_g)$ for $\Omega^1(X)$. Using the dual basis in $\Omega^1(X)^\vee$, we write

$$J(X) = \mathbb{C}^g/\Lambda,$$

where

$$\Lambda = \mathrm{Image}\left(\mathrm{H}_1(X, \mathbb{Z}) \to \mathbb{C}^g : \gamma \mapsto \int_\gamma (\omega_1, \ldots, \omega_g)\right).$$

The polarization form for $J(X)$ is given by

$$(z, w) \mapsto {}^t z \cdot (\langle \omega, \omega \rangle)^{-1}_{i,j} \cdot \overline{w}.$$

Denote by $\|\cdot\|_P$ the corresponding norm on $\mathbb{C}^g$. We also consider the standard Hermitian inner product on $\mathbb{C}^g$, which is just $(z, w) \mapsto {}^t z \cdot \overline{w}$. Here we denote the corresponding norm by $\|\cdot\|_E$. From the next two lemmas we obtain

$$\left(\lambda_{g+1} \cdots \lambda_{2g}\right)^2 \leq \left(g e^{4\pi} (pl)^{46}/\pi\right)^g$$

and hence, by Lemma 11.2.2, the estimate

$$\log(\det \Im(\tau)) = O((pl)^2 \log(pl)).$$

11.2.4 Lemma *The lattice* Λ *is generated by the subset of its elements* x
that satisfy $\|x\|_E^2 \leq g \cdot (pl)^{46}$.

Proof For the moment put $N = pl$. Following the natural surjection,

$$\Gamma_1(N) \twoheadrightarrow \Gamma_1(N)^{\mathrm{ab}} = H_1(Y_1(N), \mathbb{Z}) \twoheadrightarrow H_1(X_1(N), \mathbb{Z}),$$

any generating set for $\Gamma_1(N)$ gives a generating set for $H_1(X_1(N), \mathbb{Z})$. We
are going to take generators for $\Gamma_1(N)$ as given by Lemma 11.2.3. In particular, the absolute values of their coefficients are bounded by $N^6/4$. We have
to see now what this implies for $\|x\|_E$ for corresponding elements x of Λ.
Concretely, choose a $g = \left(\begin{smallmatrix} a & b \\ c & d \end{smallmatrix}\right)$ in $\Gamma_1(N)$. The image in Λ can be given as
follows. Choose any z in $\mathbb{H}$ and any path in $\mathbb{H}$ from z to gz. This gives us a
loop in $X_1(N)$, and the class of that loop is the image of g in $H_1(X_1(N), \mathbb{Z})$.
In order to get to Λ we compute the periods of $(\omega_1, \ldots, \omega_g)$ around this
loop. We want to get bounds for these periods. In order to do this, note first
that we can assume that $c \neq 0$. Indeed, the g with $c = 0$ are unipotent,
hence have trivial image in $H_1(X_1(N), \mathbb{Z})$. Now we make the following
choices. First, we want to take a z in $\mathbb{H}$ with $\Im(z) = \Im(gz)$. Since, for
all z in $\mathbb{H}$, $\Im(gz) = |cz + d|^{-2} \Im(z)$, the condition that $\Im(z) = \Im(gz)$ is
equivalent to $|cz + d| = 1$. We choose $z = -d/c + i/|c|$, that is, such that
$cz + d = \pm i$, depending on the sign of c. Second, the path that we take is
the straight line from z to gz. We have $|c| \geq N$ because g is in $\Gamma_1(N)$. Us-
ing furthermore that the absolute values of the coefficients of g are bounded
by $N^6/4$, we get $|gz - z| \leq |az + b| + |z| \leq N^{11}/10$ (where we have used
that $N \geq 2$). For the period of an element $\omega = f \, dq/q$ of the Atkin-Lehner
basis we obtain from this that

$$\left| \int_z^{gz} \omega \right| = \left| \int_z^{gz} f \cdot (dq)/q \right| = \left| \int_z^{gz} f \cdot 2\pi i \, dw \right| \leq \frac{2\pi N^{11}}{10} \|f\|,$$

where w denotes the standard coordinate of $\mathbb{H}$ (that is, the inclusion into $\mathbb{C}$),
and $\|f\|$ the supnorm of f on the straight line from z to gz; recall that
$q = \exp(2\pi i w)$. Writing $f = \sum_{n \geq 1} a_n(f) q^n$, Lemma 11.1.1 gives that
$|a_n(f)| \leq 2n$. Noting furthermore that that $\Im z = |c|^{-1} \geq 4/N^6$ it follows
that

$$\|f\| \leq 2 \sum_{n \geq 1} n e^{-8\pi N^{-6} n} = \frac{2r}{(1-r)^2}, \quad \text{where} \quad r = e^{-8\pi/N^6}.$$

Hence

$$\|f\| \le (N^6/4\pi)^2 \quad \text{and} \quad \left| \int_z^{gz} f(dq)/q \right| \le \frac{2\pi}{10} N^{11} \left(\frac{N^6}{4\pi} \right)^2 \le N^{23}.$$

This means that all g coordinates of our element x of $\mathbb{C}^g$ are, in absolute value, at most N^{23}. Hence $\|x\|_E^2$, being the sum of the squares of these coordinates, is at most $g \cdot N^{46} = g \cdot (pl)^{46}$. $\square$

11.2.5 Lemma *For any x in $\mathbb{C}^g$ we have the estimate*

$$\|x\|_P^2 \le (e^{4\pi}/\pi)\|x\|_E^2.$$

Proof By Lemma 11.1.4 the matrix $((\omega, \omega))^{-1}$ is almost diagonal, having diagonal elements $1/\|\omega\|^2$ corresponding to newforms ω on $X_1(pl)$, and 2-by-2 blocks corresponding to newforms ω on $X_1(l)$ and $X_1(p)$. The 2-by-2 block corresponding to a newform ω on $X_1(l)$ is

$$\frac{1}{(p-1)\|\omega\|^2((p+1)^2 - |a_p(\omega)|^2)} \begin{pmatrix} p+1 & -\overline{a_p(\omega)} \\ -a_p(\omega) & p+1 \end{pmatrix},$$

where the norm $\|\omega\|^2$ is taken on $X_1(l)$.

The 2-by-2 block corresponding to a newform ω on $X_1(p)$ is

$$\frac{1}{(l-1)\|\omega\|^2((l+1)^2 - |a_l(\omega)|^2)} \begin{pmatrix} l+1 & -\overline{a_l(\omega)} \\ -a_l(\omega) & l+1 \end{pmatrix},$$

where the norm $\|\omega\|^2$ is taken on $X_1(p)$.

A short calculation shows that for any (z_1, z_2) in $\mathbb{C}^2$ one has

$$\frac{\begin{pmatrix} z_1 & z_2 \end{pmatrix} \begin{pmatrix} p+1 & -\overline{a_p(\omega)} \\ -a_p(\omega) & p+1 \end{pmatrix} \begin{pmatrix} \overline{z_1} \\ \overline{z_2} \end{pmatrix}}{(p-1)((p+1)^2 - |a_p(\omega)|^2)} \le (|z_1|^2 + |z_2|^2),$$

and similarly for $X_1(p)$, so that all in all one gets

$$\|(z_1, \ldots, z_g)\|_P^2 \le \frac{|z_1|^2}{\|\omega_1\|^2} + \cdots + \frac{|z_g|^2}{\|\omega_g\|^2}.$$

The lemma follows by the lower bound from Lemma 11.1.2. $\square$

Next we consider the factor $\exp(-\pi\,{}^t y(\Im(\tau))^{-1}y)|\vartheta(z;\tau)|$. Since in our previous estimates the choice of τ was irrelevant, it will cause no loss of generality here if we restrict to τ lying in the so-called Siegel fundamental domain F_g, which is the set of matrices $\tau = x+iy$ satisfying the conditions:

1. for each entry x_{ij} of x one has $|x_{ij}| \leq \frac{1}{2}$;

2. for all γ in $\mathrm{Sp}(2g,\mathbb{Z})$ one has $\det \Im(\gamma\cdot\tau) \leq \det \Im(\tau)$;

3. y is Minkowski reduced, that is, for each $\xi = (\xi_1,\ldots,\xi_g)$ in $\mathbb{Z}^g$ and each i such that $\xi_i,\ldots,\xi_g$ are nonzero, one has $\xi\, y\,{}^t\xi \geq y_{ii}$ and moreover, for each $1 \leq i \leq g-1$ one has $y_{i,i+1} \geq 0$.

It is well known that F_g contains at least one representative from each $\mathrm{Sp}(2g,\mathbb{Z})$-orbit on $\mathbb{H}_g$. We claim that for τ in F_g the estimate

$$\exp(-\pi\,{}^t y(\Im(\tau))^{-1}y)|\vartheta(z;\tau)| \leq 2^{3g^3+5g}$$

holds, for all z in $\mathbb{C}^g$. Thus, this factor gives us a contribution $O((pl)^6)$. In order to prove the estimate, write $y = \Im(z) = (\Im(\tau))\cdot b$ with b in $\mathbb{R}^g$. Then it is easy to see that

$$\exp(-\pi\,{}^t y(\Im(\tau))^{-1}y)|\vartheta(z;\tau)| \leq \sum_{n\in\mathbb{Z}^g} \exp(-\pi\,{}^t(n+b)(\Im(\tau))(n+b))\,.$$

Since the $\Im(\tau)$ are Minkowski reduced, we have, for any m in $\mathbb{R}^g$ (cf. [Igu], V §4),

$$^t m\Im(\tau)m \geq c(g)\sum_{i=1}^{g} m_i^2(\Im(\tau))_{ii}, \quad c(g) = \left(\frac{4}{g^3}\right)^{g-1}\left(\frac{3}{4}\right)^{g(g-1)/2}.$$

Moreover, we have $(\Im(\tau))_{i,i} \geq \sqrt{3}/2$ for $i = 1,\ldots,g$. From this we derive:

$$\sum_{n \in \mathbb{Z}^g} \exp(-\pi\,{}^t(n+b)(\Im(\tau))(n+b))$$

$$\leq \sum_{n \in \mathbb{Z}^g} \exp\left(-\sum_{i=1}^{g} \pi c(g)(n_i + b_i)^2 (\Im(\tau))_{i,i}\right)$$

$$\leq \prod_{i=1}^{g} \sum_{n_i \in \mathbb{Z}} \exp(-\pi c(g)(n_i + b_i)^2 (\Im(\tau))_{i,i})$$

$$\leq \prod_{i=1}^{g} \frac{2}{1 - \exp(-\pi c(g)(\Im(\tau))_{ii})}$$

$$\leq 2^g \left(1 + \frac{2}{\sqrt{3}\pi c(g)}\right)^g.$$

From this and the formula for $c(g)$ the required estimate follows and the proof of Theorem 11.2.1 is finished. $\square$

11.3 UPPER BOUNDS FOR ARAKELOV GREEN FUNCTIONS ON THE CURVES $X_1(pl)$

The aim of this section is to give an upper bound for Arakelov Green functions on the curves $X_1(pl)$ that will enable us to bound from above the contributions of the intersection numbers in the right-hand side of the inequality in Theorem 9.2.5. As the $X_l(\mathbb{C})$ are compact, it is clear that for each l such an upper bound exists, but we need upper bounds that grow at most as a power of l.

In order to establish such upper bounds we will use Franz Merkl's result on Green functions on arbitrary Riemann surfaces given in Chapter 10.

Instead of using Merkl's result, we could certainly also have used recent work by Jorgenson and Kramer in [Jo-Kr]. The results of Jorgenson and Kramer date back to the same time as those of Merkl (early Spring 2004). We chose to use Merkl's results because his approach is more elementary, and we had the details earlier than those of Jorgenson and Kramer.

The following theorem gives a suitable upper bound for the Arakelov-Green functions $g_{a,\mu}$ (see (4.4.2) and Proposition 4.4.3) on the modular curves $X_1(pl)$ with p and l distinct primes.

11.3.1 Theorem *There is a real number c such that for all pairs of distinct prime numbers p and l for which the genus of $X_1(pl)$ is at least one, and for all distinct a and b on $X_1(pl)(\mathbb{C})$, we have*

$$g_{a,\mu}(b) \leq c \cdot (pl)^6.$$

Let ∞ denote the cusp ∞ on $X_1(pl)$, and let q be the standard local coordinate around ∞ given by the map $\tau \mapsto \exp(2\pi i \tau)$ from the region $\Im\tau > 1$ in $\mathbb{H}$ to $\mathbb{C}$. Then we have

$$|\log \|dq\|_{\mathrm{Ar}}(\infty)| = O((pl)^6),$$

where $\|\cdot\|_{\mathrm{Ar}}$ denotes the Arakelov metric on Ω^1 (see Section 4.4).

Proof We write for the moment N for pl. We will apply Theorem 10.1, but we will carry out the estimates on the more symmetrical modular curve $X(N)$, which for us is $\Gamma(N)\backslash(\mathbb{H} \cup \mathbb{P}^1(\mathbb{Q}))$. Let $h\colon X(N) \to X_1(N)$ be the canonical map; it has degree N. We let μ denote the Arakelov $(1,1)$-form on $X_1(N)$, and we define $\mu' = h^*\mu/N$. The characterizing properties of Green functions directly imply that

$$(11.3.2) \qquad h^* g_{a,\mu} = \sum_{h(b)=a} g_{b,\mu'},$$

where the b are counted with multiplicity.

As in Chapter 10 we fix a constant r_1 with $0 < r_1 < 1$; we take $r_1 := 3/4$. We need to construct an atlas with charts $z^{(j)}\colon U^{(j)} \to \mathbb{C}$ for $X(N)$ with all $z^{(j)}(U^{(j)})$ containing the closed unit disk and with the $U_{r_1}^{(j)}$ covering $X(N)$.

We start with a construction of a local coordinate $z\colon U \to \mathbb{C}$ in a neighborhood of the standard cusp ∞. Since $\mathrm{SL}_2(\mathbb{Z}/N\mathbb{Z})$ acts transitively on the set of cusps of $X(N)$, this construction will suffice to give the full atlas. Our initial coordinate is induced by the map z from $\mathbb{H}$ to $\mathbb{C}$ that sends τ to $e^{2\pi i \tau/N}$. As the following lemma is valid for all integers $n \geq 1$, we state it in that generality, and will apply it with $n := N$.

11.3.3 Lemma *Let n be in $\mathbb{Z}_{\geq 1}$. The subset in $\mathbb{H}$ given by the conditions $-1/2 \leq \Re\tau < n - 1/2$ and $\Im\tau > 1/n$ is mapped injectively to $X(n)(\mathbb{C})$.*

Proof First we note that for $\left(\begin{smallmatrix} a & b \\ c & d \end{smallmatrix}\right)$ in $\mathrm{SL}_2(\mathbb{Z})$ and for τ in $\mathbb{H}$ we have

$$\Im\left(\frac{a\tau + b}{c\tau + d}\right) = \frac{\Im(\tau)}{|c\tau + d|^2}.$$

Let us call D the set of $\tau \in \mathbb{H}$ that satisfy the two conditions of the lemma:

$$D = \{\tau \in \mathbb{H} \mid -1/2 \le \Re(\tau) < n - 1/2 \text{ and } \Im(\tau) > 1/n\}.$$

Let τ be in D, and let $\gamma = \left(\begin{smallmatrix} a & b \\ c & d \end{smallmatrix}\right)$ be in $\Gamma(n)$, such that $\gamma\tau \ne \tau$. If $c = 0$ then, as $ad = 1$, we have $a = d$ and $\tau' = \tau \pm b$ with b a nonzero multiple of n, and so τ' is not in D. If $c \ne 0$ then we have $|c| \ge n$ because $n|c$, and

$$\Im\left(\frac{a\tau + b}{c\tau + d}\right) = \frac{\Im(\tau)}{|c\tau + d|^2} \le \frac{\Im(\tau)}{(\Im(c\tau))^2} \le \frac{\Im(\tau)}{n^2(\Im(\tau))^2}$$

$$= \frac{1}{n^2\Im(\tau)} < \frac{1}{n^2\cdot(1/n)} = \frac{1}{n},$$

hence $\gamma\tau$ is not in D. $\square$

In particular, the region of τ with $-1/2 \le \Re\tau < N - 1/2$ and $\Im\tau > 1/2$ is mapped injectively into $X(N)$ to give an open neighborhood U of ∞. We could replace the condition "$\Im\tau > 1/2$" by "$\Im\tau > 1/N$," but that would not make the work to be done significantly easier. The map $\tau \mapsto e^{2\pi i\tau/N}$ gives an isomorphism:

(11.3.4) $z: U \longrightarrow D(0, e^{-\pi/N}) \subset \mathbb{C}.$

The region of τ with $-1/2 \le \Re\tau < N - 1/2$ and $\Im\tau > 3/4$ gives an open neighborhood V of ∞, contained in U, such that the translates of V under $\mathrm{SL}_2(\mathbb{Z}/N\mathbb{Z})$ cover $X(N)$ (note that $3/4 < \sqrt{3}/2$). The image of V under z is the disk $D(0, e^{-3\pi/2N})$. However, the quotient of the radii $e^{-\pi/N}$ and $e^{-3\pi/2N}$ tends to 1 as N tends to infinity, hence we cannot work with these disks directly.

We define a new coordinate $z' := e^{3\pi/2N} z$ to get $z'V = D(0, 1)$. Then $z'U = D(0, e^{\pi/2N})$. Let $\varepsilon := \varepsilon(N) := e^{\pi/2N} - 1$ be the difference between the two new radii. Then $\varepsilon > \pi/2N$. We can choose $O(\varepsilon^{-2})$ open disks $D(a, \varepsilon)$ with center a in $D(0, 1)$, such that the union of the $D(a, \varepsilon/2)$ contains $D(0, 1)$; we let A denote the set of these a. The $D(a, \varepsilon)$ are contained in $z'U = D(0, 1 + \varepsilon)$.

The group $\mathrm{SL}_2(\mathbb{Z}/N\mathbb{Z})$ acts transitively on the set of cusps of $X(N)$. For each cusp c, we choose a g_c in $\mathrm{SL}_2(\mathbb{Z}/N\mathbb{Z})$ such that $c = g_c \infty$. The open sets of our atlas for $X(N)$ are then the $U^{(a,c)}$ with a in A and c a cusp, defined by

$$U^{(a,c)} := g_c \cdot (z')^{-1} D(a, \varepsilon).$$

The required coordinates $z^{(a,c)}$ on the $U^{(a,c)}$ are defined by the composition of isomorphisms:

$$z^{(a,c)}: U^{(a,c)} \xrightarrow{g_c^{-1}} U^{(a,\infty)} \xrightarrow{z'} D(a,\varepsilon) \xrightarrow{-a} D(0,\varepsilon) \xrightarrow{\cdot 3/2\varepsilon} D(0, 3/2).$$

Indeed, the images $z^{(a,c)}U^{(a,c)}$ contain the unit disk, and $U_{r_1}^{(a,c)}$ corresponds via $z' \circ g_c^{-1}$ to the subdisk $D(a, \varepsilon/2)$ of $D(a, \varepsilon)$, hence the $U_{r_1}^{(a,c)}$ cover $X(N)$. The exact number $n = n(N)$ of $U^{(a,c)}$ is the cardinality of A times the number of cusps, hence $n = O(N^4)$. We choose a numbering of $A \times \{\text{Cusps}\}$ with the integers $\{1, \ldots, n\}$, and we will denote our charts as

(11.3.5) $$z^{(j)}: U^{(j)} \to D(0, 3/2) \subset \mathbb{C}.$$

11.3.6 Lemma *For the local coordinates $z^{(j)}: U^{(j)} \to \mathbb{C}$ on $X(N)$ that we have just defined, the following holds. For all j and k in $\{1, \ldots, n\}$ we have*

$$\sup_{U_1^{(j)} \cap U_1^{(k)}} \left| \frac{dz^{(j)}}{dz^{(k)}} \right| \leq M,$$

with $M = 6$.

Proof Let j and k be in $\{1, \ldots, n\}$. If j and k arise from the same cusp, then $z^{(j)}$ and $z^{(k)}$ differ by a translation, hence $dz^{(j)}/dz^{(k)} = 1$. Now suppose that j and k arise from two distinct cusps. We may suppose then, by acting with an element of $\mathrm{SL}_2(\mathbb{Z}/N\mathbb{Z})$, that k arises from the standard cusp ∞. Let x denote the cusp from which j arises. The coordinate $z^{(j)}$ is then obtained as above from an element g_x of $\mathrm{SL}_2(\mathbb{Z})$ that sends ∞ to x. Let us write $g_x^{-1} = \left(\begin{smallmatrix} a & b \\ c & d \end{smallmatrix} \right)$. Note that $c \neq 0$, hence $|c| \geq 1$. Let z be a point in $\mathbb{H}$ with $-1/2 \leq \Re z < N - 1/2$ that maps to an element in $U_1^{(j)} \cap U_1^{(k)}$. Then we know that $1/2 < \Im z < 1$ because disks given by $\Im z > 1$ around different cusps do not meet at all. Likewise, we then know that

$$\frac{1}{2} < \Im\left(\frac{az+b}{cz+d}\right) = \frac{\Im(z)}{|cz+d|^2} < 1.$$

Hence, as $(\Im z)/|cz+d|^2 \leq (\Im z)/c^2(\Im z)^2$, we have $\Im z < 2/c^2$ which gives in fact $|c| = 1$. Under these conditions, we estimate:

$$\left|\log\left|\frac{d\exp\left(2\pi i\frac{az+b}{cz+d}/N\right)}{d\exp(2\pi iz/N)}\right|\right|$$

$$= \left|\log\left|\frac{\exp\left(2\pi i\frac{az+b}{cz+d}/N\right)d\frac{az+b}{cz+d}}{\exp(2\pi iz/N)dz}\right|\right|$$

$$= \left|\log\left|\exp\left(2\pi i\frac{az+b}{cz+d}/N\right)\right| - \log|\exp(2\pi iz/N)| - \log|cz+d|^2\right|$$

$$\leq 4\pi/N + 4\pi/N + \log 4.$$

So indeed, for $N \geq 6$, we can take $M = 6$. Some explanations are perhaps in order here: as $\Im z$ and $\Im(az+b)/(cz+d)$ are between $1/2$ and 2, $|\exp(2\pi iz/N)|$ and $|\exp(2\pi i\frac{az+b}{cz+d}/N)|$ are between $\exp(-4\pi/N)$ and $\exp(-\pi/N)$. As $\Im(az+b)/(cz+d) = (\Im z)/|cz+d|^2$, we see that $|cz+d|^2$ is between $1/4$ and 4. □

Our next task is to produce a suitable bound, as in Theorem 10.1, of the type $\mu \leq c_1|dz^{(j)} \wedge d\bar{z}^{(j)}|$. We start with a bound for μ on disks around ∞ on $X_1(pl)$.

11.3.7 Lemma *Let r be a real number such that $0 < r < 1$. We map $D(0,r)$ to $X_1(pl)$ by sending $q \neq 0$ to $(\mathbb{C}^\times/q^{\mathbb{Z}}, \zeta_{pl})$. The image of this map is the image in $X_1(pl)$ of the region in $\mathbb{H}$ defined by the condition "$\Im\tau > -(\log r)/2\pi$," plus the cusp ∞. We still denote by μ the $(1,1)$-form on $D(0,r)$ induced by μ. Then we have, on $D(0,r)$,*

$$\mu \leq \frac{28e^{4\pi}}{\pi}\frac{1}{(1-r)^4}\cdot\frac{i}{2}dqd\bar{q}.$$

Proof We first bound, on the disk $D(0,r)$, and for a newform f, the functions $\sum_{n\geq 1} a_n(f)q^{n-1}$ and $\sum_{n\geq 1} a_n(f)q^{nl-1}$. We have, for $|q| < r$ by

Lemma 11.1.1,

$$\left|\sum_{n\geq 1} a_n(f)q^{n-1}\right| \leq \sum_{n\geq 1}|a_n(f)|r^{n-1} \leq 2\sum_{n\geq 1}nr^{n-1} = \frac{2}{(1-r)^2},$$

and next:

$$\left|\sum_{n\geq 1} a_n(f)q^{nl-1}\right| \leq 2\sum_{n\geq 1}nr^{nl-1} = \frac{2r^{l-1}}{(1-r^l)^2} \quad \text{for } |q| < r.$$

Now recall from Corollary 11.1.6 that for μ we have the expression

$$
\mu = \frac{i}{2g}\sum_\omega \frac{\omega \wedge \overline{\omega}}{\|\omega\|^2} + \frac{i}{2g}\sum_\omega \left(\frac{(p+1)s_p^*(\omega \wedge \overline{\omega}) + (p+1)t_p^*(\omega \wedge \overline{\omega})}{(p-1)\|\omega\|^2((p+1)^2 - |a_p(f_\omega)|^2)}\right.
$$
$$
\left. - \frac{a_p(f_\omega)s_p^*\omega \wedge \overline{t_p^*\omega} + \overline{a_p(f_\omega)}t_p^*\omega \wedge \overline{s_p^*\omega}}{(p-1)\|\omega\|^2((p+1)^2 - |a_p(f_\omega)|^2)}\right)
$$
$$
+ \frac{i}{2g}\sum_\omega \left(\frac{(l+1)s_l^*\omega \wedge \overline{s_l^*\omega} + (l+1)t_l^*\omega \wedge \overline{t_l^*\omega}}{(l-1)\|\omega\|^2((l+1)^2 - |a_l(f_\omega)|^2)}\right.
$$
$$
\left. - \frac{a_l(f_\omega)s_l^*\omega \wedge \overline{t_l^*\omega} + \overline{a_l(f_\omega)}t_l^*\omega \wedge \overline{s_l^*\omega}}{(l-1)\|\omega\|^2((l+1)^2 - |a_l(f_\omega)|^2)}\right),
$$

the first sum running over $\Omega^1(X_1(pl))^{\text{new}}$, the second sum running over $\Omega^1(X_1(l))^{\text{new}}$, and the third sum running over $\Omega^1(X_1(p))^{\text{new}}$. We bound the different terms from this expression for μ. The contribution of ω in $\Omega^1(X_1(pl))^{\text{new}}$ gives, for $|q| < r$,

$$
\frac{i}{2g\|\omega\|^2}\omega \wedge \overline{\omega} = \frac{1}{g\|\omega\|^2}\left|\sum_{n\geq 1}a_n(f_\omega)q^{n-1}\right|^2 \cdot \frac{i}{2}dqd\overline{q}
$$
$$
\leq \frac{1}{g}\frac{e^{4\pi}}{\pi}\left(\frac{2}{(1-r)^2}\right)^2 \cdot \frac{i}{2}dqd\overline{q}
$$
$$
= \frac{1}{g}\frac{e^{4\pi}}{\pi}\frac{4}{(1-r)^4} \cdot \frac{i}{2}dqd\overline{q}
$$

The contribution of an element ω of $\Omega^1(p)^{\text{new}}$ is:

$$
\frac{i}{2g} \sum_\omega \left(\frac{(l+1)s_l^*\omega \wedge \overline{s_l^*\omega} + (l+1)t_l^*\omega \wedge \overline{t_l^*\omega}}{(l-1)\|\omega\|^2((l+1)^2 - |a_l(f_\omega)|^2)} \right.
$$

$$
\left. - \frac{a_l(f_\omega)s_l^*\omega \wedge \overline{t_l^*\omega} + \overline{a_l(f_\omega)}t_l^*\omega \wedge \overline{s_l^*\omega}}{(l-1)\|\omega\|^2((l+1)^2 - |a_l(f_\omega)|^2)} \right)
$$

$$
\leq \frac{1}{g} \frac{e^{4\pi}}{\pi(l-1)^3} \left(\frac{4(l+1)}{(1-r)^4} + \frac{4(l+1)r^{2(l-1)}}{(1-r^l)^4} \right.
$$

$$
\left. + \frac{16\sqrt{l}r^{l-1}}{(1-r)^2(1-r^l)^2} \right) \cdot \frac{i}{2} dq d\overline{q}.
$$

Here one uses the Weil bounds on $a_l(f_\omega)$. Symmetrically (in p and l), the contribution to μ of an element ω of $\Omega^1(l)^{\text{new}}$ is

$$
\frac{i}{2g} \sum_\omega \left(\frac{(p+1)s_p^*\omega \wedge \overline{s_p^*\omega} + (p+1)t_p^*\omega \wedge \overline{t_p^*\omega}}{(p-1)\|\omega\|^2((p+1)^2 - |a_p(f_\omega)|^2)} \right.
$$

$$
\left. - \frac{a_p(f_\omega)s_p^*\omega \wedge \overline{t_p^*\omega} + \overline{a_p(f_\omega)}t_p^*\omega \wedge \overline{s_p^*\omega}}{(p-1)\|\omega\|^2((p+1)^2 - |a_p(f_\omega)|^2)} \right)
$$

$$
\leq \frac{1}{g} \frac{e^{4\pi}}{\pi(p-1)^3} \left(\frac{4(p+1)}{(1-r)^4} + \frac{4(p+1)r^{2(p-1)}}{(1-r^p)^4} \right.
$$

$$
\left. + \frac{16\sqrt{p}r^{p-1}}{(1-r)^2(1-r^p)^2} \right) \cdot \frac{i}{2} dq d\overline{q}.
$$

Now we sum all contributions up, over the elements of $\Omega^1(X_1(pl))^{\text{new}}$, $\Omega^1(X_1(p))^{\text{new}}$, and $\Omega^1(X_1(l))^{\text{new}}$. We get for $|q| < r$

$$
\mu \leq \frac{4e^{4\pi}}{\pi} \left(\frac{1}{(1-r)^4} + \frac{1}{(1-r)^4} + \frac{r^{2(l-1)}}{(1-r^l)^4} + \frac{r^{l-1}}{(1-r)^2(1-r^l)^2} \right.
$$

$$
\left. + \frac{1}{(1-r)^4} + \frac{r^{2(p-1)}}{(1-r^p)^4} + \frac{r^{p-1}}{(1-r)^2(1-r^p)^2} \right) \cdot \frac{i}{2} dq d\overline{q},
$$

and finally,

$$
\mu \leq \frac{28e^{4\pi}}{\pi} \frac{1}{(1-r)^4} \cdot \frac{i}{2} dq d\overline{q} \quad \text{for } |q| < r.
$$

$\square$

Our next step is to consider the disks gU in $X(pl)$, for g in $\mathrm{SL}_2(\mathbb{Z}/pl\mathbb{Z})$ and for U as in (11.3.4).

11.3.8 Lemma *Let g be in $\mathrm{SL}_2(\mathbb{Z}/pl\mathbb{Z})$ and let $z: U \to D(0, e^{-\pi/pl})$ be as in (11.3.4). Let $z_g := z \circ g^{-1}: gU \to D(0, e^{-\pi/pl})$. Then we have, for the restriction to gU of the pullback $h^*\mu$ of μ along $h: X(N) \to X_1(N)$,*

$$(h^*\mu)|_{gU} \leq c(pl)^4 |dz_g \, d\overline{z_g}|,$$

with c independent of p and l.

Proof Consider the map $h \circ g \circ z^{-1}$ from $D(0, e^{-\pi/pl})$ to $X_1(pl)$ and the pullback of μ to $D(0, e^{-\pi/pl})$. We observe that μ is invariant under all automorphisms of $X_1(pl)$. This applies in particular to the diamond operators and the Atkin-Lehner pseudo-involutions (defined in (2.5.14)). As the group generated by these automorphisms permutes the cusps of $X_1(pl)$ transitively, we can take such an automorphism α such that $\alpha \circ h \circ g \circ z^{-1}$ sends $D(0, e^{-\pi/pl})$ to a disk around the cusp ∞, where we can then apply Lemma 11.3.7. The pullbacks of μ via $h \circ g \circ z^{-1}$ and $\alpha \circ h \circ g \circ z^{-1}$ are the same. We are also free to replace the coordinate z by ζz with $\zeta \in \mathbb{C}$ such that $|\zeta| = 1$.

The map $h \circ g \circ z^{-1}$ sends a point $0 \neq q \in D(0, e^{-\pi/pl})$ to the point of $X_1(pl)$ corresponding to $(\mathbb{C}^\times/q^{pl\mathbb{Z}}, \zeta_p^a(q^l)^b, \zeta_l^c(q^p)^d)$ for certain a and b in $\mathbb{F}_p$ and c and d in $\mathbb{F}_l$ depending on g. After replacing h with h composed with a suitable diamond operator, and z by ζz with ζ a suitable element of $\mu_{pl}(\mathbb{C})$, we are in one of four cases, which we will treat one by one.

In the first case, q is mapped to $(\mathbb{C}^\times/q^{pl\mathbb{Z}}, \zeta_{pl})$. Then the map $h \circ g \circ z^{-1}$ factors as the cover $D(0, e^{-\pi/pl}) \to D(0, e^{-\pi})$ of degree pl sending q to q^{pl}, followed by the map of Lemma 11.3.7 that sends q^{pl} to the point $(\mathbb{C}^\times/q^{pl\mathbb{Z}}, \zeta_{pl})$. Then we have, on $D(0, e^{-\pi/pl})$,

$$h^*\mu \leq \frac{28e^{4\pi}}{\pi} \frac{1}{(1 - e^{-\pi})^4} \cdot \frac{i}{2} d(q^{pl}) d(\overline{q^{pl}})$$

$$\leq \frac{28e^{4\pi}}{\pi} \frac{1}{(1 - e^{-\pi})^4} (pl)^2 \cdot \frac{i}{2} dq \, d\overline{q}.$$

In the second case, q is mapped to $(\mathbb{C}^\times/q^{pl\mathbb{Z}}, q^l, \zeta_l)$. In this case, we compose it with the pseudo-involution w_{ζ_p}, which brings us to the point

$(\mathbb{C}^\times/q^l\mathbb{Z}, \zeta_p, \zeta_l)$. The map then factors as the lth power map between disks from $D(0, e^{-\pi/pl})$ to $D(0, e^{-\pi/p})$, followed by the map of Lemma 11.3.7 composed with a suitable diamond operator. We find

$$h^*\mu \leq \frac{28e^{4\pi}}{\pi} \frac{1}{(1 - e^{-\pi/p})^4} \cdot \frac{i}{2} d(q^l) d(\overline{q^l})$$

$$\leq \frac{28e^{4\pi}}{\pi} \frac{1}{(1 - e^{-\pi/p})^4} l^2 \cdot \frac{i}{2} dq d\overline{q}.$$

The third case is obtained by interchanging the roles of p and l, so we will not make it explicit.

In the fourth case, q is mapped to $(\mathbb{C}^\times/q^{pl\mathbb{Z}}, q)$. We compose with the pseudo-involution $w_{\zeta_{pl}}$, which brings us to $(\mathbb{C}^\times/q^\mathbb{Z}, \zeta_{pl})$. This is the map of Lemma 11.3.7. We find

$$h^*\mu \leq \frac{28e^{4\pi}}{\pi} \frac{1}{(1 - e^{-\pi/pl})^4} \cdot \frac{i}{2} dq d\overline{q}.$$

In these four cases, we see that the factor in front of $(i/2)dq d\overline{q}$ in the upper bound for $h^*\mu$ on $D(0, e^{-\pi/pl})$ is $O((pl)^4)$. This gives the required estimate on gU, as dq on $D(0, e^{-\pi/pl})$ corresponds to dz_g on gU. $\qquad\square$

11.3.9 Lemma *For the local coordinates $z^{(j)}$ and the real $(1, 1)$-form μ' on $X(pl)$ as defined in (11.3.5) we have*

$$\mu' \leq c_1 |dz^{(j)} \wedge d\overline{z}^{(j)}|$$

on $U_1^{(j)}$ with $c_1 = c_1(pl) = O(pl)$.

Proof First of all, we have, by definition: $\mu' = (1/pl)h^*\mu$. The definition of the $z^{(j)}$ (see 11.3.5) plus the definitions $z' = e^{3\pi/2pl}z$ and $\varepsilon = e^{\pi/2pl} - 1$ give

$$dz = \frac{2\varepsilon}{3} e^{-3\pi/2pl} \cdot dz^{(j)}.$$

If pl gets large, then the factor $e^{3\pi/2pl}$ tends to 1 and for ε we have

$$\varepsilon = (\pi/2pl)(1 + O(1/pl)).$$

Combining all this with Lemma 11.3.8 finishes the proof. $\qquad\square$

We can now finish the proof of Theorem 11.3.1. We apply Theorem 10.1 on $X(pl)$ with the $(1, 1)$-form μ'. Then we have $n = O((pl)^4)$, $M = 6$ (Lemma 11.3.6) and $c_1 = O(pl)$ (Lemma 11.3.9). We obtain from (10.2) that there exists a constant c such that $g_{b,\mu'}(b') \leq c \cdot (pl)^5$ for all distinct primes p and l, and all distinct b and b' on $X(pl)$. For distinct a and a' on $X_1(pl)$ we then have (see (11.3.2))

$$g_{a,\mu}(a') = \sum_{h(b)=a} g_{b,\mu'}(h(a')) \leq c \cdot (pl)^6.$$

The statement that $\log \|dz^{(j)}\|_{\mathrm{Ar}}(P) = O((pl)^6)$ follows from the inequality (10.3) in Theorem 10.1. Indeed, if locally we define the function f by $g_{a,\mu} = \log |z - z(a)| + f$ then $f(a) = -\log \|dz\|_{\mathrm{Ar}}(a)$. $\qquad\square$

11.4 BOUNDS FOR INTERSECTION NUMBERS ON $X_1(pl)$

In this section, we will bound the intersection numbers occurring in the right-hand side of the inequality in Theorem 9.2.5, in the situation described in Section 8.2.

11.4.1 Theorem *Let p and l be two distinct prime numbers, both at least 5, and let $\mathcal{X}$ be the semistable model over $B := \operatorname{Spec} \mathbb{Z}[\zeta_{pl}]$ that is provided by [Ka-Ma]. For two cusps P and Q (possibly equal) in $\mathcal{X}(B)$ we have*

$$(P, P) \leq 0, \quad \text{and} \quad |(P, Q)| = O((pl)^7).$$

For a cuspidal effective divisor D of degree g on $\mathcal{X}$ we have

$$|(D, D - \omega_{\mathcal{X}/B})| = O((pl)^{11}).$$

Proof As p and l are at least 5, the genus of $X_1(pl)$ is at least two. By the adjunction formula (see (4.4.8)). we have $-(P, P) = (P, \omega_{\mathcal{X}/B})$, and by [Fal1], Theorem 5, we have $(P, \omega_{\mathcal{X}/B}) \geq 0$, hence $(P, P) \leq 0$.

Let us now derive an upper bound for $(P, \omega_{\mathcal{X}/B})$. As the automorphism group of $\mathcal{X}$ over B preserves the Arakelov intersection product on $\mathcal{X}$ and acts transitively on the cusps, it suffices to do this for the standard cusp ∞. The Fourier expansion at ∞ of the rational function j on $\mathcal{X}$ is of the form $j = 1/q + f$ with $f \in \mathbb{Z}[[q]]$. Therefore, $1/j$ is regular in a neighborhood of ∞, and has a zero of order one along ∞. It follows that $d(1/j)$ generates $\infty^* \omega_{\mathcal{X}/B}$, and $d(1/j) = dq$ in $\infty^* \omega_{\mathcal{X}/B}$. By definition of the Arakelov intersection product and Theorem 11.3.1 we then have

$$(\infty, \omega_{\mathcal{X}/B}) = -[\mathbb{Q}(\zeta_{pl})) : \mathbb{Q}] \log \|dq\|_{\mathrm{Ar}}(\infty) = O((pl)^7).$$

We now know $|(P, P)| = O((pl)^7)$ for all cusps P in $\mathcal{X}(B)$. We will now show that $|(P, Q)| = O((pl)^7)$. By the Manin-Drinfeld Theorem, see [Dri], the image of the divisor $P - Q$ in $J_1(pl)(\mathbb{Q}(\zeta_{pl}))$ is of finite order. Let Φ be a vertical fractional divisor such that for any irreducible component C of a fiber of $\mathcal{X}$ over B we have $(P - Q - \Phi, C) = 0$. By [Hri] or Theorem 4 of [Fal1] we have $(P - Q - \Phi, P - Q - \Phi) = 0$. Equivalently, we have

$$2(P, Q) = (P, P) + (Q, Q) - (P - Q, \Phi).$$

The term $(P - Q, \Phi)$ can be dealt with by the method used in the proof of Lemma 9.2.6. We work this out in this special situation. We make Φ unique by demanding that its support be disjoint from P. Of course, this does not change the number $(P - Q, \Phi)$, but it makes it easier to talk about Φ. The support of Φ is contained in the reducible fibers. These are exactly the fibers in the characteristics p and l. Let us estimate the contribution at the prime p. We have to sum over the maximal ideals of the $\mathbb{F}_p$-algebra $\mathbb{F}_p[x]/(x^{l-1} + \cdots + 1)$. All residue fields of this algebra are isomorphic to a finite extension $\mathbb{F}$ of $\mathbb{F}_p$, and the number of them is $(l-1)/\dim_{\mathbb{F}_p} \mathbb{F}$. Let $\mathcal{X}_{\mathbb{F}}$ be the fiber at one of these residue fields, and let $\Phi_{\mathbb{F}}$ be the part of Φ that has support in $\mathcal{X}_{\mathbb{F}}$. Then $\mathcal{X}_{\mathbb{F}}$ is the union of two irreducible components I and I', with transversal intersection at the supersingular points. The number of $\overline{\mathbb{F}}_p$-valued supersingular points is given by

$$s := \#\mathcal{X}_{\mathbb{F}}(\overline{\mathbb{F}}_p)^{\mathrm{s.s.}} = \#\left(X_1(l)(\overline{\mathbb{F}}_p)^{\mathrm{s.s.}}\right) = \frac{(p-1)(l^2-1)}{24}.$$

The degree on I of the restriction to it of $\mathcal{O}_{\mathcal{X}}(I')$ is s, and, symmetrically, $\deg_{I'} \mathcal{O}_{\mathcal{X}}(I) = s$. As $I + I' = \mathcal{X}_{\mathbb{F}}$ is a principal divisor in a neighborhood of $\mathcal{X}_{\mathbb{F}}$, the restrictions of $\mathcal{O}_{\mathcal{X}}(I + I')$ to I and I' are trivial, and we have $\deg_I \mathcal{O}_{\mathcal{X}}(I) = -s$ and $\deg_{I'} \mathcal{O}_{\mathcal{X}}(I') = -s$. It follows that $\Phi_{\mathbb{F}}$ is one of the following fractional divisors: $\Phi_{\mathbb{F}} = 0$ if P and Q specialize to points on the same irreducible component of $\mathcal{X}_{\mathbb{F}}$; $\Phi_{\mathbb{F}} = (1/s) \cdot I$ if P specializes to a point on I' and Q to a point on I; $\Phi_{\mathbb{F}} = (1/s) \cdot I'$ if P specializes to a point on I and Q to a point on I'. If we denote by $(P - Q, \Phi)_{\mathbb{F}}$ the contribution to $(P - Q, \Phi)$ at the fiber $\mathcal{X}_{\mathbb{F}}$, we have

$$|(P - Q, \Phi)_{\mathbb{F}}| \leq (2/s) \cdot \log \#\mathbb{F}.$$

Summing this over the residue fields of $\mathbb{F}_p[x]/(x^{l-1} + \cdots + 1)$ gives, for the contribution $(P - Q, \Phi)_p$ to $(P - Q, \Phi)$ at p,

$$|(P - Q, \Phi)_p| \leq \frac{l-1}{\dim_{\mathbb{F}_p}(\mathbb{F})} \cdot \frac{2}{s} \cdot \log \#\mathbb{F} = \frac{48 \log p}{(p-1)(l+1)}.$$

Likewise, we have, for the contribution at l to $(P - Q, \Phi)$,

$$|(P - Q, \Phi)_l| \leq \frac{48 \log l}{(l-1)(p+1)}.$$

So, finally,

$$|(P - Q, \Phi)| \leq \frac{48 \log l}{(l-1)(p+1)} + \frac{48 \log p}{(p-1)(l+1)}.$$

The estimate $|(P, Q)| = O((pl)^7)$ now follows.

To get to the second statement of the theorem, note that

$$(D, D - \omega_{\mathcal{X}/B}) = (D, D + \omega_{\mathcal{X}/B}) - 2(D, \omega_{\mathcal{X}/B})$$
$$= \sum_{k \neq l} (P_k, P_l) - 2(D, \omega_{\mathcal{X}/B}),$$

where $D = P_1 + \cdots + P_g$, with repetitions allowed. By our previous estimates, we get $|(D, D - \omega_{\mathcal{X}/B})| = O(g^2(pl)^7) = O((pl)^{11})$. $\qquad \square$

We will also need a lower bound for the intersection number of two distinct points on $X_1(pl)$.

11.4.2 Theorem *There is an integer c such that for all pairs of distinct primes p and l such that $X_1(pl)$ has genus at least one, for any extension K of $\mathbb{Q}(\zeta_{5l})$, and for P and Q distinct points in $X_1(pl)(K)$ we have*

$$\frac{1}{[K : \mathbb{Q}]}(P, Q) \geq c(pl)^6,$$

where (P, Q) is the Arakelov intersection number of P and Q on the minimal regular model of $X_1(pl)$ over O_K.

Proof We have

$$(P, Q) = (P, Q)_{\text{fin}} + (P, Q)_\infty,$$

with $(P, Q)_{\text{fin}}$ the contribution from the finite places of K, and $(P, Q)_\infty$ the contribution from the infinite places. As $P \neq Q$, we have $(P, Q)_{\text{fin}} \geq 0$. On the other hand, we have

$$(P, Q)_\infty = \sum_\sigma -g_\sigma(P_\sigma, Q_\sigma).$$

By Theorem 11.3.1 we have

$$g_\sigma(P_\sigma, Q_\sigma) \leq c \cdot (pl)^6$$

for some absolute constant c. This finishes the proof. $\square$

11.5 A BOUND FOR $h(x_l'(Q))$ IN TERMS OF $h(b_l(Q))$

In this section we do what was promised at the beginning of Section 9.1, by stating and proving the following proposition and a corollary.

11.5.1 Proposition *There is a real number c such that the following holds. Let b be in $\overline{\mathbb{Q}}$, such that $b^5(b^2 + 11b - 1) \neq 0$. Let (u, v) in $\overline{\mathbb{Q}}^2$ be a torsion point on the elliptic curve E_b over $\overline{\mathbb{Q}}$ given by the equation*

$$(11.5.2) \qquad y^2 + (b + 1)xy + by = x^3 + bx^2,$$

that is, on the fiber at b of the universal elliptic curve with a point of order 5 given in Proposition 8.2.8. Then the absolute heights $h(u)$ and $h(v)$ are bounded from above by $c + 14h(b)$.

Proof Let b, u, and v in $\overline{\mathbb{Q}}$ be as in the statement of the proposition. We will now invoke known bounds for the difference between the Weil height and the Néron-Tate height on elliptic curves over number fields. Such bounds are given, for example, in [Dem], [Zim], and [Sil]. In [Sil] bounds are given for elliptic curves given by general Weierstrass equations, but under the assumption that the coefficients are algebraic integers. Therefore, for us it seems better to use the following bound in Zimmer's theorem on p. 40 of [Zim].

11.5.3 Theorem (Zimmer) *Let E be an elliptic curve over $\overline{\mathbb{Q}}$ given by a Weierstrass equation $y^2 = x^3 + Ax + B$, and let $P \in E(\overline{\mathbb{Q}})$. Let h denote the absolute Weil height on $\mathbb{P}^2(\overline{\mathbb{Q}})$, and $\hat{h}$ the absolute Néron-Tate height on E attached to h. Then one has*

$$- 2^{-1}(2^{-1}h(1 : A^3 : B^2) + 7 \log 2) \leq h(P) - \hat{h}(P)$$
$$\leq 2^{-1}h(1 : A^3 : B^2) + 6 \log 2 .$$

So, we must compare our plane elliptic curve E_b with one given by a standard Weierstrass equation. We put

$$v_1 := v + ((b + 1)u + b)/2, \quad u_1 := u + (b + (b + 1)^2/4)/3 .$$

Then (u_1, v_1) is a point on the elliptic curve E_b' given by a Weierstrass equation $y^2 = x^3 + Ax + B$, with A and B polynomials in b, with coefficients in $\mathbb{Q}$, of degrees at most 4 and 6, respectively. We note that A and B depend only on b, not on (u, v). Using Lemma 4.2.3 and writing A as $A_0 + b(A_1 + b(A_2 + b(A_3 + bA_4)))$, we see that there is a real number c_1, such that for all b we have $h(A) \leq c_1 + 4h(b)$. Similarly, there is a c_2 such that $h(B) \leq c_2 + 6h(b)$. Therefore, there is a c_3 such that $h(1 : A^3 : B^2) \leq c_3 + 24h(b)$. Zimmer's theorem 11.5.3, plus the fact that the Néron-Tate height of torsion points is zero, tells us that there is a c_4 such that for all b and for all torsion points (u, v) on E_b, we have

$$h(u_1, v_1) \leq c_4 + 12h(b) .$$

Expressing u and v in u_1 and v_1, and using again Lemma 4.2.3, we get a real number c_5 such that for all b and all (u, v) as in the proposition we are proving, we have

$$h(u) \leq c_5 + 14h(b), \quad h(v) \leq c_5 + 14h(b) .$$

This ends the proof of Proposition 11.5.1. $\qquad\qquad\qquad\qquad \square$

11.5.4 Corollary *There is a real number c such that for each l and each $Q_{x,i}$ as in the beginning of Section 9.1 we have*

$$h(x_l'(Q_{x,i})) \leq c + 14h(b_l(Q_{x,i})).$$

Proof By definition (Section 8.2), $x_1'(Q_{x,i})$ is the x-coordinate of a point of order $5l$ on the fiber at $b_l(Q_{x,i})$ of the elliptic curve E in Proposition 8.2.8. Proposition 11.5.1 gives the result. $\qquad\square$

11.6 AN INTEGRAL OVER $X_1(5l)$

In this section we will give an upper bound for the integral appearing in the estimate in Theorem 9.1.1. We recall the situation: l is a prime number with $l > 5$, and b is the regular function on $Y_1(5)_{\mathbb{Q}}$ given by Proposition 8.2.8. We also view b as a regular function b_l on $Y_1(5l)_{\mathbb{Q}}$ via pullback along the map $Y_1(5l)_{\mathbb{Q}} \to Y_1(5)_{\mathbb{Q}}$ that sends $(E/S, P_5, P_l)$ to $(E/S, P_5)$, where S is any $\mathbb{Q}$-scheme, E/S an elliptic curve, P_5 in $E(S)$ everywhere of order 5, and P_l in $E(S)$ everywhere of order l.

11.6.1 Proposition *There exist real numbers A and B such that for all primes $l > 5$,*

$$\int_{X_1(5l)(\mathbb{C})} \log(|b_l|^2 + 1)\, \mu \leq A + B \cdot l^6 ,$$

where μ is the Arakelov $(1, 1)$-form.

Proof In order to simplify the notation in the proof, we will let $X_1(5l)$ denote the Riemann surface $X_1(5l)_{\mathbb{Q}}(\mathbb{C})$ of complex points of the curve $X_1(5l)_{\mathbb{Q}}$ over $\mathbb{Q}$, and we will drop the subscript l in b_l. We will denote points of $Y_1(5l)$ as triples (E, P_5, P_l), with E a complex elliptic curve with points P_5 of order 5 and P_l of order l. Similarly, we will denote points of $Y_1(5)$ by pairs (E, P), and points on the j-line just by elliptic curves.

The equations in Proposition 8.2.8 show that the rational function b on $X_1(5)$ has exactly one pole, that it is of order one, and that at that point the function $j : X_1(5) \to \mathbb{P}^1$ has a pole of order 5. The region in $\mathbb{H}$ consisting of the τ with $\Im(\tau) > 1$ gives an embedding of the disk $D(0, e^{-2\pi})$ into the j-line, sending $q \neq 0$ to $\mathbb{C}^\times/q^{\mathbb{Z}}$. The inverse image of this disk under $j : X_1(5) \to \mathbb{P}^1$ consists of 4 disks, one around each cusp. The two disks around the cusps where j is ramified are given by the embeddings $D(0, e^{-2\pi/5}) \to X_1(5)$, sending $q \neq 0$ to $(\mathbb{C}^\times/q^{5\mathbb{Z}}, q)$ and to $(\mathbb{C}^\times/q^{5\mathbb{Z}}, q^2)$. As the integral in the proposition that we are proving does not change if we replace b by its image under a diamond operator $\langle a \rangle$ with $a \in \mathbb{F}_5^\times$, we may and do suppose that b has its pole at the center of the punctured disk

$$U = \{(\mathbb{C}^\times/q^{5\mathbb{Z}}, q) : q \in D(0, e^{-2\pi/5})^*\} \subset X_1(5).$$

The integral of $\log(|b|^2 + 1)\mu$ over the complement of the inverse image of U in $X_1(5l)$ is bounded by the supremum of $\log(|b|^2 + 1)$ on $X_1(5) - U$. This upper bound does not depend on l. Hence it is enough to prove that the integral of $\log(|b|^2 + 1)\mu$ over the inverse image of U in $X_1(5l)$ is bounded by $B \cdot l^6$, for a suitable B. Now this inverse image of U is a union of $2(l-1)$ punctured disks U_c and V_d, with c and d running through $\mathbb{F}_l^\times$ given by

$$U_c = \{(\mathbb{C}^\times/q^{5\mathbb{Z}}, q, \zeta_l^c) : q \in D(0, e^{-2\pi/5})^*\}$$

and

$$V_d = \{(\mathbb{C}^\times/q^{5l\mathbb{Z}}, q^l, q^{5d}) : q \in D(0, e^{-2\pi/5l})^*\}.$$

For each c in $\mathbb{F}_l^\times$ the map $X_1(5l) \rightarrow X_1(5)$ restricts to the isomorphism $U_c \rightarrow U$ given by $q \mapsto q$. For each d in $\mathbb{F}_l^\times$, the restriction $V_d \rightarrow U$ is given by $q \mapsto q^l$.

Around the standard unramified cusp of $X_1(5l)$ we have the punctured disk

$$W_l = \{(\mathbb{C}^\times/q^{\mathbb{Z}}, \zeta_5, \zeta_l) : q \in D(0, e^{-2\pi/5})^*\}.$$

By applying a suitable element from the group of automorphisms of $X_1(5l)$ generated by the Atkin-Lehner pseudo-involutions and the diamond operators we can establish isomorphisms of the U_c with W_l.

The point $(\mathbb{C}^\times/q^{5\mathbb{Z}}, q, \zeta_l^c)$ in U_c is then first sent to $(\mathbb{C}^\times/q^{5\mathbb{Z}}, q, \zeta_l)$ by $\langle c^{-1} \rangle$, and then to $(\mathbb{C}^\times/q^{\mathbb{Z}}, \zeta_5, \zeta_l)$ by the Atkin-Lehner pseudo-involution that divides out by the group generated by the point of order 5 (see Definition 2.5.14). The coordinate q on W_l is then identified in this way with the coordinate q on U_c. We observe that μ is invariant under each automorphism of $X_1(5l)$. Lemma 11.3.7 gives a real number C_1 such that for all l and c, the positive real $(1, 1)$-form μ on U_c can be estimated from above by

$$\mu|_{U_c} \leq C_1 \cdot \frac{i}{2} dq d\bar{q}.$$

Similarly, for each d in $\mathbb{F}_l^\times$ a suitable automorphism of $X_1(5l)$ maps V_d to the punctured disk W_l':

$$W_l' = \{(\mathbb{C}^\times/q^\mathbb{Z}, \zeta_5, \zeta_l) : q \in D(0, e^{-2\pi/5l})^*\}.$$

Lemma 11.3.7 gives a real number C_2 such that for all l and d, the positive real $(1,1)$-form μ on V_d can be estimated from above by

$$\mu|_{V_d} \leq C_2 \cdot \frac{1}{(1 - e^{-2\pi/5l})^4} \cdot \frac{i}{2} dq d\overline{q}.$$

Now the function qb on U extends to a holomorphic function on a disk containing U, hence $|qb|$ is bounded on U. Hence there is a real number $C_3 > 1$ such that $|b|^2 + 1 \leq C_3 \cdot |q^{-2}|$ on U. It follows that on all U_c we have $|b|^2 + 1 \leq C_3 \cdot |q^{-2}|$, and that on all V_d we have $|b|^2 + 1 \leq C_3 \cdot |q^{-2l}|$ (note that under $V_d \to U$, $(\mathbb{C}^\times/q^{5l\mathbb{Z}}, q^l, q^{5d})$ is sent to $(\mathbb{C}^\times/q^{5l\mathbb{Z}}, q^l)$). We remark that $1/(1 - e^{-x}) = x^{-1}(1 + O(x))$ as x tends to 0 from above. Hence there is a $C_4 \in \mathbb{R}$ such that for all l,

$$\frac{1}{(1 - e^{-2\pi/5l})^4} \leq C_4 \cdot l^4.$$

We get, for all l:

$$\int_{b^{-1}U} \log(|b|^2 + 1)\,\mu$$

$$\leq (l-1)C_1 \int_{D(0, e^{-2\pi/5})} \log(C_3 \cdot |q^{-2}|) \cdot \frac{i}{2} dq d\overline{q}$$

$$+ (l-1)C_2 \frac{1}{(1 - e^{-2\pi/5l})^4} \int_{D(0, e^{-2\pi/5l})} \log(C_3 \cdot |q^{-2l}|) \cdot \frac{i}{2} dq d\overline{q}$$

$$\leq \int_{|z|<1} \left(l C_1 \log(C_3 \cdot |z|^{-2}) + C_2 C_4 l^5 \log(C_3 \cdot |z|^{-2l}) \right) \frac{i}{2} dz d\overline{z}$$

$$\leq \int_{|z|<1} \left(C_1 l \log(C_3 \cdot |z|^{-2}) + C_2 C_4 l^6 \log(C_3 \cdot |z|^{-2}) \right) \frac{i}{2} dz d\overline{z}$$

$$= (C_1 l + C_2 C_4 l^6) \cdot (\pi \log C_3 + \pi).$$

This finishes the proof of Proposition 11.6.1   □

11.7 FINAL ESTIMATES OF THE ARAKELOV CONTRIBUTION

We will now put the estimates of the preceding sections together, in the situation of Section 8.2. We briefly recall this situation. We have a prime number $l > 5$, and let X_l denote the modular curve $X_1(5l)$ over $\mathbb{Q}$, and g_l its genus. The Jacobian variety of X_l is denoted by J_l. In $J_l(\overline{\mathbb{Q}})[l]$ we have the $\mathrm{Gal}(\overline{\mathbb{Q}}/\mathbb{Q})$-module V that realizes the representation ρ from $\mathrm{Gal}(\overline{\mathbb{Q}}/\mathbb{Q})$ to $\mathrm{GL}_2(\mathbb{F})$ attached to a surjective ring morphism $f: \mathbb{T}(1, k) \twoheadrightarrow \mathbb{F}$ such that the image of ρ contains $\mathrm{SL}_2(\mathbb{F})$. We have an effective divisor D_0 on $X_{l, \mathbb{Q}(\zeta_l)}$, of degree g_l, supported on the cusps. For every x in V there is a unique effective divisor $D_x = Q_{x,1} + \cdots + Q_{x,g_l}$ of degree g_l such that x is the class of $D_x - D_0$. We have written $D_x = D_x^{\mathrm{fin}} + D_x^{\mathrm{cusp}}$, where D_x^{cusp} is the part of D_x supported on the cusps. The numbering of the $Q_{x,i}$ is such that $D_x^{\mathrm{fin}} = Q_{x,1} + \cdots + Q_{x,d_x}$. We have morphisms b_l and x'_l from $X_{l,\mathbb{Q}}$ to $\mathbb{P}^1_{\mathbb{Q}}$ that, seen as rational functions, have their poles contained in the set of cusps of X_l.

The following proposition gives upper bounds for the absolute heights of the algebraic numbers $b_l(Q_{x,i})$ and $x'_l(Q_{x,i})$, polynomial in l. The height function h used here is as defined in (4.1.6). The proof of the proposition combines the involved arguments of the previous sections.

11.7.1 Proposition *There is an integer c such that for all x in V as above, and for all $i \in \{1, \ldots, d_x\}$, the absolute heights of $b_l(Q_{x,i})$ and $x'_l(Q_{x,i})$ are bounded from above by $c \cdot l^{12}$.*

Proof We will write just b for b_l. Let V be as in the proposition. Theorem 9.1.1 shows that for all $x \in V$, all $i \in \{1, \ldots, d_x\}$, and all number fields K containing $\mathbb{Q}(\zeta_{5l})$ over which $Q_{x,i}$ is rational, we have
(11.7.2)

$$
h(b(Q_{x,i})) \leq \frac{1}{[K : \mathbb{Q}]} \left((Q_{x,i}, b^* \infty)_x + l^2 \sum_\sigma \sup_{X_\sigma} g_\sigma \right.
$$
$$
\left. + \frac{1}{2} \sum_\sigma \int_{X_\sigma} \log(|b|^2 + 1) \mu_{X_\sigma} \right) + \frac{1}{2} \log 2.
$$

Here $\mathcal{X}$ is the minimal regular model of X_l over $B := \mathrm{Spec}\, O_K$.

Let us first concentrate on the second and third terms of the right-hand side of (11.7.2). Theorem 11.3.1 gives an integer c_1 such that for all σ we have $\sup_{X_\sigma} g_\sigma \leq c_1 \cdot l^6$, uniformly in all l. Proposition 11.6.1 gives an integer c_2 such that for all σ we have $(1/2) \int_{X_\sigma} \log(|b|^2 + 1) \mu_{X_\sigma} \leq c_2 \cdot l^6$, uniformly in all l. Hence, for all $x \in V$, all $i \in \{1, \dots, d_x\}$ and all number fields K containing $\mathbb{Q}(\zeta_{5l})$ over which $Q_{x,i}$ is rational, we have

$$(11.7.3) \qquad h(b(Q_{x,i})) \leq \frac{1}{[K : \mathbb{Q}]}(Q_{x,i}, b^* \infty)_x + c_1 \cdot l^8 + c_2 \cdot l^6,$$

uniformly in all l and V.

We now concentrate on the first term. We recall that $b^* \infty$ is an effective cuspidal divisor on X_l. Let x be in V. We apply Theorem 9.2.5, where (in the notation of that theorem) we assume that all $Q_{x,i}$ ($i \in \{1, \dots, g_l\}$) are K-rational, K contains $\mathbb{Q}(\zeta_{5l})$, and P is a cusp. We also use the obvious fact that $\log \#\mathrm{R}^1 p_* O_{\mathcal{X}}(D_x)$ is nonnegative. That gives

$$\frac{(D_x, P)}{[K : \mathbb{Q}]} \leq \frac{1}{[K : \mathbb{Q}]} \left(-\frac{1}{2}(D_0, D_0 - \omega_{\mathcal{X}/B}) + 2g_l^2 \sum_{s \in B} \delta_s \log \#k(s) \right.$$
$$+ \sum_\sigma \log \|\vartheta\|_{\sigma, \sup} + \frac{g_l}{2}[K : \mathbb{Q}] \log(2\pi)$$
$$\left. + \frac{1}{2} \deg \det p_* \omega_{\mathcal{X}/B} + (D_0, P) \right).$$

Theorem 11.4.1, applied with $B = \mathrm{Spec}(\mathbb{Z}[\zeta_{5l}])$, gives

$$\frac{1}{[K : \mathbb{Q}]}|(D_0, D_0 - \omega_{\mathcal{X}/B})| = O(l^{10}),$$

and

$$\frac{1}{[K : \mathbb{Q}]}|(D_0, P)| = O(l^8),$$

as D_0 is an effective cuspidal divisor of degree g_l and $g_l = O(l^2)$. By Theorem 11.2.1 we have

$$\frac{1}{[K : \mathbb{Q}]} \sum_\sigma \log \|\vartheta\|_{\sigma, \sup} = O(l^6).$$

By Theorem 11.1.7 we have

$$\frac{1}{[K : \mathbb{Q}]} \deg \det p_* \omega_{\mathcal{X}/B} = O(l^2 \log l).$$

Finally, we have

$$\frac{g_l^2}{[K:\mathbb{Q}]} \sum_{s \in B} \delta_s \log \#k(s) = O(l^6),$$

by the following argument. The only nontrivial contributions come from s over 5 and over l. The total contribution at 5 is independent of which extension K of $\mathbb{Q}(\zeta_5)$ we use, and for $\mathbb{Q}(\zeta_5)$ there is only one s over 5, $k(s) = \mathbb{F}_5$, and δ_s equals the number of supersingular points in $X_1(l)(\overline{\mathbb{F}}_5)$, which is $O(l^2)$. The contribution from l can be computed over $\mathbb{Q}(\zeta_l)$. Then there is one s, and $k(s) = \mathbb{F}_l$, and δ_s is the number of supersingular points in $X_1(5)(\overline{\mathbb{F}}_l)$, which is $O(l)$ (see Section 8.1).

Putting these last estimates together, we get that there is an integer c_3 such that for all l and x we have

$$(11.7.4) \qquad\qquad \frac{1}{[K:\mathbb{Q}]}(D_x, P) \leq c_3 \cdot l^{10}.$$

Now D_x is the sum of g_l points $Q_{x,i}$. In order to get upper bounds for the individual $(Q_{x,i}, P)$ we need a lower bound for these. Theorem 11.4.2 gives us a lower bound if $Q_{x,i} \neq P$ and Theorem 11.4.1 gives us one if $P = Q_{x,i}$. Putting these together, we get an integer c_4 such that for all l, x and i,

$$\frac{1}{[K:\mathbb{Q}]}(Q_{x,i}, P) \geq c_4 \cdot l^6.$$

The last two estimates together imply that there is an integer c_5 such that for all l, x and i we have

$$\frac{1}{[K:\mathbb{Q}]}(Q_{x,i}, P) \leq c_5 \cdot l^{10}.$$

As $b^*\infty$ is an effective cuspidal divisor on X_l, of degree $O(l^2)$, the previous inequality implies that there is an integer c_6 such that for all l, x and $i \in \{1, \ldots, d_x\}$, we have

$$\frac{1}{[K:\mathbb{Q}]}(Q_{x,i}, b^*\infty)_x \leq c_6 \cdot l^{12}.$$

This finishes the proof concerning the height of $b_l(Q_{x,i})$. Corollary 11.5.4 then finishes the proof. $\square$

We recall, from the end of Section 8.2, that, in the situation as described at the beginning of this section, we take a linear combination $f_l := b_l + n x_l'$ with $0 \le n \le g_l^2 (\#\mathbb{F})^4$, such that under the map $f_l : X_{l,\overline{\mathbb{Q}}} \to \mathbb{P}^1_{\overline{\mathbb{Q}}}$ the divisors $D_x^{\mathrm{fin}} = Q_{x,1} + \cdots + Q_{x,d_x}$, for $x \in V$, have distinct images $f_{l,*} D_x^{\mathrm{fin}}$. Suppose that f_l is any such linear combination. The $f_{l,*} D_x^{\mathrm{fin}}$ are then distinct effective divisors of degree d_x on $\mathbb{A}^1_{\overline{\mathbb{Q}}}$.

For each x in V, we get a polynomial $P_{D_0, f_l, x}$ with coefficients in $\overline{\mathbb{Q}}$ given by

$$P_{D_0, f_l, x}(t) = \prod_{i=1}^{d_x} (t - f_l(Q_{x,i})) \quad \text{in } \overline{\mathbb{Q}}[t],$$

and the map that sends x to $P_{D_0, f_l, x}$ is injective and $\mathrm{Gal}(\overline{\mathbb{Q}}/\mathbb{Q}(\zeta_l))$-equivariant.

We have seen that there is an integer m with $0 \le m \le g \cdot (\#\mathbb{F})^4$ such that the map

$$a_{D_0, f_l, m} : V \to \overline{\mathbb{Q}}, \quad x \mapsto P_{D_0, f_l, x}(m) = \prod_{i=1}^{d_x} (m - f_l(Q_{x,i}))$$

is injective and hence a generator of the $\mathbb{Q}(\zeta_l)$-algebra $A_{l, \mathbb{Q}(\zeta_l)}$ associated with V. Assume that m is such an integer. The following theorem gives our final upper bound for the absolute height of the coefficients of the minimal polynomial

$$(11.7.5) \quad P_{D_0, f_l, m} = \prod_{x \in V} (T - a_{D_0, f_l, m}(x)) = \sum_j P_j T^j \quad \text{in } \mathbb{Q}(\zeta_l)[T]$$

of $a_{D_0, f_l, m}$ over $\mathbb{Q}(\zeta_l)$.

11.7.6 Theorem *There exists an integer c such that for all l, V, D_0, f_l, and m as above we have, for all P_j as in (11.7.5),*

$$h(P_j) \le c \cdot l^{14} \cdot (\#\mathbb{F})^2.$$

Proof Let c_1 be an integer as given by Proposition 11.7.1. Let l, V, D_0, f_l and m be as in the theorem. For each $x \in V$ and each i in $\{1, \ldots, d_x\}$

we have, by Proposition 11.7.1, the definition of f_l, and Lemma 4.2.3,

$$
\begin{aligned}
h(f_l(Q_{x,i})) &= h(b_l(Q_{x,i}) + nx'_l(Q_{x,i})) \\
&\leq \log 2 + h(b_l(Q_{x,i})) + h(nx'_l(Q_{x,i})) \\
&\leq \log 2 + h(b_l(Q_{x,i})) + h(n) + h(x'_l(Q_{x,i})) \\
&\leq \log 2 + c_1 \cdot l^{12} + \log\left(l^2 \cdot (\#\mathbb{F})^4\right) + c_1 \cdot l^{12} \\
&\leq 2c_1 \cdot l^{12} + 2\log l + 4\log(\#\mathbb{F}) + \log 2 \\
&\leq c_2 \cdot l^{12} + 4\log(\#\mathbb{F}),
\end{aligned}
$$

(11.7.7)

for $c_2 = 2c_1 + 1$.

In order to simplify the notation during the rest of this proof, we write a for $a_{D_0, f_l, m}$ and P for $P_{D_0, f_l, m}$. Then we have, for each $x \in V$ (using $d_x \leq l^2$ and (11.7.7)),

$$
\begin{aligned}
h(a(x)) &= h\left(\prod_{i=1}^{d_x} (m - f_l(Q_{x,i}))\right) \\
&\leq \sum_{i=1}^{d_x} h(m - f_l(Q_{x,i})) \\
&\leq \sum_{i=1}^{d_x} \left(\log 2 + h(m) + h(f_l(Q_{x,i}))\right) \\
&\leq d_x \cdot \left(\log 2 + \log\left(l^2(\#\mathbb{F})^4\right) + c_2 \cdot l^{12} + 4\log(\#\mathbb{F})\right) \\
&\leq c_3 \cdot l^{14} + 8l^2 \log(\#\mathbb{F}),
\end{aligned}
$$

(11.7.8)

where $c_3 = c_2 + 1$.

Let j be in $\{0, \ldots, \#V\}$. Then P_j is, up to a sign, the value of the elementary symmetric polynomial of degree $\#V - j$ evaluated in the $a(x)$, where x ranges through V. Lemma 4.2.4, together with (11.7.8), gives us

$$
\begin{aligned}
h(P_j) &\leq \#V \cdot \log 2 + \sum_{x \in V} h(a(x)) \\
&\leq \#V \cdot \left(\log 2 + c_3 \cdot l^{14} + 8l^2 \log(\#\mathbb{F})\right) \\
&\leq c \cdot (\#\mathbb{F})^2 \cdot l^{14},
\end{aligned}
$$

where $c = c_3 + 1$, and where we have used that $\log \#\mathbb{F} \leq l \log l$ since $\mathbb{F}$ is a quotient of $\mathbb{T}(1, k)$ and $k \leq l + 1$. $\square$

A last consequence of all Arakelovian estimates is the following upper bound for the term $\log \#\mathrm{R}^1 p_* O_{\mathcal{X}}(D_x)$ in Theorem 9.2.5.

11.7.9 Theorem *There is an integer c such that for all $l > 5$ prime and all x in V, we have*

$$\frac{1}{[K : \mathbb{Q}]} \log \#\mathrm{R}^1 p_* O_{\mathcal{X}}(D_x) \leq c \cdot l^{10}.$$

Proof We have already seen, in the proof of Proposition 11.7.1, that the right-hand side of the inequality in Theorem 9.2.5, divided by $[K : \mathbb{Q}]$, is bounded from above by a constant times l^{10}. We have also seen in Theorem 11.4.2 that the term $(D_x, P)/[K : \mathbb{Q}]$ on the left-hand side is bounded from below by a constant times l^8 (recall that D_x is of degree $O(l^2)$). This proves the inequality. $\square$

This upper bound will be very useful for us, as the next interpretation shows. Recall that X_l has good reduction over $\mathbb{Z}[1/5l]$.

11.7.10 Theorem *There is an integer c with the following property. Let l, X_l, V, and D_0 be as in the beginning of this section. A prime number $p \nmid 5l$ is said to be V-good, if for all x in $V - \{0\}$ the following two conditions are satisfied:*

1. *at all places v of $\overline{\mathbb{Q}}$ over p the specialization $D_{x,\overline{\mathbb{F}}_p}$ at v is the unique effective divisor on the reduction $X_{l,\overline{\mathbb{F}}_p}$ such that the difference with $D_{0,\overline{\mathbb{F}}_p}$ represents the specialization of x;*

2. *the specializations of the noncuspidal part D_x^{fin} of D_x at all v above p are disjoint from the cusps.*

Then we have

$$\sum_{p \text{ not } V\text{-good}} \log p \leq c l^{12} \cdot (\#\mathbb{F})^2.$$

Proof First of all, a prime number p satisfies conditions (1) and (2) for all x in $V - \{0\}$ if and only if it satisfies one of them, as $\mathrm{Gal}(\overline{\mathbb{Q}}/\mathbb{Q}(\zeta_l))$ acts transitively on $V - \{0\}$ by assumption.

We take K to be the extension of $\mathbb{Q}(\zeta_l)$ that corresponds to the transitive $\mathrm{Gal}(\overline{\mathbb{Q}}/\mathbb{Q}(\zeta_l))$-set $V - \{0\}$; this is the field of definition of one x in $V - \{0\}$. Then $[K : \mathbb{Q}] = O(l \cdot (\#\mathbb{F})^2)$. We define $S(V)$ to be the image in $\mathrm{Spec}(\mathbb{Z}[1/5l])$ of the support in $\mathrm{Spec}(O_K)$ of the finite O_K-module $R^1 p_* O_\mathcal{X}(D_x)$. By Theorem 11.7.9 we have

$$\log \#(\mathbb{Z}[1/5l] \otimes R^1 p_* O_\mathcal{X}(D'_x)) = O(l^{11} \cdot (\#\mathbb{F})^2),$$

and hence

$$\sum_{p \in S(V)} \log p = O(l^{11} \cdot (\#\mathbb{F})^2).$$

We claim that the p in $S(V)$ are precisely the primes $p \notin \{5, l\}$ such that condition (1) is not satisfied for x. To see this, we first note that for a morphism $O_K \to \overline{\mathbb{F}}_p$ the canonical map from $\overline{\mathbb{F}}_p \otimes_{O_K} R^1 p_* O_\mathcal{X}(D_x)$ to $H^1(X_{l,\overline{\mathbb{F}}_p}, O_{\mathcal{X}_{\overline{\mathbb{F}}_p}}(D_{x,\overline{\mathbb{F}}_p}))$ is an isomorphism (see Theorem III.12.1 of [Hart]; base change and cohomology in top dimension commute). The divisor $D_{x,\overline{\mathbb{F}}_p}$ is the unique effective divisor in its linear equivalence class if and only if $h^0(X_{l,\overline{\mathbb{F}}_p}, O_{\mathcal{X}_{\overline{\mathbb{F}}_p}}(D_{x,\overline{\mathbb{F}}_p})) = 1$, which, by Riemann-Roch, is equivalent to $h^1(X_{l,\overline{\mathbb{F}}_p}, O_{\mathcal{X}_{\overline{\mathbb{F}}_p}}(D_{x,\overline{\mathbb{F}}_p})) = 0$.

Now we let $T(V)$ denote the set of primes $p \notin \{5, l\}$ such that at least one specialization of D_x^{fin} at a place of K above p is not disjoint from the cusps. Taking into account that $[K : \mathbb{Q}] = O(l \cdot (\#\mathbb{F})^2)$, equation (11.7.4) gives us an upper bound

$$(D_x, P) = O(l^{11} \cdot (\#\mathbb{F})^2),$$

As the degree of D_x^{cusp} is at most $O(l^2)$, Theorem 11.4.1 gives us

$$|(D_x^{\mathrm{cusp}}, P)| = O(l^7 \cdot (\#\mathbb{F})^2), \quad \text{hence} \quad (D_x^{\mathrm{fin}}, P) = O(l^{11} \cdot (\#\mathbb{F})^2).$$

As the divisor Cusps has degree $O(l)$, we have

$$(D_x^{\mathrm{fin}}, \mathrm{Cusps}) = O(l^{12} \cdot (\#\mathbb{F})^2).$$

The intersection number $(D_x^{\mathrm{fin}}, \mathrm{Cusps})$ is the sum of $(D_x^{\mathrm{fin}}, \mathrm{Cusps})_{\mathrm{fin}}$, the contribution of the finite places, and $(D_x^{\mathrm{fin}}, \mathrm{Cusps})_\infty$, the contribution of the infinite places. We have

$$(D_x^{\mathrm{fin}}, \mathrm{Cusps})_\infty = \sum_{i,P,\sigma} -g_\sigma(Q_{x,i}, P),$$

where the sum is taken over the i with $1 \leq i \leq d_x$, over the cusps P, and the $\sigma\colon K \to \mathbb{C}$. Then Theorem 11.3.1 gives the upper bound

$$(D_x^{\mathrm{fin}}, \mathrm{Cusps})_{\mathrm{fin}} = (D_x^{\mathrm{fin}}, \mathrm{Cusps}) - (D_x^{\mathrm{fin}}, \mathrm{Cusps})_\infty = O(l^{12} \cdot (\#\mathbb{F})^2).$$

By the definition of our set $T(V)$ and the definition of $(D_x^{\mathrm{fin}}, \mathrm{Cusps})_{\mathrm{fin}}$, we get

$$\sum_{p \in T(V)} \log p \leq (D_x^{\mathrm{fin}}, \mathrm{Cusps})_{\mathrm{fin}} = O(l^{12} \cdot (\#\mathbb{F})^2).$$

The proof of the theorem is then finished by noticing that the set of primes $p \notin \{5, l\}$ that are not V-good is precisely the union of $S(V)$ and $T(V)$. $\square$

Chapter Twelve

Approximating V_f over the complex numbers

by Jean-Marc Couveignes

In this chapter, we address the problem of computing torsion divisors on modular curves with an application to the explicit calculation of modular representations. We assume that we are given an even integer $k > 2$, a prime integer $l > 6(k - 1)$, a finite field $\mathbb{F}$ with characteristic l, and a ring epimorphism $f : \mathbb{T}(1, k) \to \mathbb{F}$. We want to compute the associated Galois representation $\rho_f : \mathrm{Gal}(\overline{\mathbb{Q}}/\mathbb{Q}) \to \mathrm{GL}_2(\mathbb{F})$. This representation lies in the Jacobian variety of the modular curve $X_1(l)$. Indeed, let $f_2 : \mathbb{T}(l, 2) \to \mathbb{F}$ be the unique ring homomorphism such that $f_2(T_m) = f(T_m)$ for every positive integer m. Let V_f be the subgroup of $J_1(l)(\overline{\mathbb{Q}})[l]$ cut out by the kernel of f_2. This is a dimension 2 vector space over $\mathbb{F}$. Given a finite generating set $(t_1, \ldots, t_r)$ for $\ker(f_2)$, where $r = (l^2 - 1)/6$ as in Theorem 2.5.13, we can rewrite V_f as a finite intersection

$$(12.0.11) \qquad V_f = \bigcap_{1 \leq i \leq r} \ker\left(t_i, J_1(l)(\overline{\mathbb{Q}})[l]\right).$$

Then V_f realizes ρ_f and we may write ρ_f as a morphism

$$\rho_f : \mathrm{Gal}(\overline{\mathbb{Q}}/\mathbb{Q}) \to \mathrm{GL}(V_f).$$

We will assume that the image of ρ_f contains $\mathrm{SL}(V_f)$; otherwise ρ_f would be reducible according to Theorem 2.5.20. The reducible case is treated in Section 14.2.

We want to compute the splitting field K_f of ρ_f as an extension of $\mathbb{Q}$. Computing V_f using general algorithms from computer algebra, like Buchberger's algorithm, seems difficult in this situation because V_f is defined as a subset of the l-torsion subgroup inside $J_1(l)$. A naive algebraic description of V_f would lead us to write down an equation for $J_1(l)[l]$, something similar to an l-division polynomial for $J_1(l)$. The degree of such a polynomial would be the cardinality of the group of torsion points, that is, $l^{2g(X_1(l))}$, where the genus $g(X_1(l))$ of $X_1(l)$ grows quadratically in l. Such a degree is far too large for us: we are looking for an algorithm with polynomial time complexity in the size $\#V_l$ of V_l.

We describe below an algorithm for computing elements in V_f. This is a deterministic algorithm, and the running time is polynomial in $\#V_l$. We shall work with the Jacobian variety $J_1(5l)$ rather than with $J_1(l)$. We set

$$X = X_1(5l),$$

and we denote by g the genus of the latter curve. We note that the conditions above imply $k \geq 4$, $l \geq 19$, and $g \geq 289$.

Using the map $B_{5l,l,1} : X_1(5l) \to X_1(l)$ defined in Section 2.2, and the associated morphism $B_{5l,l,1}^* : J_1(l) \to J_1(5l)$ between Jacobian varieties, we can see the subspace V_f as an $\mathbb{F}_l$-subspace inside the l-torsion of the Jacobian variety

$$J = J_1(5l).$$

To avoid confusion we set

$$W_f := B_{5l,l,1}^*(V_f) \subset J_1(5l),$$

and we call W_f the *Ramanujan subspace* associated with f. Elements in J are represented by divisors on the curve X. For every class in W_f, we compute a sharp enough approximation for some divisor in this class. It will be explained in Chapter 14 how one can compute the field of definition of this divisor from such a complex approximation, using the upper bounds for the naive height of torsion divisors on modular curves proved in Section 11.7.

This chapter is organized as follows. In Section 12.1 we recall how points on X are represented using standard coordinates taking values in the complex unit disk. Section 12.2 recalls state of the art algorithms for computing the lattice of periods Λ of the Jacobian variety J of the modular curve X. Section 12.3 describes an algebraic model for $X_1(5l)$ and relates it to the analytic model $\Gamma_1(5l)\backslash\mathbb{H}^*$.

The next four sections collect useful intermediate results. Section 12.4 provides explicit inequalities relating coefficients and values of converging power series. In Section 12.5 we prove formal identities relating Jacobian and Wronskian determinants. These identities are useful for the local study of the Jacobi integration map. Section 12.6 collects simple quantitative facts about the Jacobi integration map. Section 12.7 relates several natural norms on the space of parabolic modular forms of weight 4.

A point on J can be represented in two different ways. We may consider it as a class $x + \Lambda$ in $\mathbb{C}^g/\Lambda = J(\mathbb{C})$. We may also fix a degree g divisor Ω on X and represent an element in $J(\mathbb{C})$ by a divisor $Q - \Omega$ in the corresponding linear equivalence class, where Q is an effective degree g divisor on X. In Sections 12.8 and 12.9 we adopt this latter point of view and we show that it is very convenient for computational purposes and leads to polynomial time algorithms. Unfortunately, the points we are interested in (the x belonging to the Ramanujan subspace W_f) are rather difficult to characterize and compute in this form. However, assuming the divisor Ω has been chosen correctly (e.g., take for Ω the divisor D_0 constructed in Section 8.1), to every x in W_f there corresponds a unique divisor Q_x such that $Q_x - \Omega$ lies in the class represented by $x + \Lambda \in J(\mathbb{C})$. Such a Q_x will be called a *Ramanujan divisor*. Computing $x \in \mathbb{C}^g$ is not too difficult because the defining equations of V_f given in equation (12.0.11) become linear in the analytic model $\mathbb{C}^g/\Lambda$. The difficulty then is to compute Q_x, once given x. This is a typical example of the inverse Jacobi problem. Section 12.10 provides a partial general solution for this problem: it explains how, given Ω and some $x + \Lambda$, one can find a divisor Q such that $Q - \Omega$ lies in the corresponding linear equivalence class. Since we are only working with approximations, we must control the error made in computing Q from x. The output divisor Q' is hopefully close to Q but most likely not equal to it. We call x' the image of Q' by the Jacobi integration map. Statements in Section 12.10 control the difference between x and x'. We cannot hope a much better result in full generality because, in general, the divisor Q is not even unique, because x could lie in the singular locus of the Jacobi map. To relate the distance between Q' and Q and the distance between x' and x, we need some information about the local behavior of the Jacobi map at $Q - \Omega$. Using results from Arakelov theory proved in Section 11.7,

we show in Sections 12.11, 12.12, and 12.13 that when x lies in W_f then
the error on x and the error on Q_x are nicely related. This finishes the proof
of the main Theorem 12.10.7 in this chapter. Section 12.14 provides a more
algebraic variant of this theorem.

In this chapter, we shall use several times the main statements in Chapter 5, in particular Lemma 5.4.22 and Theorem 5.4.2. These statements
basically say that it is possible to compute efficiently sharp approximations
of zeros of power series, provided we don't prospect near the boundary of
the disk of convergence. In particular these zeros are well conditioned: they
are not dramatically affected by a small perturbation of the series. We sug-
gest that the reader look at the first pages and main statements in Chapter 5
before going further in this chapter.

Notation: in this chapter, the symbol Θ stands for a positive effective
absolute constant. So any statement containing this symbol becomes true
if the symbol is replaced in every occurrence by some large enough real
number. The letter i stands for the square root of -1 in $\mathbb{C}$ having positive
imaginary part.

12.1 POINTS, DIVISORS, AND COORDINATES ON X

In this section we recall how points, functions, forms, and divisors are represented on a modular curve. We first recall in Section 12.1.1 how to define
convenient local analytic parameters on X. Section 12.1.6 is devoted to
simple divisors (divisors with all multiplicities equal to 1). We introduce
a quantified definition for simpleness that will be useful in Section 12.8.
Section 12.1.7 introduces a natural and convenient pseudo-distance on X.

12.1.1 Local analytic parameters

We denote by $\mathcal{F}$ the classical fundamental domain for the action of $\mathrm{SL}_2(\mathbb{Z})$
on the Poincaré upper half-plane $\mathbb{H}$. We set $D = \bar{\mathcal{F}} \cup \{\infty\}$. We set

$$T = \begin{pmatrix} 1 & 1 \\ 0 & 1 \end{pmatrix}.$$

For every positive integer w we set

$$D_w = \bigcup_{0 \leq k \leq w-1} T^k(D).$$

We denote by F_w the image of D_w by the map $z \mapsto \exp(2iw^{-1}\pi z)$. This is a compact subset of the open disk $D(0,1) \in \mathbb{C}$. It is even contained in $D(0, \exp(-\pi/w))$. Any cusp on $X(\mathbb{C})$ can be written $\gamma(\infty)$ for some $\gamma \in SL_2(\mathbb{Z})$. These γ can be chosen once for all with entries $\leq l^\Theta$ in absolute value. We denote by Ξ the set of all these chosen γ. The set Ξ parametrizes the cusps of X. We assume that the identity belongs to Ξ. It parametrizes the cusp ∞ itself. We write the topological space $X(\mathbb{C})$ as a union

$$X(\mathbb{C}) = \bigcup_{\gamma \in \Xi} \gamma(D_{w_\gamma}) = \bigcup_{\gamma \in \Xi} \bigcup_{0 \leq k \leq w_\gamma - 1} \gamma(T^k(D)),$$

where w_γ is the width of the cusp $\gamma(\infty)$. We say that the γT^k arising in this union form a *standard system of right cosets representatives* for $\Gamma_1(5l)$ in $SL_2(\mathbb{Z})$.

Every point on X is represented by a complex number z in $\gamma(D_{w_\gamma}) \subset \mathbb{H}^*$ for some γ in Ξ. But $\gamma^{-1}(z) \in D_{w_\gamma}$ will often be more convenient. And

(12.1.2) $$q_\gamma = \exp(2iw_\gamma^{-1}\pi\gamma^{-1}(z))$$

is even more convenient. So most of the time a complex point on X will be given as a pair (γ, q), where $\gamma \in \Xi$ and $q \in F_{w_\gamma} \subset D(0, \exp(-\pi/w_\gamma))$ is the value of q_γ at this point. We set

$$D_\gamma = \gamma(D_{w_\gamma}) \subset \Gamma_1(5l)\backslash\mathbb{H}^* = X_1(5l).$$

When γ is the identity Id, we sometimes write q instead of q_{Id}. The parameter q traditionally plays a more important role. Functions and forms on $X_1(5l)$ are often identified with their q-expansions. The function field $\mathbb{C}(X_1(5l))$ can thus be identified with a subfield of the field of Puiseux series $\mathbb{C}\{\{q\}\}$.

Given a complex number q in the open disk $D(0,1)$, let $z \in \mathbb{H} \cup \{\infty\}$ be such that

$$\exp(2iw_\gamma^{-1}\pi\gamma^{-1}(z)) = q.$$

In the special case $q = 0$ we set $z = \gamma(\infty)$. Such a z may not be unique. But two such z are mapped onto each other by some power of $\gamma \times T^{w_\gamma} \times \gamma^{-1}$. Since the latter lies in $\Gamma_1(5l)$ we have defined a map

$$\mu_\gamma : D(0, 1) \to X_1(5l).$$

This is the parametrization associated with γ. It sends F_{w_γ} onto D_γ. Any form (resp. function) on $X_1(5l)$ can be lifted to $D(0, 1)$ along the map μ_γ. For example, Klein's modular function $j(z)$ is usually given as a function of $q = q_{\mathrm{Id}}$. There exists a Laurent series $J(x)$ in the indeterminate x such that $j(z) = J(q)$. Further, there are rational integers $c(k)$, for $k \in \mathbb{Z}_{\geq 1}$, such that

(12.1.3) $$J(x) = \frac{1}{x} + 744 + \sum_{k \geq 1} c(k) x^k \quad \text{in } \mathbb{Z}((x)).$$

It can be checked easily that the expansion of j at the cusp $\gamma(\infty)$ is given by

(12.1.4) $$j(z) = J(q_\gamma^{w_\gamma}),$$

where w_γ is the width of the cusp $\gamma(\infty)$. It is a consequence of a famous theorem by Petersson and Rademacher that

(12.1.5) $$|c(k)| \leq \Theta^{\sqrt{k}}.$$

12.1.6 Simple divisors

Let d be a positive integer and let $Q = Q_1 + Q_2 + \cdots + Q_d$ be a degree d effective divisor on X. Let ϵ be a nonnegative real number. We say that Q is ϵ-simple if the following conditions hold true.

1. For every integer k such that $1 \leq k \leq d$, the point Q_k belongs to D_{γ_k} for a unique γ_k in Ξ. So $Q_k = (\gamma_k, q_k)$. We ask that q_k lies in the interior of F_{w_k} where w_k is the width of the cusp $\gamma_k(\infty)$.

2. The distance between q_k and the boundary of F_{w_k} is $> \epsilon$.

3. If $1 \leq k_1 < k_2 \leq d$ and $\gamma_{k_1} = \gamma_{k_2} = \gamma$, we write $Q_{k_1} = (\gamma, q_1)$ and $Q_{k_2} = (\gamma, q_2)$ and we ask that $|q_2 - q_1| > \epsilon$.

Not every divisor Q is ϵ-simple, but if $\epsilon < 1/(d\Theta)$ there exists an ϵ-simple divisor $Q' = Q'_1 + Q'_2 + \cdots + Q'_d$ such that for every $1 \leq k \leq d$ we have $Q'_k = (\gamma_k, q'_k)$ and $|q'_k - q_k| \leq \Theta d\epsilon$.

12.1.7 A pseudo-distance on X

Given two points P_1 and P_2 on X, we would like to measure how far P_2 is from P_1. We pick some γ in Ξ and try to write P_1 as (γ, q_1) and P_2 as (γ, q_2) for some q_1 and q_2 in $D(0, \exp(-\pi/w_\gamma))$. If this is not possible, we set $d_\gamma(P_1, P_2) = +\infty$. Otherwise, we define $d_\gamma(P_1, P_2)$ to be the minimum of $|q_2 - q_1|$ for all possible q_1 and q_2. So

$$d_\gamma(P_1, P_2) = \min_{\substack{q_1, q_2 \in D(0, \exp(-\pi/w_\gamma)) \\ \mu_\gamma(q_1) = P_1 \\ \mu_\gamma(q_2) = P_2}} |q_2 - q_1|.$$

We set

$$d_X(P_1, P_2) = \min_{\gamma \in \Xi} d_\gamma(P_1, P_2).$$

The function $d_X : X \times X \to \mathbb{R}^+ \cup \{\infty\}$ is symmetric and separating. We avoid calling it a distance because it does not fulfill the triangle inequality.

12.2 THE LATTICE OF PERIODS

This section is devoted to the explicit calculation of the lattice of periods of X. All the algorithms in this section are detailed in the two books by Cremona [Cre] and Stein [Ste2] and in Bosman's thesis [Bos2]. See also Chapter 6.

We need a complex analytic description of the torus $J(\mathbb{C})$ as $\mathbb{C}^g/\Lambda$, where Λ is the lattice of periods. We first compute an explicit description of the first singular cohomology group

$$H_1^{\text{sing}}(X_1(5l), \mathbb{Z}).$$

Using Manin-Shokurov theory we find a basis $\mathcal{B}_1^{\text{sing}}$ for this $\mathbb{Z}$-module. Every element in this basis is an integer linear combination of Manin symbols

$$(12.2.1) \qquad \sum_\gamma c_\gamma \{\gamma(0), \gamma(\infty)\}.$$

The γ in the sum (12.2.1) runs over the standard system of right cosets representatives for $\Gamma_1(5l)$ in $SL_2(\mathbb{Z})$. The integer coefficients c_γ can be chosen to be such that

$$|c_\gamma| \leq \exp(l^\Theta).$$

We also need a basis $\mathcal{B}_{DR}^1$ of the space of holomorphic differentials

$$\mathcal{H}^1 = H_{DR}^{1,0}(X_1(5l)),$$

or equivalently a basis of the space $S_2(\Gamma_1(5l))$ of weight two cusp forms. We shall use the standard basis made of normalized newforms of level $5l$ together with normalized newforms of level l lifted to level $5l$ by the two degeneracy maps.

Let $f = \sum_{k \geq v} f_k q^k$ be a form in this basis. The q-valuation v of f is 1 or 5. The first nonzero coefficient f_v in the q-expansion of f is 1. The coefficients f_k in the q-expansion of f are algebraic integers. The absolute value of f_k is $\leq k^\Theta$. One can compute an approximation of f_k within $\exp(-m)$ in deterministic polynomial time $(klm)^\Theta$.

The action of Atkin-Lehner involutions is expressed in the basis $\mathcal{B}_{DR}^1$ by Theorem 2 of [Asa]. The action of the diamond operators is known also because every form in $\mathcal{B}_{DR}^1$ is an eigenform for the Hecke algebra $\mathbb{T}^{(5l)}$ generated by the operators T_n for n prime to $5l$.

We also need the expansion of every form $f(q)$ in $\mathcal{B}_{DR}^1$ at every cusp $\gamma(\infty)$. More precisely, $f(q)q^{-1}dq$ should be rewritten as $h(q_\gamma)q_\gamma^{-1}dq_\gamma$ for every γ in Ξ. Since the level $5l$ is square-free, the group generated by the Atkin-Lehner involutions and the diamond operators acts simply transitively on the cusps. So there is an automorphism in this group that sends ∞ to $\gamma(\infty)$. This automorphism can be represented by a matrix W_γ in $GL_2(\mathbb{Q})$ having integer entries, as explained in Section 6.2 of Chapter 6. If

$$\gamma = \begin{pmatrix} a & b \\ c & d \end{pmatrix}$$

then the width w_γ of the cusp $\gamma(\infty)$ is

$$w_\gamma = \frac{5l}{\gcd(5l, c)}.$$

Let r be the unique integer in $[0, w_\gamma[$ such that $d \equiv cr \bmod w_\gamma$. Set $b' = b - ar$ and $d' = (d - cr)/w_\gamma$ and $c' = c/\gcd(5l, c)$. Then

$$(12.2.2) \qquad W_\gamma = \begin{pmatrix} aw_\gamma & b' \\ 5lc' & w_\gamma d' \end{pmatrix},$$

and the product

$$(12.2.3) \qquad W_\gamma^{-1} \times \gamma = \begin{pmatrix} w_\gamma^{-1} & rw_\gamma^{-1} \\ 0 & 1 \end{pmatrix}$$

fixes ∞ and acts on Fourier expansions like the substitution $q \mapsto \zeta q^{1/w_\gamma}$ for some root of unity $\zeta = \exp(\frac{2ri\pi}{w_\gamma})$. Since the action of W_γ on forms is known, we can compute the expansion of all forms in $\mathcal{B}_{\mathrm{DR}}^1$ at all cusps in deterministic polynomial time $(klm)^\Theta$, where k is the q-adic accuracy and m the complex absolute accuracy of coefficients.

Once we have computed bases for both the singular homology and the de Rham cohomology we can compute the lattice Λ of periods. Since we are given a basis $\mathcal{B}_{\mathrm{DR}}^1$ of holomorphic differentials, the lattice Λ is well defined inside $\mathbb{C}^{\mathcal{B}_{\mathrm{DR}}^1}$ as the image of the integration map $\mathrm{H}_1^{\mathrm{sing}}(X_1(5l), \mathbb{Z}) \to \mathbb{C}^{\mathcal{B}_{\mathrm{DR}}^1}$ sending a cycle c onto the vector $(\int_c \omega)_{\omega \in \mathcal{B}_{\mathrm{DR}}^1}$. The image of the basis $\mathcal{B}_1^{\mathrm{sing}}$ by the integration map is a basis $\mathcal{B}_{\mathrm{per}}$ of the lattice Λ of periods. So we now have three bases

$$\mathcal{B}_1^{\mathrm{sing}}, \ \mathcal{B}_{\mathrm{DR}}^1, \ \text{and} \ \mathcal{B}_{\mathrm{per}} \ \text{ for } \ \mathrm{H}_1^{\mathrm{sing}}(X_1(5l), \mathbb{Z}), \ \mathrm{H}_{\mathrm{DR}}^{1,0}(X_1(5l)), \ \text{ and } \Lambda,$$

respectively. The so-called matrix of periods has entries $\int_c \omega$, where c is a cycle in the basis $\mathcal{B}_1^{\mathrm{sing}}$ of $\mathrm{H}_1^{\mathrm{sing}}(X_1(5l), \mathbb{Z})$ and $\omega = f(q)q^{-1}dq$ is a form in the basis $\mathcal{B}_{\mathrm{DR}}^1$ of $\mathrm{H}_{\mathrm{DR}}^{1,0}(X_1(5l))$. Computing these periods reduces to evaluating integrals of the form $\int_\alpha^\beta f(q)q^{-1}dq$, where α and β are two cusps. We first cut this integral in two pieces

$$\int_\alpha^\beta f(q)q^{-1}dq = \int_i^\beta f(q)q^{-1}dq - \int_i^\alpha f(q)q^{-1}dq.$$

Since the group generated by Atkin-Lehner involutions and diamond operators acts transitively on the cusps, we can reduce to the computation of integrals of the form $\int_\alpha^\infty f(q)q^{-1}dq$ where $\alpha = (a + bi)/c$ and a, b and c are integers bounded by l^Θ in absolute value. Since the coefficients f_k

in the q-expansion of f are bounded by k^Θ, we can compute approximations of the entries in the period matrix within $\exp(-m)$ in deterministic
polynomial time $(lm)^\Theta$.

We note that the L^∞ norm on $\mathbb{C}^{\mathcal{B}^1_{\mathrm{DR}}}$ induces a distance d_J on the quotient
$\mathbb{C}^{\mathcal{B}^1_{\mathrm{DR}}}/\Lambda = J(\mathbb{C})$

$$(12.2.4) \qquad d_J(x + \Lambda, y + \Lambda) = \min_{z \in \Lambda} |x - y - z|_\infty.$$

This distance will be useful when evaluating rounding errors in the course
of numerical computations. We denote by

$$\phi : X \to J$$

the Jacobi integration map. This map is well defined once we have chosen
a degree one divisor on X as origin. The integration map ϕ is Lipschitz in
the following sense. Assume that $P_1 = \mu_\gamma(q_1)$ and $P_2 = \mu_\gamma(q_2)$ for some
γ in Ξ and some q_1 and q_2 in $D(0, \exp(-\pi/w_\gamma))$. Then

$$d_J(\phi(P_2), \phi(P_1)) \le l^\Theta \times |q_2 - q_1|.$$

This will be proved in Section 12.6.1; see equation (12.6.3). We deduce that
for any two points P_1 and P_2 on X one has

$$d_J(\phi(P_2), \phi(P_1)) \le l^\Theta \times d_X(P_1, P_2).$$

For every positive integer k, we also denote by ϕ the integration map
$\phi : X^k \to J$. We denote by

$$\phi^\star : \mathrm{Div}(X) \to J$$

the map induced by ϕ on the group of divisors on X. The restriction of $\phi^\star$
to the subgroup $\mathrm{Div}^0(X)$ of degree zero divisors is independent of the origin
we have chosen.

12.3 MODULAR FUNCTIONS

Since we plan to compute the field of definition of some very special divisors on the modular curve $X = X_1(5l)$ we must be able to evaluate some
well chosen modular functions of weight 0 and level 5 or $5l$ at a given point

$z \in \mathbb{H}$. In this section we describe algebraic models for $X_1(5)$ and $X_1(5l)$ and we explain how to compute the expansions of the involved modular functions at every cusp. We first consider the curve $X_1(5)$ in Section 12.3.1. Section 12.3.19 introduces a plane singular model for $X_1(5l)$. The singularities of this model are studied in Section 12.3.21. We finally recall in Section 12.3.26 how to compute the zeta function of $X_1(5l)$.

12.3.1 The modular curve $X_1(5)$

In this section we recall the definition of several classical level 5 modular functions and show how to compute their expansions at each of the four cusps of $X_1(5)$. Let b be an indeterminate, and consider the elliptic curve E_b in Tate normal form with equation

$$(12.3.2) \qquad y^2 + (1+b)xy + by = x^3 + bx^2.$$

The point $P = (0,0)$ has order 5 and its multiples are $2P = (-b, b^2)$, $3P = (-b, 0), 4P = (0, -b)$. Call $\mathbb{P}_b^1$ the projective line with parameter b. The modular invariant of E_b is

$$(12.3.3) \qquad j(b) = -\frac{(b^4 + 12b^3 + 14b^2 - 12b + 1)^3}{b^5(b^2 + 11b - 1)}.$$

Let

$$s = -\frac{11 + 5\sqrt{5}}{2}$$

and $\bar{s}$ be the two complex roots of $b^2 + 11b - 1$. We call $A_\infty, A_0, A_s,$ and $A_{\bar{s}}$ the points on $\mathbb{P}_b^1$ corresponding to the values ∞, 0, s, and $\bar{s}$ of the parameter b. The elliptic curve $E_b \to \mathbb{P}_b^1 - \{A_\infty, A_0, A_s, A_{\bar{s}}\}$ is the universal elliptic curve with one point of order 5. So there exists a unique isomorphism between the modular curve $X_1(5) = \Gamma_1(5l)\backslash\mathbb{H}^*$ and $\mathbb{P}_b^1$ that is compatible with the moduli structure on either side. We want to compute this isomorphism. More precisely we compute the expansions of b at every cusp of $X_1(5)$. To this end we compare the curve in equation (12.3.2) and the Tate curve [Tat2] with equation

$$(12.3.4) \qquad y'^2 + x'y' = x'^3 + a_4 x' + a_6,$$

where

$$a_4 = -5 \sum_{n \geq 1} \sigma_3(n) q^n,$$

$$a_6 = - \sum_{n \geq 1} \frac{5\sigma_3(n) + 7\sigma_5(n)}{12} q^n,$$

and q is a formal parameter. We note that the coefficients in the expansions above are integers and we have

$$a_4 = \frac{1 - E_4(q)}{48},$$

$$a_6 = \frac{1 - 3E_4(q) + 2E_6(q)}{1728},$$

where

$$E_4(q) = 1 + 240 \sum_{n \geq 1} \sigma_3(n) q^n,$$

$$E_6(q) = 1 - 504 \sum_{n \geq 1} \sigma_5(n) q^n.$$

The modular invariant of the Tate curve in equation (12.3.4) is

$$(12.3.5) \qquad j(q) = \frac{1728 E_4^3(q)}{E_4^3(q) - E_6^2(q)} = \frac{1}{q} + 744 + 196884q + \cdots .$$

Any isomorphism between the two Weierstrass curves in equations (12.3.2) and (12.3.4) must take the form

$$(12.3.6) \qquad \begin{aligned} x &= u^2 x' + r, \\ y &= u^3 y' + su^2 x' + t. \end{aligned}$$

Straightforward calculation gives the following necessary and sufficient conditions for the affine transform in (12.3.6) to induce an isomorphism of Weierstrass curves:

(12.3.7)
$$u^2 = -\frac{E_4(q)}{E_6(q)} \times \frac{(b^2+1)(b^4+18b^3+74b^2-18b+1)}{(b^4+12b^3+14b^2-12b+1)},$$

$$r = \frac{u^2-b^2-6b-1}{12},$$

$$s = \frac{u-b-1}{2},$$

$$t = \frac{b^3+7b^2-(5+u^2)b+1-u^2}{24},$$

$$\frac{1728E_4^3(q)}{E_4^3(q)-E_6^2(q)} = -\frac{(b^4+12b^3+14b^2-12b+1)^3}{b^5(b^2+11b-1)}.$$

We can simplify these expressions a bit. The last one just means

$$j(q) = j(b).$$

From the classical ([Sch3], Proposition 7.1) identities

$$\left(\frac{q\,dj}{dq}\right)^2 = j(j-1728)E_4$$

and

$$\left(\frac{q\,dj}{dq}\right)^3 = -j^2(j-1728)E_6$$

we deduce

$$\left(\frac{q\,db}{dq}\right) = \frac{E_6(q)}{E_4(q)} \times \frac{b(b^2+11b-1)(b^4+12b^3+14b^2-12b+1)}{5(b^2+1)(b^4+18b^3+74b^2-18b+1)}.$$

So the expression for u^2 can be written

(12.3.8)
$$u^2 = -\frac{b(b^2+11b-1)}{5\left(\frac{q\,db}{dq}\right)}.$$

The expansion of j as a series in q has integer coefficients and can be computed using equation (12.3.5). We deduce the expansion of q as a series in j^{-1}

(12.3.9) $$q = j^{-1} + 744j^{-2} + 750420j^{-3} + \cdots.$$

It has integer coefficients and one can compute it from the expansion of j as a series in q using any reasonable algorithm for the reversion of a power series: brute force linear algebra or the more efficient algorithms in [Br-Ku] or the quasi-optimal algorithms in [Ke-Um].

We first study the situation locally at A_∞. A local parameter for $\mathbb{P}^1_b$ at A_∞ is b^{-1}. The expansion

$$(12.3.10) \qquad j^{-1} = -b^{-5} + 25b^{-6} + \cdots$$

of j^{-1} as a series in b^{-1} has integer coefficients and can be computed using equation (12.3.3) and standard algorithms for polynomial arithmetic. We substitute (12.3.10) in (12.3.9) and find

$$q = -b^{-5} + 25b^{-6} + \cdots .$$

One more reversion gives the expansion of b^{-1} as a series in $q^{\frac{1}{5}}$.

$$(12.3.11) \qquad b^{-1} = -q^{\frac{1}{5}} + 5q^{\frac{2}{5}} + \cdots .$$

This expansion defines an embedding of the local field at A_∞ inside the field of Puiseux series $\mathbb{C}\{\{q\}\}$ in the formal parameter q. We deduce the expansion of u^2 from equations (12.3.8) and (12.3.11)

$$(12.3.12) \qquad u^2 = q^{\frac{-2}{5}} + 4q^{\frac{-1}{5}} + 4 - 10q^{\frac{1}{5}} - 30q^{\frac{2}{5}} + \cdots .$$

We also find

$$(12.3.13) \qquad r = -1 - 5q^{\frac{1}{5}} - 10q^{\frac{2}{5}} + 35q^{\frac{4}{5}} + 45q + \cdots .$$

The coordinates of the 5-torsion point P on E_b are $x_P = 0$ and $y_P = 0$. The image of P on the Tate curve has x'-coordinate

$$x'_P = -ru^{-2} = q^{\frac{2}{5}} + q^{\frac{3}{5}} + 2q^{\frac{4}{5}} - 2q + \cdots .$$

Since on the Tate curve we have

$$(12.3.14) \qquad x'(w, q) = \sum_{n \in \mathbb{Z}} \frac{wq^n}{(1 - wq^n)^2} - 2\sum_{n \geq 1} \frac{nq^n}{1 - q^n},$$

we deduce that the value of the parameter w at the 5-torsion point P is

$$w(P) = q^{\pm \frac{2}{5}} \bmod \langle q \rangle.$$

We may take either sign in the exponent above because we may choose either isomorphism corresponding to either possible values for u. We decide that

(12.3.15) $$w(P) = q^{\frac{2}{5}} \bmod \langle q \rangle.$$

So the limit curve when $q \to 0$, or equivalently when $b \to \infty$, is a 5-gon, with a 5-torsion point lying on the second component after the one carrying the origin. So let

$$\gamma = \begin{pmatrix} a & b \\ c & d \end{pmatrix}$$

be any matrix in $\mathrm{SL}_2(\mathbb{Z})$ such that c is 2 modulo 5 and d is 0 modulo 5. For example, take

$$\gamma = \begin{pmatrix} 1 & 2 \\ 2 & 5 \end{pmatrix}.$$

In particular, the cusp $\gamma(\infty) = 1/2$ has width 5. We set

$$q_\gamma(z) = \exp(2i\pi\gamma^{-1}(z)/5).$$

If we replace $q^{1/5}$ by q_γ in equation (12.3.11) we turn b^{-1} into a modular function with weight 0 and level 5. Equation (12.3.11) with $q^{1/5}$ replaced by q_γ then gives the expansion of this modular function at the cusp $\gamma(\infty) = 1/2$. By construction, b has no pole outside the cusps.

If we replace $q^{1/5}$ by q_γ in equation (12.3.12), then $u^{-2}q_\gamma^{-1}dq_\gamma$ defines a differential on $\Gamma_1(5l)\backslash\mathbb{H}^*$, or equivalently a modular function with weight 2 and level 5. Equation (12.3.12) with $q^{1/5}$ replaced by q_γ then gives the expansion of this modular function at the cusp $\gamma(\infty) = 1/2$. By construction, u^{-2} has no zero and no pole outside the cusps.

We now study the situation locally at A_0. A local parameter at A_0 is b. We find $j^{-1} = b^5 + 25b^6 + \cdots$ and $q = b^5 + 25b^6 + \cdots$ and we fix an embedding of the local field at A_0 inside $\mathbb{C}\{\{q\}\}$ by setting

(12.3.16) $$b = q^{\frac{1}{5}} - 5q^{\frac{2}{5}} + 15q^{\frac{3}{5}} - 30q^{\frac{4}{5}} + \cdots.$$

Using equation (12.3.8) we deduce

(12.3.17) $$u^2 = 1 - 6q^{\frac{1}{5}} + 19q^{\frac{2}{5}} - 40q^{\frac{3}{5}} + 55q^{\frac{4}{5}} + \cdots,$$

and

$$r = -q^{\frac{1}{5}} + 4q^{\frac{2}{5}} - 10q^{\frac{3}{5}} + 15q^{\frac{4}{5}} + \cdots .$$

So the coordinate x'_P of the 5-torsion point P is

$$x'_P = q^{\frac{1}{5}} + 2q^{\frac{2}{5}} + 3q^{\frac{3}{5}} + 5q^{\frac{4}{5}} + 3q + O(q^{\frac{6}{5}}),$$

and the parameter w at P can be taken to be $w(P) = q^{\frac{1}{5}}$ mod $\langle q \rangle$ this time. The limit curve when $q \to 0$ or equivalently when $b \to 0$, is a 5-gon, with a 5-torsion point lying on the first component after the one carrying the origin. So let

$$\gamma = \begin{pmatrix} a & b \\ c & d \end{pmatrix}$$

be any matrix in $\mathrm{SL}_2(\mathbb{Z})$ such that c is 1 modulo 5 and d is 0 modulo 5. For example, take

$$\gamma = \begin{pmatrix} 0 & -1 \\ 1 & 0 \end{pmatrix}.$$

In particular, the cusp $\gamma(\infty) = 0$ has width 5. We set

$$q_\gamma(z) = \exp(2i\pi\gamma^{-1}(z)/5).$$

If we replace $q^{1/5}$ by q_γ in equations (12.3.16) and (12.3.17) we obtain the expansions of b and $u^{-2}q_\gamma^{-1}dq_\gamma$ at the cusp $\gamma(\infty) = 0$. The latter differential is the same as the one defined in the previous paragraph. So it defines the same modular function of weight 2 and level 5.

We now study the situation locally at A_s. A local parameter at A_s is $b - s$ and

$$j^{-1} = \left(-\frac{1}{2} + \frac{11\sqrt{5}}{50}\right)(b-s) + \left(-\frac{45443}{125} + \frac{4064\sqrt{5}}{25}\right)(b-s)^2 + \cdots ,$$

$$b - s = -\frac{125 + 55\sqrt{5}}{2}q - (375 + 170\sqrt{5})q^2 - \frac{3375 + 1495\sqrt{5}}{2}q^3 + \cdots ,$$

$$u^2 = -\frac{25 + 11\sqrt{5}}{2} - (200 + 90\sqrt{5})q - \frac{3575 + 1595\sqrt{5}}{2}q^2 + \cdots ,$$

$$r = -\frac{7 + 3\sqrt{5}}{2} - (100 + 45\sqrt{5})q - (1300 + 580\sqrt{5})q^2 + \cdots ,$$

and

$$x'_P = -\frac{1}{2} + \frac{\sqrt{5}}{10} - \frac{5 + \sqrt{5}}{2}q + \frac{-15 + \sqrt{5}}{2}q^2 + (-10 + \sqrt{5})q^3 + \cdots .$$

We note that the coordinate $x'(P)$ of the 5-torsion point P is

$$x'(P) = \frac{w}{(1 - w)^2} + O(q),$$

where $w = \exp(\frac{4i\pi}{5}) = \zeta_5^2$. So the parameter w at P can be taken to be $w(P) = \zeta_5^2 \bmod \langle q \rangle$ this time. The limit curve when $b \to s$ is thus a 1-gon equipped with the 5-torsion point ζ_5^2 in its smooth locus $\mathbb{G}_m$. Let

$$\gamma = \begin{pmatrix} a & b \\ c & d \end{pmatrix}$$

be any matrix in $\mathrm{SL}_2(\mathbb{Z})$ such that c is 0 modulo 5 and d is 2 modulo 5. For example, take

$$\gamma = \begin{pmatrix} 3 & 1 \\ 5 & 2 \end{pmatrix}.$$

In particular, the cusp $\gamma(\infty) = 3/5$ has width 1. We set

$$q_\gamma(z) = \exp(2i\pi\gamma^{-1}(z)).$$

If we replace q by q_γ in the expansions above we obtain the expansions of b and $u^{-2}q_\gamma^{-1}dq_\gamma$ at the cusp $\gamma(\infty) = 3/5$. The latter differential is the same as the one defined in the two previous paragraphs. So it defines the same modular function of weight 2 and level 5.

We finally study the situation locally at $A_{\bar{s}}$. A local parameter at A_s is $b - \bar{s}$ and

$$j^{-1} = \left(-\frac{1}{2} - \frac{11\sqrt{5}}{50} \right)(b - \bar{s}) - \left(\frac{45443}{125} + \frac{4064\sqrt{5}}{25} \right)(b - \bar{s})^2 + \cdots ,$$

$$b - \bar{s} = \frac{-125 + 55\sqrt{5}}{2}q + (-375 + 170\sqrt{5})q^2 + \frac{-3375 + 1495\sqrt{5}}{2}q^3 + \cdots ,$$

$$u^2 = \frac{-25 + 11\sqrt{5}}{2} + (-200 + 90\sqrt{5})q + \frac{-3575 + 1595\sqrt{5}}{2}q^2 + \cdots,$$

$$r = \frac{-7 + 3\sqrt{5}}{2} + (-100 + 45\sqrt{5})q + (-1300 + 580\sqrt{5})q^2 + \cdots,$$

and

$$x'_P = -\frac{1}{2} - \frac{\sqrt{5}}{10} + \frac{-5 + \sqrt{5}}{2}q - \frac{15 + \sqrt{5}}{2}q^2 - (10 + \sqrt{5})q^3 + \cdots.$$

We note that the coordinate $x'(P)$ of the 5-torsion point P is

$$x'(P) = \frac{w}{(1 - w)^2} + O(q),$$

where $w = \exp(\frac{2i\pi}{5}) = \zeta_5$. So the parameter w at P can be taken to be $w(P) = \zeta_5 \mod \langle q \rangle$ this time. The limit curve when $b \to \bar{s}$ is thus a 1-gon equipped with the 5-torsion point ζ_5 in its smooth locus $\mathbb{G}_m$. Let

$$\gamma = \begin{pmatrix} a & b \\ c & d \end{pmatrix}$$

be any matrix in $\mathrm{SL}_2(\mathbb{Z})$ such that c is 0 modulo 5 and d is 1 modulo 5. For example take

$$\gamma = \mathrm{Id} = \begin{pmatrix} 1 & 0 \\ 0 & 1 \end{pmatrix}.$$

In particular, the cusp $\gamma(\infty) = \infty$ has width 1. We set

$$q_{\mathrm{Id}}(z) = \exp(2i\pi z).$$

If we replace q by q_{Id} in the expansions above, we obtain the expansions of b and $u^{-2}q_{\mathrm{Id}}^{-1}dq_{\mathrm{Id}}$ at the cusp ∞. The latter differential is the same as the one defined in the three previous paragraphs. So it defines the same modular function of weight 2 and level 5.

Altogether we have proved the following.

12.3.18 Lemma (computing expansions of b and u^2) *There exists a deterministic algorithm that given an integer $k \geq 1$ computes the k first terms in the expansions of b and u^2 at each of the four cusps of $X_1(5)$, at the expense of k^{Θ} elementary operations.*

We recall that the functions b and u^2 are defined in equation (12.3.7). Further b is a weight zero and level 5 modular function having no pole outside the cusps. The weight 2 and level 5 modular function u^{-2} has no zero and no pole outside the cusps. We also notice that the coefficients in the expansions of b and u^2 lie in $\mathbb{Z}$ or $\mathbb{Z}[\frac{1+\sqrt{5}}{2}]$. Lemma 12.4.15 implies that there exists a positive constant Θ such that for every integer $k \geq 1$, the kth coefficient in these expansions has absolute value $\leq \exp(\Theta\sqrt{k})$.

12.3.19 A plane model for $X_1(5l)$

Let $l > 5$ be a prime. We call $X_l = X_1(5l)$ or just X the moduli of elliptic curves with one point of order $5l$. The genus of X_l is $g_l = (l-2)^2$. In this section we define and study a natural projective plane singular model C_l for this curve. In particular, we enumerate the geometric points on X_l above every singularity of C_l and we explain how to compute series expansions for affine coordinates at every such branch. Finally we recall how to compute the zeta function of the function field $\mathbb{F}_p(X_l)$ for every prime integer $p \notin \{5, l\}$.

Let b be an indeterminate and let E_b be the elliptic curve given in equation (12.3.2). The field $\mathbb{Q}(b)$ of rational fractions in b is the function field $\mathbb{Q}(X_1(5))$ of the modular curve $X_1(5)$ over $\mathbb{Q}$. The map

$$B_{5l,5,1} : X_l \to X_1(5)$$

introduced in Section 2.2 defines a degree $l^2 - 1$ extension $\mathbb{Q}(X_l)/\mathbb{Q}(b)$. We construct an explicit model for this extension. The multiplication by l isogeny

$$[l] : E_b \to E_b$$

induces a degree l^2 rational function on x-coordinates:

$$x \mapsto \frac{N(x)}{D(x)},$$

where $N(x)$ is a monic degree l^2 polynomial in $\mathbb{Q}(b)[x]$. Recursion formulae for division polynomials (see [Eng], Section 3.6) provide a quick algorithm for computing this polynomial, and also show that the coefficients of $N(x)$ actually lie in $\mathbb{Z}[b]$. The roots of $N(x)$ are the x-coordinates of the points Q on E_b such that $[l]Q$ is $P = (0, 0)$. If l is congruent to ± 1 modulo 5, then $lP = \pm P$ and x divides $N(x)$. Otherwise $N(x)$ is divisible by $x - b$. Call $T_l(b, x)$ the quotient of $N(x)$ by x or $x - b$, accordingly. This is a monic polynomial in $\mathbb{Z}[b][x]$ with degree $l^2 - 1$ in x. As a polynomial in x we have

(12.3.20) $$T_l(b, x) = \sum_{0 \le k \le l^2 - 1} a_{l^2 - 1 - k}(b) x^k$$

where $a_0(b) = 1$. We call d be the total degree of T_l.

Let F be a field extension of $\mathbb{Q}(b)$ where $T_l(b, x) \in \mathbb{Q}(b)[x]$ has a root. A suitable twist of the Tate curve E_b given by equation (12.3.2) has a point of order $5l$ defined over F. This proves that the function field extension $\mathbb{Q}(X_l)/\mathbb{Q}(b)$ can be embedded in $F/\mathbb{Q}(b)$. Since the degree of $T_l(b, x)$ in x is equal to the degree of the extension $\mathbb{Q}(X_l)/\mathbb{Q}(b)$ we deduce that the polynomial T_l is irreducible in $\mathbb{Q}(b)[x]$ and the quotient field $\mathbb{Q}(b)[x]/T_l$ is isomorphic to $\mathbb{Q}(X_l)$. Since the latter field is a regular extension of $\mathbb{Q}(b) = \mathbb{Q}(X_1(5))$ we deduce that T_l is absolutely irreducible.

We call $C_l \subset \mathbb{P}^2$ be the projective curve with homogeneous equation $T_l(\frac{b}{z}, \frac{x}{z}) z^d$ in the variables b, x, and z. The map $B_{5l,5,1} : X_1(5l) \to X_1(5)$ is unramified except at $b \in \{0, \infty, s, \bar{s}\}$. So for every point R on X_l such that $b(R) \notin \{0, s, \bar{s}, \infty\}$, the function $b - b(R)$ is a uniformizing parameter at R. Let $\mathcal{U}$ be the affine open set with equation

$$zb(b^2 - 11bz + z^2) \ne 0.$$

Every point on $C_l \cap \mathcal{U}$ is smooth and all points on X_l above points in $C_l - \mathcal{U}$ are cusps in the modular sense (i.e., the modular invariant has a pole at these points). The smooth point $R = (b_R, x_R)$ on $C_l \cap \mathcal{U}$ is the moduli of the curve E_{b_R} equipped with the unique $5l$-torsion point Q having x-coordinate x_R and such that $[l]Q = P = (0, 0) \in E_{b_R}$.

12.3.21 The singularities of C_l

We study the cusps of $X_1(5l)$ that are mapped onto A_∞ by $B_{5l,5,1}$. Set $\zeta_l = \exp(\frac{2i\pi}{l})$. Let α and β be integers such that $0 \le \alpha, \beta \le l - 1$. Let also $\tilde{\alpha}$ and $\tilde{\beta}$ be integers such that $0 \le \tilde{\alpha}, \tilde{\beta} \le l - 1$, and

$$\tilde{\alpha} \equiv \alpha/5 \bmod l,$$

and

$$\tilde{\beta} \equiv (\beta - 2)/5 \text{ mod } l.$$

We set $w_Q = \zeta_l^{\tilde{\alpha}} q^{\frac{\tilde{\beta}}{l}} q^{\frac{2}{5l}}$ and observe that

$$w_Q^5 = \zeta_l^{\alpha} q^{\frac{\beta}{l}} \text{ mod } \langle q \rangle$$

and

$$w_Q^l = q^{\frac{2}{5}} \text{ mod } \langle q \rangle = w_P,$$

according to equation (12.3.15). We denote by Q the point on the Tate curve in (12.3.4), having w-coordinate w_Q. The isomorphism given by equations (12.3.6) maps Q onto a $5l$-torsion point on the curve E_b. This point is called Q also. The couple (E_b, Q) defines a point on $X_1(5l)$ that is mapped onto (E_b, P) by $B_{5l,5,1}$. We substitute w by w_Q in expansion (12.3.14) and find

$$x_Q' = \zeta_l^{\tilde{\alpha}} q^{\frac{\tilde{\beta}}{l}} q^{\frac{2}{5l}} (1 + O(q^{\frac{1}{5l}}))$$

if $0 \leq \tilde{\beta} \leq \frac{l-1}{2}$ and

$$x_Q' = \zeta_l^{-\tilde{\alpha}} q^{\frac{l-\tilde{\beta}}{l} - \frac{2}{5l}} (1 + O(q^{\frac{1}{5l}}))$$

if $\frac{l+1}{2} \leq \tilde{\beta} \leq l - 1$. Using equation (12.3.6) and the expansions in equations (12.3.12) and (12.3.13) we find

(12.3.22) $$\qquad x_Q + 1 = \zeta_l^{\tilde{\alpha}} q^{\frac{\tilde{\beta}}{l} + \frac{2}{5l} - \frac{2}{5}} (1 + O(q^{\frac{1}{5l}}))$$

if $0 \leq \tilde{\beta} \leq \frac{l-1}{2}$ and

(12.3.23) $$\qquad x_Q + 1 = \zeta_l^{-\tilde{\alpha}} q^{\frac{l-\tilde{\beta}}{l} - \frac{2}{5l} - \frac{2}{5}} (1 + O(q^{\frac{1}{5l}}))$$

if $\frac{l+1}{2} \leq \tilde{\beta} \leq l - 1$.

Let

$$\gamma = \begin{pmatrix} a & b \\ c & d \end{pmatrix}$$

be any matrix in $SL_2(\mathbb{Z})$ such that c is $5\tilde{\beta} + 2$ modulo $5l$ and d is $5\tilde{\alpha}$ modulo $5l$. In particular, the cusp $\gamma(\infty) = a/c$ has width $w = \frac{5l}{\gcd(5l,c)}$. So w is 5 if $\beta = 0$ and $5l$ otherwise. We set

$$q_\gamma(z) = \exp(2i\pi\gamma^{-1}(z)/w).$$

If we replace $q^{1/w}$ by q_γ in equation (12.3.22) or (12.3.23) above we obtain the expansion of x at the cusp $\gamma(\infty) = a/c$. The same method applies to cusps of $X_1(5l)$ that are mapped onto A_0, A_s or $A_{\tilde{s}}$ by $B_{5l,5,1}$.

12.3.24 Lemma (computing expansions of x and b) *Let $l > 5$ be a prime integer. Let $C_l \subset \mathbb{P}^2$ be the singular plane model of $X_1(5l)$ given by equation (12.3.20). The only singularities of C_l correspond to cusps of $X_1(5l)$. There exists a deterministic algorithm that given an integer $k \geq 1$ computes the k first terms in the expansions of x and b at each of the cusps of $X_1(5l)$, at the expense of $(kl)^\Theta$ elementary operations. The coefficient of q_γ^k in these expansions is bounded from above by $\exp(l^\Theta \sqrt{|k| + 1})$.*

So we have a very accurate description of the singularities of C_l since for every branch of $X_1(5l)$ above such a singularity we can compute expansions of both coordinates b and x as series in the local parameter q_γ. We notice that the coefficients in these expansions lie in $\mathbb{Z}[\zeta_{5l}]$. The bound on the coefficients in the q_γ-expansions given in Lemma 12.3.24 results from Lemma 12.4.15 and equation (12.3.6). An interesting consequence of this bound is that we can bound from above the functions $x \circ \mu_\gamma$, $b \circ \mu_\gamma$ and their derivatives with respect to q_γ.

12.3.25 Lemma (size and variation of the functions $j, b,$ and x) *For every positive constant K_1 there exists a positive constant K_2 such that the following holds true. Let $l > 5$ be a prime and let $\gamma \in \Xi$ correspond to a cusp $\gamma(\infty)$ of $X_1(5l)$. Let w_γ be the width of this cusp. Let q_1 and q_2 belong to $D(0, \exp(-\pi/w_\gamma))$ and assume that both $|q_1|$ and $|q_2|$ are $\geq \exp(-l^{K_1})$. Let $P_1 = (\gamma, q_1) = \mu_\gamma(q_1)$ be the point on $X_1(5l)$ associated with q_1. Let $P_2 = (\gamma, q_2) = \mu_\gamma(q_2)$. Then $j(P_1), b(P_1)$ and $x(P_1)$ are bounded in absolute value by $\exp(l^{K_2})$. Similarly the absolute values of $j(P_2) - j(P_1), b(P_2) - b(P_1),$ and $x(P_2) - x(P_1)$ are bounded from above* *by* $\exp(l^{K_2}) \times |q_1 - q_1|$. *One can compute* $j(P_1)$, $b(P_1)$ *and* $x(P_1)$ *from* q_1 *in time polynomial in* l *and the required absolute accuracy.*

This lemma says that computing x and b is a well-conditioned problem as long as we don't get too close to a cusp.

12.3.26 Computing the zeta function of X

We shall also need the following result due to Manin, Shokurov, Merel, and Cremona [Man1, Mer, Cre, Fre].

12.3.27 Lemma (Manin, Shokurov, Merel, Cremona) *For* $l \geq 7$ *a prime and* $p \notin \{5, l\}$ *another prime, the zeta function of* X_l *mod* p *can be computed in deterministic polynomial time in* l *and* p.

Proof We first compute the action of the Hecke operator T_p on the space of Manin symbols for the congruence group $\Gamma_1(5l)$ associated with X_l. Then, from the Eichler-Shimura identity $T_p = F_p + p < p > / F_p$ we deduce the characteristic polynomial of the Frobenius F_p. $\qquad\square$

12.4 POWER SERIES

In this section we give some notation and state a few useful elementary properties of power series in one or several variables. We are mainly interested in relating the size of coefficients in the expansions and the size of the values of the series where it converges. In the course of our calculations we shall encounter three kinds of power series. Expansions of normalized eigenforms have rather small coefficients, according to the Ramanujan conjecture. To deal with such expansions we introduce in Definition 12.4.2 the *type* of a power series. We shall also consider modular functions having no pole outside the cusps. The coefficients in the expansions of such functions may be larger, but they are controlled by the Petersson and Rademacher inequality as explained in Lemma 12.4.15 below. Even more general modular functions may have quite big coefficients. To deal with this case, we introduce in Definition 12.4.19 the exp-type of a power series.

12.4.1 The type of a power series

Let $g \geq 1$ be an integer. The L^∞ norm of a vector $\mathbf{x} = (x_1, \ldots, x_g) \in \mathbb{C}^g$ is $\max_k |x_k|$, and it is denoted $|\mathbf{x}|_\infty$. The L^1 norm is $|\mathbf{x}|_1 = \sum_k |x_k|$ and the L^2 norm is $|\mathbf{x}|_2 = \sqrt{\sum_k |x_k|^2}$. We set

$$\|\mathbf{x}\| = (|x_1|, \ldots, |x_g|).$$

If $\mathbf{y} = (y_1, \ldots, y_g)$ is another vector in $\mathbb{C}^g$, we denote by

$$\mathbf{x} \star \mathbf{y} = (x_1 y_1, \ldots, x_g y_g)$$

the componentwise product. If $\mathbf{n} = (n_1, \ldots, n_g) \in \mathbb{N}^g$, we write

$$\mathbf{n}! = n_1! n_2! \cdots n_g! \quad \text{and} \quad \mathbf{x}^\mathbf{n} = x_1^{n_1} \cdots x_g^{n_g}.$$

We set $\mathbf{0}_g = (0, \ldots, 0)$, and $\mathbf{1}_g = (1, \ldots, 1)$, and $\mathbf{2}_g = (2, \ldots, 2)$. We denote $P(\mathbf{x}, \mathbf{r}) = \prod_{k=1}^g D(x_k, r_k) \subset \mathbb{C}^g$, the polydisk with center $\mathbf{x}$ and polyradius $\mathbf{r}$. When $\mathbf{r} = (r, r, \ldots, r)$ we just write $P(\mathbf{x}, r)$.

If $\mathbf{x} = (x_1, \ldots, x_g) \in \mathbb{R}^g$, we say that $\mathbf{x} \geq \mathbf{0}_g$ if and only if $x_k \geq 0$ for every k. We say that $\mathbf{x} > \mathbf{0}_g$ if and only if $x_k > 0$ for every k. A formal power series f in the g variables $x_1, \ldots, x_g$ is a formal sum $f = \sum_\mathbf{k} f_\mathbf{k} \mathbf{x}^\mathbf{k}$, where the index $\mathbf{k}$ runs over $\mathbb{N}^g$.

12.4.2 Definition (type of a power series in several variables) Let A be a real number ≥ 1 and $\mathbf{n} = (n_1, \ldots, n_g) \in \mathbb{N}^g$ such that $\mathbf{n} \geq \mathbf{1}_g$. We say that f is of type $(A, \mathbf{n})$ if for every $\mathbf{k} \geq \mathbf{0}_g$ we have

$$|f_\mathbf{k}| \leq A(\mathbf{k} + \mathbf{1}_g)^\mathbf{n} = A \prod_{1 \leq m \leq g} (k_m + 1)^{n_m}.$$

For every $\mathbf{z} \in P(\mathbf{0}_g, 1)$ we deduce an upper bound for the value of f at $\mathbf{z}$.

$$|f(\mathbf{z})| \leq \sum_{\mathbf{k} \geq \mathbf{0}_g} A(\mathbf{k} + \mathbf{1}_g)^\mathbf{n} |\mathbf{z}^\mathbf{k}|$$

$$\leq A \prod_{1 \leq m \leq g} \sum_{k_m \geq 0} (k_m + 1)^{n_m} |z_m|^{k_m}$$

(12.4.3)

$$\leq \frac{\mathbf{n}! A}{\prod_m (1 - |z_m|)^{n_m + 1}}$$

$$= \frac{\mathbf{n}! A}{(\mathbf{1}_g - \|\mathbf{z}\|)^{\mathbf{n} + \mathbf{1}_g}}.$$

We check that if f is of type $(A, \mathbf{a})$ and h is of type $(B, \mathbf{b})$, then the product fh is of type

(12.4.4) $$(AB, \mathbf{a} + \mathbf{b} + 1_g).$$

12.4.5 Refocusing a power series

Given a formal power series f of type $(A, \mathbf{n})$ and a vector

$$\mathbf{c} = (c_1, \ldots, c_g) \in P(\mathbf{0}_g, 1),$$

we set

$$F_\mathbf{c}(\mathbf{y}) = F_\mathbf{c}(y_1, \ldots, y_g)$$
$$= f(\mathbf{c} + \mathbf{y} \star (1_g - \|\mathbf{c}\|)) = f((c_m + y_m(1 - |c_m|))_m).$$

We call $F_\mathbf{c}$ the *refocused* series of f at $\mathbf{c}$. According to [Cou], Lemme 16, this is a series of type $(A_\mathbf{c}, \mathbf{n} + 1_g)$, where

(12.4.6) $$A_\mathbf{c} = \mathbf{n}! A \exp(g + |\mathbf{n}|_1) 2^{g + |\mathbf{n}|_1} (1_g - \|\mathbf{c}\|)^{-\mathbf{n} - 2g}.$$

In particular, it converges for $\mathbf{y} \in P(\mathbf{0}_g, 1)$.

12.4.7 Bounding the remainder

For any integer $u \geq 0$ we denote by $R_u(\mathbf{x})$, the remainder of order u of the series $f(\mathbf{x})$. So

$$f(\mathbf{x}) = \sum_{|\mathbf{k}|_1 \leq u-1} f_\mathbf{k} \mathbf{x}^\mathbf{k} + R_u(\mathbf{x}).$$

According to [Cou], Lemme 17, if f is of type $(A, \mathbf{n})$ and if r is a real in $]0, 1[$ and if $|\mathbf{z}| \leq r$, then

(12.4.8) $$|R_u(\mathbf{z})| \leq B(u + 1)^{(n+1)g} |\mathbf{z}|_\infty^u$$

where $n = |\mathbf{n}|_\infty$ and

$$B = \frac{\mathbf{n}! 2Ag}{(1 - r)^{g + |\mathbf{n}|_1}}.$$

Further, if κ is a real in $]0, 1[$ and if

$$u \geq \max \left(\frac{16(ng)^2}{(\log r)^2}, \frac{2(\log \kappa - \log B)}{\log r} \right),$$

then

$$|R_u(\mathbf{z})| \le \kappa \text{ for } |\mathbf{z}| \le r.$$

12.4.9 The type of a quotient

Let $f = \sum_{\mathbf{k}} f_{\mathbf{k}} \mathbf{x}^{\mathbf{k}}$ be a formal power series in the $g \ge 2$ variables $x_1, \ldots,$ x_g. Assume that f has type $(A, \mathbf{n})$ for some $A \ge 1$ and $\mathbf{n} \ge \mathbf{1}_g$. Assume that f is divisible by a polynomial $P(x_1, \ldots, x_g)$. So there exists a formal power series $h = \sum_{\mathbf{k}} h_{\mathbf{k}} \mathbf{x}^{\mathbf{k}}$ such that $f = Ph$. We want to estimate the size of coefficients in h. We shall only need the case when $P = x_2 - x_1$. So we restrict to this special case. Let $\mathbf{k} = (k_1, \ldots, k_g)$. For every integer m such that $2 \le m \le g$ we set

$$u_m = \frac{k_m + 1}{k_m + n_m + 1}.$$

We set

$$\hat{u}_1 = \frac{k_1 + 1}{k_1 + n_1 + 1}.$$

If $|\hat{u}_1 - u_2| < |\frac{\hat{u}_1 + 1}{2} - u_2|$ we set $u_1 = \frac{\hat{u}_1 + 1}{2}$. Otherwise we set $u_1 = \hat{u}_1$. In any case

$$\frac{1}{|1 - u_1|} \le \frac{2(k_1 + n_1 + 1)}{n_1},$$

$$\frac{1}{u_1} \le \frac{k_1 + n_1 + 1}{k_1 + 1},$$

and

$$\frac{1}{|u_2 - u_1|} \le \frac{3(k_2 + n_2 + 1)}{n_2}.$$

From Cauchy's integral $h_{\mathbf{k}}$ is equal to

$$\frac{1}{(2\pi i)^g} \int_{|z_1| = u_1} \cdots \int_{|z_g| = u_g} \frac{f(z_1, z_2, \ldots, z_g)}{(z_2 - z_1) \prod_{1 \le m \le g} z_m^{k_m + 1}} dz_1 dz_2 \ldots dz_g.$$

Using equation (12.4.3) we find

$$|f(z_1, z_2, \ldots, z_g)| \leq \frac{\mathbf{n}! A}{\prod_{1 \leq m \leq g} (1 - u_m)^{n_m + 1}}$$

$$\leq \mathbf{n}! 2^{n_1 + 1} A \prod_{1 \leq m \leq g} \left(\frac{k_m + n_m + 1}{n_m} \right)^{n_m + 1}.$$

So

$$|h_\mathbf{k}| \leq \frac{3\mathbf{n}! 2^{n_1 + 1} A(k_2 + n_2 + 1)}{n_2} \prod_{1 \leq m \leq g} \frac{(k_m + n_m + 1)^{k_m + n_m + 1}}{(k_m + 1)^{k_m} (n_m)^{n_m + 1}}.$$

So h has type $(12\mathbf{n}! 2^{n_1 + |n|_1 + g} A \exp(|n|_1), \mathbf{n} + 2 \times \mathbf{1}_g)$.

Now if we manage to divide f by K polynomials of the form $x_{j_1} - x_{j_2}$ we obtain a series of type

(12.4.10) $\qquad (A \exp(8gK(|\mathbf{n}|_\infty + 2K)^2), \mathbf{n} + 2K \times \mathbf{1}_g).$

12.4.11 The type of derivatives

If f is a formal power series in one variable of type (A, n), then the derivative f' of f is of type

$$(A2^n, n + 1).$$

So the the dth derivative of f is of type

(12.4.12) $\qquad (A2^{dn + \frac{d(d-1)}{2}}, n + d).$

12.4.13 The Petersson-Rademacher inequality

The modular functions that appear in Lemma 12.3.18 are not modular forms: they have poles at the cusps. Since we plan to evaluate these functions at well chosen points z in the Poincaré upper half-plane, we must control the size of the coefficients in the expansions of these functions.

Let $f : \mathbb{H} \to \mathbb{C}$ be a holomorphic periodic function with integer period e. So $f(z + e) = f(z)$ for every $z \in \mathbb{H}$. We assume that f is meromorphic at ∞. So f can be written as a series

$$f(z) = \mathbf{F}(q_e) = \sum_{\frac{k}{e} \geq v} a_{\frac{k}{e}} q_e^k,$$

where

$$q_e = \exp\left(\frac{2i\pi z}{e}\right),$$

$v \in \frac{1}{e}\mathbb{Z}$ is the valuation of f at ∞, the $a_{\frac{k}{e}}$ are the coefficients in the Puiseux expansion of f at ∞, and the series $\mathbf{F}$ converges in the unit disk.

Conversely, to every Puiseux series

$$\sum_{\frac{k}{e} \geq v} a_{\frac{k}{e}} q^{\frac{k}{e}}$$

with radius of convergence ≥ 1 we can associate a holomorphic function $f : \mathbb{H} \to \mathbb{C}$ having period e, and meromorphic at ∞.

By abuse of notation we say that f is a Puiseux series with radius ≥ 1 and period e. The set of all such functions is a ring. An example of such a series is provided by Klein's modular invariant $j(z) = J(q)$ where J is the series in equation (12.1.3). In view of the Petersson-Rademacher inequality (12.1.5) it is natural to state the following lemma.

12.4.14 Lemma (The Petersson and Rademacher's property) *Let*

$$f(z) = \sum_{\frac{k}{e} \geq v} a_{\frac{k}{e}} q^{\frac{k}{e}}$$

be a Puiseux series with radius of convergence ≥ 1. Then the two properties below are equivalent.

1. *There exist two positive constants K_1 and K_3 such that for every integer $k \geq K_3$ we have*

$$\left|a_{\frac{k}{e}}\right| \leq \exp\left(K_1\sqrt{\frac{k}{e}}\right).$$

2. *There exist two positive constants K_2 and K_4 such that for every $z = x + iy \in \mathbb{H}$ with $y^{-1} \geq K_4$ we have*

$$|f(z)| \leq \exp\left(\frac{K_2}{y}\right).$$

We say that such a Puiseux series is PR (like Petersson and Rademacher). The set of PR series is a ring that is integrally closed in the ring of Puiseux series with radius of convergence ≥ 1.

This lemma is easily proved: one implication results from Cauchy's formula and the other implication is trivial. Using Lemma 12.4.14 and the Petersson-Rademacher inequality (12.1.5) we prove the following lemma.

12.4.15 Lemma (integral modular functions are PR) *Let $N \geq 1$ be an integer. Let $f : \mathbb{H} \to \mathbb{C}$ be a modular function of weight 0 for the group $\Gamma_1(N)$. Assume that f has no pole outside the cusps. So f is holomorphic on $\mathbb{H}$, or equivalently it belongs to the ring $\mathbb{C}[Y_1(N)]$ of integral functions on $Y_1(N)$. Then the expansion of f at any cusp is PR.*

So the Petersson-Rademacher inequality for the Fourier coefficients of j holds true for any integral function on $Y_1(N)$. We also prove the following lemma concerning the discriminant form.

12.4.16 Lemma (the discriminant and its inverse are PR) *Let*

$$\Delta(q) = q \prod_{n \geq 1} (1 - q^n)^{24} = \sum_{n \geq 1} \tau(n) q^n$$

be the discriminant form. The inverse Δ^{-1} of Δ is a PR series in q.

Proof This results from the product formula for Δ. $\qquad\qquad\qquad$ □

We deduce that Lemma 12.4.15 extends to modular functions of arbitrary weight.

12.4.17 Lemma (integral modular functions are PR) *Let $N \geq 1$ be an integer. Let $f : \mathbb{H} \to \mathbb{C}$ be a modular function of weight $k \in 2\mathbb{Z}$ for the group $\Gamma_1(N)$. Assume that f has no pole outside the cusps. Then the expansion of f at any cusp is PR.*

12.4.18 The exp-type of a power series

We shall have to deal with series in one variable having bigger coefficients than the ones introduced in Sections 12.4.1 and 12.4.13. The successive derivatives appearing in Lemma 12.11.6 are a good example. We no longer care about convergence then. We just want to control the size (of the logarithms) of the coefficients.

12.4.19 Definition (exp-type of a power series in one variable) Let n be an integer ≥ 2 and let A and B be two real numbers ≥ 1. We say that a formal power series $f = \sum_{k \geq 0} f_k x^k$ is of exp-type (A, B, n) if for every $k \geq 0$ we have

$$|f_k| \leq \exp((Ak + B)^n).$$

If f_1 is of exp-type (A_1, B_1, n_1) and f_2 is of exp-type (A_2, B_2, n_2), then the product $f_1 f_2$ is of exp-type

$$(12.4.20) \qquad\qquad (A_1 + A_2, B_1 + B_2 + 1, n),$$

where n is the maximum of n_1 and n_2.

If f is of exp-type (A, B, n) and $k \geq 1$ is an integer, then the kth power f^k is of exp-type $(kA, kB + k - 1, n)$. The derivative $f' = df/dx$ of f is of exp-type

$$(A, A + B + 1, n).$$

If f is of exp-type (A, B, n) and $g(x) = 1/(1 - xf(x))$, then g is of exp-type

$$(12.4.21) \qquad\qquad (\sqrt{A + B + 1}, 0, 2n).$$

12.5 JACOBIAN AND WRONSKIAN DETERMINANTS OF SERIES

In this section we state and prove an algebraic identity relating Wronskian and Jacobian determinants. This identity will be useful to control the local behavior of the Jacobi integration map. We first state this identity in its simplest and most natural form in Section 12.5.2. We then state a more general identity in Section 12.5.7. The proofs are given in Sections 12.5.14 and 12.5.19. For any integer $g \geq 2$ we set

$$(12.5.1) \qquad\qquad g\dagger = \prod_{1 \leq m \leq g-1} m! = \prod_{1 \leq m \leq g} m^{g-m}.$$

12.5.2 A special case

We assume that $g \geq 2$ is an integer and we consider g formal power series $f_1(x)$, $f_2(x)$, ..., $f_g(x)$ in one variable x, with coefficients in $\mathbb{C}$. The *Wronskian determinant* associated with $\mathbf{f} = (f_1, \ldots, f_g)$ is the determinant

$$(12.5.3) \qquad W_{\mathbf{f}}(x) = \frac{1}{g\dagger} \times \begin{vmatrix} f_1(x) & \cdots & f_g(x) \\ f_1'(x) & \cdots & f_g'(x) \\ \vdots & & \vdots \\ f_1^{(g-1)}(x) & \cdots & f_g^{(g-1)}(x) \end{vmatrix}.$$

We may also introduce g indeterminates y_1, y_2, ..., y_g and define the *Jacobian determinant* associated with $\mathbf{f}$ to be the determinant

$$\mathcal{J}_{\mathbf{f}} = \begin{vmatrix} f_1(y_1) & \cdots & f_g(y_1) \\ f_1(y_2) & \cdots & f_g(y_2) \\ \vdots & & \vdots \\ f_1(y_g) & \cdots & f_g(y_g) \end{vmatrix}.$$

Let now $D = \prod_{k<m}(y_m - y_k)$ be the reduced discriminant. Then the Jacobian determinant $\mathcal{J}_{\mathbf{f}}$ lies in the ring $\mathbb{C}[[y_1, \ldots, y_g]]$ and is divisible by the reduced discriminant in this ring. Further, the quotient $\mathcal{J}_{\mathbf{f}}/D$ is congruent to $W_{\mathbf{f}}(0)$ modulo the maximal ideal of $\mathbb{C}[[y_1, \ldots, y_g]]$:

$$(12.5.4) \qquad \frac{\mathcal{J}_{\mathbf{f}}}{D} \equiv W_{\mathbf{f}}(0) \bmod (y_1, y_2, \ldots, y_g)\mathbb{C}[[y_1, \ldots, y_g]].$$

A proof of this identity is given in Sections 12.5.14 and 12.5.19.

Now assume that all the series $f_k(x)$ have type (A, n). Then the Jacobian determinant $\mathcal{J}_{\mathbf{f}}$ has type

$$(12.5.5) \qquad \left(g! A^g, n1_g\right).$$

We deduce from equations (12.4.12) and (12.4.4) that the Wronskian determinant $W_{\mathbf{f}}$ has type

$$(12.5.6) \qquad \left(A^g \exp(\Theta n g^2 + \Theta g^3), gn - 1 + \frac{g(g+1)}{2}\right).$$

12.5.7 A more general identity

Let $g \geq 2$ be an integer. Let $\mathbb{K}$ be any field with characteristic zero. Let $f_1(x), \ldots, f_g(x)$ be g formal power series in one variable x having coefficients in $\mathbb{K}$. Let $\mathbf{f} = (f_1, \ldots, f_g)$ be the corresponding vector. Let $1 \leq n \leq g$ be an integer and consider the n first derivatives

$$\mathbf{f}^{(0)}, \mathbf{f}^{(1)}, \ldots, \mathbf{f}^{(n-1)}$$

of $\mathbf{f}$ with respect to the variable x. The exterior product

$$(12.5.8) \qquad \mathcal{W}_{\mathbf{f},n} = \frac{\mathbf{f}^{(0)}}{0!} \wedge \frac{\mathbf{f}^{(1)}}{1!} \wedge \cdots \wedge \frac{\mathbf{f}^{(n-1)}}{(n-1)!} \in \bigwedge^n (\mathbb{K}[[x]])^g$$

is a sort of *partial Wronskian determinant* associated with the vector $\mathbf{f}$.

Now let $S \geq 1$ be an integer and let

$$n = m_1 + m_2 + \cdots + m_S$$

be a partition of n in S parts. In particular m_s is a positive integer for every $1 \leq s \leq S$. Let $y_1, y_2, \ldots, y_S$ be S distinct indeterminates and consider the corresponding *partial Jacobian determinant*

$$(12.5.9) \qquad \mathcal{J}_{\mathbf{f},(m_s)_{1 \leq s \leq S}} = \mathcal{W}_{\mathbf{f},m_1}(y_1) \wedge \mathcal{W}_{\mathbf{f},m_2}(y_2) \cdots \wedge \mathcal{W}_{\mathbf{f},m_S}(y_S)$$

in

$$\bigwedge^n (\mathbb{K}[[y_1, \ldots, y_S]])^g .$$

Let

$$D_{(m_s)_{1 \leq s \leq S}} = \prod_{1 \leq s_1 < s_2 \leq S} (y_{s_2} - y_{s_1})^{m_{s_1} m_{s_2}}$$

be the corresponding *partial weighted discriminant*. The following identity is a partial generalization of equation (12.5.4)

$$(12.5.10) \qquad \frac{\mathcal{J}_{\mathbf{f},(m_s)_{1 \leq s \leq S}}}{D_{(m_s)_{1 \leq s \leq S}}} \equiv \mathcal{W}_{\mathbf{f},n}(0) \bmod (y_1, y_2, \ldots, y_S).$$

This means in particular that $\mathcal{J}_{\mathbf{f},(m_s)_{1 \leq s \leq S}}$ is divisible by the weighted discriminant $D_{(m_s)_{1 \leq s \leq S}}$ in the $\mathbb{K}[[y_1, \ldots, y_S]]$-module

$$\bigwedge^n (\mathbb{K}[[y_1, \ldots, y_S]])^g .$$

A proof of this identity is given in Sections 12.5.14 and 12.5.19.

Now let $K \geq 1$ be an integer and let

$$g = n_1 + n_2 + \cdots + n_K$$

be a partition of the dimension g. So n_k is a positive integer for each $1 \leq k \leq K$. We consider K vectors $\mathbf{f}_1, \ldots, \mathbf{f}_K$ in $(\mathbb{K}[[x]])^g$. We introduce K indeterminates $x_1, \ldots, x_K$ and following equation (12.5.8) we define the *total Wronskian determinant* to be

$$(12.5.11) \quad \mathcal{W}_{(\mathbf{f}_k, n_k)_{1 \leq k \leq K}} = \bigwedge_{1 \leq k \leq K} \mathcal{W}_{\mathbf{f}_k, n_k}(x_k) \in \mathbb{K}[[x_1, x_2, \ldots, x_K]].$$

For every $1 \leq k \leq K$ let S_k be a positive integer and let

$$n_k = m_{k,1} + m_{k,2} + \cdots + m_{k,S_k}$$

be a partition of n_k into S_k parts. For every $1 \leq k \leq K$ and $1 \leq s \leq S_k$ we introduce the new indeterminate $y_{k,s}$, and following equation (12.5.9) we define the *total Jacobian determinant* to be

$$(12.5.12) \quad \mathcal{J}_{(\mathbf{f}_k, (m_{k,s})_{1 \leq s \leq S_k})_{1 \leq k \leq K}} = \bigwedge_{1 \leq k \leq K} \mathcal{J}_{\mathbf{f}_k, (m_{k,s})_{1 \leq s \leq S_k}}$$

in

$$\mathbb{K}[[(y_{k,s})_{1 \leq k \leq K; \, 1 \leq s \leq S_k}]].$$

We set

$$\mathbf{n} = (n_k)_{1 \leq k \leq K},$$

$$\mathbf{m}_k = (m_{k,s})_{1 \leq s \leq S_k},$$

and

$$\mathbf{m} = (m_{k,s})_{1 \leq k \leq K; \, 1 \leq s \leq S_k}.$$

Both $\mathbf{m}$ and $\mathbf{n}$ are partitions of g, and $\mathbf{m}$ is a refinement of $\mathbf{n}$. We define the discriminant relative to $\mathbf{m}$ and $\mathbf{n}$ to be

$$D_{\mathbf{m}, \mathbf{n}} = \prod_{1 \leq k \leq K} D_{\mathbf{m}_k} = \prod_{1 \leq k \leq K} \prod_{1 \leq s_1 < s_2 \leq S_k} (y_{k,s_2} - y_{k,s_1})^{m_{k,s_1} m_{k,s_2}}.$$

Collecting K equations like (12.5.10) we obtain

(12.5.13)
$$\frac{\mathcal{J}_{(\mathbf{f}_k, \mathbf{m}_k)_{1 \le k \le K}}}{D_{\mathbf{m}, \mathbf{n}}} \equiv \mathcal{W}_{(\mathbf{f}_k, \mathbf{n}_k)_{1 \le k \le K}}(0) \bmod (y_{k,s})_{1 \le k \le K; \, 1 \le s \le S_k}.$$

This generalization of equation (12.5.4) will be useful in Section 12.13.6 when studying the Jacobi map. We now prove it using a formal analogue of the Jacobi map.

12.5.14 Proof of equation (12.5.10) in a special case

In this section assume that $S = n$ and $m_1 = m_2 = \cdots = m_S = 1$. We write $\mathbf{f}$ as a series in x with coefficients in $\mathbb{K}^g$

$$\mathbf{f} = \Psi_0 + \Psi_1 x + \Psi_2 x^2 + \cdots \in \mathbb{K}^g[[x]]$$

where

$$\Psi_k = \frac{\mathbf{f}^{(k)}}{k!}(0) \in \mathbb{K}^g$$

for every $k \ge 0$. We consider the formal integration

$$\mathbf{F} = \int_0^x \mathbf{f}(x)dx = \Psi_0 x + \frac{\Psi_1}{2}x^2 + \frac{\Psi_2}{3}x^3 + \cdots \in \mathbb{K}^g[[x]].$$

We introduce S new indeterminates $y_1, \ldots, y_S$, and we set

(12.5.15) $\Phi(y_1, \ldots, y_S) = \mathbf{F}(y_1) + \cdots + \mathbf{F}(y_S) \in \mathbb{K}^g[[y_1, \ldots, y_S]].$

We denote by

$$\mathfrak{m} = (y_1, \ldots, y_S)\mathbb{K}[[y_1, \ldots, y_S]]$$

the maximal ideal in $\mathbb{K}[[y_1, \ldots, y_S]]$. For every $k \ge 1$ we call

$$v_k = y_1^k + \cdots + y_S^k$$

the kth Newton's power sum. We check that

$$\Phi(y_1, \ldots, y_S) = \Psi_0 v_1 + \frac{\Psi_1}{2}v_2 + \cdots + \frac{\Psi_{S-1}}{S}v_S + R,$$

where the remainder R is a vector in $(\mathbb{K}[[y_1, \ldots, y_S]])^g$ whose coefficients are symmetric functions in the $(y_s)_{1 \leq s \leq S}$ and belong to $\mathfrak{m}^{S+1}$. So these coefficients belong to $\mathbb{K}[[\nu_1, \ldots, \nu_S]]$ and for every $1 \leq k \leq S$ the partial derivative $\frac{\partial R}{\partial \nu_k}$ is zero modulo

$$\mathfrak{n} = \mathfrak{m} \cap \mathbb{K}[[\nu_1, \ldots, \nu_S]] = (\nu_1, \ldots, \nu_S)\mathbb{K}[[\nu_1, \ldots, \nu_S]].$$

We deduce that for every $1 \leq k \leq S$

$$(12.5.16) \qquad \frac{\partial \Phi}{\partial \nu_k} \equiv \frac{\Psi_{k-1}}{k} \mod \mathfrak{n}.$$

On the other hand, it is clear from the definition of Φ in equation (12.5.15) that for every $1 \leq s \leq S$,

$$(12.5.17) \qquad \frac{\partial \Phi}{\partial y_s} = \frac{d\mathbf{F}}{dx}(y_s) = \mathbf{f}(y_s).$$

Finally, for every $1 \leq k \leq S$ and $1 \leq s \leq S$ we have

$$\frac{\partial \nu_k}{\partial y_s} = k y_s^{k-1}$$

and the determinant

$$(12.5.18) \qquad \left| \frac{\partial \nu_k}{\partial y_s} \right|_{k,s} = S! \prod_{1 \leq s_1 < s_2 \leq S} (y_{s_2} - y_{s_1}).$$

Equation (12.5.10) then follows from (12.5.16), (12.5.17) and (12.5.18) applying the chain rule for derivatives.

12.5.19 Proof of equation (12.5.10) in general

We introduce the n indeterminates $x_{s,j}$ for $1 \leq s \leq S$ and $1 \leq j \leq m_s$. We put the lexicographic order on these indeterminates and we apply equation (12.5.10) to the series $\mathbf{f}$ and partition $n = 1 + 1 + \cdots + 1$. We obtain

$$
(12.5.20) \qquad
\begin{aligned}
\mathbf{f}(x_{1,1}) \wedge \cdots \wedge \mathbf{f}(x_{S,m_S}) \\
= \left(\mathcal{W}_{\mathbf{f},n}(0) + O(x_{1,1}, \ldots, x_{S,m_S}) \right) \\
\times \prod_{(1,1) \leq (s_1, j_1) < (s_2, j_2) \leq (S, m_S)} (x_{s_2, j_2} - x_{s_1, j_1})
\end{aligned}
$$

where $O(x_{1,1}, \ldots, x_{S,m_S})$ stands for any element in the ideal generated by $x_{1,1}, \ldots, x_{S,m_S}$ in $\mathbb{K}[[x_{1,1}, \ldots, x_{S,m_S}]]$.

We introduce S new indeterminates $y_1, \ldots, y_S$. We also introduce n indeterminates $z_{s,j}$ for $1 \le s \le S$ and $1 \le j \le m_s$. For every $1 \le s \le S$ we consider $\mathbf{f}(y_s + z)$ as a series in z with coefficients in $\mathbb{K}[[y_s]]$. We apply equation (12.5.10) to the series $\mathbf{f}(y_s + z)$ and partition $m_s = 1 + 1 + \cdots + 1$. We obtain

(12.5.21)
$$\mathbf{f}(y_s + z_{s,1}) \wedge \cdots \wedge \mathbf{f}(y_s + z_{s,m_s})$$
$$= \left(\mathcal{W}_{\mathbf{f},m_s}(y_s) + O(z_{s,1}, \ldots, z_{s,m_s}) \right) \times \prod_{1 \le j_1 < j_2 \le m_s} (z_{s,j_2} - z_{s,j_1}).$$

We now replace $x_{s,j}$ by $y_s + z_{s,j}$ in equation (12.5.20) and we obtain

(12.5.22)
$$\mathbf{f}(y_1 + z_{1,1}) \wedge \cdots \wedge \mathbf{f}(y_S + z_{S,m_S})$$
$$= \left(\mathcal{W}_{\mathbf{f},n}(0) + O(y_1, \ldots, y_S, z_{1,1}, \ldots, z_{S,m_S}) \right)$$
$$\times \prod_{(1,1) \le (s_1,j_1) < (s_2,j_2) \le (S,m_S)} (y_{s_2} - y_{s_1} + z_{s_2,j_2} - z_{s_1,j_1}).$$

We now notice that the left-hand side of equation (12.5.22) is the wedge product of the left-hand sides of the S equations like (12.5.21). Further, the discriminants on the right-hand sides of the S equations like (12.5.21) divide the discriminant in the right-hand side of equation (12.5.22). So we equate the right-hand side of equation (12.5.22) and the wedge product of the right-hand sides of the S equations like (12.5.21). We then divide by the product of the S small discriminants. We then reduce modulo the ideal generated by the variables $z_{s,j}$ and obtain

$$\bigwedge_{1 \le s \le S} \mathcal{W}_{\mathbf{f},m_s}(y_s) = \left(\mathcal{W}_{\mathbf{f},n}(0) + O(y_1, \ldots, y_s) \right)$$
$$\times \prod_{1 \le s_1 < s_2 \le S} (y_{s_2} - y_{s_1})^{m_{s_1} m_{s_2}},$$

as was to be proved.

12.6 A SIMPLE QUANTITATIVE STUDY OF THE JACOBI MAP

In this section we prove some upper and lower bounds for the Jacobi map. Upper bounds are rather trivial but important to control the complexity of the algorithms. Lower bounds are not very surprising either. But they play an important role in the proof of Theorem 12.10.5.

12.6.1 Upper bounds for the Jacobi map

We first prove that the Jacobi map is Lipschitz with constant l^{Θ}. Let $\gamma \in \Xi$ and call $\phi_\gamma = \phi \circ \mu_\gamma$ the composition of the Jacobi map $\phi : X \to J(\mathbb{C})$ with the modular parametrization $\mu_\gamma : D(0, 1) \to X$. So

$$\phi_\gamma : q_\gamma \mapsto \left(\int^{q_\gamma} \omega \right)_{\omega \in \mathcal{B}^1_{DR}} ;$$

we don't need to specify the origin of the integral here.

Every $\omega = f(q)q^{-1}dq$ in $\mathcal{B}^1_{DR}$ can be written $h(q_\gamma)q_\gamma^{-1}dq_\gamma$, where $h(q_\gamma)$ is the expansion of the modular form f at the cusp $\gamma(\infty)$. It is a consequence of Ramanujan's conjecture proved by Deligne and the explicit formulae by Asai for pseudo-eigenvalues that the series $h(q_\gamma)q_\gamma^{-1}$ is of type

(12.6.2) $\qquad\qquad\qquad (\Theta, \Theta).$

Therefore if q_1 and q_2 belong to $D(0, \exp(-\pi/w_\gamma)) \supset F_{w_\gamma}$ the integral $\int_{q_1}^{q_2} h(q_\gamma)q_\gamma^{-1}dq_\gamma$ is bounded in absolute value by $|q_2 - q_1|$ times l^{Θ} according to equation (12.4.8). So the Jacobi map is Lipschitz with constant

(12.6.3) $\qquad\qquad\qquad \leq l^{\Theta}.$

We now consider some vector $\vec{\gamma} = (\gamma_k)_{1 \leq k \leq g} \in \Xi^g$, and we call

$$\phi_{\vec{\gamma}} : D(0, 1)^g \to \mathbb{C}^{\mathcal{B}^1_{DR}}/\Lambda$$

the composition of the Jacobi map $\phi : X^g \to J(\mathbb{C})$ with the product of the modular parametrizations

$$\mu_{\vec{\gamma}} = \prod_{1 \leq k \leq g} \mu_{\gamma_k} : D(0, 1)^g \to X^g.$$

For every $1 \leq k \leq g$ we set $w_k = w_{\gamma_k}$. Let $\mathbf{q} = (q_1, \ldots, q_g)$ with $q_k \in D(0, \exp(-\pi/w_k))$ for every $1 \leq k \leq g$. We study the map $\phi_{\vec{\gamma}}$ locally at $\mathbf{q}$. The tangent space to $D(0, 1)^g \subset \mathbb{C}^g$ at $\mathbf{q}$ is identified with $\mathbb{C}^g$, and we denote by $(\delta_k)_{1 \leq k \leq g}$ its canonical basis. The underlying $\mathbb{R}$-vector space has basis $(\delta_1, \ldots, \delta_g, i\delta_1, \ldots, i\delta_g)$. For $1 \leq k \leq g$ we set $\delta_{k+g} = i\delta_k$.

Similarly we call $(e_k)_{1 \le k \le g}$ the canonical basis of $\mathbb{C}^g = \mathbb{C}^{\mathcal{B}^1_{DR}}$ and set $e_{k+g} = i e_k$ for $1 \le k \le g$. So $(e_k)_{1 \le k \le 2g}$ is a basis of the $\mathbb{R}$-vector space underlying $\mathbb{C}^g$. Let $D_{\mathbf{q}} \phi_{\vec{\gamma}}$ be the differential of $\phi_{\vec{\gamma}}$ at $\mathbf{q}$. For $1 \le k \le 2g$ let ρ_k be the image of δ_k by this differential. The determinant of $D_{\mathbf{q}} \phi_{\vec{\gamma}}$

$$\frac{\rho_1 \wedge \ldots \wedge \rho_{2g}}{e_1 \wedge \ldots \wedge e_{2g}}$$

is the square of the absolute value of the Jacobian determinant

(12.6.4)
$$\mathcal{J}_{\vec{\gamma}}(\mathbf{q}) = \left| \frac{\omega}{dq_{\gamma_k}}(q_k) \right|_{1 \le k \le g, \, \omega \in \mathcal{B}^1_{DR}}.$$

As a series in the g indeterminates $q_1, q_2, \ldots, q_g$, the Jacobian determinant $\mathcal{J}_{\vec{\gamma}}$ has type

(12.6.5)
$$(\exp(\Theta g^2), \Theta 1_g).$$

This results from equations (12.6.2) and (12.5.5) and the definition of $\mathcal{J}_{\vec{\gamma}}$ in equation (12.6.4). If $\vec{\gamma} = (\gamma, \gamma, \ldots, \gamma)$ is the repetition of g times the same γ in Ξ, we write $\mathcal{J}_{\gamma}$ for $\mathcal{J}_{\vec{\gamma}}$ and we denote by $\mathcal{W}_{\gamma}(q)$ the Wronskian determinant associated with $\mathcal{J}_{\gamma}$. This is a series in one variable q. We deduce from equations (12.5.6) and (12.6.2) that the series $\mathcal{W}_{\gamma}(q)$ has type

(12.6.6)
$$(\exp(\Theta g^3), \Theta g^2).$$

12.6.7 Lower bounds for the Jacobi map

We now bound from below the Wronskian determinant $\mathcal{W}_{\gamma}(q)$ and the Jacobian determinant $\mathcal{J}_{\gamma}(\mathbf{q})$ for some special values of q and $\mathbf{q}$. So we assume that $\vec{\gamma} = (\gamma, \gamma, \ldots, \gamma)$ is the repetition of g times the same γ in Ξ and we study the Jacobi map $\phi : X^g \to J$ in the neighborhood of $(\gamma(\infty), \ldots, \gamma(\infty))$, where $\gamma(\infty)$ is the cusp associated with γ. The parameter at the cusp $\gamma(\infty)$ is the q_{γ} from equation (12.1.2).

We first treat the case when $\gamma = \text{Id}$ and we write $\mathcal{J}_{\infty}$ (resp. $\mathcal{W}_{\infty}$) for $\mathcal{J}_{\text{Id}}$ (resp. $\mathcal{W}_{\text{Id}}$). We recall that the denominator in the definition of the Wronskian determinant given in equation (12.5.3) is the integer $g\dagger$ defined in equation (12.5.1). The expansions in $q = q_{\text{Id}}$ of the $\frac{\omega}{dq}$ for $\omega \in \mathcal{B}^1_{DR}$ are power series with algebraic integer coefficients; and they are permuted by the absolute Galois group of $\mathbb{Q}$. So the series $(g\dagger)^2 \times \mathcal{W}_\infty(q)^2$ has coefficients in $\mathbb{Z}$.

We set $\eta = \frac{g(g+1)}{2}$. The product $\mathcal{W}_\infty(q)(dq)^\eta$ is a degree η holomorphic form on X. Therefore it has $2(g-1)\eta$ zeros counting multiplicities. Since the q-valuation v of $\mathcal{W}_\infty(q)$ is the multiplicity of the cusp ∞ in the divisor of $\mathcal{W}_\infty(q)(dq)^\eta$, we deduce that $v \leq 2(g-1)\eta$. So the series $(g\dagger)^2 \times \mathcal{W}_\infty(q)^2$ has valuation $2v \leq 2(g-1)g(g+1)$ and rational integer coefficients. We deduce from equations (12.6.6) and (12.4.4) that the type of the series $\mathcal{W}_\infty(q)^2$ is

(12.6.8) $\qquad\qquad (\exp(\Theta g^3), \Theta g^2)$.

We write

$$\mathcal{W}_\infty(q)^2 = \frac{c}{(g\dagger)^2} \times q^{2v} + R_{2v+1}(q),$$

where c is a nonzero rational integer and R_{2v+1} is the remainder of order $2v + 1$. We can bound this remainder using equations (12.4.8) and (12.6.8).

$$|R_{2v+1}(q)| \leq \exp(\Theta l^6) \times |q|^{2v+1}.$$

So if $|q| \leq \exp(-\Theta l^6)$ we have $|\mathcal{W}_\infty(q)^2| \geq \frac{1}{2(g\dagger)^2} \times |q|^{2(g-1)g(g+1)}$ and

(12.6.9)
$$|\mathcal{W}_\infty(q)| \geq \frac{1}{2g\dagger} \times |q|^{(g-1)g(g+1)} \geq \exp(-\Theta l^5) \times |q|^{(g-1)g(g+1)}.$$

So we fix such a q. For example, we take

(12.6.10) $\qquad\qquad q = 10^{-K_1 l^6}$

for some large enough positive constant K_1. We set

(12.6.11) $\qquad\qquad \mathbf{q} = q\mathbf{1}_g = (q, \ldots, q)$

and $\mathbf{x} = (x_1, \ldots, x_g)$, where $x_1, \ldots, x_g$ are new indeterminates. The Jacobian determinant

$$\mathcal{J}_\infty(\mathbf{q} + \mathbf{x} \star (\mathbf{1}_g - \|\mathbf{q}\|))$$

is a power series in the g variables $x_1, \ldots, x_g$. This is indeed the Jacobian determinant associated with the g series

$$(f(q + x(1 - q))/(q + x(1 - q)))_{f(q)q^{-1} dq \in B^1_{DR}}$$

in the variable x. Equation (12.5.4) gives us the first nonzero term in the expansion of this series at $\mathbf{x} = \mathbf{0}_g$:

$$\mathcal{J}_\infty(\mathbf{q} + \mathbf{x} \star (\mathbf{1}_g - \|\mathbf{q}\|)) = \mathcal{W}_\infty(q)(1-q)^{\frac{g(g-1)}{2}} \prod_{k<m} (x_m - x_k)$$

$$+ R_{\frac{g(g-1)}{2}+1}(\mathbf{x}).$$

The type of the Jacobian determinant $\mathcal{J}_\infty(\mathbf{q})$ as a series in $\mathbf{q}$ is given by equation (12.6.5). We deduce from equation (12.4.6) that the refocused series $\mathcal{J}_\infty(\mathbf{q}+\mathbf{x}\star(\mathbf{1}_g-\|\mathbf{q}\|))$ is a series in $\mathbf{x}$ of type $(\exp(\Theta g^2), \Theta \mathbf{1}_g)$. Using equation (12.4.8) we deduce that for $\mathbf{x}$ in $P(\mathbf{0}_g, \exp(-\pi))$,

$$\left| R_{\frac{g(g-1)}{2}+1}(\mathbf{x}) \right| \le \exp(\Theta g^2) |\mathbf{x}|_\infty^{\frac{g(g-1)}{2}+1}.$$

We set $s = |\mathbf{x}|_\infty$ and we assume that $\mathbf{x}$ takes the special form

$$(12.6.12) \qquad \mathbf{x} = \left(\frac{s}{g}, \frac{2s}{g}, \dots, \frac{(g-1)s}{g}, s \right)$$

and $s \le \exp(-\pi)$. Then

$$\left| \mathcal{W}_\infty(q)(1-q)^{\frac{g(g-1)}{2}} \prod_{k<m} (x_m - x_k) \right| \ge |\mathcal{W}_\infty(q)| \left(\frac{s(1-q)}{g} \right)^{\frac{g(g-1)}{2}}$$

$$\ge \Theta^{-1} |\mathcal{W}_\infty(q)| \left(\frac{s}{g} \right)^{\frac{g(g-1)}{2}}.$$

We take

$$(12.6.13) \qquad\qquad\qquad s = 10^{-K_2 l^{12}}$$

for some large enough positive constant K_2. Using equations (12.6.10) and (12.6.9) we obtain the following lower bound for the Jacobian determinant:

$$(12.6.14) \qquad |\mathcal{J}_\infty(\mathbf{q} + \mathbf{x}(1-q))| \ge \exp(-\Theta l^{16}),$$

where q, $\mathbf{q}$, and $\mathbf{x}$ are given by equations (12.6.10), (12.6.11), (12.6.12), and (12.6.13). In particular $|\mathbf{q} + \mathbf{x}(1-q)|_\infty$ can be assumed to be $\le \exp(-2\pi)$. So

$$(12.6.15) \qquad \max_{\mathbf{q} \in \bar{D}(0,\exp(-2\pi))^g} |\mathcal{J}_\infty(\mathbf{q})| \ge \exp(-\Theta l^{16}).$$

In order to bound from below $\mathcal{J}_\gamma$ for any $\gamma \in \Xi$, we observe that $\mathcal{J}_\gamma$ and $\mathcal{J}$ are closely related. If w is the width of the cusp $\gamma(\infty)$, there exists a wth root of unity ζ_γ and an algebraic number λ_γ of absolute value 1 (the product of all pseudo-eigenvalues) such that the following formal identity in $\mathbb{C}[[q_1, \ldots, q_g]]$ holds true

$$\mathcal{J}_\gamma(q_1, \ldots, q_g) = \lambda_\gamma \mathcal{J}_\infty(\zeta_\gamma q_1, \ldots, \zeta_\gamma q_g).$$

So the lower bound in equation (12.6.15) is also valid for every $\mathcal{J}_\gamma$.

12.7 EQUIVALENCE OF VARIOUS NORMS

The main algorithm in this text (the one in Section 12.8) uses a subroutine that computes the complex roots of an analytic function on a compact set. This problem is well conditioned according to Lemma 5.4.22 of Chapter 5, provided we have a decent lower bound for the maximum of the function in question. In our situation, the analytic functions are derived from quadratic differentials on X. We need simple conditions for these functions not to be uniformly small in absolute value in the neighborhood of any cusp. The second inequality in equation (12.7.3) below provides such a condition. In order to prove this inequality we study Jacobian determinants associated with weight 4 cusp forms on X, locally at every cusp.

12.7.1 Space of quadratic differentials

In this section we shall make use of parabolic modular forms of weight 4 on X. To every such form $f(q)$ one can associate a quadratic differential $\omega = f(q)q^{-2}(dq)^2$. The divisor of ω is related to the divisor of f by the relation

$$\mathrm{Div}(\omega) = \mathrm{Div}(f) - 2\mathrm{Cusps},$$

where Cusps is the sum of all cusps. Note that $X_1(5l)$ has no elliptic point. The map $f(q) \mapsto f(q)q^{-2}(dq)^2$ defines a bijection between the space of weight 4 parabolic forms and the space

$$\mathcal{H}^2(\mathrm{Cusps}) = \mathrm{H}^0(X, \Omega_{X/\mathbb{C}}^{\otimes 2} \otimes \mathcal{O}_X(\mathrm{Cusps}))$$

of quadratic differentials with divisor $\geq$ $-$ Cusps. We denote by g_2 the dimension of the latter space. This is $3g - 3$ plus the degree of Cusps (number of cusps).

We shall need a basis $\mathcal{B}_{\mathrm{quad}}$ for the space $\mathcal{H}^2(\mathrm{Cusps})$ or equivalently a basis for the space $S_4(\Gamma_1(5l))$ of weight 4 cusp forms. We shall again use the standard basis made of normalized newforms of level $5l$ together with normalized newforms of level l lifted to level $5l$ by the two degeneracy maps. We also need the expansion of every form in $\mathcal{B}_{\mathrm{quad}}$ at every cusp $\gamma(\infty)$ for $\gamma \in \Xi$. More precisely, $f(q)q^{-2}(dq)^2$ should be rewritten as $h(q_\gamma)q_\gamma^{-2}(dq_\gamma)^2$ for every γ in Ξ. As for degree one forms, and using the same methods, we can compute the expansion of all quadratic forms in $\mathcal{B}_{\mathrm{quad}}$ at all cusps in deterministic polynomial time $(klm)^\Theta$, where k is the q-adic accuracy and m the complex absolute accuracy of coefficients.

We now define several important norms on the space $\mathcal{H}^2(\mathrm{Cusps})$. If ω is a form in $\mathcal{H}^2(\mathrm{Cusps})$, we denote by $|\omega|_\infty$ the L^∞ norm in the basis $\mathcal{B}_{\mathrm{quad}}$. To every cusp $\gamma(\infty)$ with $\gamma \in \Xi$ we associate a norm on $\mathcal{H}^2(\mathrm{Cusps})$. We define $|\omega|_\gamma$ to be the maximum of the absolute value of the function $\omega q_\gamma (dq_\gamma)^{-2}$ for $|q_\gamma| \leq 1/2$.

$$(12.7.2) \qquad |\omega|_\gamma = \max_{|q_\gamma| \leq 1/2} \left| \frac{q_\gamma \, \omega}{(dq_\gamma)^2} \right|.$$

Any two such norms are of course equivalent: their ratios are bounded by a constant. More interestingly, the logarithm of this constant factor is polynomial in the level $5l$ of X: for any γ in Ξ and any ω in $\mathcal{H}^2(\mathrm{Cusps})$ we have

$$(12.7.3) \qquad l^{-\Theta} \times |\omega|_\gamma \leq |\omega|_\infty \leq \exp(l^\Theta) \times |\omega|_\gamma.$$

These inequalities will be proved in Section 12.7.17.

12.7.4 Jacobian determinant of weight 4 cusp forms

Recall that we have constructed in Section 12.7.1 a basis $\mathcal{B}_{\mathrm{quad}}$ for the space $\mathcal{H}^2(\mathrm{Cusps})$ of quadratic differential forms. If $\vec{\gamma} \in \Xi^{g_2}$ and $\mathbf{q} = (q_k)_{1 \leq k \leq g_2}$ we define the quadratic Jacobian determinant

$$(12.7.5) \qquad \mathcal{J}_{\vec{\gamma}}^{\mathrm{quad}}(\mathbf{q}) = \left| \frac{q_{\gamma_k} \omega}{(dq_{\gamma_k})^2}(q_k) \right|_{1 \leq k \leq g_2, \, \omega \in \mathcal{B}_{\mathrm{quad}}}.$$

It is a consequence of Ramanujan's conjectures proved by Deligne and the explicit formulae by Asai for pseudo-eigenvalues that the series $\frac{q_\gamma \omega}{(dq_\gamma)^2}$ is of type

$$(12.7.6) \qquad\qquad (\Theta, \Theta).$$

for every ω in $\mathcal{B}_{\text{quad}}$ and every γ in Ξ. We deduce from equation (12.5.5) that the series in g_2 variables defined by equation (12.7.5) has type

$$(12.7.7) \qquad\qquad (\exp(\Theta l^4), \Theta \mathbf{1}_{g_2}).$$

Further, if $\vec{\gamma} = (\gamma, \dots, \gamma) \in \Xi^{g_2}$ is the repetition of g_2 times the same γ, we write $\mathcal{J}_\gamma^{\text{quad}}$ for $\mathcal{J}_{\vec{\gamma}}^{\text{quad}}$ and we denote by $\mathcal{W}_\gamma^{\text{quad}}$ the corresponding Wronskian determinant. We deduce from equation (12.5.6) that this series in one variable has type

$$(12.7.8) \qquad\qquad (\exp(\Theta l^6), \Theta l^4).$$

We need a similar estimate to equation (12.6.15) for these quadratic Jacobian determinants $\mathcal{J}_\gamma^{\text{quad}}$. We first treat the case when $\gamma = \text{Id}$, and we write $\mathcal{J}_\infty^{\text{quad}}$ (resp. $\mathcal{W}_\infty^{\text{quad}}$) for $\mathcal{J}_{\text{Id}}^{\text{quad}}$ (resp. $\mathcal{W}_{\text{Id}}^{\text{quad}}$). We recall that the denominator in the definition of $\mathcal{W}_\infty^{\text{quad}}$ is $g_2\dagger$ with the notation introduced in equation (12.5.1).

The expansions in $q = q_{\text{Id}}$ of the $\frac{q\omega}{(dq)^2}$ for $\omega \in \mathcal{B}_{\text{quad}}$ are power series with algebraic integer coefficients and they are permuted by the absolute Galois group of $\mathbb{Q}$. So the series $(g_2\dagger)^2 \times \mathcal{W}_\infty^{\text{quad}}(q)^2$ has coefficients in $\mathbb{Z}$. We set $\eta_2 = \frac{g_2(g_2+3)}{2}$. The product $\mathcal{W}_\infty^{\text{quad}}(q)q^{-g_2}(dq)^{\eta_2}$ is a degree η_2 form on X; it is holomorphic outside Cusps. More precisely, it belongs to

$$\mathcal{H}^{\eta_2}(g_2\text{Cusps}) = \text{H}^0(X, \Omega_{X/\mathbb{C}}^{\otimes \eta_2} \otimes \mathcal{O}_X(g_2\text{Cusps})).$$

Therefore it has

$$2(g-1)\eta_2 + g_2 \deg(\text{Cusps}) \leq \Theta l^6$$

zeros counting multiplicities. Therefore the q-valuation v_2 of $\mathcal{W}_\infty^{\text{quad}}(q)$ is $\leq \Theta l^6$. So the series $(g_2\dagger)^2 \times \mathcal{W}_\infty^{\text{quad}}(q)^2$ has valuation $\leq \Theta l^6$ and rational integer coefficients. We deduce from equations (12.7.8) and (12.4.4) that the type of the series $\mathcal{W}_\infty^{\text{quad}}(q)^2$ is

$$(12.7.9) \qquad\qquad (\exp(\Theta l^6), \Theta l^4).$$

We write

$$W_\infty^{\text{quad}}(q)^2 = \frac{c}{(g_2\dagger)^2} q^{2v_2} + R_{2v_2+1}(q),$$

where c is a nonzero rational integer and R_{2v_2+1} is the remainder of order $2v_2+1$. We can bound this remainder using equations (12.4.8) and (12.7.9):

$$|R_{2v_2+1}(q)| \le \exp(\Theta l^6)|q|^{2v_2+1}.$$

So if $|q| \le \exp(-\Theta l^6)$ we have $|W_\infty^{\text{quad}}(q)^2| \ge \frac{1}{2(g_2\dagger)^2}|q|^{2v_2}$ and

$$(12.7.10) \qquad\qquad |W_\infty^{\text{quad}}(q)| \ge \exp(-\Theta l^5)|q|^{\Theta l^6}.$$

So we fix such a q. For example, we take

$$(12.7.11) \qquad\qquad\qquad q = 10^{-K_3 l^6},$$

where K_3 is a large enough positive constant. We set

$$(12.7.12) \qquad\qquad\qquad \mathbf{q} = q\mathbf{1}_{g_2} = (q, \ldots, q)$$

and $\mathbf{x} = (x_1, \ldots, x_{g_2})$, where $x_1, \ldots, x_{g_2}$ are new indeterminates. The Jacobian determinant

$$\mathcal{J}_\infty^{\text{quad}}(\mathbf{q} + \mathbf{x} \star (\mathbf{1}_{g_2} - \|\mathbf{q}\|))$$

is a power series in the g_2 variables $x_1, \ldots, x_{g_2}$. This is indeed the Jacobian determinant associated with the g_2 series

$$(f(q + x(1-q))/(q + x(1-q)))_{f(q)q^{-2}(dq)^2 \in \mathcal{B}_{\text{quad}}}$$

in the variable x. Equation (12.5.4) gives us the first nonzero term in the expansion of this series at $\mathbf{x} = \mathbf{0}_{g_2}$:

$$\mathcal{J}_\infty^{\text{quad}}(\mathbf{q} + \mathbf{x} \star (\mathbf{1}_{g_2} - \|\mathbf{q}\|)) = W_\infty^{\text{quad}}(q)(1-q)^{\frac{g_2(g_2-1)}{2}} \prod_{k<m} (x_m - x_k)$$

$$+ R_{\frac{g_2(g_2-1)}{2}+1}(\mathbf{x}).$$

The type of the Jacobian determinant $\mathcal{J}_\infty^{\text{quad}}(\mathbf{q})$ as a series in $\mathbf{q}$ is given by equation (12.7.7). We deduce from equation (12.4.6) that the refocused series $\mathcal{J}_\infty^{\text{quad}}(\mathbf{q} + \mathbf{x} \star (\mathbf{1}_{g_2} - \|\mathbf{q}\|))$ is a series in $\mathbf{x}$ of type $(\exp(\Theta l^4), \Theta \mathbf{1}_{g_2})$. Using equation (12.4.8) we deduce that for $\mathbf{x}$ in $P(\mathbf{0}_{g_2}, \exp(-\pi))$

$$\left| R_{\frac{g_2(g_2-1)}{2}+1}(\mathbf{x}) \right| \leq \exp(\Theta l^4) |\mathbf{x}|_\infty^{\frac{g_2(g_2-1)}{2}+1}.$$

We set $s = |\mathbf{x}|_\infty$ and we assume that $\mathbf{x}$ takes the special form

$$(12.7.13) \qquad \mathbf{x} = \left(\frac{s}{g_2}, \frac{2s}{g_2}, \dots, \frac{(g_2-1)s}{g_2}, s \right)$$

and $s \leq \exp(-\pi)$. Then

$$\left| \mathcal{W}_\infty^{\text{quad}}(q)(1-q)^{\frac{g_2(g_2-1)}{2}} \prod_{k<m}(x_m - x_k) \right| \geq \left| \mathcal{W}_\infty^{\text{quad}}(q) \right| \left(\frac{s(1-q)}{g_2} \right)^{\frac{g_2(g_2-1)}{2}}$$

$$\geq \Theta^{-1} \left| \mathcal{W}_\infty^{\text{quad}}(q) \right| \left(\frac{s}{g_2} \right)^{\frac{g_2(g_2-1)}{2}}$$

We take

$$(12.7.14) \qquad\qquad s = 10^{-K_4 l^{12}},$$

where K_4 is a large enough positive constant. Using equations (12.7.11) and (12.7.10), we obtain the following lower bound for the quadratic Jacobian determinant

$$(12.7.15) \qquad |\mathcal{J}_\infty^{\text{quad}}(\mathbf{q} + \mathbf{x}(1 - q))| \geq \exp(-\Theta l^{16})$$

when q, $\mathbf{q}$, and $\mathbf{x}$ are given by equations (12.7.11), (12.7.12), (12.7.13), and (12.7.14). In particular $|\mathbf{q} + \mathbf{x}(1 - q)|_\infty$ can be assumed to be $\leq 1/2$. So

$$(12.7.16) \qquad \max_{\mathbf{q} \in \bar{D}(0,1/2)^{g_2}} \left| \mathcal{J}_\infty^{\text{quad}}(\mathbf{q}) \right| \geq \exp(-\Theta l^{16}).$$

In order to bound from below $\mathcal{J}_\gamma^{\text{quad}}$ for any $\gamma \in \Xi$ we observe that if w is the width of the cusp $\gamma(\infty)$, there exists a wth root of unity ζ_γ and an algebraic number $\lambda_\gamma^{\text{quad}}$ of absolute value 1 (the product of all pseudo-eigenvalues) such that the following formal identity in $\mathbb{C}[[q_1, \dots, q_g]]$ holds true:

$$\mathcal{J}_\gamma^{\text{quad}}(q_1, \dots, q_{g_2}) = \lambda_\gamma^{\text{quad}} \mathcal{J}_\infty^{\text{quad}}(\zeta_\gamma q_1, \dots, \zeta_\gamma q_{g_2}).$$

So the lower bound in equation (12.7.16) is also valid for every $\mathcal{J}_\gamma^{\text{quad}}$.

12.7.17 Equivalence of norms on $\mathcal{H}^2$(Cusps)

In Section 12.7.1 we defined various norms on the space of quadratic differential forms $\mathcal{H}^2$(Cusps). If

$$\omega = f(q)q^{-2}(dq)^2$$

is a quadratic differential form in $\mathcal{H}^2$(Cusps), the norm $|\omega|_\infty$ is the L^∞ norm associated with the basis $\mathcal{B}_{\text{quad}}$. For every $\gamma \in \Xi$, the norm $|\omega|_\gamma$ is defined by equation (12.7.2). We write

$$\omega = f(q)q^{-2}(dq)^2 = h(q_\gamma)q_\gamma^{-2}(dq_\gamma)^2.$$

We must prove both inequalities in equation (12.7.3). The first inequality is a trivial consequence of equations (12.7.6) and (12.4.8).

We denote by $\mathcal{M}_\gamma^{\text{quad}}$ the matrix occurring in the definition of the Jacobian determinant $\mathcal{J}_\gamma^{\text{quad}}$. So

$$\mathcal{M}_\gamma^{\text{quad}}(q_1, q_2, \ldots, q_{g_2}) = \left(\frac{q_\gamma \omega}{(dq_\gamma)^2}(q_k) \right)_{1 \le k \le g_2, \, \omega \in \mathcal{B}_{\text{quad}}}$$

$$= \left(\frac{h(q_k)}{q_k} \right)_{1 \le k \le g_2, \, h(q_\gamma)q_\gamma^{-2}(dq_\gamma)^2 \in \mathcal{B}_{\text{quad}}}.$$

In particular, $\mathcal{J}_\gamma^{\text{quad}}$ is the determinant of $\mathcal{M}_\gamma^{\text{quad}}$.

Now let q, $\mathbf{q}$ and $\mathbf{x}$ be given by equations (12.7.11), (12.7.12), (12.7.13), and (12.7.14). We set

$$\mathbf{r} = \mathbf{q} + \mathbf{x}(1 - q) = (r_1, r_2, \ldots, r_{g_2}) \in \bar{D}(0, 1/2)^{g_2}$$

and denote by $\mathcal{M}_\gamma^{\text{quad}}(\mathbf{r})$ the evaluation of $\mathcal{M}_\gamma^{\text{quad}}$ at $\mathbf{r}$.

The entries in $\mathcal{M}_\gamma^{\text{quad}}(\mathbf{r})$ are bounded above by l^Θ in absolute value. Using equation (12.7.15) we deduce that the entries in the inverse matrix of $\mathcal{M}_\gamma^{\text{quad}}(\mathbf{r})$ are bounded in absolute value by $\exp(\Theta l^{16})$.

Let ω be a form in $\mathcal{H}^2$(Cusps) and let $\mathbf{c} \in \mathbb{C}^{g_2}$ be the coordinate vectors of $\omega = h(q_\gamma)q_\gamma^{-2}(dq_\gamma)^2$ in the basis $\mathcal{B}_{\text{quad}}$. For every $1 \le k \le g_2$ set $v_k = h(r_k)/r_k$ and let $\mathbf{v} = (v_1, \ldots, v_{g_2})$ be the corresponding vector. We have

$$\mathbf{v}^t = \mathcal{M}_\gamma^{\text{quad}}(\mathbf{r}) \times \mathbf{c}^t,$$

where $\mathbf{v}^t$ is the transposed vector of $\mathbf{v}$ and $\mathbf{c}^t$ is the transposed vector of $\mathbf{c}$. So

$$|\omega|_\infty = |\mathbf{c}|_\infty \leq g_2 \exp(\Theta l^{16}) \times |\mathbf{v}|_\infty,$$

and this is $\leq \exp(\Theta l^{16}) \times |\omega|_\gamma$ from the definition of $|\omega|_\gamma$ given in equation (12.7.2).

12.8 AN ELEMENTARY OPERATION IN THE JACOBIAN VARIETY

An important prerequisite for the explicit computation in the Jacobian variety J is to be able to compute the linear space associated with some divisor on X. In this section, we describe an algorithm that solves the following elementary problem: given $3g - 4$ points $P_1, P_2, \ldots, P_{3g-4}$ in $X(\mathbb{C})$, find g points $Q_1, \ldots, Q_g$ in $X(\mathbb{C})$ such that

$$Q_1 + \cdots + Q_g \sim 2\mathcal{K} - (P_1 + \cdots + P_{3g-4}),$$

where $\sim$ stands for linear equivalence of divisors and $\mathcal{K}$ is the canonical class. This elementary problem will be used as a building block for explicit arithmetic operations in the Jacobian variety J of X. We observe that the solution is not always unique. However, the image of $Q_1 + \cdots + Q_g$ by the Jacobi integration map $\phi^\star : \mathrm{Sym}^g X \to J$ is well defined. When doing numerical approximations, it will be convenient to measure the error in $J(\mathbb{C}) = \mathbb{C}/\Lambda$ in terms of the distance d_J defined in equation (12.2.4).

We shall solve the above problem in two steps. We set

$$P = P_1 + \cdots + P_{3g-4},$$

and we first look for a differential quadratic form ω in the linear space $\mathcal{H}^2(-P) \subset \mathcal{H}^2(\mathrm{Cusps})$ using our explicit knowledge of the latter space and linear algebra algorithms. We then compute the divisor (ω) of ω and output the (effective) difference $(\omega) - P$. We now provide details for these two steps.

We denote by $T = P + \mathrm{Cusps}$ the divisor obtained by adding the cusps to P. The degree of T is $g_2 - 1$. We write $T = T_1 + T_2 + \cdots + T_{g_2-1}$, where $T_k = (\gamma_k, q_k)$ for every $1 \leq k \leq g_2 - 1$. Let $\epsilon_1 = \exp(-m_1)$ be a positive real number. We assume that $m_1 \geq \Theta l$. As explained in Section 12.1.6 we can find an ϵ_1-simple divisor $T' = T'_1 + T'_2 + \cdots + T'_{g_2-1}$ such that for every $1 \leq k \leq g_2 - 1$ we have $T'_k = (\gamma_k, q'_k)$ and

$$|q'_k - q_k| \leq \Theta(g_2 - 1)\epsilon_1.$$

We look for a quadratic form in $\mathcal{H}^2$(Cusps) having divisor $\geq -$ Cusps $+ T'$. The space of such forms can be described as the kernel of a matrix $\mathcal{M}$. Each of the $g_2 - 1$ rows of $\mathcal{M}$ corresponds to a point T'_k for some $1 \leq k \leq g_2-1$. The g_2 columns of $\mathcal{M}$ correspond to the g_2 forms in the basis $\mathcal{B}_{\text{quad}}$. The entry of $\mathcal{M}$ at the line corresponding to the point $T'_k = (\gamma_k, q'_k)$ and column corresponding to the form ω in $\mathcal{B}_{\text{quad}}$ is obtained in the following way. We consider the expansion of ω in the variable q_{γ_k}

$$\omega = h(q_{\gamma_k})q_{\gamma_k}^{-2}(dq_{\gamma_k})^2,$$

and we evaluate the function $h(q_{\gamma_k})q_{\gamma_k}^{-1}$ at the value q'_k of q_{γ_k} corresponding to T'_k. The entries in $\mathcal{M}$ are bounded by l^Θ in absolute value according to equations (12.6.2) and (12.4.8).

We can't compute $\mathcal{M}$ exactly. Instead of that, we fix a positive real $\epsilon_2 = \exp(-m_2)$ and we compute a matrix $\mathcal{M}'$ with decimal entries in $\mathbb{Z}[i, 1/10]$ such that the difference $\mathcal{M}' - \mathcal{M}$ has L^∞ norm $\leq \epsilon_2$. The entries in this matrix $\mathcal{M}'$ can be chosen to have numerators and denominators bounded in absolute value by $\exp(\Theta(l + m_2))$. We find a nonzero vector in the kernel of $\mathcal{M}'$, having coefficients in $\mathbb{Z}[i]$ and bounded in absolute value by $\exp(\Theta l^2(l + m_2))$. We divide this vector by its largest coefficient an obtain a vector $v = (v_k)_{1 \leq k \leq g_2}$ with L^∞ norm equal to 1. This vector may not lie in the kernel of $\mathcal{M}$ but $\mathcal{M}v$ has coefficients bounded by $g_2 \exp(-m_2)$. We call v the quadratic differential form in $\mathcal{H}^2$(Cusps) having coordinate vector v in the basis $\mathcal{B}_{\text{quad}}$. By definition we have

(12.8.1) $$|v|_\infty = 1.$$

Using Lemma 5.4.22 of Chapter 5 together with equations (12.7.3) and (12.8.1) we show that if $m_2 \geq l^\Theta$, then for every $1 \leq k \leq g_2 - 1$ the form v has a zero $T''_k = (\gamma_k, q''_k)$ such that

$$|q'_k - q''_k| \leq \exp(-m_2 \times l^{-\Theta}).$$

If $m_2 \geq l^{\Theta} \times m_1$ then these $g_2 - 1$ zeros must be pairwise distinct, because each of them is close to some T'_k and the latter points form an ϵ_1-simple divisor. The divisor of ν can be written

$$(\nu) = T'' - \text{Cusps} + Q',$$

where $T'' = \sum_{1 \leq k \leq g_2-1} T''_k$ and Q' is a degree g effective divisor. We rewrite (ν) as

$$(\nu) = P + Q' + \sum_{1 \leq k \leq g_2-1} \left(T''_k - T_k \right).$$

The image of the error term $\sum_{1 \leq k \leq g_2-1} \left(T''_k - T_k \right)$ by $\phi^{\star}$ is small in the torus $J(\mathbb{C}) = \mathbb{C}/\Lambda$. More precisely,

$$d_J \left(0, \phi^{\star} \left(\sum_{1 \leq k \leq g_2-1} \left(T''_k - T_k \right) \right) \right) \leq \exp(-m_1/\Theta),$$

provided $m_1 \geq l^{\Theta}$ and $m_2 \geq m_1 \times l^{\Theta}$.

So Q' is a good approximation for the solution Q to the original problem. Using the algorithm in Theorem 5.4.2 of Chapter 5, we compute an approximation of the divisor of ν and output the corresponding approximation Q'' of Q'.

12.8.2 Lemma (an elementary operation) *There is a deterministic algorithm that on input a degree $3g - 4$ effective divisor $P = P_1 + \cdots + P_{3g-4}$ on $X_1(5l)$, returns a degree g effective divisor $Q = Q_1 + \cdots + Q_g$ such that*

$$Q_1 + \cdots + Q_g \sim 2\mathcal{K} - (P_1 + \cdots + P_{3g-4}),$$

where $\mathcal{K}$ is the canonical class on $X_1(5l)$. The running time is $(lm)^{\Theta}$, where $5l$ is the level and m the required absolute accuracy of the result.

Recall that the accuracy of the result in the above statement is measured in the torus $\mathbb{C}/\Lambda$ using the distance d_J introduced in equation (12.2.4).

12.9 ARITHMETIC OPERATIONS IN THE JACOBIAN VARIETY

We fix a degree g effective divisor Ω on X. We also need an effective degree $g - 4$ auxiliary divisor Π. For example, we may choose a point O as origin for the Jacobi integration map (e.g., O could be the cusp at infinity), and set $\Omega = gO$ and $\Pi = (g - 4)O$. An element in $\mathrm{Pic}^0(X)$ is given as the class of a divisor $Q - \Omega$, where Q is a degree g effective divisor. Let R be another degree g effective divisor. In order to add the class of $Q - \Omega$ and the class of $R - \Omega$ we apply Lemma 12.8.2 twice. We first apply it to the divisor $Q + R + \Pi$. This is indeed a degree $3g - 4$ effective divisor. We obtain a degree g effective divisor T such that $T \sim 2\mathcal{K} - Q - R - \Pi$. We again apply Lemma 12.8.2 to the divisor $T + \Omega + \Pi$ this time. And we obtain a degree g effective divisor U such that $U + \Omega \sim Q + R$. So the class of $U - \Omega$ is the sum of the classes of $Q - \Omega$ and $R - \Omega$.

If we now wish to compute the opposite of the class $Q - \Omega$, we apply Lemma 12.8.2 to the divisor $2\Omega + \Pi$ and obtain a degree g effective divisor R, linearly equivalent to $2\mathcal{K} - \Pi - 2\Omega$. We apply Lemma 12.8.2 to the divisor $R + Q + \Pi$ and obtain a degree g effective divisor T such that $T - \Omega \sim -(Q - \Omega)$.

12.9.1 Theorem (arithmetic operations in $J_1(5l)$) *Addition and subtraction in the Jacobian variety of $X_1(5l)$ can be computed in deterministic time $(lm)^\Theta$, where $5l$ is the level and m the required absolute accuracy of the result.*

Again, the accuracy of the result is measured in the torus $\mathbb{C}/\Lambda$ using the distance d_J introduced in equation (12.2.4). In particular the error belongs to a *group*, and when chaining operations in the Jacobian variety, the successive errors add to each other: the error on the result is the sum of the errors on either input plus the error introduced in the current calculation. This observation is particularly useful in conjunction with the fast exponentiation algorithm of Section 5.1: if we multiply a divisor $Q - \Omega$ by a positive integer N, assuming that every elementary operation introduces and error $\leq \epsilon$, then the error on the final result is $\leq \Theta \times \epsilon \times N \log N$, so the loss of accuracy is $\leq \Theta \log N$.

12.9.2 Theorem (fast exponentiation in $J_1(5l)$) *There is a deterministic algorithm that, on input two degree g effective divisors Ω and Q on $X_1(5l)$ and a positive integer N, outputs a degree g effective divisor R such that*

$$R - \Omega \sim N(Q - \Omega).$$

The algorithm runs in time $(lm \log N)^{\Theta}$, where $5l$ is the level and m the required absolute accuracy of the result.

12.10 THE INVERSE JACOBI PROBLEM

In this section we are given a degree g effective origin divisor Ω on X and an element x in $\mathbb{C}^{\mathcal{B}_{DR}^1}$, and we want to solve the inverse Jacobi problem for

$$x + \Lambda \in \mathbb{C}^{\mathcal{B}_{DR}^1}/\Lambda = J(\mathbb{C}).$$

So we look for a degree g effective divisor $P = P_1 + \cdots + P_g$ on X such that $\phi^\star(P - \Omega) = x + \Lambda$. We note that the solution might not be unique.

The main idea is the following: we start from a family of $2g$ classes $b_1 + \Lambda, b_2 + \Lambda, \ldots, b_{2g} + \Lambda$ in J for which the inverse Jacobi problem is already solved: for every $1 \le k \le 2g$, we know a degree 0 divisor B_k such that $\phi^\star(B_k) = b_k + \Lambda$. We try to approximate x by an integer combination $\sum_{1 \le k \le 2g} N_k b_k$. This should not be too difficult if the b_k are very small and $\mathbb{R}$-linearly independent: we compute the coordinates of $x \in \mathbb{C}^g$ in the $\mathbb{R}$-basis made of the b_k and we round each of these coordinates to the closest integer. Once we have found the N_k we note that the divisor $\sum_k N_k B_k$ would be a nice solution to the problem if it were a difference between two effective degree g divisors. This is not the case, of course, but using the algorithms in Theorems 12.9.2 and 12.9.1 we find a degree g effective divisor P such that $P - \Omega$ is linearly equivalent to $\sum_k N_k B_k$. We output P and we are done.

There remains to explain how to find the b_k and the corresponding B_k. For every $1 \le k \le 2g$, we set $B_k = R'_k - R_k$, where R_k and R'_k are two points on X that are very close. More precisely, we choose the g first points

$$R_k = (\gamma_k, q_k), \qquad \text{for } 1 \le k \le g,$$

and we set $R_{k+g} = R_k$. We also choose a positive integer χ and we set $\Upsilon = \exp(-\chi)$. We assume that $\chi \geq l^\Theta$ so Υ is small. For $1 \leq k \leq g$, we set

$$R'_k = (\gamma_k, q_k + \Upsilon), \qquad \text{and} \qquad R'_{k+g} = (\gamma_k, q_k + i\,\Upsilon).$$

We set

$$b_k = \left(\int_{R_k}^{R'_k} \omega \right)_{\omega \in \mathcal{B}^1_{DR}} \in \mathbb{C}^{\mathcal{B}^1_{DR}}.$$

These integrals can be computed efficiently using the same method as for period integrals. We now want to quantify the condition that these b_k should be $\mathbb{R}$-linearly independent. So let $(e_k)_{1 \leq k \leq g}$ be the canonical basis of $\mathbb{C}^g$ and set $e_{k+g} = i e_k$ for $1 \leq k \leq g$. So $(e_k)_{1 \leq k \leq 2g}$ is a basis of the $\mathbb{R}$-vector space underlying $\mathbb{C}^g$. We shall need a lower bound for the determinant

(12.10.1)
$$\frac{b_1 \wedge \ldots \wedge b_{2g}}{e_1 \wedge \ldots \wedge e_{2g}}.$$

Since Υ is going to be small, we derive such a lower bound from the local study of the Jacobi integration map. We call $\vec{\gamma} \in \Xi^g$ the vector $(\gamma_k)_{1 \leq k \leq g}$, and we call

$$\phi_{\vec{\gamma}} : D(0,1)^g \to \mathbb{C}^{\mathcal{B}^1_{DR}}/\Lambda$$

the composition of the Jacobi map $\phi : X^g \to J(\mathbb{C})$ with the product of the modular parametrizations

$$\mu_{\vec{\gamma}} = \prod_{1 \leq k \leq g} \mu_{\gamma_k} : D(0,1)^g \to X^g.$$

For every $1 \leq k \leq g$ we set $w_k = w_{\gamma_k}$. The value q_k of the coordinate q_{γ_k} at the point R_k belongs to $D(0, \exp(-\pi/w_k))$. We set

$$\mathbf{q} = (q_1, \ldots, q_g).$$

We study the map $\phi_{\vec{\gamma}}$ locally at $\mathbf{q}$. The tangent space to $D(0,1)^g \subset \mathbb{C}^g$ at $\mathbf{q}$ is identified with $\mathbb{C}^g$, and we denote by $(\delta_k)_{1 \leq k \leq g}$ its canonical basis. The underlying $\mathbb{R}$-vector space has basis $(\delta_1, \ldots, \delta_g, i\delta_1, \ldots, i\delta_g)$. For

$1 \leq k \leq g$ we set $\delta_{k+g} = i\delta_k$. Let $D_\mathbf{q}\phi_{\vec{\gamma}}$ be the differential of $\phi_{\vec{\gamma}}$ at $\mathbf{q}$. For $1 \leq k \leq 2g$ let ρ_k be the image of δ_k by this differential. equations (12.6.2) and (12.4.8) imply that the L^∞ norm of ρ_k in the basis $(e_m)_{1 \leq m \leq 2g}$ satisfies

$$(12.10.2) \qquad |\rho_k|_\infty \leq l^\Theta.$$

The determinant of $D_\mathbf{q}\phi_{\vec{\gamma}}$

$$\frac{\rho_1 \wedge \ldots \wedge \rho_{2g}}{e_1 \wedge \ldots \wedge e_{2g}}$$

is the square of the absolute value of the Jacobian determinant

$$(12.10.3) \qquad \mathcal{J}_{\vec{\gamma}}(\mathbf{q}) = \left| \frac{\omega}{dq_{\gamma_k}}(q_k) \right|_{1 \leq k \leq g, \, \omega \in \mathcal{B}^1_{DR}}.$$

We denote λ the opposite of the logarithm of the absolute value of this determinant, and we call it the *illconditioning* of $\mathbf{q}$. We shall see that the inverse Jacobi problem is well conditioned unless the illconditioning is large.

We first observe that we can bound from below the norm of every ρ_k in terms of λ. Indeed, the determinant of $D_\mathbf{q}\phi_{\vec{\gamma}}$ is bounded from above by the product of the L^2 norms $\prod_{1 \leq k \leq 2g} |\rho_k|_2$, so

$$\exp(-2\lambda) \leq |\rho_k|_2 \times \prod_{j \neq k} |\rho_j|_2 \leq |\rho_k|_2 \times \exp(\Theta l^3)$$

using equation (12.10.2). So

$$(12.10.4) \qquad |\rho_k|_2 \geq \exp(-2\lambda - \Theta l^3).$$

We now can bound from below the determinant of equation (12.10.1) in terms of the illconditioning λ. For every $1 \leq k \leq 2g$ we notice that $\Upsilon\rho_k$ is the first order approximation of b_k. We deduce from equations (12.6.2) and (12.4.8) that

$$|b_k - \Upsilon\rho_k| \leq l^\Theta \Upsilon^2.$$

Using the lower bound (12.10.4) we deduce that

$$\frac{\Upsilon}{2}|\rho_k|_2 \leq |b_k|_2 \leq \frac{3\Upsilon}{2}|\rho_k|_2 \leq l^\Theta\Upsilon$$

provided $\chi \geq l^\Theta + 2\lambda$.

Using multilinearity of the determinant we can bound the difference

$$\frac{b_1 \wedge \ldots \wedge b_{2g}}{e_1 \wedge \ldots \wedge e_{2g}} - \Upsilon^{2g} \frac{\rho_1 \wedge \ldots \wedge \rho_{2g}}{e_1 \wedge \ldots \wedge e_{2g}}$$

by

$$2^{2g} \Upsilon^{2g-1} \left(\max_{1 \leq k \leq 2g} |\rho_k|_2 \right)^{2g-1} \max_{1 \leq k \leq 2g} |b_k - \Upsilon \rho_k|_2$$

$$\leq 2^{2g} l^{\Theta(2g-1)} l^\Theta \Upsilon^{2g+1}$$

and this is less than half of $\Upsilon^{2g} |\frac{\rho_1 \wedge \cdots \wedge \rho_{2g}}{e_1 \wedge \cdots \wedge e_{2g}}| = \exp(-2\lambda) \Upsilon^{2g}$ as soon as

$$\chi \geq \Theta l^3 + 2\lambda.$$

We deduce that

$$\left| \frac{b_1 \wedge \cdots \wedge b_{2g}}{e_1 \wedge \cdots \wedge e_{2g}} \right| \geq \frac{1}{2} \exp(-2\lambda - 2g\chi).$$

So we have a lower bound for the determinant of the transition matrix between the basis $(e_k)_{1 \leq k \leq 2g}$ and the basis $(b_k)_{1 \leq k \leq 2g}$. Further, the entries in this matrix are bounded by $l^\Theta \Upsilon$ in absolute value. Therefore the entries in the inverse matrix are bounded by

$$2 \exp(2\lambda + 2g\chi) l^{\Theta g} \Upsilon^{2g-1} \leq \exp(2\lambda + \chi + \Theta l^3)$$

in absolute value. We thus can compute this inverse matrix in time polynomial in l, λ, χ and the required absolute accuracy.

In Section 12.2 we constructed a basis $\mathcal{B}_{per}$ for the lattice of periods, consisting of vectors with coordinates bounded by $\exp(l^\Theta)$ in absolute value in the basis $(e_k)_{1 \leq k \leq 2g}$. The coordinates of these periods in the basis $(b_k)_{1 \leq k \leq 2g}$ are bounded by

$$\exp(l^\Theta + 2\lambda + \chi)$$

in absolute value.

Every point in the fundamental parallelogram associated with the basis $\mathcal{B}_{per}$ (i.e., having coordinates in $[0, 1]$ in this basis) has coordinates

$$\leq \exp(2\lambda + \chi + l^\Theta)$$

in absolute value in the basis $(b_k)_{1 \leq k \leq 2g}$. When we replace the latter coordinates by the closest integer, the induced error is bounded by $l^{\Theta} \Upsilon$ according to the L^2 norm for the canonical basis $(e_k)_{1 \leq k \leq 2g}$.

According to equation (12.6.14) there exists a vector **q** with illconditioning

$$\lambda \leq l^{\Theta}.$$

This finishes the proof of the following theorem.

12.10.5 Theorem (inverse Jacobi problem) *There exists a deterministic algorithm that takes as input*

- *a prime integer l,*

- *an element x in the tangent space $\mathbb{C}^{\mathcal{B}_{DR}^1}$ to $J_1(5l)$ at the origin (where $\mathcal{B}_{DR}^1$ is the basis of $H_{DR}^{1,0}$ made of normalized newforms of level $5l$ together with normalized newforms of level l lifted to level $5l$ by the two degeneracy maps),*

- *a degree g effective (origin) divisor Ω on $X_1(5l)$,*

and returns an approximation of a degree g effective divisor

$$P = P_1 + \cdots + P_g$$

on $X_1(5l)$ such that $\phi^*(P - \Omega) = x + \Lambda$.

The running time is $(l \times \log(2 + |x|_\infty) \times m)^{\Theta}$, where $5l$ is the level, m is the required absolute accuracy of the result, and $\log(2 + |x|_\infty)$ is the size of x, that is, the logarithm of its L^∞ norm in the canonical basis of $\mathbb{C}^{\mathcal{B}_{DR}^1}$.

We insist that the absolute accuracy in this theorem be measured in the space $\mathbb{C}^{\mathcal{B}_{DR}^1}$ using the L^∞ norm, or equivalently in the Jacobian variety $J = J_1(5l)$ using the distance d_J. However, when x belongs to the Ramanujan subspace W_f and the degree g origin divisor Ω is the cuspidal divisor D_0 manufactured in Section 8.1, then there is a unique effective degree g divisor Q such that $\phi^*(Q - \Omega) = x + \Lambda$. We call it the *Ramanujan divisor* associated with x. In that case, we can and must control the error in X^g. We use the pseudo-distance d_X introduced in Section 12.1 to define a pseudo-distance on $\text{Sym}^g X$. The pseudo-distance $d_{\text{Sym}^g X}(Q, Q')$ between $Q = \sum_{1 \le k \le g} Q_k$ and $Q' = \sum_{1 \le k \le g} Q'_k$ is defined to be

$$(12.10.6) \qquad d_{\text{Sym}^g X}(Q, Q') = \min_{\sigma \in S_g} \max_{1 \le k \le g} d_X(Q_{\sigma(k)}, Q'_k).$$

It is symmetric and separating but does not fulfill the triangle inequality, so we cannot call it a distance. Sections 12.11 to 12.13 will be mainly devoted to the proof of the theorem below.

12.10.7 Theorem (approximating V_f) *There exists a deterministic algorithm that takes as input a positive integer m, an even integer $k > 2$, a prime integer $l > 6(k-1)$, a finite field $\mathbb{F}$ with characteristic l, a ring epimorphism $f : \mathbb{T}(1, k) \to \mathbb{F}$, and a cuspidal divisor Ω on $X_1(5l)$ like the divisor D_0 constructed in Section 8.1, and computes a complex approximation within $\exp(-m)$ for every element in $W_f \subset J_1(5l)$, the image of $V_f \subset J_1(l)$ by $B^*_{5l,l,1}$. Here $V_f \subset J_1(l)$ is defined by equation (12.0.11) and we assume that the image of the Galois representation ρ_f associated with f contains $\text{SL}(V_f)$. The algorithm returns for every element x in W_f a complex approximation*

$$Q'_x = \sum_{1 \le n \le g} Q'_{x,n}$$

of the corresponding Ramanujan divisor (the unique effective degree g divisor Q_x such that $Q_x - \Omega$ lies in the class represented by x) such that

$$d_{\text{Sym}^g X}(Q_x, Q'_x) \le \exp(-m).$$

Every point $Q'_{x,n}$ is given as a couple (γ, q), where $\gamma \in \Xi$ and

$$q \in F_{w_\gamma} \subset D(0, \exp(-\pi/w_\gamma))$$

with the notation introduced in Section 12.1.1. The running time of the algorithm is $\le (m \times \#V_f)^\Theta$ for some absolute constant Θ. Here $\#V_f$ is the cardinality of the Galois representation V_f and m is the required absolute accuracy.

There are two differences between Theorems 12.10.5 and 12.10.7. While Theorem 12.10.5 controls the error in $J_1(5l)$, Theorem 12.10.7 controls the error in $X_1(5l)^g$. Unfortunately, Theorem 12.10.7 only applies to special divisors like the $Q_x - \Omega$. For these divisors, one can prove that the inverse Jacobi problem is reasonably well conditioned. Results in Section 8.1 prove that Q_x is well defined, and using Proposition 11.7.1 in Section 11.7 one can show that computing Q_x from x is a well-conditioned problem. This will be the purpose of the next three sections.

Another remark concerning notation. We denote by Ω the divisor D_0 introduced in Section 8.1. And we write Q_x rather than D_x. The only reason for this slight change in notation is that many things are already called D in this chapter, and we want to avoid any possible confusion.

12.11 THE ALGEBRAIC CONDITIONING

An important feature of Theorems 12.9.1, 12.9.2, and 12.10.5 is that the error in all these statements is measured in the torus $J(\mathbb{C})$. We have seen that this helps control the accumulation of errors when we chain computations. However, when solving the inverse Jacobi problem, we want to control the error in $\mathrm{Sym}^g X$, at least for the final result of the computation. So we need a theoretical estimate for the error in $\mathrm{Sym}^g X$ in terms of the error in $J(\mathbb{C})$. This will be the main concern of this and the following two sections.

So assume that we are given a degree g effective origin divisor Ω on X and a vector x in $\mathbb{C}^{B_{\mathrm{DR}}^1}$, and we look for an effective degree g divisor P such that $\phi^\star(P - \Omega) = x + \Lambda$, where Λ is the lattice of periods. We write

$$(12.11.1) \qquad P = n_1 P_1 + n_2 P_2 + \cdots + n_K P_K,$$

where the $(P_k)_{1 \leq k \leq K}$ are pairwise distinct points on X and the $(n_k)_{1 \leq k \leq K}$ are positive integers such that

$$n_1 + n_2 + \cdots + n_K = g.$$

We shall make two assumptions.

The *first assumption* is rather essential: we assume that the divisor P is *nonspecial*, or equivalently that

$$(12.11.2) \qquad \dim(H^0(X, \mathcal{O}_X(P))) = 1.$$

A first interesting consequence of this assumption is that the answer to the inverse Jacobi problem is unique, and the error can be defined as the distance to the unique solution.

The *second assumption* is more technical. We assume that Klein's modular function j does not take the values 0 or 1728 at any of the points $(P_k)_{1 \leq k \leq K}$. So

$$j(P_k) \notin \{0, 1728\}.$$

Removing the second assumption would only result in a heavier presentation. By contrast, the first assumption plays a crucial role in the forthcoming calculations. Its meaning is that the Jacobi map $\phi^\star : \mathrm{Sym}^g X \to J$ is a local diffeomorphism at $P - \Omega$. Our first task is to reformulate this first assumption in a more algebraic setting. We look for an algebraic variant of the Jacobi determinant in equation (12.10.3). Knowing that such an algebraic quantity is nonzero when P is a Ramanujan divisor, we will bound it from below in that case in Section 12.12.

We assume that we are given a basis $\mathcal{B}_{\mathbb{Z}}^1$ of $\mathrm{H}_{\mathrm{DR}}^{1,0}(X_1(5l))$ such that for every $\omega = f(q)q^{-1}dq$ in $\mathcal{B}_{\mathbb{Z}}^1$, the associated modular form $f(q)$ has rational integer coefficients and type $(\exp(\Theta l^4), \Theta)$. The existence of such a basis is granted by Lemma 12.11.5 below. We stress that we don't try to compute such a basis. We are just happy to know that it exists.

We also need an algebraic uniformizing parameter t_Q at every point Q on X such that $j(Q) \notin \{0, 1728\}$. When Q is not a cusp either, the differential dj of Klein's function j has no pole or zero at Q. So $j - j(Q)$ is a uniformizing parameter at Q. So we set

$$t_Q = j - j(Q)$$

in that case. When $Q = \gamma(\infty)$ is a cusp and $\gamma \in \Xi$, we set

$$j_\gamma = j \circ W_\gamma^{-1},$$

where W_γ is given in equation (12.2.2). We deduce from equation (12.2.3) that the automorphism W_γ^{-1} maps the cusp $\gamma(\infty)$ to the cusp ∞. So j_γ has a simple pole at $\gamma(\infty)$ and j_γ^{-1} is a uniformizing parameter at $\gamma(\infty)$. So we set

$$t_Q = j_\gamma^{-1}$$

in that case. We notice that j_γ only depends on the width of $\gamma(\infty)$. We call $j_1 = j$, j_5, j_l, and j_{5l} the four corresponding functions. Thus t_Q is one of the following functions: $j - j(Q)$, $1/j$, $1/j_5$, $1/j_l$, or $1/j_{5l}$.

Let ω be a form in $\mathcal{B}^1_{\mathbb{Z}}$ and let k be an integer such that $1 \leq k \leq K$, where K is the number of distinct points in the divisor P of equation (12.11.1). Let $t_k = t_{P_k}$ be the algebraic uniformizing parameter at P_k and consider the Taylor expansion of ω/dt_k at P_k,

(12.11.3)
$$\omega/dt_k = \xi^\omega_{k,0} + \xi^\omega_{k,1} \times \frac{t_k}{1!} + \xi^\omega_{k,2} \times \frac{t_k^2}{2!} + \cdots$$
$$+ \xi^\omega_{k,n_k-1} \times \frac{t_k^{n_k-1}}{(n_k - 1)!} + O(t_k^{n_k}).$$

We only need the first n_k terms in this expansion, where n_k is the multiplicity of P_k in the divisor P. We form the matrix

(12.11.4)
$$\mathcal{M}^{\text{alg}}_P = \left(\xi^\omega_{k,m} \right)_{1 \leq k \leq K, \, 0 \leq m \leq n_k-1; \, \omega \in \mathcal{B}^1_{\mathbb{Z}}}.$$

The rows in $\mathcal{M}^{\text{alg}}_P$ are indexed by pairs (k, m), where $1 \leq k \leq K$ and $0 \leq m \leq n_k - 1$. The columns in $\mathcal{M}^{\text{alg}}_P$ are indexed by forms ω in $\mathcal{B}^1_{\mathbb{Z}}$. We call $\mathcal{M}^{\text{alg}}_P$ the *algebraic Wronskian matrix* at P. The determinant of $\mathcal{M}^{\text{alg}}_P$ will play an important role in the sequel. We call it the *algebraic conditioning*.

An important feature of the matrix $\mathcal{M}^{\text{alg}}_P$ is that its entries are algebraic functions evaluated at the points P_k for $1 \leq k \leq K$. Indeed, let

$$t_k = t_{P_k} \in \{j - j(P_k), 1/j, 1/j_5, 1/j_l, 1/j_{5l}\}$$

be the chosen algebraic uniformizing parameter at the point P_k and set $\omega^{(0)} = \omega/dt_k$. For every integer $m \geq 0$ set $\omega^{(m+1)} = d\omega^{(m)}/dt_k$. Then

$$\xi^\omega_{k,m} = \omega^{(m)}(P_k).$$

Lemma 12.11.6 below provides useful information about the algebraic dependency between the derivatives $\omega^{(m)}$ and Klein's function j. It is an important consequence of our first assumption in equation (12.11.2) that the determinant of the algebraic Wronskian matrix is nonzero:

$$\det \mathcal{M}^{\text{alg}}_P \neq 0.$$

In Section 12.12 we shall prove a lower bound for this determinant in the special case when P is a Ramanujan divisor, using the theory of heights and Proposition 11.7.1. To finish this section, there remains to state and prove Lemmas 12.11.5 and 12.11.6. We first construct the basis $\mathcal{B}^1_{\mathbb{Z}}$.

12.11.5 Lemma (a rational basis) *There is a basis $\mathcal{B}^1_{\mathbb{Z}}$ of $H^{1,0}_{DR}(X_1(5l))$ such that for every $\omega = f(q)q^{-1}dq$ in $\mathcal{B}^1_{\mathbb{Z}}$, the associated modular form $f(q)$ has rational integer coefficients and type*

$$(\exp(\Theta l^4), \Theta).$$

The transition matrix from the basis $\mathcal{B}^1_{DR}$ to the basis $\mathcal{B}^1_{\mathbb{Z}}$ has algebraic integer entries bounded in absolute value by $\exp(\Theta l^4)$, and its determinant is the square root of a nonzero rational integer.

Proof We construct $\mathcal{B}^1_{\mathbb{Z}}$ from $\mathcal{B}^1_{DR}$ using a descent process.

Let ω be a differential form in the basis $\mathcal{B}^1_{DR}$, and let

$$f(q) = \omega q(dq)^{-1} = \sum_{m \geq 1} f_m q^m$$

be the corresponding modular form. Let $\mathbb{Z}_f$ be the ring generated by the coefficients of $f(q)$. It is a consequence of Theorem 2.5.11 that, as a $\mathbb{Z}$-module, $\mathbb{Z}_f$ is generated by the f_m for $m \leq 4l^2$. Let $\mathbb{K}_f$ be the fraction field of $\mathbb{Z}_f$. Let $\mathbb{L}$ be a strict subfield of $\mathbb{K}_f$. Let $\mathbf{a} = (a_m)_{1 \leq m \leq 4l^2}$ be a vector with rational integer coefficients. The associated linear combination $\sum_{1 \leq m \leq 4l^2} a_m f_m$ belongs to $\mathbb{L}$ if and only if $\mathbf{a}$ belongs to a submodule of $\mathbb{Z}^{4l^2}$ with rank $< 4l^2$. The degree d_f of $\mathbb{K}_f$ over $\mathbb{Q}$ is bounded above by $2g$. So the number of strict subfields of $\mathbb{K}_f$ is $< 2^{2g}$. So there exist rational integers $(a_m)_{1 \leq m \leq 4l^2}$ such that $0 \leq a_m < 2^{2g}$ and $\theta = \sum_{1 \leq m \leq 4l^2} a_m f_m$ does not belong to any strict subfield of $\mathbb{K}_f$. This θ is an algebraic integer that generates $\mathbb{K}_f$ over $\mathbb{Q}$. For $0 \leq k \leq d_f - 1$ we set

$$\mathrm{Tr}(\theta^k f) = \sum_{m \geq 1} \mathrm{Tr}(\theta^k f_m)q^m,$$

where $\mathrm{Tr} : \mathbb{Z}_f \to \mathbb{Z}$ is the trace map. Since $|f_m| \leq \Theta(m+1)^{\Theta}$ and $|\theta| \leq \exp(\Theta l^2)$ we deduce that $|\mathrm{Tr}(\theta^k f_m)| \leq \exp(\Theta l^4)(m+1)^{\Theta}$. So the series $\mathrm{Tr}(\theta^k f)$ for $0 \leq k \leq d_f - 1$ has type $(\exp(\Theta l^4), \Theta)$.

We do the same construction for every Galois orbit in $\mathcal{B}^1_{DR}$. We collect all the forms thus obtained. This makes a basis $\mathcal{B}^1_{\mathbb{Z}}$ of $H^{1,0}_{DR}(X_1(5l))$ consisting of forms $f(q)q^{-1}dq$, where $f(q)$ is a series with integer coefficients and of type $(\exp(\Theta l^4), \Theta)$. □

Now let us prove some quantitative statements about the algebraic dependency between the successive derivatives $\omega^{(m)}$ and j.

12.11.6 Lemma (an algebraic relation) *Let t be one of the functions*

$$j, 1/j, 1/j_5, 1/j_l, 1/j_{5l}.$$

Let ω be a form in $\mathcal{B}^1_{\mathbb{Z}}$, the basis given by Lemma 12.11.5. Let $m \geq 0$ be an integer. Set $\omega^{(0)} = \omega/dt$ and $\omega^{(m+1)} = d\omega^{(m)}/dt$. There exists a nonzero irreducible polynomial $E(X, Y) \in \mathbb{Z}[X, Y]$ such that

$$E(\omega^{(m)}, t) = 0.$$

Its degree in either variable is

$$\leq (lm)^\Theta,$$

and its coefficients are bounded in absolute value by

$$\exp((lm)^\Theta).$$

Proof All the functions involved belong to the field of level $5l$ and weight 0 modular functions having Puiseux expansion in $\mathbb{Q}\{\{q\}\}$ at the cusp ∞. This field is the field of functions of $X_\mu(5l)/\mathbb{Q}$, the twist of $X_1(5l)$ parametrizing μ_{5l}-structures on elliptic curves. This twist splits over $\mathbb{Q}(\zeta_{5l})$. The field $\mathbb{Q}(X_\mu(5l))$ is a regular extension of $\mathbb{Q}(j)$.

We assume that $t = j$ since the other cases are quite similar. The differential dj has $4(l^2 - 1)$ zeros of multiplicity 2 (the points in the fiber of j above 0) and $6(l^2 - 1)$ zeros of multiplicity 1 (the points in the fiber of j above 1728). So $\omega^{(0)} = \omega/dj$ has degree $\leq 14l^2$ and at most $10l^2$ poles. When we differentiate $\omega^{(0)}$ we increase by one the multiplicity of each pole. So $d\omega^{(0)}$ has the same poles as $\omega^{(0)}$ and the total multiplicity of these poles is at most $24l^2$. When we divide by dj we don't add poles, but we increase the multiplicities by 1 (in the fiber of j above 1728) or

2 (in the fiber of j above 0). The degree of the polar divisor of $\omega^{(1)}$ is thus $\leq 38l^2$. We go on like that and we prove that the degree of $\omega^{(m)}$ is $\leq (14 + 24m)l^2 \leq 24(m + 1)l^2$. The degree of j is $12(l^2 - 1)$. So there is an irreducible polynomial $E(x, y)$ in $\mathbb{Z}[x, y]$ such that $E(\omega^{(m)}, j^{-1}) = 0$, and $\deg_x E \leq 12(l^2 - 1)$, and $\deg_y E \leq 24(m + 1)l^2$.

In order to bound the coefficients in $E(x, y)$ we consider the expansions of j^{-1} and $\omega^{(m)}$ at the cusp ∞. This is a $\mathbb{Q}$-rational point on $X_\mu(5l)$. Remember that

$$j(q) = \frac{1}{q} + 744 + \sum_{k \geq 1} c(k)q^k,$$

so

$$-q^2 \frac{dj}{dq} = 1 - q \sum_{k \geq 1} kc(k)q^k.$$

Equation (12.1.5) implies that $\sum_{k \geq 1} kc(k)q^k$ has exp-type $(\Theta, 0, 2)$ in the sense of Section 12.4.18. Now equation (12.4.21) implies that $(-q^2 \frac{dj}{dq})^{-1}$ has exp-type $(K_1, 0, 4)$ where $K_1 \geq 1$ is an absolute constant. We write

$$\omega^{(0)} = -q^2 \frac{\omega}{dq} \left(-q^2 \frac{dj}{dq} \right)^{-1}.$$

The series ω/dq has type $(\exp(\Theta l^4), \Theta)$ and exp-type $(\Theta, \Theta l^2, 2)$. Using equation (12.4.20) we deduce that $\omega^{(0)}$ has exp-type $(K_2, l^2 K_3, 4)$ for some absolute constants $K_2 \geq 1$ and $K_3 \geq 1$. A simple iteration shows that $\omega^{(m)}$ has exp-type

$$\left(K_2 + mK_1, l^2 K_3 + mK_2 + \frac{m(m - 1)}{2} K_1 + 2m, 4 \right).$$

So if a is an integer such that $0 \leq a \leq 12(l^2 - 1)$ then $(\omega^{(m)})^a$ has exp-type

$$(\Theta l^2(m + 1), \Theta l^2(l + m + 1)^2, 4).$$

On the other hand, j^{-1} has exp-type $(\Theta, 0, 4)$ and if b is an integer such that $0 \leq b \leq 24(m + 1)l^2$ then j^{-b} has exp-type

$$(\Theta(m + 1)l^2, \Theta(m + 1)l^2, 4).$$

So all the monomials $(\omega^{(m)})^a j^{-b}$ arising in the equation $E(\omega^{(m)}, j^{-1}) = 0$ have exp-type

$$(\Theta l^2(m+1), \Theta l^2(l+m+1)^2, 4),$$

and the coefficients in their q-expansions up to order

$$2 \deg(\omega^{(m)}) \times \deg(j^{-1}) \le \Theta(m+1)l^4$$

are rational integers bounded in absolute value by

$$\exp(\Theta l^{24}(m+1)^8).$$

Since the coefficients in $E(x, y)$ are solutions of the homogeneous system given by these truncated q-expansions, they are bounded in absolute value by

$$\exp(\Theta l^{28}(m+1)^9).$$

$\square$

12.12 HEIGHTS

In this section we recall basic facts about heights of algebraic numbers and deduce upper and lower bounds for the determinant of the algebraic Wronskian matrix in equation (12.11.4) when the divisor P is a Ramanujan divisor. Let $\bar{\mathbb{Q}} \subset \mathbb{C}$ be the algebraic closure of $\mathbb{Q}$ in $\mathbb{C}$. Let $\alpha \in \bar{\mathbb{Q}}$ be an algebraic number. The *degree* d_α of α is the degree of the field extension $\mathbb{Q}(\alpha)/\mathbb{Q}$. Let

$$f(x) = a_{d_\alpha} x^{d_\alpha} + a_{d_\alpha-1} x^{d_\alpha-1} + \cdots + a_0$$

be the unique irreducible polynomial in $\mathbb{Z}[x]$ such that $f(\alpha) = 0$ and $a_{d_\alpha} > 0$. We say that a_{d_α} is the *denominator* of α and we denote it $\mathfrak{d}_\alpha$.

Let $\mathbb{K}$ be a number field containing α. The *multiplicative height* of α with respect to $\mathbb{K}$ is

$$H_{\mathbb{K}}(\alpha) = \prod_\sigma \max(1, |\sigma(\alpha)|) \prod_v \max(1, |\alpha|_v),$$

where the σ in the first product runs over the set of embeddings of $\mathbb{K}$ into $\mathbb{C}$ and the v in the second product runs over the non-Archimedean places of $\mathbb{K}$, counting multiplicities. In the special case $\mathbb{K} = \mathbb{Q}(\alpha)$ we have

$$H_{\mathbb{Q}(\alpha)}(\alpha) = \eth_\alpha \prod_{1 \leq k \leq d_\alpha} \max(1, |\alpha_k|),$$

where the α_k are the d_α roots of $f(x)$.

The *logarithmic height* of α with respect to $\mathbb{K}$ is

$$h_{\mathbb{K}}(\alpha) = \log H_{\mathbb{K}}(\alpha),$$

and the *absolute (logarithmic) height* of α is

$$h(\alpha) = \frac{h_{\mathbb{K}}(\alpha)}{\deg(\mathbb{K}/\mathbb{Q})} = \frac{h_{\mathbb{Q}(\alpha)}(\alpha)}{d_\alpha}.$$

Knowing the degree d_α and absolute height $h(\alpha)$ of a nonzero algebraic number α, we deduce the following upper and lower bounds for the absolute value $|\alpha|$:

$$\exp\left(-d_\alpha \times h(\alpha)\right) \leq |\alpha| \leq \exp\left(d_\alpha \times h(\alpha)\right).$$

Let $F(x)$ be a nonzero and degree d_F polynomial in $\mathbb{Z}[x]$ and assume that all coefficients in $F(x)$ are bounded by H_F in absolute value. Let $\mathbb{K}$ be a number field and let $\alpha \in \mathbb{K}$ be an algebraic number. We set $\beta = F(\alpha)$.

If σ is any embedding of $\mathbb{K}$ into $\mathbb{C}$ we have $\sigma(\beta) = F(\sigma(\alpha))$, so

$$\max(1, |\sigma(\beta)|) \leq (d_F + 1) H_F \max(1, |\sigma(\alpha)|)^{d_F}.$$

Now, let v be a non-Archimedean valuation of $\mathbb{K}$. We have

$$\max(1, |\beta|_v) \leq \max(1, |\alpha|_v)^{d_F}.$$

Forming the product over all σ and all v we find that the absolute logarithmic height of β is

(12.12.1) $$\leq d_F \times h(\alpha) + \log(d_F + 1) + \log H_F.$$

Now assume α and β belong to a number field $\mathbb{K}$ and let $E(x, y) \in \mathbb{Z}[x, y]$ be a polynomial such that $E(\alpha, y) \neq 0$ and $E(\alpha, \beta) = 0$. Assume that all coefficients in $E(x, y)$ are bounded by H_E in absolute value. Call d_x (resp. d_y) the degree of $E(x, y)$ with respect to the variable x (resp. y).

We write

$$E(x, y) = \sum_{0 \leq k \leq d_y} E_k(x)y^k.$$

We deduce from inequality (12.12.1) that every $E_k(\alpha)$ has absolute height

(12.12.2) $\qquad \leq d_x \times h(\alpha) + \log(d_x + 1) + \log H_E.$

Let K be the largest k such that $E_k(\alpha) \neq 0$. Set

$$F(y) = E(\alpha, y) = \sum_{0 \leq k \leq K} E_k(\alpha)y^k.$$

If σ is any embedding of $\mathbb{K}$ into $\mathbb{C}$, we call

$$^{\sigma}F(y) = E(\sigma(\alpha), y) = \sum_{0 \leq k \leq K} E_k(\sigma(\alpha))y^k$$

the polynomial obtained by applying σ to all the coefficients in $F(y)$. Applying Landau's inequality to $^{\sigma}F(y)$ we find that

$$\max(1, |\sigma(\beta)|) \leq \frac{\sqrt{d_y + 1} \times \max_{0 \leq k \leq K}(|E_k(\sigma(\alpha))|)}{|E_K(\sigma(\alpha))|},$$

and this is

$$\leq \sqrt{d_y + 1} \times (d_x + 1)H_E \times \max(1, |\sigma(\alpha)|)^{d_x} \times \max(1, |E_K(\sigma(\alpha))|^{-1}).$$

Now, let v be a non-Archimedean valuation of $\mathbb{K}$. Applying Gauss's lemma to $F(y)$ we find that

$$\max(1, |\beta|_v) \leq |E_K(\alpha)|_v^{-1} \times \max(1, |\alpha|_v)^{d_x}$$
$$\leq \max(1, |E_K(\alpha)|_v^{-1}) \times \max(1, |\alpha|_v)^{d_x}.$$

Forming the product over all σ and all v we find that the absolute logarithmic height of β is bounded from above by

$$\frac{\log(d_y + 1)}{2} + \log(d_x + 1) + \log H_E + d_x \times h(\alpha) + h(E_K(\alpha)).$$

Using equation (12.12.2) we deduce that

$$h(\beta) \leq \frac{\log(d_y + 1)}{2} + 2\log(d_x + 1) + 2\log H_E + 2d_x \times h(\alpha).$$

12.12.3 Lemma (relating heights of algebraic numbers) *Let* $\alpha, \beta \in \overline{\mathbb{Q}}$
be two algebraic numbers, and let $E(x, y) \in \mathbb{Z}[x, y]$ *be a polynomial such
that* $E(\alpha, y) \neq 0$ *and* $E(\alpha, \beta) = 0$. *Assume that all coefficients in* $E(x, y)$
are bounded by H_E *in absolute value. Call* d_x *(resp.* d_y) *the degree of*
$E(x, y)$ *with respect to the variable* x *(resp.* y). *Then the absolute heights
of* α *and* β *are related by the following inequality:*

$$h(\beta) \leq \frac{\log(d_y + 1)}{2} + 2\log(d_x + 1) + 2\log H_E + 2d_x \times h(\alpha).$$

We now can bound the heights of the entries in the algebraic Wronskian
of Section 12.11, at least in the cases we are interested in. Let

$$x + \Lambda \in W_f \subset J(\mathbb{C}) = \mathbb{C}^{\mathcal{B}^1_{\mathrm{DR}}}/\Lambda$$

be a nonzero vector in the Ramanujan subspace W_f of Theorem 12.10.7.
Assume that the degree g origin divisor Ω on $X_1(5l)$ is the cuspidal divisor
D_0 manufactured in Section 8.1. Let P be the unique degree g divisor such
that

$$\phi^\star(P - \Omega) = x + \Lambda.$$

It is a consequence of Proposition 8.2.2 and equation (8.2.3) in Section 8.1
that the two assumptions of Section 12.11 are satisfied in that case: the divisor $P = n_1 P_1 + n_2 P_2 + \cdots + n_K P_K$ is *nonspecial* and Klein's modular function j does not take the values 0 or 1728 at any of the points
$(P_k)_{1 \leq k \leq K}$. Proposition 11.7.1 in Section 11.7 implies that the absolute
height of every $j(P_k)$ is $\leq l^\Theta$. Lemmas 12.11.6 and 12.12.3 imply that
every entry in the algebraic Wronskian matrix of equation (12.11.4) has absolute height $\leq l^\Theta$. Using Lemma 4.2.6 we deduce that

(12.12.4) $$h(\det(\mathcal{M}_P^{\mathrm{alg}})) \leq l^\Theta$$

in that case.

Note also that, as an algebraic integer, the algebraic conditioning has degree at most $8(l-1) \times (\#V_f - 1)$ because the square of it lies in the definition
field of $x + \Lambda$ over $\mathbb{Q}(\zeta_{5l})$ and the latter field is a degree $\leq \#V_f - 1$ extension
of $\mathbb{Q}(\zeta_{5l})$.

We deduce that for all primes p but a finite number bounded by $(\#V_f)^\Theta$, the divisor P remains nonspecial when we reduce modulo any place $\mathfrak{p}$ above p. Indeed, we can assume that $p \notin \{5, l\}$ so $X_1(5l)$ has good reduction modulo p. We also can assume that $j(P_k) \notin \{0, 1728\}$ mod $\mathfrak{p}$ because both the degree and the absolute height of $j(P_k)$ are $\leq (\#V_f)^\Theta$. We also can assume that the P_k remain pairwise distinct modulo $\mathfrak{p}$ for the same reason: we just need to exclude less than $(\#V_f)^\Theta$ primes p. We can also assume that every P_k which is not a cusp does not reduce modulo $\mathfrak{p}$ onto a cusp. We also assume that p is larger than the genus g of $X_1(5l)$ so that the Taylor expansion in equation (12.11.3) remains valid modulo $\mathfrak{p}$. So the algebraic Wronskian matrix reduces modulo $\mathfrak{p}$ to the algebraic Wronskian matrix. Excluding a few more primes, but no more than $(\#V_f)^\Theta$, we can assume that the algebraic conditioning does not vanish modulo $\mathfrak{p}$. So the divisor P remains nonspecial.

12.12.5 Lemma (reduction modulo p of a Ramanujan divisor) *Call Ω the cuspidal divisor D_0 on $X_1(5l)$ introduced in Section 8.1 and let W_f be the Ramanujan subspace W_f of Theorem 12.10.7. Let*

$$x \in W_f \subset J_1(5l),$$

and let P be the unique divisor on $X_1(5l)$ such that $P - \Omega$ lies in the class defined by x. Then P is nonspecial, and for all primes p but a finite number bounded by $(\#V_f)^\Theta$, the divisor P remains nonspecial modulo any place $\mathfrak{p}$ above p.

We stress that the above lemma is very similar to Theorem 11.7.10.

12.13 BOUNDING THE ERROR IN X^g

In this section we fill the gap between Theorems 12.10.5 and 12.10.7. For $x + \Lambda \in W_f \subset J(\mathbb{C})$ and P the corresponding Ramanujan divisor, we relate the error on P and the error on x. We need some control on the Jacobi integration map locally at P. A first step in this direction is the upper bound for the height of the algebraic conditioning, given in equation (12.12.4). In Section 12.13.1 we deduce a lower bound for the determinant of the differential of the Jacobi map at P. This bound implies that the Jacobi map is nonsingular on a reasonably large neighborhood of P, as we show in Section 12.13.6. In Section 12.13.11 we deduce that the inverse of the Jacobi map is well defined and Lipschitz (with reasonably small constant) on a (reasonably large) neighborhood of $x + \Lambda$. We shall use the identities between Wronskian and Jacobian determinants of power series proved in Section 12.5. The reason for these algebraic complications is the following: we need an equation for the singular locus in X^g of the Jacobi integration map. The space X^g is stratified by the diagonals. The strata correspond to partitions of $\{1, 2, 3, \ldots, g\}$. We obtain a different equation for the singular locus on every stratum. These various equations are related by algebraic identities between Jacobian and Wronskian determinants.

12.13.1 The analytic conditioning of a divisor

Let

$$(12.13.2) \qquad P = n_1 P_1 + \cdots + n_K P_K$$

be a degree g effective divisor on X where the n_k are positive integers for $1 \leq k \leq K$. For every k we write

$$P_k = (\gamma_k, q_k),$$

where $\gamma_k \in \Xi$, and $q_k \in F_{w_k} \subset D(0, \exp(-\pi/w_k))$, and w_k is the width of the cusp $\gamma_k(\infty)$. Let

$$\phi_{(n_k)_{1 \leq k \leq K}} : \prod_{1 \leq k \leq K} \mathrm{Sym}^{n_k} X \to J(\mathbb{C}) = \mathbb{C}^{\mathcal{B}^1_{\mathrm{DR}}}/\Lambda$$

be the relevant Jacobi integration map in this context. We stress that this map is different from the maps introduced before. Its initial set is a sort of *semisymmetric product*, that is, something between the Cartesian product X^g and the full symmetric product $\mathrm{Sym}^g X$. In order to write down the differential of this map at the divisor P, we need a local system of coordinates on $\prod_{1 \leq k \leq K} \mathrm{Sym}^{n_k} X$. We shall use partial Newton power sums. Let

$$R = (\gamma, q_R) = \mu_\gamma(q_R)$$

be a point on X. For every $n \geq 1$ and $m \geq 1$ we define the mth power sum $\nu_{R,n,m}$ to be the function that takes the value

$$\left(q_\gamma(Q_1) - q_R\right)^m + \left(q_\gamma(Q_2) - q_R\right)^m + \cdots + \left(q_\gamma(Q_n) - q_R\right)^m$$

at $\{Q_1, \ldots, Q_n\} \in \mathrm{Sym}^n X$.

It is well defined on a neighborhood of $\{R, R, \ldots, R\}$ in $\mathrm{Sym}^n X$ because μ_γ is locally invertible at R and its inverse q_γ defines a local parameter there. Now let us come back to the divisor P in equation (12.13.2). The functions

$$\left(\nu_{P_k, n_k, m}\right)_{1 \leq k \leq K; \, 1 \leq m \leq n_k}$$

form a local system of coordinates on $\prod_{1 \leq k \leq K} \mathrm{Sym}^{n_k} X$ at the divisor P. The matrix of the differential at P of the integration map $\phi_{(n_k)_{1 \leq k \leq K}}$ in the bases $\left(d\nu_{P_k, n_k, m}\right)_{1 \leq k \leq K; \, 1 \leq m \leq n_k}$ and $\mathcal{B}_{\mathrm{DR}}^1$ is

$$(12.13.3) \qquad \mathcal{M}_P^{\mathrm{ana}} = \left(\psi_{k,m}^\omega / (m+1)!\right)_{1 \leq k \leq K, \, 0 \leq m \leq n_k - 1; \, \omega \in \mathcal{B}_{\mathrm{DR}}^1},$$

where $\psi_{k,m}^\omega$ is the mth derivative of ω / dq_{γ_k} with respect to q_{γ_k}, evaluated at $P_k = (\gamma_k, q_k)$. The determinant of this matrix is called the *analytic conditioning* of the divisor P. This analytic conditioning and the algebraic conditioning defined in Section 12.11 after equation (12.11.4) differ by simple factors we should not be afraid of.

Now, we assume that we are in the context of Theorem 12.10.7. We call Ω the cuspidal divisor D_0 constructed in Section 8.1 and we assume that the class of $P - \Omega$ corresponds to a point x in the Ramanujan subspace $W_f \subset J$. Equation (12.12.4) implies that the algebraic conditioning of P is $\geq \exp(-(\#V_f)^\Theta)$. In order to obtain a similar lower bound for the analytic conditioning, we must bound from below the complementary factors.

The factor $1 / \prod_{1 \leq k \leq K} \prod_{1 \leq m \leq n_k} m!$ is $\geq \exp(-\Theta l^7)$.

Lemma 12.11.5 implies that the factor coming from the change of bases from $\mathcal{B}_{\mathrm{DR}}^1$ to $\mathcal{B}_{\mathbb{Z}}^1$ is $\geq \exp(-l^\Theta)$.

There are also factors due to the change of coordinates. Assume, for example, that P_k is not a cusp. So the algebraic parameter at P_k is $j - j(P_k)$. The analytic parameter at P_k is $q_{\gamma_k} - q_k$. The extra factor in the analytic conditioning is thus

$$\left(\frac{dj}{dq_{\gamma_k}}(P_k)\right)^{\frac{n_k(n_k+1)}{2}}.$$

Lemma 12.13.5 below ensures that there exists a constant K_1 such that this factor is $\geq \exp(-(\#V_f)^{K_1})$ provided there exists a constant K_2 such that $|j(P_k)|$ and $|j(P_k) - 1728|$ are both $\geq \exp(-(\#V_f)^{K_2})$. But this latter condition is met because $j(P_k)$ is not in $\{0, 1728\}$, and both its degree and logarithmic height are $\leq (\#V_f)^{\Theta}$.

Assume now that P_k is a cusp. Then the algebraic parameter at P is the function $1/j_{w_k}$ introduced in Section 12.11, where $w_k \in \{1, 5, l, 5l\}$ is the width of the cusp $P_k = \gamma_k(\infty)$. And the derivative $d(1/j_{w_k})/dq_{\gamma_k}$ is just 1. So the analytic conditioning at a Ramanujan l-torsion divisor is

(12.13.4) $\hspace{3cm} \geq \exp(-(\#V_f)^{\Theta}).$

To finish this section, there remains to state and prove Lemma 12.13.5.

12.13.5 Lemma (lower bounds for dJ/dx) *For every positive real number K_1 there exists a positive real number K_2 such that the following statement is true.*

Let $L \geq 2$ be an integer and let $J(x)$ be Klein's series given in equation (12.1.3). Let $q \in D(0, 1)$ be a complex number in the unit disk such that $1 - |q| \geq L^{-K_1}$, and $|J(q)| \geq \exp(-L^{K_1})$, and $|J(q) - 1728| \geq \exp(-L^{K_1})$. Then $|\frac{dJ}{dx}(q)| \geq \exp(-L^{K_2})$.

Proof We first note that $\frac{dJ}{dx}(q)$ only vanishes if $J(q) \in \{0, 1728\}$. So we just want to prove that if $\frac{dJ}{dx}$ is small at q then q is close to a zero of it. We would like to apply Lemma 5.4.22 to the series $-x^2 J'(x) = -x^2 \frac{dJ}{dx}$. Unfortunately this series is not of type (A, n) because its coefficients are a bit too large. So we set $R = (1 + |q|)/2$ and $\mathfrak{J}(x) = -x^2 R^2 \times J'(Rx)$. From the hypothesis in the lemma there exists a positive real K_3 such that $\log R^n \leq -n \times L^{-K_3}$. From equation (12.1.5) there exists an absolute constant K_4 such that the coefficient of x^n in $-x^2 \times J'(x)$ is $\leq \exp(K_4 \sqrt{n})$. This implies that the series $\mathfrak{J}(x)$ has type $(\exp(L^{K_5}), K_5)$ for some $K_5 \geq 1$. We apply Lemma 5.4.22 of Chapter 5 to the series $\mathfrak{J}(x)$ at q/R and we are done. $\hspace{1cm} \square$

12.13.6 The neighborhood of a nonspecial divisor

The singular locus of the Jacobi integration map $\phi : X^g \to J$ is a strict closed subset. So every nonspecial effective degree g divisor has a neighborhood in $\mathrm{Sym}^g X$ consisting of nonspecial divisors. In this section, we provide a quantified version of this statement for Ramanujan divisors. So we assume that we are in the context of Theorem 12.10.7. We call Ω the cuspidal divisor D_0 constructed in Section 8.1 and we assume that

$$(12.13.7) \qquad P = n_1 P_1 + \cdots + n_K P_K$$

is a Ramanujan divisor corresponding to the point x in the Ramanujan subspace $W_f \subset J$. We first need to define simple neighborhoods of the divisor P in equation (12.13.7). For every $1 \le k \le K$ we write

$$(12.13.8) \qquad P_k = (\gamma_k, q_k)$$

for $\gamma_k \in \Xi$ and $q_k \in F_{w_k} \subset D(0, \exp(-\pi/w_k))$. Let ϵ be a positive real number. We call $\mathcal{P}_\epsilon$ the set of degree g effective divisors P' that can be written

$$P' = \sum_{1 \le k \le K} \sum_{1 \le m \le n_k} P'_{k,m},$$

where $P'_{k,m} = (\gamma_k, q'_{k,m}) = \mu_{\gamma_k}(q'_{k,m})$ and $|q'_{k,m} - q_k| \le \epsilon$. In particular, every such P' satisfies $d_{\mathrm{Sym}^g X}(P, P') \le \epsilon$ provided $\epsilon \le 1/(\Theta l)$.

We know that P is nonspecial. So if ϵ is small enough, then P' is nonspecial as well. In order to write down the analytic conditioning of P' we must take multiplicities into account. So we rewrite P' as

$$P' = \sum_{1 \le k \le K} \sum_{1 \le s \le S_k} m_{k,s} P'_{k,s},$$

and, for every k, we denote $\mathbf{m}_k = (m_{k,s})_{1 \le s \le S_k}$ the corresponding partition of n_k into S_k nonempty parts. In particular,

$$n_k = m_{k,1} + m_{k,2} + \cdots + m_{k,S_k}.$$

We set $\mathbf{n} = (n_k)_{1 \le k \le K}$ and $\mathbf{m} = (m_{k,s})_{1 \le k \le K; 1 \le s \le S_k}$. These are partitions of g, and $\mathbf{m}$ is a refinement of $\mathbf{n}$. For every $1 \le k \le K$ we call

$$x_k = q_{\gamma_k} - q_k$$

the local analytic parameter at P_k. We call $\mathbf{f}_k$ the vector in $(\mathbb{C}[[x_k]])^{\mathcal{B}_{\mathrm{DR}}^1}$ defined by

$$(12.13.9) \qquad \mathbf{f}_k = \left(\mathrm{Taylor}\left(\frac{\omega}{dq_{\gamma_k}}, q_k \right) \right)_{\omega \in \mathcal{B}_{\mathrm{DR}}^1} \in (\mathbb{C}[[x_k]])^{\mathcal{B}_{\mathrm{DR}}^1},$$

where $\mathrm{Taylor}(\frac{\omega}{dq_{\gamma_k}}, q_k)$ is the Taylor expansion of $\frac{\omega}{dq_{\gamma_k}}$ at q_k in the parameter x_k.

For every $1 \le k \le K$ and $1 \le s \le S_k$ we write

$$P'_{k,s} = (\gamma_k, q'_{k,s}),$$

and we set $y_{k,s} = q'_{k,s} - q_k$.

The analytic conditioning of P is the Wronskian in equation (12.5.11) evaluated at $(0, 0, \ldots, 0) \in \mathbb{C}^K$ and divided by $\mathbf{n}! = \prod_{1 \le k \le K} n_k!$.

The analytic conditioning of P' is the Jacobian determinant in equation (12.5.12) divided by $\mathbf{m}! = \prod_{1 \le k \le K} \prod_{1 \le s \le S_k} m_{k,s}!$ More precisely, the Jacobian determinant in equation (12.5.12) is $\mathbf{m}!$ times the Taylor expansion of the analytic conditioning in the parameters $(y_{k,s})_{1 \le k \le K; 1 \le s \le S_k}$.

These two conditionings are related by equation (12.5.13). The conditioning at P' is divisible by the determinant $D_{\mathbf{m},\mathbf{n}}$ relating the two partitions. And the quotient specializes to the conditioning at P times $\mathbf{n}!/\mathbf{m}!$

The conditioning $\mathcal{J}_{(\mathbf{f}_k,\mathbf{m}_k)_{1 \le k \le K}}$ is a series in the $\sum_{1 \le k \le K} S_k$ variables $(y_{k,s})_{1 \le k \le K; 1 \le s \le S_k}$. Using equation (12.6.2) about the size of coefficients in modular forms, the type of refocused series as described in Section 12.4.5, the type of derivatives given in Section 12.4.11, and the type of quotient series given in Section 12.4.9, we can prove that the series

$$\frac{\mathcal{J}_{(\mathbf{f}_k,\mathbf{m}_k)_{1 \le k \le K}}}{D_{\mathbf{m},\mathbf{n}}}$$

has type $(\exp(l^{\Theta}), l^{\Theta}\mathbf{1}_g)$.

We denote by λ the opposite of the logarithm of the absolute value of the analytic conditioning of P. We call it the *analytic illconditioning* of P. It generalizes the illconditioning introduced in Section 12.10. The only difference is that here we take multiplicities into account. Using equation (12.5.13) and inequality (12.4.8) on bounding the remainder, we prove that if

$$(12.13.10) \qquad\qquad -\log \epsilon \ge l^{\Theta} + \lambda,$$

then P' is a nonspecial divisor too. According to equation (12.13.4) the analytic illconditioning λ is $\leq (\#V_f)^\Theta$. So any divisor in the neighborhood $\mathcal{P}_\epsilon$ is nonspecial, provided $\epsilon \leq \exp(-(\#V_f)^\Theta)$.

12.13.11 Relating the direct and inverse errors

We continue the local study of the Jacobi integration map at a Ramanujan divisor $P \in \text{Sym}^g X$ as in equations (12.13.7) and (12.13.8). We can associate to the divisor $P \in \text{Sym}^g X$ a point

$$Q = (n_1 P_1, n_2 P_2, \ldots, n_K P_K)$$

on the semisymmetric product $\prod_{1 \leq k \leq K} \text{Sym}^{n_k} X$. We call $\mathcal{S}_{(n_k)_{1 \leq k \leq K}}$ or just $\mathcal{S}$ the map

$$\mathcal{S} : X^g \rightarrow \prod_{1 \leq k \leq K} \text{Sym}^{n_k} X$$

that sends $(R_1, \ldots, R_g)$ onto

$$(\{R_1, \ldots, R_{n_1}\}, \{R_{n_1+1}, \ldots, R_{n_1+n_2}\}, \ldots, \{R_{n_1+\cdots+n_{K-1}+1}, \ldots, R_g\}).$$

We check that the following diagram commutes:

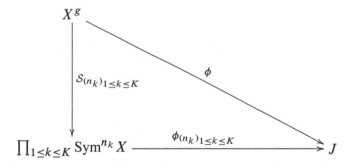

We call $U = (P_1, \ldots, P_1, P_2, \ldots, P_2, \ldots, P_K, \ldots, P_K) \in X^g$ the unique g-uple such that

$$\mathcal{S}(U) = Q.$$

We call $x = \phi(U)$ the point in J corresponding to the divisor P. We know that P is a nonspecial divisor. The map $\phi_{(n_k)_{1 \leq k \leq K}}$ is thus a local diffeomorphism at Q. However, the map ϕ is not a local diffeomorphism at U because $\mathcal{S}$ is not, unless the partition $\mathbf{n}$ is $(1, 1, \ldots, 1)$. We shall allow ourselves to write $\phi^\bullet$ instead of $\phi_{(n_k)_{1\le k\le K}}$ in this section. So $x = \phi(U) = \phi^\bullet(Q) = \phi^\star(P)$. We need an explicit analytic description of the maps ϕ at U. We have local coordinates $(z_{k,m})_{1\le k\le K;\, 1\le m\le n_k}$ at $U \in X^g$. Let

(12.13.12) $$R = (R_{k,m})_{1\le k\le K;\, 1\le m\le n_k}$$

be a g-uple in X^g, where

$$R_{k,m} = (\gamma_k, q_{k,m}) = \mu_\gamma(q_{k,m}).$$

We set

$$z_{k,m} = q_{k,m} - q_k.$$

For every $1 \le k \le K$ we define the series $\mathbf{F}_k \in (\mathbb{C}[[x_k]])^{\mathcal{B}^1_{\mathrm{DR}}}$ to be the formal integral

$$\mathbf{F}_k(x_k) = \int_0^{x_k} \mathbf{f}_k(x_k)dx_k = \mathbf{F}_{k,1} \times x_k + \mathbf{F}_{k,2} \times x_k^2 + \cdots,$$

where $\mathbf{f}_k$ is the vector defined in equation (12.13.9). This is the Taylor expansion of the Jacobi map at the point P_k. The Taylor expansion of ϕ at U is

$$\mathrm{Taylor}(\phi, U) = \sum_{1\le k\le K} \sum_{1\le m\le n_k} \mathbf{F}_k(z_{k,m}),$$

and it lies in

$$\left(\mathbb{C}[[(z_{k,m})_{1\le k\le K;\, 1\le m\le n_k}]]\right)^{\mathcal{B}^1_{\mathrm{DR}}}.$$

We stress that for every ω in $\mathcal{B}^1_{\mathrm{DR}}$, the corresponding coordinate in the above Taylor expansion is a series of type

(12.13.13) $$(l^\Theta, \Theta \times \mathbf{1}_g)$$

because it is mostly the expansion of a modular form at a point which is not too close to the boundary of the unit disk.

We now split this Taylor expansion in two pieces. We write

$$\mathrm{Taylor}(\phi, U) = T_1 + T_2,$$

where T_1 is a sort of principal part

$$T_1 = \sum_{1 \le k \le K} \sum_{1 \le m \le n_k} \sum_{1 \le j \le n_k} \mathbf{F}_{k,j} \times z_{k,m}^j$$

(12.13.14)
$$= \sum_{1 \le k \le K} \sum_{1 \le j \le n_k} \mathbf{F}_{k,j} \sum_{1 \le m \le n_k} z_{k,m}^j$$

$$= \sum_{1 \le k \le K} \sum_{1 \le j \le n_k} \mathbf{F}_{k,j} \times v_{k,j},$$

where

$$v_{k,j} = \sum_{1 \le m \le n_k} z_{k,m}^j.$$

We call $T_2 = \mathrm{Taylor}(\phi, U) - \phi_1$ the corresponding remainder. It is clear from equation (12.13.14) that T_1 can be written

$$T_1 = T_1^\bullet \circ \mathcal{S},$$

where $T_1^\bullet$ is the first order term (the differential) in the Taylor expansion of $\phi^\bullet$ at $Q = \mathcal{S}(U)$. We write $T_2 = T_2^\bullet \circ \mathcal{S}$, where $T_2^\bullet$ is the corresponding remainder. The maps T_1 and T_2 are defined on a neighborhood of U and take values in the torus J. We notice that T_1 and T_2 can be defined on the polydisk $\prod_{1 \le k \le K} D(q_k, 1 - |q_k|)^{n_k}$ and take values in $\mathbb{C}^{B_{\mathrm{DR}}^1}$.

We set $S = \mathcal{S}(R)$ and we prove that if R is close enough to U then the principal term $T_1^\bullet$ dominates the remainder $T_2^\bullet$. In order to control how close is R to U, we set

$$\epsilon = \max_{1 \le k \le K;\, 1 \le m \le n_k} |z_{k,m}|$$

and

$$\eta = \max_{1 \le k \le K;\, 1 \le j \le n_k} |v_{k,j}|.$$

We prove that if η is $\le \exp(-l^\Theta)$ then $T_1^\bullet$ dominates $T_2^\bullet$.

We first bound the remainder $T_2(R) = T_2^\bullet(S)$ from above in terms of η. Using the Buckholtz inequality (5.3.11) we can bound the coordinates

$$z_{k,m} = q_{k,m} - q_k$$

for every $1 \leq k \leq K$ and $1 \leq m \leq n_k$ in terms of the bound η on Newton power functions:

$$|z_{k,m}| \leq 5\eta^{1/n_k},$$

so

$$|z_{k,m}|^{n_k+1} \leq 5^{g+1} \times \eta^{(n_k+1)/n_k} \leq 5^{g+1} \times \eta^{1+1/g}.$$

Using the type estimate in equations (12.13.13) and (12.4.8) we deduce that

$$(12.13.15) \qquad \left|T_2^{\bullet}(S)\right|_{\infty} \leq \exp(l^{\Theta}) \times \eta^{1+1/g}.$$

We now bound from below $T_1(R) = T_1^{\bullet}(S)$ in terms of η. We note that $T_1^{\bullet}$ is a linear map and its matrix in the bases $(v_{k,j})_{1 \leq k \leq K; 1 \leq j \leq n_k}$ and $\mathcal{B}_{\mathrm{DR}}^1$ is the matrix $\mathcal{M}_P^{\mathrm{ana}}$ of equation (12.13.3). The coefficients in this matrix are $\leq l^{\Theta}$ in absolute value. The determinant of this matrix is the analytic conditioning at P. According to equation (12.13.4) the analytic conditioning at the Ramanujan l-torsion divisor P is $\geq \exp(-(\#V_f)^{\Theta})$. We deduce that

$$(12.13.16) \qquad \left|T_1^{\bullet}\right|_{\infty} \geq \exp(-(\#V_f)^{\Theta}) \times \eta.$$

We deduce from equations (12.13.16) and (12.13.15) that

$$\left|T_1^{\bullet}\right|_{\infty} > 2\left|T_2^{\bullet}\right|_{\infty}$$

as soon as $\epsilon \leq \exp(-(\#V_f)^{\Theta})$.

Using Lemma 12.13.17 below we deduce that there exist two absolute positive constants K_1 and K_2 such that for every y in $J(\mathbb{C})$ such that

$$d_J(x, y) \leq \exp(-(\#V_f)^{K_1})$$

there exists a g-uple R as in equation (12.13.12) such that $\phi(R) = y$ and the corresponding ϵ satisfies

$$\epsilon \leq \exp((\#V_f)^{K_2}) \times d_J(x, y).$$

Using again that the conditioning of P is $\geq \exp(-(\#V_f)^{\Theta})$ together with the estimates in Section 12.13.6, for example, equation (12.13.10), we deduce that if the constant K_1 has been chosen big enough, then the degree g effective divisor associated with R is nonspecial.

So every degree g effective divisor $P' \in \mathrm{Sym}^g X$ such that

$$d_J(\phi^\star(P'), \phi^\star(P)) \leq \exp(-(\#V_f)^{K_1})$$

is nonspecial and provides a good approximation of $P \in \mathrm{Sym}^g X$ in the sense that

$$d_{\mathrm{Sym}^g X}(P', P) \leq \exp((\#V_f)^{K_2}) \times d_J(\phi^\star(P'), \phi^\star(P)).$$

This relation between the error in J and the error in $\mathrm{Sym}^g X$ finishes the proof of Theorem 12.10.7. To finish this section there remains to state and prove Lemma 12.13.17.

12.13.17 Lemma (perturbation of a nonsingular linear map) *Let $g \geq 1$ be a positive integer and let $\bar{P}(0_g, 1_g) \in \mathbb{C}^g$ be the closed polydisk with polyradius 1_g and centered at the origin. Let $T : \bar{P}(0_g, 1_g) \to \mathbb{C}^g$ be a continuous function and let $T_1 : \mathbb{C}^g \to \mathbb{C}^g$ be a linear function. Set $T_2 = T - T_1$ and assume that for every $R = (R_1, \ldots, R_g)$ in $\bar{P}(0_g, 1_g)$ we have $|T_1(R)|_\infty > 2|T_2(R)|_\infty$. Let $K > 0$ be a positive real number such that for every R in $\mathbb{C}^g$*

$$|T_1(R)|_\infty \geq K \times |R|_\infty.$$

Then the image by T of the polydisk $\bar{P}(0_g, 1_g)$ contains the polydisk

$$\bar{P}\left(0_g, \frac{K}{2} \times 1_g\right)$$

with polyradius $K/2 \times 1_g$.

Proof For every real number $t \in [0, 1]$ we denote by $\mathcal{S}_t$ the image by $T_1 + t T_2$ of the L^∞ sphere with radius 1. Then $\mathcal{S}_t$ is contained in

$$\mathbb{C}^g - \bar{P}\left(0_g, \frac{K}{2} \times 1_g\right)$$

for every $t \in [0, 1]$. So $\mathcal{S}_1$ is homologous to $\mathcal{S}_0$ in $\mathbb{C}^g - \bar{P}(0_g, \frac{K}{2} \times 1_g)$ and its class in

$$H_{2g-1}\left(\mathbb{C}^g - \bar{P}\left(0_g, \frac{K}{2} \times 1_g\right), \mathbb{Z}\right)$$

is nonzero.

So for every x in $\bar{P}(0_g, \frac{K}{2} \times 1_g)$, the class of $\mathcal{S}_1$ in $H_{2g-1}(\mathbb{C}^g - \{x\}, \mathbb{Z})$ is nonzero either. Assume now that x does not belong to the image by T of the polydisk $\bar{P}(0_g, 1_g)$. Then $\mathcal{S}_1$ is the boundary of $T(\bar{P}(0_g, 1_g)) \subset \mathbb{C}^g - \{x\}$ and its class in $H_{2g-1}(\mathbb{C}^g - \{x\}, \mathbb{Z})$ is trivial. A contradiction. $\square$

12.14 FINAL RESULT OF THIS CHAPTER

We now state the final result of this chapter.

12.14.1 Theorem (approximating V_f) *There exists a deterministic algorithm that takes as input an even integer $k > 2$, a prime integer $l > 6(k-1)$, a finite field $\mathbb{F}$ with characteristic l, a ring epimorphism $f : \mathbb{T}(1, k) \to \mathbb{F}$, and a cuspidal divisor Ω on $X_1(5l)$ like the divisor D_0 constructed in Section 8.1, and computes complex approximations for every element in*

$$W_f \subset J_1(5l),$$

the image of $V_f \subset J_1(l)$ by $B^*_{5l,l,1}$. Here $V_f \subset J_1(l)$ is defined by equation (12.0.11) and we assume that the image of the Galois representation ρ_f associated with f contains $\mathrm{SL}(V_f)$. The algorithm returns for every element x in W_f the unique degree g effective divisor Q_x such that $Q_x - \Omega$ lies in the class represented by x. More precisely, the algorithm returns the cuspidal part Q_x^{cusp} of Q_x and a complex approximation of its finite part $Q_x^{\mathrm{fin}} = \sum_{1 \le n \le d_x} Q_{x,n}$. Every point $Q_{x,n}$ is given by complex approximations of its affine coordinates $(b(Q_{x,n}), x(Q_{x,n}))$ in the plane model C_l of $X_1(5l)$ given in Section 12.3.19. The running time of the algorithm is $\le (m \times \#V_f)^\Theta$ for some absolute constant Θ. Here $\#V_f$ is the cardinality of the Galois representation V_f and m is the required absolute accuracy.

There are only two differences between this Theorem 12.14.1 and the previous Theorem 12.10.7. First, we claim that we can separate the cuspidal and the finite part of Q_x. Second, we return algebraic coordinates b and x for the points $Q_{x,n}$ rather than analytic ones.

Indeed, the algorithm in Theorem 12.10.7 returns for every point P in the support of Q_x an approximation P' of P such that $d_X(P, P') \le \exp(-m_1)$ for the m_1 we have chosen. We know that both the degree and the logarithmic height of $j(P)$ are $\le (\#V_f)^\Theta$. So there exists an absolute constant

$K_1 \geq 1$ such that if $1/j(P)$ is $\leq \exp(-(\#V_f)^{K_1})$ then it is zero. But the Petersson-Rademacher inequality tells us that there exists an $\epsilon > 0$ such that if x is a complex number bounded by ϵ in absolute value then

$$(12.14.2) \qquad\qquad |J(x)| \geq \frac{1}{2|x|},$$

where $J(x)$ is Klein's series given in equation (12.1.3).

Using equation (12.1.4) we deduce that there exists an absolute constant K_2 such that if $P = (\gamma, q)$ and q is $\leq \exp(-(\#V_f)^{K_2})$ then $q = 0$ and the point $P = (\gamma, q)$ is the cusp $\gamma(\infty)$.

Now remember that P' is given as (γ', q'), where $\gamma' \in \Xi$ and

$$q' \in F_{w'} \subset D(0, \exp(-\pi/w'))$$

and w' is the width of $\gamma'(\infty)$. Assume that P is the cusp $\gamma(\infty)$ for some γ in Ξ. The definition of the distance d_X implies that $\gamma' = \gamma$ and q' has absolute value $\leq \exp(-m_1)$. If we have chosen $m_1 \geq (\#V_f)^{K_2} + 1$ then we can decide if the point P in question is a cusp from the returned point P'.

Now assume that P is not a cusp. Since $d_X(P', P) \leq \exp(-m_1)$ there exist a γ in Ξ and $q, q' \in D(0, \exp(-\pi/w_\gamma))$ such that

$$P = \mu_\gamma(q), P' = \mu_\gamma(q') \text{ and } |q' - q| \leq \exp(-m_1).$$

Since P is not a cusp we have $|q| \geq \exp(-(\#V_f)^{K_2})$. If we have chosen $m_1 \geq (\#V_f)^{K_2} + 1$, then $|q'| \geq \exp(-(\#V_f)^{K_2} - 1)$. Lemma 12.3.25 implies that there exists a constant K_3 such that $j(P') - j(P)$, $b(P') - b(P)$ and $x(P') - x(P)$ are bounded in absolute value by

$$\exp(l^{K_3}) \times |q' - q| \leq \exp(l^{K_3} - m_1).$$

Assuming $m_1 \geq l^{K_3} + 1$ we deduce that

$$(12.14.3) \qquad\qquad |j(P)| \geq |j(P')| - 1.$$

Further, P' is given as (γ', q''), where $\gamma' \in \Xi$,

$$q'' \in F_{w'} \subset D(0, \exp(-\pi/w')),$$

and w' is the width of $\gamma'(\infty)$. We deduce from equations (12.14.2) and (12.14.3) that

$$|q''| \geq \min(\epsilon, \frac{1}{2(1 + |j(P)|)}) \geq \exp(-(\#V_f)^{K_1} - 1).$$

In order to compute $b(P')$ and $x(P')$ we just substitute $q_{\gamma'}$ by q'' in the expansions for x and b at $\gamma'(\infty)$ computed in Section 12.3. So this last step is polynomial time in l and the required accuracy.

Chapter Thirteen

Computing V_f modulo p

by Jean-Marc Couveignes

In this chapter we address the problem of computing in the group of l^k-torsion rational points in the Jacobian variety of algebraic curves over finite fields, with an application to computing modular representations.

Let p be a prime and let $\mathbb{F}_p = \mathbb{Z}/p\mathbb{Z}$ be the field with p elements. Let $\overline{\mathbb{F}}_p$ be an algebraic closure of $\mathbb{F}_p$. For any power q of p we call $\mathbb{F}_q \subset \overline{\mathbb{F}}_p$ the field with q elements. Let $\mathbb{A}^2 \subset \mathbb{P}^2$ be the affine and projective planes over $\mathbb{F}_q$. Let $C \subset \mathbb{P}^2$ be a plane projective geometrically integral curve over $\mathbb{F}_q$. Let X be its smooth projective model and let J be the Jacobian variety of X. We note g the genus of X and d the degree of C. We assume that we are given the numerator of the zeta function of the function field $\mathbb{F}_q(X)$. So we know the characteristic polynomial of the Frobenius endomorphism F_q of J. This is a monic degree $2g$ polynomial $\chi(x)$ with integer coefficients.

Let $l \neq p$ be a prime integer and let $k \geq 1$ be an integer. We look for a *nice generating set* for the group $J[l^k](\mathbb{F}_q)$ of l^k-torsion points in $J(\mathbb{F}_q)$. By *nice* we mean that the generating set $(g_i)_{1 \leq i \leq I}$ should induce a decomposition of $J[l^k](\mathbb{F}_q)$ as a direct product $\prod_{1 \leq i \leq I} < g_i >$ of cyclic subgroups with nondecreasing orders. Given such a generating set and an $\mathbb{F}_q$-endomorphism of J, we also want to describe the action of this endomorphism on $J[l^k](\mathbb{F}_q)$ by an $I \times I$ integer matrix.

By an algorithm in this chapter we usually mean a probabilistic Las Vegas algorithm. In some places we shall give deterministic or probabilistic Monte Carlo algorithms, but this will be stated explicitly. See Section 5.1 for a reminder of computational complexity theory. The main reason for using probabilistic Turing machines is that we shall need to construct generating sets for the Picard group of curves over finite fields. Solving such a problem in the deterministic world is out of reach at this time; see Section 5.1.

Our unique goal is to prove, as quickly as possible, that the problems studied in this chapter can be solved in probabilistic polynomial time. So we shall never try to be optimal. A detailed description of more efficient algorithms can be found in [Cou2], from which this chapter is inspired. Further improvements can be found in Peter Bruin's PhD thesis [Bru].

In Section 13.1 we recall how to compute in the Picard group $J(\mathbb{F}_q)$. Section 13.2 gives a naive algorithm for picking random elements in this group. Pairings are useful when looking for relations between divisor classes. So we recall how to compute pairings in Section 13.3. Section 13.4 is concerned with characteristic subspaces for the action of Frobenius inside the l^∞-torsion of $J(\overline{\mathbb{F}}_p)$. In Section 13.5 we look for a convenient surjection from $J(\mathbb{F}_q)$ onto its l^k-torsion subgroup. We use the Kummer exact sequence and the structure of the ring generated by the Frobenius endomorphism. In Section 13.6 we give an algorithm that, on input a degree d plane projective curve over $\mathbb{F}_q$, plus some information on its singularities, and the zeta function of its function field, returns a nice generating set for the group of l^k-torsion points inside $J(\mathbb{F}_q)$ in probabilistic polynomial time in $\log q$, d, and l^k. In Section 13.7 we apply the general algorithms of Section 13.6 to the modular curve $X_1(5l)$ in order to compute explicitly a modular representations V_f modulo l. Such a representation modulo l can be realized as a subgroup W_f inside the l-torsion of $J_1(5l)/\mathbb{Q}$. The idea is to compute the reduction modulo p of the group scheme W_f as a subgroup of $J_1(5l)/\mathbb{F}_p$, for many small primes p. One can then lift using the Chinese remainder theorem, as will be explained in Section 14.8.

Notation: The symbol Θ in this chapter stands for a positive effective absolute constant. So any statement containing this symbol becomes true if the symbol is replaced in every occurrence by some large enough real number.

13.1 BASIC ALGORITHMS FOR PLANE CURVES

In this section, we recall elementary results about computing in the Picard group of an algebraic curve over a finite field. See [Hac, Vol, Poo2, Die] for a more detailed treatment.

13.1.1 Finite fields

We should first explain how finite fields are represented. The prime field $\mathbb{F}_p$ is just $\mathbb{Z}/p\mathbb{Z}$, so elements in it are represented as integers in $[0, p - 1[$. The base field $\mathbb{F}_q$ is given as $\mathbb{F}_p[x]/f(x)$, where $f(x)$ is an irreducible monic polynomial with degree a in $\mathbb{F}_p[x]$, where p is the characteristic and $q = p^a$. A finite extension of $\mathbb{F}_q$ is given as $\mathbb{F}_q[y]/h(y)$, where $h(y)$ is a monic irreducible polynomial in $\mathbb{F}_q[y]$. We shall never use two extensions of $\mathbb{F}_q$ simultaneously. Recall that polynomial factoring in $\mathbb{F}_q[x]$ is Las Vegas probabilistic polynomial time in $\log q$ and the degree of the polynomial to be factored (see [Ga-Ge], Chapter 14).

13.1.2 Plane projective curves and their smooth model

We now explain how curves are represented in this chapter. To start with, a projective plane curve C over $\mathbb{F}_q$ is given by a degree d homogeneous polynomial $E(\mathrm{x}, \mathrm{y}, \mathrm{z})$ in the three variables x, y, and z, with coefficients in $\mathbb{F}_q$. The curve C is assumed to be absolutely integral. By a *point* on C we mean a geometric point: an element of $C(\overline{\mathbb{F}}_q)$, where $\overline{\mathbb{F}}_q = \overline{\mathbb{F}}_p$ is the algebraic closure of $\mathbb{F}_p$ fixed in the introduction. Any $\overline{\mathbb{F}}_q$-point on C can be represented by its affine or projective coordinates.

Let X be a smooth model of C and let $X \to C$ be the desingularization map. If $P \in X(\overline{\mathbb{F}}_q)$ is a geometric point on X above a singular point S on C, we say that P is a *singular branch*. The *conductor* $\mathfrak{C}$ is an effective divisor on X. It is the closed subscheme of X defined by the sheaf $\mathrm{Ann}_{\mathcal{O}_C}(\mathcal{O}_X/\mathcal{O}_C)$. Every multiplicity in $\mathfrak{C}$ is even. Some authors call $\mathfrak{C}$ the *adjunction divisor*. Its support consists of all singular branches. The conductor expresses the local behavior of the map $X \to C$; see IV.1 in [Ser1] and [Gor]. We have $\deg(\mathfrak{C}) = 2\delta$, where

$$\delta = \frac{(d-1)(d-2)}{2} - g$$

is the difference between the arithmetic genus of C and the geometric genus of X. Since $\delta \le (d-1)(d-2)/2$, the support of $\mathfrak{C}$ contains at most $(d-1)(d-2)/2$ geometric points in $X(\overline{\mathbb{F}}_q)$. So the field of definition of any singular branch on X is an extension of $\mathbb{F}_q$ with degree

$$\leq (d-1)(d-2)/2.$$

A modern reference for singularities of plane curves is Section 5.8 of [Cas].

The smooth model X of C is not given as a projective variety. Indeed, we shall only need a nice local description of X above every singularity of C. This means that we need a list of all singular points on C, and a list (a labeling) of all points in $X(\overline{\mathbb{F}}_q)$ lying above every singularity of C (the singular branches), and a uniformizing parameter at every such branch. We also need the Laurent series expansions of affine plane coordinates in terms of all these uniformizing parameters.

More precisely, let $P \in X(\overline{\mathbb{F}}_q)$ be a geometric point above a singular point S, and let v be the corresponding valuation. The field of definition of P is an extension field $\mathbb{F}_P$ of $\mathbb{F}_q$ with degree $\leq (d-1)(d-2)/2$. Let x_S and y_S be affine coordinates that vanish at the singular point S on C. We need a local parameter t_P at P and expansions $x_S = \sum_{k \geq v(x_S)} a_k t_P^k$ and $y_S = \sum_{k \geq v(y_S)} b_k t_P^k$ with coefficients in $\mathbb{F}_P$.

Because these expansions are not finite, we just assume that we are given a black box that on input a positive integer n returns the first n terms in all these expansions. In all the cases we shall be interested in, this black box will contain a Turing machine that answers in time polynomial in n, $\log q$, and the genus g. This is the case for curves with ordinary multiple points for example. We have shown in Section 12.3 that this is also the case for the standard plane model of modular curves $X_1(5l)$. Using general normalization and factorization algorithms [Po-Za] one can show, (Section 2 in [Die]), that this is indeed the case for all plane curves without any restriction, but this is beyond the scope of this text.

We may also assume that we are given the conductor $\mathfrak{C}$ of C as a combination of singular branches with even coefficients. The following algorithms still work if the conductor is replaced by any divisor $\mathfrak{D}$ that is greater than the conductor and has polynomial degree in d. Such a divisor can be found easily: the singular branches on X are supposed to be known already, and the multiplicities are bounded above by $(d-1)(d-2)/2$.

Given a smooth point $P = (x_P : y_P : z_P)$ on X, we call $\mathbb{F}_P$ be the field of definition of P. Assume, for example, that $z_P \neq 0$. We set $x = \mathrm{x}/\mathrm{z}$

and $y = y/z$ and we rewrite P as $(x_P : y_P : 1)$. At least one among the two functions $x - x_P$ and $y - y_P$ is a uniformizing parameter at P. We choose one and call it t_P. One can compute the expansions of x and y as power series in t_P using Newton iteration. This takes times polynomial in d, $\log \# \mathbb{F}_P$ and the expected t_P-adic accuracy.

13.1.3 Divisors, forms, and functions

Smooth $\overline{\mathbb{F}}_q$-points on C are represented by their affine or projective coordinates. Labeling for the branches above singular points is given in the description of X. So we know how to represent divisors on X. For any integer $h \geq 0$ the $\mathbb{F}_q$-linear space $H^0(\mathbb{P}^2, \mathcal{O}_{\mathbb{P}^2}(h))$ of degree h homogeneous polynomials in x, y, and z has dimension $(h + 1)(h + 2)/2$. A basis for it is made of all monomials of the form $x^a y^b z^c$ with $a, b, c \in \mathbb{N}$ and $a + b + c = h$. We denote by $\mathcal{O}_X(h)$ the pullback of $\mathcal{O}_{\mathbb{P}^2}(h)$ to X. Let F be a degree h form on $\mathbb{P}^2$ having nonzero pullback F_X on X. Let $\Delta = (F_X)$ be the divisor of this restriction. The map $f \mapsto \frac{f}{F_X}$ is a bijection from $H^0(X, \mathcal{O}_X(h))$ to the linear space $H^0(X, \mathcal{O}_X(\Delta))$ associated with Δ.

We assume that we are given a divisor $\mathfrak{D}$ bigger than the conductor $\mathfrak{C}$. We assume that the degree of $\mathfrak{D}$ is $\leq d^{\ominus}$. We have explained in Section 13.1.2 how to find such a divisor.

The dimension of $H^0(X, \mathcal{O}_X(h)(-\mathfrak{D}))$ is at least $dh + 1 - g - \deg(\mathfrak{D})$, and is equal to this number when it exceeds $g - 1$. This is the case if

$$h \geq d + \frac{\deg \mathfrak{D}}{d}.$$

The dimension of $H^0(X, \mathcal{O}_X(h)(-\mathfrak{D}))$ is greater than $2g$ if

$$h \geq \frac{3d}{2} + \frac{\deg \mathfrak{D}}{d}.$$

We take h to be the smallest integer fulfilling this condition. The composite map $X \twoheadrightarrow C \hookrightarrow \mathbb{P}^2$ induces a map

$$\rho_h : H^0(\mathbb{P}^2, \mathcal{O}_{\mathbb{P}^2}(h)) \to H^0(X, \mathcal{O}_X(h)).$$

The image of ρ_h contains $H^0(X, \mathcal{O}_X(h)(-\mathfrak{D}))$. This is known as Noether's residue theorem, Theorem 7 in [Gor].

It will be convenient to describe $H^0(X, \mathcal{O}_X(h)(-\mathfrak{D}))$ as a quotient,

$$0 \to \operatorname{Ker} \rho_h \to \rho_h^{-1}(H^0(X, \mathcal{O}_X(h)(-\mathfrak{D}))) \xrightarrow{\rho_h} H^0(X, \mathcal{O}_X(h)(-\mathfrak{D})) \to 0.$$

We need linear equations for

$$\rho_h^{-1}(H^0(X, \mathcal{O}_X(h)(-\mathfrak{D}))) \subset H^0(\mathbb{P}^2, \mathcal{O}_{\mathbb{P}^2}(h)).$$

We consider a generic homogeneous form

$$F(\mathrm{x}, \mathrm{y}, \mathrm{z}) = \sum_{a+b+c=h} \epsilon_{a,b,c} \mathrm{x}^a \mathrm{y}^b \mathrm{z}^c$$

of degree h in x, y, and z. For every branch P above a singular point $S \in C$ (assuming for example that S has nonzero z-coordinate) we replace in $F(\frac{\mathrm{x}}{\mathrm{z}}, \frac{\mathrm{y}}{\mathrm{z}}, 1)$ the affine coordinates $x = \frac{\mathrm{x}}{\mathrm{z}}$ and $y = \frac{\mathrm{y}}{\mathrm{z}}$ by their expansions as series in the local parameter t_P at this branch. We ask the resulting series in t_P to have valuation at least the multiplicity of P in the divisor $\mathfrak{D}$. Every singular branch thus produces linear equations in the $\epsilon_{a,b,c}$. We obtain similar equations at every smooth point in the support of $\mathfrak{D}$. The collection of all such equations defines the subspace $\rho_h^{-1}(H^0(X, \mathcal{O}_X(h)(-\mathfrak{D})))$.

A basis for the subspace $\operatorname{Ker} \rho_h$ of $\rho_h^{-1}(H^0(X, \mathcal{O}_X(h)(-\mathfrak{D})))$ consists of all $\mathrm{x}^a \mathrm{y}^b \mathrm{z}^c E(\mathrm{x}, \mathrm{y}, \mathrm{z})$ with $a + b + c = h - d$. We fix a supplementary space M_C to $\operatorname{Ker} \rho_h$ in $\rho_h^{-1}(H^0(X, \mathcal{O}_X(h)(-\mathfrak{D})))$ and we assimilate $H^0(X, \mathcal{O}_X(h)(-\mathfrak{D}))$ to it.

Given a homogeneous form in three variables, one can compute its divisor on X using resultants and the given expansions of affine coordinates in terms of the local parameters at every singular branch. A function is given as a quotient of two forms.

13.1.4 The Brill-Noether algorithm

The linear space M_C computed in the previous paragraph is isomorphic to $H^0(X, \mathcal{O}_X(h)(-\mathfrak{D}))$ via the map ρ_h. This space allows us to compute in the group $J(\mathbb{F}_q)$ of $\mathbb{F}_q$-points in the Jacobian variety of X. We fix an effective $\mathbb{F}_q$-divisor Ω with degree g on X. This Ω will play the role of origin: a point $\alpha \in J(\mathbb{F}_q)$ is represented by a divisor $A - \Omega$ in the corresponding linear equivalence class, where A is an effective $\mathbb{F}_q$-divisor on X with degree g. Given another point $\beta \in J(\mathbb{F}_q)$ by a similar divisor $B - \Omega$, we can compute the space $H^0(X, \mathcal{O}_X(h)(-\mathfrak{D} - A - B))$, which is nontrivial, and pick a nonzero form F_1 in it. The divisor of F_1 is $(F_1) = A + B + \mathfrak{D} + R$, where R is an effective divisor with degree $hd - 2g - \deg(\mathfrak{D})$. The linear space $H^0(X, \mathcal{O}_X(h)(-\mathfrak{D} - R - \Omega))$ has dimension at least 1. We pick a nonzero form F_2 in it. It has divisor $(F_2) = \mathfrak{D} + R + \Omega + D$, where D is effective with degree g. Then $D - \Omega$ is linearly equivalent to $A - \Omega + B - \Omega$.

In order to invert the class α of $A - \Omega$, we pick a nonzero form F_1 in $H^0(X, \mathcal{O}_X(h)(-\mathfrak{D} - 2\Omega))$. The divisor of F_1 is $(F_1) = 2\Omega + \mathfrak{D} + R$, where R is an effective divisor with degree $hd - 2g - \deg(\mathfrak{D})$. The linear space $H^0(X, \mathcal{O}_X(h)(-\mathfrak{D} - R - A))$ has dimension at least 1. We pick a nonzero form F_2 in it. It has divisor $(F_2) = \mathfrak{D} + R + A + B$, where B is effective with degree g. And $B - \Omega$ is linearly equivalent to $-(A - \Omega)$.

13.1.5 Lemma (arithmetic operations in the Jacobian variety) *Let C be a degree d plane projective absolutely integral curve over $\mathbb{F}_q$. Let g be the geometric genus of C. Assume that we are given the smooth model X of C and a $\mathbb{F}_q$-divisor with degree g on X, denoted Ω. We assume that Ω is given as a difference between two effective divisors with degrees bounded by d^Θ. This Ω serves as an origin. Arithmetic operations in the Picard group $\mathrm{Pic}^0(X/\mathbb{F}_q)$ can be performed in time polynomial in $\log q$ and d. This includes addition, subtraction and comparison of divisor classes.*

Proof If Ω is not effective, we use Lemma 13.1.6 below to compute a nonzero function f in $H^0(X, \mathcal{O}_X(\Omega))$ and we write $\Omega' = (f) + \Omega$. This is an effective divisor with degree g. We replace Ω by Ω' and finish as in the paragraph before Lemma 13.1.5 □

We now recall the principle of the Brill-Noether algorithm for computing complete linear series. Recall that functions in $\mathbb{F}_q(X)$ are represented as quotients of forms.

13.1.6 Lemma (Brill-Noether) *There exists an algorithm that on input a degree d plane projective absolutely integral curve $C/\mathbb{F}_q$ and the smooth model X of C and two effective $\mathbb{F}_q$-divisors A and B on X, computes a basis for $H^0(X, \mathcal{O}_X(A - B))$ in time polynomial in d, $\log q$, and the degrees of A and B.*

Proof We assume that $\deg(A) \geq \deg(B)$; otherwise $H^0(X, \mathcal{O}_X(A - B))$ is equal to $\{0\}$. Let a be the degree of A. We let h be the smallest integer such that

$$hd - g + 1 > a + \deg \mathfrak{D}.$$

The space $H^0(X, \mathcal{O}_X(h)(-\mathfrak{D} - A))$ is nonzero. It is contained in the image of the map

$$\rho_h : H^0(\mathbb{P}^2, \mathcal{O}_{\mathbb{P}^2}(h)) \to H^0(X, \mathcal{O}_X(h)),$$

so that we can represent it as a subspace of $H^0(\mathbb{P}^2, \mathcal{O}_{\mathbb{P}^2}(h))$.

We pick a nonzero form f in $H^0(X, \mathcal{O}_X(h)(-\mathfrak{D} - A))$ and compute its divisor $(f) = \mathfrak{D} + A + D$. The space $H^0(X, \mathcal{O}_X(h)(-\mathfrak{D} - B - D))$ is contained in the image of the map ρ_h so that we can represent it as a subspace of $H^0(\mathbb{P}^2, \mathcal{O}_{\mathbb{P}^2}(h))$.

We compute forms $\gamma_1, \gamma_2, \ldots, \gamma_k$ in $H^0(\mathbb{P}^2, \mathcal{O}_{\mathbb{P}^2}(h))$ such that their images by ρ_h provide a basis for $H^0(X, \mathcal{O}_X(h)(-\mathfrak{D} - B - D))$. A basis for $H^0(X, \mathcal{O}_X(A - B))$ is made of the functions $\frac{\gamma_1}{f}, \frac{\gamma_2}{f}, \ldots, \frac{\gamma_k}{f}$. $\qquad\square$

We deduce an explicit moving lemma for divisors.

13.1.7 Lemma (moving divisor lemma I) *There exists an algorithm that on input a degree d plane projective absolutely integral curve $C/\mathbb{F}_q$ and the smooth model X of C and a degree zero divisor $D = D^+ - D^-$ and an effective divisor A with degree $< q$ on X, computes a divisor $E = E^+ - E^-$ linearly equivalent to D and disjoint to A in time polynomial in d and $\log q$ and the degrees of D^+, and A. Further the degree of E^+ and E^- can be taken to be $\leq 2gd$.*

Proof Let O be an $\mathbb{F}_q$-rational divisor on X such that $1 \leq \deg(O) \leq d$ and whose support is disjoint from that of A. We may take O to be a well chosen fiber of some plane coordinate function on X. We compute the linear space $H^0(X, \mathcal{O}_X(D^+ - D^- + 2gO))$.

The subspace of $H^0(X, \mathcal{O}_X(D^+ - D^- + 2gO))$ consisting of functions f such that $(f) + D^+ - D^- + 2gO$ is not disjoint to A is contained in a union of at most $\deg(A) < q$ hyperplanes. We conclude invoking Lemma 13.1.8 below. $\qquad\square$

There remains to state and prove the

13.1.8 Lemma (solving inequalities) *Let q be a prime power, $d \geq 2$, and $n \geq 1$ two integers, and let $H_1, \ldots, H_n$ be hyperplanes inside $V = \mathbb{F}_q^d$, each given by a linear equation. Assume that $n < q$. There exists a deterministic algorithm that finds a vector in $U = V - \bigcup_{1 \leq k \leq n} H_k$ in time polynomial in $\log q$, d, and n.*

Proof This is proved by lowering the dimension d. For $d = 2$ we pick any affine line L in V not containing the origin. We observe that there are at least $q - n$ points in $U \cap L = L - \bigcup_{1 \leq k \leq n} L \cap H_k$. We enumerate points in L until we find one which is not in any H_k. This requires at most $n + 1$ trials.

Assume now that d is bigger than 2. Hyperplanes in V are parametrized by the projective space $\mathbb{P}(\hat{V})$, where $\hat{V}$ is the dual of V. We enumerate points in $\mathbb{P}(\hat{V})$ until we find a hyperplane K distinct from every H_k. We compute a basis for K and an equation for every $H_k \cap K$ in this basis. This way, we have lowered the dimension by 1. $\qquad \square$

We can strengthen the moving divisor algorithm a bit by removing the condition that A has degree $< q$. Indeed, in case this condition is not met, we call α the smallest integer such that $q^\alpha > \deg(A)$ and we set $\beta = \alpha + 1$. We apply Lemma 13.1.7 after base change to the field with q^α elements and find a divisor E_α. We call e_α the norm of E_α from $\mathbb{F}_{q^\alpha}$ to $\mathbb{F}_q$. It is equivalent to αD. We similarly construct a divisor e_β that is equivalent to $(\alpha + 1)D$. We return the divisor $E = e_\beta - e_\alpha$. We observe that we can take $\alpha \leq 1 + \log_q \deg(A)$ so the degree of the positive part E^+ of E is $\leq 4gd(\log_q(\deg(A)) + 2)$.

13.1.9 Lemma (moving divisor lemma II) *There exists an algorithm that on input a degree d plane projective absolutely integral curve $C/\mathbb{F}_q$ and the smooth model X of C and a degree zero $\mathbb{F}_q$-divisor $D = D^+ - D^-$ and an effective divisor A on X, computes a divisor $E = E^+ - E^-$ linearly equivalent to D and disjoint to A in time polynomial in d, $\log q$, and the degrees of D^+ and A. Further, the degree of E^+ and E^- can be taken to be $\leq 4gd(\log_q(\deg(A)) + 2)$.*

13.2 A FIRST APPROACH TO PICKING RANDOM DIVISORS

Given a finite field $\mathbb{F}_q$ and a plane projective absolutely integral curve C over $\mathbb{F}_q$ with projective smooth model X, we call J the Jacobian variety of X and we consider two related problems: picking a random element in $J(\mathbb{F}_q)$ with (close to) uniform distribution and finding a generating set for (a large subgroup of) $J(\mathbb{F}_q)$. Let g be the genus of X. We assume that we are given a degree 1 divisor $O = O^+ - O^-$, where O^+ and O^- are effective, $\mathbb{F}_q$-rational, and have degree bounded by Θg^Θ for some positive constant Θ.

We know from Theorem 2 in [Mu-S-T] that the group $\mathrm{Pic}^0(X/\mathbb{F}_q)$ is generated by the classes $[\mathfrak{p} - \deg(\mathfrak{p})O]$, where $\mathfrak{p}$ runs over the set of prime divisors of degree $\leq 1 + 2\log_q(4g - 2)$. For the convenience of the reader we quote this result as a lemma.

13.2.1 Lemma (Müller, Stein, Thiel) *Let K be an algebraic function field of one variable over $\mathbb{F}_q$. Let $N \geq 0$ be an integer. Let g be the genus of K. Let $\chi : \mathrm{Div}(K) \to \mathrm{Pic}(K) \to \mathbb{C}^*$ be a character of finite order which is nontrivial when restricted to Div^0. Assume that $\chi(\mathfrak{p}) = 1$ for every prime divisor $\mathfrak{p}$ of degree $\leq N$. Then*

$$N < 2\log_q(4g - 2).$$

If $q < 4g^2$, the number of prime divisors of degree $\leq 1 + 2\log_q(4g - 2)$ is bounded by Θg^Θ. So we can easily compute a small generating set for $J(\mathbb{F}_q)$. In the rest of this section, we will assume that the size q of the field is greater than or equal to $4g^2$. This condition ensures the existence of a $\mathbb{F}_q$-rational point.

Picking efficiently and provably random elements in $J(\mathbb{F}_q)$ with uniform distribution seems difficult to us. We first give here an algorithm for efficiently constructing random divisors with a distribution that is far from uniform but still sufficient to construct a generating set for a large subgroup of $J(\mathbb{F}_q)$. Once given generators, picking random elements becomes much easier.

Let r be the smallest prime integer bigger than 30, $2g - 2$, and d. We note that r is less than $\max(4g - 4, 2d, 60)$. The set $\mathcal{P}(r, q)$ of $\mathbb{F}_q$-places with degree r on X has cardinality

$$\#\mathcal{P}(r,q) = \frac{\#X(\mathbb{F}_{q^r}) - \#X(\mathbb{F}_q)}{r}.$$

So

$$(1 - 10^{-2})\frac{q^r}{r} \leq \#\mathcal{P}(r,q) \leq (1 + 10^{-2})\frac{q^r}{r}.$$

Indeed, $\left|\#X(\mathbb{F}_{q^r}) - q^r - 1\right| \leq 2gq^{\frac{r}{2}}$ and $\left|\#X(\mathbb{F}_q) - q - 1\right| \leq 2gq^{\frac{1}{2}}$. So

$$\left|\#\mathcal{P}(r,q) - \frac{q^r}{r}\right| \leq \frac{4g+3}{r}q^{\frac{r}{2}} \leq 8q^{\frac{r}{2}},$$

and $8rq^{\frac{-r}{2}} \leq r2^{3-\frac{r}{2}} \leq 10^{-2}$ since $r \geq 31$.

Since we are given a degree d plane model C for the curve X, we have a degree d map $x : X \to \mathbb{P}^1$. This is the composition of the desingularization map $X \to C$ with the restriction to C of the rational map $[x, y, z] \to [x, z]$. Since $d < r$, the function x maps $\mathcal{P}(r,q)$ to the set $\mathcal{U}(r,q)$ of monic prime polynomials of degree r over $\mathbb{F}_q$. The cardinality of $\mathcal{U}(r,q)$ is $\frac{q^r - q}{r}$, so

$$(1 - 10^{-9})\frac{q^r}{r} \leq \#\mathcal{U}(r,q) \leq \frac{q^r}{r}.$$

The fibers of the map $x : \mathcal{P}(r,q) \to \mathcal{U}(r,q)$ have cardinality between 0 and d. We can pick a random element in $\mathcal{U}(r,q)$ with uniform distribution in the following way: we pick a random monic polynomial of degree r with coefficients in $\mathbb{F}_q$, with uniform distribution. We check whether it is irreducible. If it is, we output it. Otherwise we start again. This is polynomial time in r and $\log q$.

Given a random element in $\mathcal{U}(r,q)$ with uniform distribution, we can compute the fiber of $x : \mathcal{P}(r,q) \to \mathcal{U}(r,q)$ above it and, provided this fiber is nonempty, pick a random element in it with uniform distribution. If the fiber is empty, we pick another element in $\mathcal{U}(r,q)$ until we find a nonempty fiber. At least one in every $d \times (0.99)^{-1}$ fibers is nonempty. We thus define a distribution μ on $\mathcal{P}(r,q)$ and prove the following.

13.2.2 Lemma (a very rough measure) *There is a unique measure μ on $\mathcal{P}(r,q)$ such that all nonempty fibers of the map $x : \mathcal{P}(r,q) \to \mathcal{U}(r,q)$ have the same measure, and all points in a given fiber have the same measure.*

There exists a probabilistic algorithm that picks a random element in $\mathcal{P}(r, q)$ with distribution μ in time polynomial in d and $\log q$. For every subset Z of $\mathcal{P}(r, q)$ the measure $\mu(Z)$ is related to the uniform measure $\frac{\#Z}{\#\mathcal{P}(r,q)}$ by

$$\frac{\#Z}{d\#\mathcal{P}(r, q)} \le \mu(Z) \le \frac{d\#Z}{\#\mathcal{P}(r, q)}.$$

Now let $\mathcal{D}(r, q)$ be the set of effective $\mathbb{F}_q$-divisors with degree r on X. Since we have assumed that $q \ge 4g^2$, we know that X has at least one $\mathbb{F}_q$-rational point. Let Ω be a degree r effective divisor on $X/\mathbb{F}_q$. We associate to every α in $\mathcal{D}(r, q)$ the class of $\alpha - \Omega$ in $J(\mathbb{F}_q)$. This defines a surjection

$$\phi_r : \mathcal{D}(r, q) \to J(\mathbb{F}_q)$$

with all its fibers having cardinality $\#\mathbb{P}^{r-g}(\mathbb{F}_q)$. So the set $\mathcal{D}(r, q)$ has cardinality $\frac{q^{r-g+1}-1}{q-1}\#J(\mathbb{F}_q)$. So

$$\#\mathcal{P}(r, q) \le \#\mathcal{D}(r, q) \le q^{r-g}\frac{1 - \frac{1}{q^{r-g+1}}}{1 - \frac{1}{q}}q^g\left(1 + \frac{1}{\sqrt{q}}\right)^{2g}.$$

Since $q \ge 4g^2$ we have

$$\#\mathcal{D}(r, q) \le 2eq^r.$$

Assume that G is a finite group and ψ an epimorphism of groups

$$\psi : J(\mathbb{F}_q) \to G.$$

We look for some divisor $\Delta \in \mathcal{D}(r, q)$ such that

$$\psi(\phi_r(\Delta)) \ne 0 \in G.$$

Since all the fibers of $\psi \circ \phi_r$ have the same cardinality, the fiber above 0 has at most

$$\frac{2eq^r}{\#G}$$

elements. So the number of prime divisors $\Delta \in \mathcal{P}(r, q)$ such that $\psi(\phi_r(\Delta))$ is not 0 is at least

$$q^r\left(\frac{0.99}{r} - \frac{2e}{\#G}\right).$$

We assume that #G is at least $12r$. Then at least half of the divisors in $\mathcal{P}(r, q)$ are not mapped onto 0 by $\psi \circ \phi_r$. The μ-measure of the subset consisting of these elements is at least $\frac{1}{2d}$. So if we pick a random Δ in $\mathcal{P}(r, q)$ with μ-measure as in Lemma 13.2.2, the probability of success is at least $\frac{1}{2d}$. If we make $2d$ trials, the probability of success is $\geq 1 - \exp(-1) \geq \frac{1}{2}$.

13.2.3 Lemma (finding nonzero classes) *There is a Monte Carlo probabilistic algorithm that takes as input*

1. *a degree d and geometric genus g plane projective absolutely integral curve C over $\mathbb{F}_q$, such that $q \geq 4g^2$,*

2. *the smooth model X of C,*

3. *a degree 1 divisor $O = O^+ - O^-$, where O^+ and O^- are effective, $\mathbb{F}_q$-rational, and have degree bounded by Θg^{Θ} for some positive constant Θ,*

4. *an epimorphism $\psi : \mathrm{Pic}^0(X/\mathbb{F}_q) \to G$ (that need not be computable) such that the cardinality of G is at least $\max(48g, 24d, 720)$,*

and outputs a sequence of $2d$ elements in $\mathrm{Pic}^0(X/\mathbb{F}_q)$ such that at least one of them is not in the kernel of ψ with probability $\geq \frac{1}{2}$. The algorithm is polynomial time in d and $\log q$.

As a special case we take $G = G_0 = J(\mathbb{F}_q)$ and $\psi = \psi_0$ the identity. Applying Lemma 13.2.3 we find a sequence of elements in $J(\mathbb{F}_q)$ of which one at least is nonzero (with high probability). We take G_1 to be the quotient of G by the subgroup generated by these elements and ψ_1 the quotient map. Applying the lemma again we construct another sequence of elements in $J(\mathbb{F}_q)$ of which one at least is not in G_0 (with high probability). We go on like that and produce a sequence of subgroups in $J(\mathbb{F}_q)$ that increase with constant probability until the index in $J(\mathbb{F}_q)$ becomes smaller than $\max(48g, 24d, 720)$. Note that every step in this method is probabilistic: it succeeds with some probability that can be made very high (exponentially close to 1) while keeping a polynomial overall complexity.

13.2.4 Lemma (finding an almost generating set) *There exists a probabilistic (Monte Carlo) algorithm that takes as input*

1. a degree d and geometric genus g plane projective absolutely integral curve C over $\mathbb{F}_q$, such that $q \geq 4g^2$,

2. the smooth model X of C,

3. a degree 1 divisor $O = O^+ - O^-$, where O^+ and O^- are effective, $\mathbb{F}_q$-rational and have degree bounded by Θg^Θ for some positive constant Θ,

and outputs a sequence of elements in $\mathrm{Pic}^0(X/\mathbb{F}_q)$ that generate a subgroup of index at most

$$\max(48g, 24d, 720)$$

with probability $\geq \frac{1}{2}$. The algorithm is polynomial time in d and $\log q$.

Note that we do not catch the whole group $J(\mathbb{F}_q)$ of rational points but a subgroup $\mathcal{G}$ with index at most $\iota = \max(48g, 24d, 720)$. This is a small but annoying gap. In the sequel we shall try to compute the l-torsion of the group $J(\mathbb{F}_q)$ of rational points. Because of the small gap in the above lemma, we may miss some l-torsion points if l is smaller than ι. However, let k be an integer such that $l^k > \iota$, and let x be a point of order l in $J(\mathbb{F}_q)$. Assume that there exists a point y in $J(\mathbb{F}_q)$ such that $x = l^{k-1}y$. The group $<y>$ generated by y and the group $\mathcal{G}$ have nontrivial intersection because the product of their orders is bigger than the order of $J(\mathbb{F}_q)$. Therefore x belongs to $\mathcal{G}$.

Our strategy for computing $J(\mathbb{F}_q)[l]$ will be to find a minimal field extension $\mathbb{F}_Q$ of $\mathbb{F}_q$ such that all points in $J(\mathbb{F}_q)[l]$ are divisible by l^{k-1} in $J(\mathbb{F}_Q)$. We then shall apply the above lemma to $J(\mathbb{F}_Q)$. To finish, we have to compute $J(\mathbb{F}_q)$ as a subgroup of $J(\mathbb{F}_Q)$. To this end, we shall use the Weil pairing.

13.3 PAIRINGS

Let n be a prime to p integer and J a Jacobian variety over $\mathbb{F}_q$. The Weil pairing relates the full n-torsion subgroup $J(\overline{\mathbb{F}}_q)[n]$ with itself. It can be defined using Kummer theory and is geometric in nature. The Tate-Lichtenbaum-Frey-Rück pairing is more arithmetic and relates the n-torsion

$J(\mathbb{F}_q)[n]$ in the group of $\mathbb{F}_q$-rational points and the quotient $J(\mathbb{F}_q)/nJ(\mathbb{F}_q)$. In this section, we quickly review the definitions and algorithmic properties of these pairings, following work by Weil, Lang, Menezes, Okamoto, Vanstone, Frey, and Rück.

We first recall the definition of Weil pairing following [Lan3]. Let k be an algebraically closed field with characteristic p. For every Abelian variety A over k, we denote by $Z_0(A)_0$ the group of 0-cycles with degree 0, and by $S : Z_0(A)_0 \to A$ the summation map that associates to every 0-cycle of degree 0 the corresponding sum in A. Let V and W be two projective nonsingular integral varieties over k, and let $\alpha : V \to A$ and $\beta : W \to B$ be the canonical maps into their Albanese varieties. Let D be a divisor on $V \times W$. Let $n \geq 2$ be a prime to p integer. Let $\mathfrak{a}$ (resp. $\mathfrak{b}$) be a 0-cycle of degree 0 on V (resp. W) and let $a = S(\alpha(\mathfrak{a}))$ (resp. $b = S(\beta(\mathfrak{b}))$) be the associated point in A (resp. B). Assume that $na = nb = 0$. The Weil pairing $e_{n,D}(a, b)$ is defined in [Lan3], Chapter VI, Section 4. It is an nth root of unity in k and it is linear in a, b, and D.

Assume now that $V = W = X$ is a smooth projective integral curve over k. Assume that $A = B = J$ is its Jacobian variety and that

$$\alpha = \beta = \phi : X \to J$$

is the Jacobi map (once an origin on X has been chosen). If we take D to be the diagonal on $X \times X$, we define a pairing $e_{n,D}(a, b)$ that will be denoted $e_n(a, b)$ or $e_{n,X}(a, b)$. It is nondegenerate and does not depend on the origin for the Jacobi map.

The Jacobian variety J comes with a principal polarization, that is, an isomorphism $\lambda : J \to \hat{J}$ between J and its dual $\hat{J}$. If α is an endomorphism $\alpha : J \to J$, we denote by $^t\alpha$ its transpose $^t\alpha : \hat{J} \to \hat{J}$. If D is a divisor on J that is algebraically equivalent to zero, the image by $^t\alpha$ of the linear equivalence class of D is the linear equivalence class of the inverse image $\alpha^{-1}(D)$; see [Lan3], Chapter V, Section 1. The Rosati dual of α is defined to be $\alpha^* = \lambda^{-1} \circ {}^t\alpha \circ \lambda$. The map $\alpha \to \alpha^*$ is an involution, and α^* is the adjoint of α for the Weil pairing

(13.3.1) $$e_{n,X}(a, \alpha(b)) = e_{n,X}(\alpha^*(a), b)$$

according to [Lan3], Chapter VII, Section 2, Proposition 6.

If Y is another smooth projective integral curve over k, K its Jacobian variety, $f : X \to Y$ a nonconstant map with degree d, and $f^* : K \to J$ the associated map between Jacobian varieties, then for a and b of order dividing n in K one has

$$e_{n,X}(f^*(a), f^*(b)) = e_{n,Y}(a,b)^d.$$

The Frey-Rück pairing can be constructed from the Lichtenbaum version of Tate's pairing [Lic] as was shown in [Fr-Ru]. Let q be a power of p. Let again $n \geq 2$ be a prime to p integer and let X be a smooth projective absolutely integral curve over $\mathbb{F}_q$. Let g be the genus of X. We assume that n divides $q - 1$. Let J be the Jacobian variety of X. The Frey-Rück pairing

$$\{,\}_n : J(\mathbb{F}_q)[n] \times J(\mathbb{F}_q)/nJ(\mathbb{F}_q) \to \mathbb{F}_q^*/(\mathbb{F}_q^*)^n$$

is defined as follows. We take a class of order dividing n in $J(\mathbb{F}_q)$. Such a class can be represented by an $\mathbb{F}_q$-divisor D with degree 0. We take a class in $J(\mathbb{F}_q)$ and pick a degree zero $\mathbb{F}_q$-divisor E in this class, which we assume to be disjoint to D. The pairing evaluated at the classes $[D]$ and $[E]$ mod n is

$$\{[D], [E] \bmod n\}_n = f(E) \bmod (\mathbb{F}_q^*)^n,$$

where f is any function with divisor nD. This is a nondegenerate pairing.

We now explain how one can compute the Weil pairing, following work by Menezes, Okamoto, Vanstone, Frey, and Rück. The Tate-Lichtenbaum-Frey-Rück pairing can be computed similarly. As usual, we assume that we are given a degree d plane model C for X. Assume that $\mathfrak{a}$ and $\mathfrak{b}$ have disjoint support (otherwise we may replace $\mathfrak{a}$ by some linearly equivalent divisor using the explicit moving Lemma 13.1.7) We compute a function $f_\mathfrak{a}$ with divisor $n\mathfrak{a}$. We similarly compute a function $f_\mathfrak{b}$ with divisor $n\mathfrak{b}$. Then

$$e_n(a,b) = \frac{f_\mathfrak{b}(\mathfrak{a})}{f_\mathfrak{a}(\mathfrak{b})}.$$

This algorithm is polynomial in the degree d of C and the order n of the divisors, provided the initial divisors $\mathfrak{a}$ and $\mathfrak{b}$ are given as differences between effective divisors with polynomial degree in d.

Using an idea that appears in a paper by Menezes, Okamoto, and Van-stone [Me-Ok-Va] in the context of elliptic curves, and in [Fr-Ru] for general curves, one can make this algorithm polynomial in $\log n$ in the following way. We write $\mathfrak{a} = \mathfrak{a}_0 = \mathfrak{a}_0^+ - \mathfrak{a}_0^-$, where $\mathfrak{a}_0^+$ and $\mathfrak{a}_0^-$ are effective divisors. Let $f_\mathfrak{a}$ be the function computed in the above simpleminded algorithm. One has $(f) = n\mathfrak{a}_0^+ - n\mathfrak{a}_0^-$. We want to express $f_\mathfrak{a}$ as a product of small degree functions. We use a variant of fast exponentiation. Using Lemma 13.1.7 we compute a divisor $\mathfrak{a}_1 = \mathfrak{a}_1^+ - \mathfrak{a}_1^-$ and a function f_1, such that $\mathfrak{a}_1$ is disjoint to $\mathfrak{b}$ and $(f_1) = \mathfrak{a}_1 - 2\mathfrak{a}_0$ and the degrees of $\mathfrak{a}_1^+$ and $\mathfrak{a}_1^-$ are $\leq 4gd(\log_q(\deg(\mathfrak{b})) + 2)$. We go on and compute, for $k \geq 1$ an integer, a divisor $\mathfrak{a}_k = \mathfrak{a}_k^+ - \mathfrak{a}_k^-$ and a function f_k, such that $\mathfrak{a}_k$ is disjoint to $\mathfrak{b}$ and $(f_k) = \mathfrak{a}_k - 2\mathfrak{a}_{k-1}$ and the degrees of $\mathfrak{a}_k^+$ and $\mathfrak{a}_k^-$ are $\leq 4gd(\log_q(\deg(\mathfrak{b})) + 2)$. We write the base 2 expansion of $n = \sum_k \epsilon_k 2^k$ with $\epsilon_k \in \{0, 1\}$. We compute the function Φ with divisor $\sum_k \epsilon_k \mathfrak{a}_k$. We claim that the function $f_\mathfrak{a}$ can be written as a product of the f_k, for $k \leq \log_2 n$, and Φ with suitable integer exponents bounded by n in absolute value. Indeed we write $F_1 = f_1$, $F_2 = f_2 f_1^2$, $F_3 = f_3 f_2^2 f_1^4$, and so on. We have $(F_k) = \mathfrak{a}_k - 2^k \mathfrak{a}$, and

$$\Phi \prod_k F_k^{-\epsilon_k}$$

has divisor $n\mathfrak{a}$, so it is the $f_\mathfrak{a}$ we were looking for.

13.3.2 Lemma (computing the Weil pairing) *There is an algorithm that on input a prime to q integer $n \geq 2$, a degree d absolutely integral plane projective curve C over $\mathbb{F}_q$, its smooth model X, and two $\mathbb{F}_q$-divisors on X, denoted $\mathfrak{a} = \mathfrak{a}^+ - \mathfrak{a}^-$ and $\mathfrak{b} = \mathfrak{b}^+ - \mathfrak{b}^-$, with degree 0, and order dividing n in the Jacobian variety, computes the Weil pairing $e_n(\mathfrak{a}, \mathfrak{b})$ in time polynomial in d, $\log q$, $\log n$, and the degrees of $\mathfrak{a}^+$, $\mathfrak{a}^-$, $\mathfrak{b}^+$, $\mathfrak{b}^-$, the positive and negative parts of $\mathfrak{a}$ and $\mathfrak{b}$.*

13.3.3 Lemma (the Tate-Lichtenbaum-Frey-Rück pairing) *There exists an algorithm that on input an integer $n \geq 2$ dividing $q - 1$, a degree d absolutely integral plane projective curve C over $\mathbb{F}_q$, its smooth model X, and two $\mathbb{F}_q$-divisors on X, denoted $\mathfrak{a} = \mathfrak{a}^+ - \mathfrak{a}^-$ and $\mathfrak{b} = \mathfrak{b}^+ - \mathfrak{b}^-$, with degree 0, and such that the class of $\mathfrak{a}$ has order dividing $n \geq 2$ in the Jacobian,* *computes the Tate-Lichtenbaum-Frey-Rück pairing $\{a, b\}_n$ in time polynomial in d, $\log q$, $\log n$, and the degrees of a^+, a^-, b^+, b^-, the positive and negative parts of a and b.*

13.4 DIVISIBLE GROUPS

Let $\mathbb{F}_q$ be a finite field with characteristic p and let X be a projective smooth absolutely integral algebraic curve over $\mathbb{F}_q$. Let g be the genus of X, and let $l \neq p$ be a prime integer. We assume that $g \geq 1$. Let J be the Jacobian of X, and let $\mathrm{End}(J/\mathbb{F}_q)$ be the ring of endomorphisms of J over $\mathbb{F}_q$. Let F_q be the Frobenius endomorphism. In this section we study the action of F_q on l^k-torsion points of J. We first consider the whole l^k-torsion group. We then restrict to some well-chosen subgroups where this action is more amenable.

Let $\chi(x)$ be the characteristic polynomial of $F_q \in \mathrm{End}(J/\mathbb{F}_q)$. The Rosati dual to F_q is q/F_q. Let

$$\mathcal{O} = \mathbb{Z}[x]/\chi(x)$$

and $\mathcal{O}_l = \mathbb{Z}_l[x]/\chi(x)$. We set

$$\varphi_q = x \bmod \chi(x) \in \mathcal{O}.$$

Mapping φ_q onto F_q defines an epimorphism from the ring $\mathcal{O}$ onto $\mathbb{Z}[F_q]$. In order to control the degree of the field of definition of l^k-torsion points, we shall bound the order of φ_q in $(\mathcal{O}/l^k\mathcal{O})^*$. We set

$$\mathcal{U}_1 = (\mathcal{O}/l\mathcal{O})^* = (\mathbb{F}_l[x]/\chi(x))^*.$$

Let the prime factorization of $\chi(x) \bmod l$ be $\prod_i \chi_i(x)^{e_i}$ with $\deg(\chi_i) = f_i$. The order of $\mathcal{U}_1$ is $\prod_i l^{(e_i-1)f_i}(l^{f_i} - 1)$. Let γ be the smallest integer such that l^γ is greater than or equal to $2g$. Then the exponent of the group $\mathcal{U}_1$ divides

$$A_1 = l^\gamma \prod_i (l^{f_i} - 1).$$

We set

$$B_1 = \prod_i (l^{f_i} - 1)$$

and

$$C_1 = l^\gamma.$$

There is a unique polynomial $M_1(x) \in \mathbb{Z}[x]$ with degree $< 2g$ such that

$$\frac{\varphi_q^{A_1} - 1}{l} = M_1(\varphi_q) \in \mathcal{O}.$$

Now for every positive integer k, the element φ_q belongs to the unit group

$$\mathcal{U}_k = (\mathcal{O}/l^k\mathcal{O})^*$$

of the quotient algebra $\mathcal{O}/l^k\mathcal{O} = \mathbb{Z}[x]/(l^k, \chi(x))$. The prime factorization of $\chi(x) \bmod l$ is lifted modulo l^k as $\prod_i \Xi_i(x)$ with Ξ_i monic and $\deg(\Xi_i) = e_i f_i$, and the order of $\mathcal{U}_k$ is $\prod_i l^{f_i(ke_i-1)}(l^{f_i} - 1)$. The exponent of the latter group divides

$$A_k = A_1 l^{k-1}.$$

So we set

$$B_k = B_1 = \prod_i (l^{f_i} - 1)$$

and

$$C_k = C_1 l^{k-1} = l^{k-1+\gamma}.$$

There is a unique polynomial $M_k(x) \in \mathbb{Z}[x]$ with degree $< \deg(\chi)$ such that

$$\frac{\varphi_q^{A_k} - 1}{l^k} = M_k(\varphi_q) \in \mathcal{O}.$$

For every integer $N \geq 2$ we can compute $M_k(x) \bmod N$ from $\chi(x)$ in probabilistic polynomial time in $\log q$, $\log l$, $\log N$, k, g. Indeed we first factor $\chi(x) \bmod l$, then compute the χ_i and the e_i and f_i. We compute

$$x^{A_k} \bmod (\chi(x), l^k N)$$

using the fast exponentiation algorithm presented in Section 5.1. We remove 1 and divide by l^k.

13.4.1 Lemma (Frobenius and l-torsion) *Let k be a positive integer and $l \neq p$ a prime. Let $\chi(x)$ be the characteristic polynomial of the Frobenius F_q of $J/\mathbb{F}_q$. Let e_i and f_i be the multiplicities and inertias in the prime decomposition of $\chi(x)$ mod l. Let γ be the smallest integer such that l^γ is bigger than or equal to $2g$. Let $B = \prod_i (l^{f_i} - 1)$. Let $C_k = l^{k-1+\gamma}$ and let $A_k = BC_k$. The l^k-torsion in J splits completely over the degree A_k extension of $\mathbb{F}_q$. There is a degree $< 2g$ polynomial $M_k(x) \in \mathbb{Z}[x]$ such that*

$$F_q^{A_k} = 1 + l^k M_k(F_q).$$

For every integer N one can compute such a $M_k(x)$ mod N from $\chi(x)$ in probabilistic polynomial time in $\log q$, $\log l$, $\log N$, k, g.

   In order to state sharper results it is convenient to introduce l-divisible subgroups inside the l^∞-torsion of a Jacobian variety J, that may or may not correspond to subvarieties. We now see how to define such subgroups and control their rationality properties.

13.4.2 Lemma (divisible group) *Let $\Pi : J[l^\infty] \to J[l^\infty]$ be a group homomorphism whose restriction to its image $\mathbb{G}$ is a bijection. Multiplication by l is then a surjection from $\mathbb{G}$ onto itself. We denote by $\mathbb{G}[l^k]$ the l^k-torsion in $\mathbb{G}$. There is an integer w such that $\mathbb{G}[l^k]$ is a free $\mathbb{Z}/l^k\mathbb{Z}$ module of rank w for every k. We assume that Π commutes with the Frobenius endomorphism F_q. We then say $\mathbb{G}$ is the divisible group associated with Π. From Tate's theorem [Tat1] Π is induced by some endomorphism in $\mathrm{End}(J/\mathbb{F}_q) \otimes_{\mathbb{Z}} \mathbb{Z}_l$, and we can define Π^* the Rosati dual of Π and denote by $\mathbb{G}^* = \mathrm{Im}(\Pi^*)$ the associated divisible group, which we call the adjoint of $\mathbb{G}$.*

Remark. The dual $\mathbb{G}^*$ does not only depend on $\mathbb{G}$. It may depend on Π also. This will not be a problem for us.

Remark. We may equivalently define Π^* as the dual of Π for the Weil pairing; see equation (13.3.1).

We now give an example of a divisible group. Let $F(x) = F_1(x)$ and $G(x) = G_1(x)$ be two monic coprime polynomials in $\mathbb{F}_l[x]$, such that

$$\chi(x) = F_1(x)G_1(x) \bmod l.$$

According to Bezout's theorem we have two polynomials $H_1(x)$ and $K_1(x)$ in $\mathbb{F}_l[x]$, such that

$$F_1 H_1 + G_1 K_1 = 1$$

and $\deg(H_1) < \deg(G_1)$ and $\deg(K_1) < \deg(F_1)$.

From Hensel's lemma, for every positive integer k there exist four polynomials $F_k(x)$, $G_k(x)$, $H_k(x)$, and $K_k(x)$ in $(\mathbb{Z}/l^k\mathbb{Z})[x]$ such that F_k and G_k are monic,

$$\chi(x) = F_k(x)G_k(x) \bmod l^k,$$

$$F_k H_k + G_k K_k = 1 \bmod l^k,$$

$\deg(H_k) < \deg(G_1), \deg(K_k) < \deg(F_1), F_1 = F_k \bmod l, G_1 = G_k \bmod l, H_1 = H_k \bmod l$, and $K_1 = K_k \bmod l$.

The sequences $(F_k)_k$, $(G_k)_k$, $(H_k)_k$, $(K_k)_k$ converge in $\mathbb{Z}_l[x]$ to F_0, G_0, H_0, K_0. If we substitute F_q for x in $F_0 H_0$ we obtain a map

$$\Pi_G : J[l^\infty] \to J[l^\infty]$$

and similarly, if we substitute F_q for x in $G_0 K_0$ we obtain a map Π_F. It is clear that $\Pi_F^2 = \Pi_F$, $\Pi_G^2 = \Pi_G$, $\Pi_F + \Pi_G = 1$, and $\Pi_F \Pi_G = 0$. We call $\mathbb{G}_F = \mathrm{Im}(\Pi_F)$ and $\mathbb{G}_G = \mathrm{Im}(\Pi_G)$ the associated supplementary l-divisible groups.

13.4.3 Definition (characteristic subspaces) For every nontrivial monic factor $F(x)$ of $\chi(x) \bmod l$ such that the cofactor $G = \chi/F \bmod l$ is prime to F, we write $\chi = F_0 G_0$, the corresponding factorization in $\mathbb{Z}_l[x]$. The l-divisible group

$$\mathbb{G}_F = \mathrm{Im}(\Pi_F)$$

is called the F_0-torsion in $J[l^\infty]$ and is denoted $J[l^\infty, F_0]$. It is the characteristic subspace of F_q associated with the factor F. If $F = (x-1)^e$ is the largest power of $x - 1$ dividing $\chi(x)$ mod l, we abbreviate $\mathbb{G}_{(x-1)^e} = \mathbb{G}_1$. If $F = (x - q)^e$, then we write similarly $\mathbb{G}_{(x-q)^e} = \mathbb{G}_q$.

We notice that there exists a unit u in $\mathrm{End}(J/\mathbb{F}_q) \otimes_{\mathbb{Z}} \mathbb{Z}_l$ such that the Rosati dual Π_1^* of Π_1 is

$$\Pi_1^* = \Pi_q \circ u.$$

Therefore

$$\mathbb{G}_q = \mathbb{G}_1^*,$$

and the restriction of the Weil pairing to $\mathbb{G}_1[l^k] \times \mathbb{G}_q[l^k]$ is nondegenerate for any integer $k \geq 1$.

We now compute fields of definitions for torsion points inside such divisible groups. The action of F_q on the l^k-torsion $\mathbb{G}_F[l^k] = J[l^k, F_0]$ inside $\mathbb{G}_F$ factors through the ring $\mathcal{O}_l/(l^k, F_0(\varphi_q)) = \mathbb{Z}_l[x]/(l^k, F_0)$. We deduce the following.

13.4.4 Lemma (Frobenius and F_0-torsion) *Let k be a positive integer and $l \neq p$ a prime. Let $\chi(x)$ be the characteristic polynomial of the Frobenius F_q of J. Let $\chi = FG$ mod l with F and G monic coprime. Let e_i and f_i be the multiplicities and inertias in the prime decomposition of $F(x)$ mod l. Let γ be the smallest integer such that l^γ is greater than or equal to $2g$. Let $B(F) = \prod_i (l^{f_i} - 1)$. Let $C_k(F) = l^{k-1+\gamma}$ and $A_k(F) = B(F)C_k(F)$. The l^k-torsion in $\mathbb{G}_F$ splits completely over the degree $A_k(F)$ extension of $\mathbb{F}_q$. There is a degree $< \deg(F)$ polynomial $M_k(x) \in \mathbb{Z}_l[x]$ such that*

$$\Pi_F \circ F_q^{A_k(F)} = \Pi_F + l^k \Pi_F \circ M_k(F_q).$$

For every power N of l, one can compute such an $M_k(x)$ modulo N from $\chi(x)$ and $F(x)$ in probabilistic polynomial time in $\log q$, $\log l$, $\log N$, k, g.

If we take for F the largest power of $x - 1$ dividing $\chi(x)$ mod l in the above lemma, we can take $B(F) = 1$ so $A_k(F)$ is an l power $\leq 2gl^k$.

If we take for F the largest power of $x - q$ dividing $\chi(x)$ mod l in the above lemma, we have $B(F) = l - 1$ so $A_k(F)$ is $\leq 2g(l-1)l^k$.

So the characteristic spaces associated with the eigenvalues 1 and q split completely over small degree extensions of $\mathbb{F}_q$.

13.5 THE KUMMER MAP

Let X be a smooth projective absolutely integral curve over $\mathbb{F}_q$ of genus g and J the Jacobian variety of X. Let $n \geq 2$ be an integer dividing $q - 1$. We assume that $g \geq 1$. In this section, we construct a convenient epimorphism of groups from $J(\mathbb{F}_q)$ to $J(\mathbb{F}_q)[n]$. In the next section, we will apply this epimorphism to a random element in $J(\mathbb{F}_q)$ and obtain a random element in $J(\mathbb{F}_q)[n]$.

If P is in $J(\mathbb{F}_q)$ we take some $R \in J(\overline{\mathbb{F}}_q)$ such that $nR = P$ and form the 1-cocycle $(^\sigma R - R)_\sigma$ in $\mathrm{H}^1(\mathbb{F}_q, J[n])$. Using the Weil pairing we deduce an element

$$S \mapsto (e_n(^\sigma R - R, S))_\sigma$$

in

$$\mathrm{Hom}(J[n](\mathbb{F}_q), \mathrm{H}^1(\mathbb{F}_q, \mu_n)) = \mathrm{Hom}(J[n](\mathbb{F}_q), \mathrm{Hom}(\mathrm{Gal}(\mathbb{F}_q), \mu_n)).$$

The map that sends $P \bmod nJ(\mathbb{F}_q)$ to $S \mapsto (e_n(^\sigma R - R, S))_\sigma$ is injective because the Frey-Rück pairing is nondegenerate. We observe that $\mathrm{Hom}(\mathrm{Gal}(\mathbb{F}_q), \mu_n)$ is isomorphic to μ_n because giving an homomorphism from $\mathrm{Gal}(\mathbb{F}_q)$ to μ_n is equivalent to giving the image of the topological generator F_q. We obtain a bijection

$$T_{n,q} : J(\mathbb{F}_q)/nJ(\mathbb{F}_q) \rightarrow \mathrm{Hom}(J[n](\mathbb{F}_q), \mu_n)$$

that we call the *Tate map*. It maps P onto $S \mapsto e_n(^{F_q}R - R, S)$. If $J[n]$ splits completely over $\mathbb{F}_q$ we set $K_{n,q}(P) = {}^{F_q}R - R$ and define a bijection

$$K_{n,q} : J(\mathbb{F}_q)/nJ(\mathbb{F}_q) \rightarrow J[n](\mathbb{F}_q) = J[n]$$

that we call the *Kummer map*.

13.5.1 Definition (the Kummer map) Let $J/\mathbb{F}_q$ be a Jacobian variety and $n \geq 2$ a prime to p integer. Assume that $J[n]$ splits completely over $\mathbb{F}_q$. For P in $J(\mathbb{F}_q)$ we choose any R in $J(\overline{\mathbb{F}}_q)$ such that $nR = P$ and set $K_{n,q}(P) = {}^{F_q}R - R$. This defines a bijection

$$K_{n,q} : J(\mathbb{F}_q)/nJ(\mathbb{F}_q) \rightarrow J[n](\mathbb{F}_q) = J[n].$$

We now assume that

$$n = l^k$$

is a power of some prime integer $l \neq p$. We still make the (strong !) assumption that $J[n]$ splits completely over $\mathbb{F}_q$. We want to compute the Kummer map $K_{n,q}$ explicitly. The difficult point is to find a point R such that $nR = P$. Division in an algebraic group of large dimension is a difficult task. Let P be an $\mathbb{F}_q$-rational point in J. Let R be such that $nR = P$. Since $F_q - 1$ annihilates $J[n]$, there is an $\mathbb{F}_q$-endomorphism κ of J such that $F_q - 1 = n\kappa$. We note that κ belongs to $\mathbb{Z}[F_q] \otimes_{\mathbb{Z}} \mathbb{Q} = \mathbb{Q}[F_q]$ and therefore commutes with F_q. We have

$$\kappa(P) = (F_q - 1)(R) = K_{n,q}(P),$$

and $\kappa(P)$ is $\mathbb{F}_q$-rational. So we can compute $K_{n,q}(P)$ without computing R. We don't need to divide P by n. We shall try to divide $F_q - 1$ by n instead.

The Kummer map will be very useful, but its definition requires that $J[n]$ splits completely over $\mathbb{F}_q$. If this is not the case, we must base change to some extension of $\mathbb{F}_q$.

Let $\chi(x)$ be the characteristic polynomial of F_q and let $B = \prod_i (l^{f_i} - 1)$ where the f_i are the degrees of the irreducible factors of $\chi(x)$ mod l. Let l^γ be the smallest power of l that is greater than or equal to $2g$.

Let $C_k = l^{\gamma+k-1}$ and $A_k = BC_k$. Set $Q = q^{A_k}$. From Lemma 13.4.1 there is a polynomial $M_k(x)$ such that

$$F_Q = 1 + l^k M_k(F_q).$$

So for P an $\mathbb{F}_Q$-rational point in J and R such that $nR = P$, the Kummer map $K_{n,Q}$ applied to P is

$$M_k(F_q)(P) = (F_Q - 1)(R) = K_{n,Q}(P),$$

and this is an $\mathbb{F}_Q$-rational point. This finishes the proof of the following lemma.

13.5.2 Lemma (computing the Kummer map) *Let $J/\mathbb{F}_q$ be a Jacobian and let $g \geq 1$ its dimension. Let $l \neq p$ be a prime integer and $n = l^k$*

a power of l. Let $\chi(x)$ be the characteristic polynomial of F_q, and let $B = \prod_i (l^{f_i} - 1)$, where the f_i are the degrees of the irreducible factors of $\chi(x)$ mod l. Let l^γ be the smallest power of l that is greater than or equal to $2g$. Let $C_k = l^{\gamma+k-1}$ and $A_k = BC_k$. Set $Q = q^{A_k}$ and observe that n divides $Q - 1$ because $J[n]$ splits completely over $\mathbb{F}_Q$. There exists an endomorphism $\kappa \in \mathbb{Z}[F_q]$ of J such that $n\kappa = F_Q - 1$, and for every $\mathbb{F}_Q$-rational point P and any R with $nR = P$ one has

$$\kappa(P) = (F_Q - 1)(R) = K_{n,Q}(P).$$

This endomorphism κ induces a bijection between $J(\mathbb{F}_Q)/nJ(\mathbb{F}_Q)$ and $J[n](\mathbb{F}_Q) = J[n]$. Given $\chi(x)$ and a positive integer N, one can compute κ mod N as a polynomial in F_q with coefficients in $\mathbb{Z}/N\mathbb{Z}$ in probabilistic polynomial time in g, $\log l$, $\log q$, k, $\log N$.

    This lemma is not of much use in practice because the field $\mathbb{F}_Q$ is too big. On the other hand, we may not be interested in the whole n-torsion in J but just a small piece in it, namely the n-torsion of a given divisible group.

    So let $l \neq p$ be a prime integer and $\mathbb{G}$ an l-divisible group in $J[l^\infty]$. Let

$$\Pi \in \text{End}(J/\mathbb{F}_q) \otimes_{\mathbb{Z}} \mathbb{Z}_l$$

be a projection onto $\mathbb{G}$. So

$$\Pi : J[l^\infty] \to \mathbb{G}.$$

We assume that $\Pi^2 = \Pi$. Let $n = l^k$ and let Q be a power of q such that $\mathbb{G}[n]$ splits completely over $\mathbb{F}_Q$. Let P be an $\mathbb{F}_Q$-rational point in $\mathbb{G}$. Let $R \in \mathbb{G}(\overline{\mathbb{F}}_q)$ be such that $nR = P$. We set

$$K_{\mathbb{G},n,Q}(P) = {}^{F_Q}R - R$$

and define an isomorphism

$$K_{\mathbb{G},n,Q} : \mathbb{G}(\mathbb{F}_Q)/n\mathbb{G}(\mathbb{F}_Q) \to \mathbb{G}(\mathbb{F}_Q)[n] = \mathbb{G}[n].$$

In order to make this construction explicit, we now assume that there exists some $\kappa \in \mathbb{Z}_l[F_q]$ such that

$$(F_Q - 1 - n\kappa)\Pi = 0.$$

Lemma 13.4.4 provides us with such a Q and κ when $\mathbb{G} = J[l^\infty, F_0]$ is some characteristic subspace.

We now can compute this new Kummer map $K_{\mathbb{G},n,Q}$. Let P be an $\mathbb{F}_Q$-rational point in $\mathbb{G}$. Let $R \in \mathbb{G}$ be such that $nR = P$. From

$$(F_Q - 1 - n\kappa)\Pi(R) = 0 = (F_Q - 1 - n\kappa)(R)$$

we deduce that $K_{\mathbb{G},n,Q}(P) = \kappa(P)$. Hence the following lemma.

13.5.3 Lemma (the Kummer map for a divisible group) *Let $J/\mathbb{F}_q$ be a Jacobian. Let g be its dimension. Let $l \neq p$ be a prime integer and $n = l^k$ a power of l. We assume that $g \geq 1$. Let $\chi(x)$ be the characteristic polynomial of F_q. Assume that $\chi(x) = F(x)G(x) \bmod l$ with F and G monic coprime polynomials in $\mathbb{F}_l[x]$, and let $\mathbb{G}_F$ be the associated divisible group. Let $B = \prod_i (l^{f_i} - 1)$, where the f_i are the degrees of the irreducible factors of $F(x) \bmod l$. Let l^γ be the smallest power of l that is $\geq 2g$. Let $C_k = l^{k-1+\gamma}$ and $A_k = BC_k$. Set $Q = q^{A_k}$. From Lemma 13.4.4 there exists an endomorphism $\kappa \in \mathbb{Z}_l[F_q]$ such that*

$$\Pi_F(n\kappa - F_Q + 1) = 0,$$

and for every $\mathbb{F}_Q$-rational point $P \in \mathbb{G}_F$ and any $R \in \mathbb{G}_F$ with $nR = P$ one has

$$\kappa(P) = (F_Q - 1)(R) = K_{\mathbb{G},n,Q}(P).$$

This endomorphism κ induces a bijection between $\mathbb{G}_F(\mathbb{F}_Q)/n\mathbb{G}_F(\mathbb{F}_Q)$ and $\mathbb{G}_F[n](\mathbb{F}_Q) = \mathbb{G}_F[n]$. Given $\chi(x)$ and $F(x)$ and a power N of l, one can compute $\kappa \bmod N$ as a polynomial in F_q with coefficients in $\mathbb{Z}/N\mathbb{Z}$ in probabilistic polynomial time in g, $\log l$, $\log q$, k, $\log N$. Any $N \geq (4Q)^g$ suffices for the purpose of computing $\kappa(P)$.

13.6 LINEARIZATION OF TORSION CLASSES

Let C be a degree d plane projective absolutely integral curve C over $\mathbb{F}_q$ with geometric genus $g \geq 1$, and assume that we are given the smooth model X of C.

We also assume that we are given a degree 1 divisor $O = O^+ - O^-$ where O^+ and O^- are effective, $\mathbb{F}_q$-rational and have degree bounded by Θg^Θ for some constant Θ.

Let J be the Jacobian variety of X. We assume that $l \neq p$ is a prime integer that divides $\#J(\mathbb{F}_q)$. Let $n = l^k$ be a power of l. In this section we give a probabilistic algorithm that returns a description of $J(\mathbb{F}_q)[l^k]$ by generators and relations.

If $x_1, x_2, \ldots, x_I$ are elements in a finite commutative group G, we let $\mathcal{R}$ be the kernel of the map $\xi : \mathbb{Z}^I \to G$ defined by $\xi(a_1, \ldots, a_I) = \sum_i a_i x_i$. We call $\mathcal{R}$ the *lattice of relations* between the x_i. We first give a very general and rough algorithm for computing relations in any finite commutative group.

13.6.1 Lemma (finding relations in blackbox groups) *Let G be a finite and commutative group and let $x_1, x_2, \ldots, x_I$ be elements in G. A basis for the lattice of relations between the x_i can be computed at the expense of $3I\#G$ operations (or comparisons) in G.*

Proof We first compute and store all the multiples of x_1. So we list 0, x_1, $2x_1, \ldots$ until we find the first multiple $e_1 x_1$ that is equal to zero. This gives us the relation $r_1 = (e_1, 0, \ldots, 0) \in \mathcal{R}$. This first step requires at most $o = \#G$ operations in G and o comparisons.

We then compute successive multiples of x_2 until we find the first one, $e_2 x_2$, that is in $L_1 = \{0, x_1, \ldots, (e_1-1)x_1\}$. This gives us a second relation r_2. The couple (r_1, r_2) is a basis for the lattice of relations between x_1 and x_2. Using this lattice, we compute the list L_2 of elements in the group generated by x_1 and x_2. This second step requires at most $2o$ operations and $e_1 e_2 \leq o$ comparisons.

We then compute successive multiples of x_3 until we find the first one, $e_3 x_3$, that is in L_2. This gives us a third relation r_3. The triple (r_1, r_2, r_3) is a basis for the lattice of relations between x_1, x_2 and x_3. Using this lattice, we compute the list L_3 of elements in the group generated by x_1, x_2, and x_3. This third step requires at most $2o$ operations and o comparisons. And we go on like this. $\square$

We stress that the algorithm above is far from efficient and will not be very useful unless the group G is very small.

We now come back to the computation of generators and relations for $J(\mathbb{F}_q)[l^k]$. Let $B = l - 1$. Let l^γ be the smallest power of l that is $\geq 2g$, and let $A_k = Bl^{\gamma+k-1}$. We set $Q_k = q^{A_k}$. Let $F(x)$ be the largest power of $x - 1$ dividing the characteristic polynomial $\chi(x)$ of F_q in the ring $\mathbb{F}_l[x]$. Definition 13.4.3 and Lemma 13.5.3 provide us with two surjective maps,

$$\Pi_1 : J(\mathbb{F}_{Q_k})[l^\infty] \to \mathbb{G}_1(\mathbb{F}_{Q_k})$$

and

$$K_{\mathbb{G}_1, l^k, Q_k} : \mathbb{G}_1(\mathbb{F}_{Q_k}) \to \mathbb{G}_1[l^k].$$

If we now take for $F(x)$ the largest power of $x - q$ dividing $\chi(x)$ in the ring $\mathbb{F}_l[x]$, Definition 13.4.3 and Lemma 13.5.3 give two surjective maps

$$\Pi_q : J(\mathbb{F}_{Q_k})[l^\infty] \to \mathbb{G}_q(\mathbb{F}_{Q_k}),$$

and

$$K_{\mathbb{G}_q, l^k, Q_k} : \mathbb{G}_q(\mathbb{F}_{Q_k}) \to \mathbb{G}_q[l^k].$$

Remember that $\mathbb{G}_q = \mathbb{G}_1^*$ and that the restriction of the Weil pairing to $\mathbb{G}_1[l^k] \times \mathbb{G}_q[l^k]$ is nondegenerate. We use this pairing to build a presentation for $\mathbb{G}_1[l^k]$ and $\mathbb{G}_q[l^k]$ simultaneously. The motivation for this approach is that generators for $\mathbb{G}_1[l^k]$ provide relations for $\mathbb{G}_q[l^k]$, and conversely.

If $Q_k \geq 4g^2$, we use Lemma 13.2.4 to produce a sequence $\gamma_1, \ldots, \gamma_I$ of elements in $J(\mathbb{F}_{Q_k})$ that generate (with high probability) a subgroup of index at most

$$\iota = \max(48g, 24d, 720).$$

If $Q_k \leq 4g^2$ we use Lemma 13.2.1 to produce a sequence $\gamma_1, \ldots, \gamma_I$ of elements in $J(\mathbb{F}_{Q_k})$ that generate it.

Let N be the largest divisor of $\#J(\mathbb{F}_{Q_k})$ that is prime to l. We set

$$\alpha_i = K_{\mathbb{G}_1, l^k, Q_k}(\Pi_1(N\gamma_i))$$

and

$$\beta_i = K_{\mathbb{G}_q, l^k, Q_k}(\Pi_q(N\gamma_i)).$$

The group $\mathcal{A}_k$ generated by the α_i has index at most ι in $\mathbb{G}_1[l^k]$. The group $\mathcal{B}_k$ generated by the β_i has index at most ι in $\mathbb{G}_q[l^k]$. Let l^δ be smallest power of l that is bigger than ι and assume that $k > \delta$. Then

$$\mathbb{G}_1[l^{k-\delta}] \subset \mathcal{A}_k.$$

We now explain how to compute the lattice of relations between given elements $\rho_1, \ldots, \rho_J$ in $\mathbb{G}_1[l^k]$. We denote this lattice by $\mathcal{R}$. We recall that the restriction of the Weil pairing to $\mathbb{G}_1[l^k] \times \mathbb{G}_q[l^k]$ is a nondegenerate pairing

$$e_{l^k} : \mathbb{G}_1[l^k] \times \mathbb{G}_q[l^k] \to \mu_{l^k}.$$

We fix an isomorphism between the group

$$\mu_{l^k}(\overline{\mathbb{F}}_q) = \mu_{l^k}(\mathbb{F}_{Q_k})$$

of l^kth roots of unity and $\mathbb{Z}/l^k\mathbb{Z}$. With the preimage of 1 mod l^k chosen, computing this isomorphism is a problem called *discrete logarithm*. We can compute this discrete logarithm by exhaustive search at the expense of $O(l^k)$ operations in $\mathbb{F}_{Q_k}$. There are more efficient algorithms, but we don't need them for our complexity estimates.

We regard the matrix $(e_{l^k}(\rho_j, \beta_i))$ as a matrix with I rows, J columns, and coefficients in $\mathbb{Z}/l^k\mathbb{Z}$. This matrix defines a morphism from $\mathbb{Z}^J$ to $(\mathbb{Z}/l^k\mathbb{Z})^I$ whose kernel is a lattice $\mathcal{R}'$ that contains $\mathcal{R}$. The index of $\mathcal{R}$ in $\mathcal{R}'$ is at most ι. Indeed, $\mathcal{R}'/\mathcal{R}$ is isomorphic to the orthogonal subspace to $\mathcal{B}_k$ inside $< \rho_1, \ldots, \rho_J > \subset \mathbb{G}_1[l^k]$. So it has order $\leq \iota$. We then compute a basis of $\mathcal{R}'$. This boils down to computing the kernel of an $I \times (J + I)$ integer matrix with entries bounded by l^k. This can be done by putting this matrix in Hermite normal form (see [Coh], 2.4.3). The complexity is polynomial in I, J and $k \log l$. See [Hav], [Coh], and [Kal].

Once given a basis of $\mathcal{R}'$, the sublattice $\mathcal{R}$ can be computed using the algorithm given in the proof of Lemma 13.6.1 at the expense of $\leq 3J\iota$ operations.

We apply this method to the generators $(\alpha_i)_i$ of $\mathcal{A}_k$. Once given the lattice $\mathcal{R}$ of relations between the α_i it is a matter of linear algebra to find a basis $(b_1, \ldots, b_w)$ for $\mathcal{A}_k[l^{k-\delta}] = \mathbb{G}_1[l^{k-\delta}]$. The latter group is a rank w free module over $\mathbb{Z}/l^{k-\delta}\mathbb{Z}$ and is acted on by the q-Frobenius F_q. For every b_j we can compute the lattice of relations between $F_q(b_j)$, b_1, b_2,

..., b_w and deduce the matrix of F_q with respect to the basis $(b_1, \ldots, b_w)$. From this matrix we deduce a nice generating set for the kernel of $F_q - 1$ in $\mathbb{G}_1[l^{k-\delta}]$. This kernel is $J[l^{k-\delta}](\mathbb{F}_q)$. We deduce the following.

13.6.2 Theorem (computing the l^k-torsion in the Picard group) *There is a probabilistic Monte Carlo algorithm that on input*

1. *a degree d and geometric genus g plane projective absolutely integral curve C over $\mathbb{F}_q$,*

2. *the smooth model X of C,*

3. *a degree 1 divisor $O = O^+ - O^-$, where O^+ and O^- are effective, $\mathbb{F}_q$-rational, and have degree bounded by Θg^Θ for some positive constant Θ,*

4. *a prime l different from the characteristic p of $\mathbb{F}_q$ and a power $n = l^k$ of l,*

5. *the zeta function of X,*

outputs a set $g_1, \ldots, g_I$ of divisor classes in the Picard group of $X/\mathbb{F}_q$, such that the l^k-torsion $\mathrm{Pic}(X/\mathbb{F}_q)[l^k]$ is the direct product of the $< g_i >$, and the orders of the g_i form a nondecreasing sequence. Every class g_i is given by a divisor $G_i - gO$ in the class, where G_i is a degree g effective $\mathbb{F}_q$-divisor on X.

The algorithm runs in probabilistic polynomial time in d, $\log q$ and l^k. It outputs the correct answer with probability $\geq \frac{1}{2}$. Otherwise, it may return either nothing or a strict subgroup of $\mathrm{Pic}(X/\mathbb{F}_q)[l^k]$.

If one is given a degree zero $\mathbb{F}_q$-divisor $D = D^+ - D^-$ of order dividing l^k, one can compute the coordinates of the class of D in the basis $(g_i)_{1 \leq i \leq I}$ in polynomial time in d, $\log q$, l^k, and the degree of D^+. These coordinates are integers x_i such that $\sum_{1 \leq i \leq I} x_i g_i = [D]$.

13.7 COMPUTING V_f MODULO p

In this section, we apply the general algorithm given in Section 13.6 to the plane curve C_l constructed in Section 12.3.19, and we compute Ramanujan divisors modulo p. These divisors are defined at the beginning of Chapter 12. So we assume that we are given an even integer $k > 2$, a prime integer $l > 6(k-1)$, a finite field $\mathbb{F}$ with characteristic l, and a ring epimorphism $f : \mathbb{T}(1, k) \to \mathbb{F}$. More precisely, we are given the images $f(T_i)$ for $i \leq k/12$. We want to compute the associated Galois representation $V_f \subset J_1(l)$, or rather its image $W_f \subset J_1(5l)$ by $B_{5l,l,1}^* : J_1(l) \to J_1(5l)$. We assume that the image of ρ_f contains $\mathrm{SL}(V_f)$. We set $X_l = X_1(5l)$ and $J_l = J_1(5l)$, and we denote by g_l the genus of X_l.

Let $p \notin \{5, l\}$ a prime integer. We explain how to compute divisors on $X_l/\bar{\mathbb{F}}_p$ associated to every $\bar{\mathbb{F}}_p$-point x in $W_f \subset J_l$. The definition field $\mathbb{F}_q$ for such divisors can be predicted from the characteristic polynomial of the Frobenius endomorphism F_p acting on V_f. So the strategy is to pick random $\mathbb{F}_q$-points in the l-torsion of the Jacobian J_l and to project them onto W_f using Hecke operators.

The covering map $B_{5l,l,1} : X_l \to X_1(l)$ has degree 24. We call it π. It induces two morphisms $\pi^* : J_1(l) \to J_l$ and $\pi_* : J_l \to J_1(l)$ such that the composite map $\pi_* \circ \pi^*$ is multiplication by 24 in $J_1(l)$. We denote by $K_l \subset J_l$ the image of π^*. This is a subvariety of J_l isogenous to $J_1(l)$. The restriction of $\pi^* \circ \pi_*$ to K_l is multiplication by 24. The maps π^* and π_* induce Galois equivariant bijections between the N-torsion subgroups $J_1(l)[N]$ and $K_l[N]$ for every integer N which is prime to 6.

Using Theorem 2.5.13 we derive from f a finite set $(t_1, \ldots, t_r)$ of elements in $\mathbb{T}(l, 2)$ with $r = (l^2 - 1)/6$ and

$$(13.7.1) \qquad V_f = \bigcap_{1 \leq i \leq r} \ker\left(t_i, J_1(l)(\overline{\mathbb{Q}})[l]\right).$$

and $W_f \subset K_l \subset J_l$ is the image of V_f by π^*.

We choose an integer s such that $24s$ is congruent to 1 modulo l. For every integer $n \geq 2$ we note $T_n \in \mathbb{T}(l, 2)$ the nth Hecke operator with weight 2 and level l. We can see T_n as an endomorphism of $J_1(l)$. We set $\hat{T}_n = [s] \circ \pi^* \circ T_n \circ \pi_*$. We notice that

$$\hat{T}_n \circ \pi^* = \pi^* \circ T_n$$

on $J_1(l)[l]$. Thus, the map $\pi^* : J_1(l) \to J_l$ induces a Galois equivariant bijection of $\mathbb{T}(l, 2)$-modules between $J_1(l)[l]$ and $K_l[l]$, and $W_f = \pi^*(V_f)$

is the subspace in $K_l[l]$ cut out by all $\hat{T}_n - \tau(n)$. We notice that π^*, π_*, T_n, and $\hat{T}_n$ can be seen as correspondences as well as morphisms between Jacobians. The following lemma states that the Hecke action on divisors can be efficiently computed.

13.7.2 Lemma (computing the Hecke action) *Let $l \geq 7$ and p be primes such that $p \notin \{5, l\}$. Let $n \geq 2$ be an integer. Let q be a power of p and let D be an effective $\mathbb{F}_q$-divisor of degree $\deg(D)$ on X_l mod p. The divisor $\pi^* \circ \pi_*(D)$ can be computed in polynomial time in l, $\deg(D)$ and $\log q$. The divisor $\pi^* \circ T_n \circ \pi_*(D)$ can be computed in polynomial time in l, $\deg(D)$, n, and $\log q$.*

Proof If n is prime to l, we define the Hecke operator $T(n, n)$ as an element in the ring of correspondences on $X_1(l)$ tensored by $\mathbb{Q}$. See [Lan4], Chapter VII, Section 2. From his Theorem 2.1 we have $T_{l^i} = (T_l)^i$ and $T_{n^i} = T_{n^{i-1}}T_n - nT_{n^{i-2}}T(n, n)$ if n is prime and $n \neq l$. And of course $T_{n_1}T_{n_2} = T_{n_1 n_2}$ if n_1 and n_2 are coprime. So it suffices to explain how to compute T_l and also T_n and $T(n, n)$ for n prime and $n \neq l$.

Let $x = (E, u)$ be a point on $Y_1(l) \subset X_1(l)$ representing an elliptic curve E with one l-torsion point u. Let n be an integer. The Hecke operator T_n maps x onto the sum of all $(E_I, I(u))$, where $I : E \to E_I$ runs over the set of all isogenies of degree n from E such that $I(u)$ still has order l. If n is prime to l, the Hecke operator $T(n, n)$ maps x onto $\frac{1}{n^2}$ times (E, nu). So we can compute the action of these Hecke correspondences on points $x = (E, u)$ using Vélu's formulae [Vel].

There remains to treat the case of cusps.

We call σ_β for $1 \leq \beta \leq \frac{l-1}{2}$ the cusp on $X_1(l)$ corresponding to the l-gon equipped with an l-torsion point on the βth component. The corresponding Tate curves $\mathbb{C}^*/q$ have an l-torsion point $w = \zeta_l^\star q^{\frac{\beta}{l}}$, where the star runs over the set of all residues modulo l. There are l Tate curves at every such cusp.

We call μ_α for $1 \leq \alpha \leq \frac{l-1}{2}$ the cusps on $X_1(l)$ corresponding to a 1-gon equipped with the l-torsion point ζ_l^α in its smooth locus $\mathbb{G}_m$. The Tate curve at μ_α is the Tate curve $\mathbb{C}^*/q$ with l-torsion point $w = \zeta_l^\alpha$. One single Tate curve here: no ramification.

For n prime and $n \neq l$ we have

$$T_n(\sigma_\beta) = \sigma_\beta + n\sigma_{n\beta}$$

and

$$T_n(\mu_\alpha) = n\mu_\alpha + \mu_{n\alpha},$$

where $n\alpha$ in $\mu_{n\alpha}$ (resp. $n\beta$ in $\sigma_{n\beta}$) should be understood as a class in $(\mathbb{Z}/l\mathbb{Z})^*/\{1,-1\}$.

Similarly

$$T_l(\sigma_\beta) = \sigma_\beta + 2l \sum_{1 \leq \alpha \leq \frac{l-1}{2}} \mu_\alpha$$

and

$$T_l(\mu_\alpha) = l\mu_\alpha.$$

And, of course, if n is prime to l, then

$$T(n,n)(\sigma_\beta) = \frac{1}{n^2}\sigma_{n\beta}$$

and

$$T(n,n)(\mu_\alpha) = \frac{1}{n^2}\mu_{n\alpha}.$$

Altogether, one can compute the effect of T_n on cusps for all n. For the sake of completeness, we also give the action of the diamond operator $\langle n \rangle$ on cusps. If n is prime to l then $\langle n \rangle(\sigma_\beta) = \sigma_{n\beta}$ and $\langle n \rangle(\mu_\alpha) = \mu_{n\alpha}$.   □

We can now state the following theorem.

13.7.3 Theorem (computing V_f modulo p) *There is a probabilistic (Las Vegas) algorithm that takes as input an even integer $k > 2$, a prime integer $l > 6(k-1)$, a finite field $\mathbb{F}$ with characteristic l, a ring epimorphism $f : \mathbb{T}(1,k) \to \mathbb{F}$, a cuspidal divisor Ω on $X_1(5l)$ as constructed in Section 8.1, and a prime $p \notin \{5,l\}$, and computes the reduction modulo p of $W_f \subset J_1(5l)$. Here $V_f \subset J_1(l)$ is defined by equation (13.7.1) and W_f is the image of $V_f \subset J_1(l)$ by $B^*_{5l,l,1}$, and we assume that the image of the Galois representation ρ_f associated with f contains $\mathrm{SL}(V_f)$. The algorithm* *returns a finite extension* $\mathbb{F}_q$ *of* $\mathbb{F}_p$ *such that* W_f *splits over* $\mathbb{F}_q$. *It returns also a collection of* $\#V_f$ *effective degree* g_l *divisors* $(Q_i)_{1 \le i \le \#V_f}$ *such that the classes* $[Q_i - \Omega]$ *represent all* $\mathbb{F}_q$-*points of* W_f. *The running time of the algorithm is* $\le (p \times \#V_f)^{\Theta}$ *for some absolute constant* Θ.

Remark. It is proved in Section 8.1 that the divisor Q_x on $X_1(5l)/\mathbb{Q}$ associated to $x \in W_f \subset J_1(5l)(\bar{\mathbb{Q}})[l]$ is nonspecial. According to Theorem 11.7.10 and Lemma 12.12.5, for every prime p but a finite number bounded by l^{Θ}, all the divisors Q_x remain nonspecial modulo any place above p, and their reductions coincide with the divisors Q_i returned by the algorithm above.

Proof To prove Theorem 13.7.3 we notice that Section 12.3.19 gives us a plane model for X_l mod p and a resolution of its singularities. From Lemma 12.3.27 we obtain the zeta function of X_l mod p. The characteristic polynomial of F_p acting on the two-dimensional $\mathbb{F}$-vector space V_f is $X^2 - f(T_p)X + p^{k-1}$ mod l. Since we know $f(T_i)$ for $2 \le i \le k/12$, we deduce $f(T_p)$ using Manin-Drinfeld-Shokurov theory. Knowing the characteristic polynomial of F_p, we deduce the order of F_p acting on V_f mod p. We deduce some small enough splitting field $\mathbb{F}_q$ for V_f mod p. We then apply Theorem 13.6.2 and obtain a basis for the l-torsion in the Picard group of $X_l/\mathbb{F}_q$. The same theorem allows us to compute the matrix of the endomorphism $\pi^* \circ \pi_*$ in this basis. We deduce a $\mathbb{F}_l$-basis for the image $K_l[l](\mathbb{F}_q)$ of $\pi^* \circ \pi_*$. Using Theorem 13.6.2 again, we now write down the matrices of the Hecke operators $\hat{T}_n$ in this basis for all $n \le (l^2 - 1)/6$. It is then a matter of linear algebra to compute a basis for the intersection of the kernels of all $(t_i)_{1 \le i \le (l^2-1)/6}$ in $K_l[l](\mathbb{F}_q)$. The algorithm is Las Vegas rather than Monte Carlo because we can check the result, the group W_f having known cardinality $(\#\mathbb{F})^2$. $\square$

Chapter Fourteen

Computing the residual Galois representations

by Bas Edixhoven

In this chapter we first combine the results of Chapters 11 and 12 in order to work out the strategy of Chapter 3 in the setup of Section 8.2. This gives the main result, Theorem 14.1.1: a deterministic polynomial time algorithm, based on computations with complex numbers. The crucial transition from approximations to exact values is done in Section 14.4, and the proof of Theorem 14.1.1 is finished in Section 14.7. In Section 14.8 we replace the complex computations with the computations over finite fields from Chapter 13, and give a probabilistic (Las Vegas type) polynomial time variant of the algorithm in Theorem 14.1.1.

14.1 MAIN RESULT

For positive integers k and n we have defined, in Section 2.4, $\mathbb{T}(n, k)$ as the $\mathbb{Z}$-algebra in $\mathrm{End}_{\mathbb{C}}(S_k(\Gamma_1(n)))$ generated by the Hecke operators T_m ($m \geq 1$) and the $\langle a \rangle$ (a in $(\mathbb{Z}/n\mathbb{Z})^\times$). Theorem 2.5.11 says that $\mathbb{T}(n, k)$ is generated as $\mathbb{Z}$-module by the T_i with $1 \leq i \leq k \cdot [\mathrm{SL}_2(\mathbb{Z}) : \Gamma_1(n)]/12$. In particular, $\mathbb{T}(1, k)$ is generated as $\mathbb{Z}$-module by the T_i with $i \leq k/12$. For each k and n, each surjective ring morphism $\mathbb{T}(n, k) \to \mathbb{F}$ gives rise to a Galois representation $\rho_m \colon \mathrm{Gal}(\overline{\mathbb{Q}}/\mathbb{Q}) \to \mathrm{GL}_2(\mathbb{F})$. Theorem 2.5.20 says that if $n = 1$ and the characteristic l of $\mathbb{F}$ satisfies $l > 6(k - 1)$, then ρ_m is reducible, or has image containing $\mathrm{SL}_2(\mathbb{F})$.

14.1.1 Theorem *There is a deterministic algorithm that on input a positive integer k, a finite field $\mathbb{F}$, and a surjective ring morphism $f \colon \mathbb{T}(1, k) \to \mathbb{F}$ such that the associated Galois representation $\rho \colon \mathrm{Gal}(\overline{\mathbb{Q}}/\mathbb{Q}) \to \mathrm{GL}_2(\mathbb{F})$ is* reducible or has image containing $\mathrm{SL}_2(\mathbb{F})$, computes ρ in time polynomial in k and $\#\mathbb{F}$. The morphism f is given by the images of $T_1, \ldots, T_{\lfloor k/12 \rfloor}$. More explicitly, the algorithm gives:

1. a Galois extension K of $\mathbb{Q}$, given as a $\mathbb{Q}$-basis e and the products $e_i e_j$ (that is, the $a_{i,j,k}$ in $\mathbb{Q}$ such that $e_i e_j = \sum_k a_{i,j,k} e_k$ are given);

2. a list of the elements σ of $\mathrm{Gal}(K/\mathbb{Q})$, where each σ is given as its matrix with respect to e;

3. an injective morphism ρ from $\mathrm{Gal}(K/\mathbb{Q})$ into $\mathrm{GL}_2(\mathbb{F})$, making $\mathbb{F}^2$ into a semisimple representation of $\mathrm{Gal}(\overline{\mathbb{Q}}/\mathbb{Q})$,

such that K is unramified outside l, with l the characteristic of $\mathbb{F}$, and such that for all prime numbers p different from l we have

$$\mathrm{trace}(\rho(\mathrm{Frob}_p)) = f(T_p) \quad \text{and} \quad \det(\rho(\mathrm{Frob}_p)) = p^{k-1} \quad \text{in } \mathbb{F}.$$

14.1.2 Remark Of course, we do not only prove *existence* of such an algorithm, but we actually *describe* one in the proof. We cannot claim that we really *give* such an algorithm because we did not make all constants in our estimates explicit, for example, those in Theorems 11.7.6 and 11.7.10.

Proof As this proof is rather long, we divide it into sections.

14.2 REDUCTION TO IRREDUCIBLE REPRESENTATIONS

Let k, $\mathbb{F}$, l, and f be as in Theorem 14.1.1. By definition, the associated representation $\rho \colon \mathrm{Gal}(\overline{\mathbb{Q}}/\mathbb{Q}) \to \mathrm{GL}_2(\mathbb{F})$ is semisimple, unramified outside $\{l\}$, and has $\det \circ \rho = \chi_l^{k-1}$, with $\chi_l \colon \mathrm{Gal}(\overline{\mathbb{Q}}/\mathbb{Q}) \to \mathbb{F}_l^\times$ the mod l cyclotomic character. Hence ρ is reducible if and only if it is of the form $\chi_l^i \oplus \chi_l^j$, for some i and j in $\mathbb{Z}/(l-1)\mathbb{Z}$ with $i+j = k-1$ in $\mathbb{Z}/(l-1)\mathbb{Z}$. The following result gives us an effective way to decide if ρ is reducible, and to determine ρ in that case. More precisely, using the standard algorithms based on modular symbols this proposition reduces the proof of Theorem 14.1.1 to the case where ρ is irreducible.

14.2.1 Proposition *In this situation, if $l = 2$, then $\rho \cong 1 \oplus 1$, and if $l = 3$, then $\rho \cong 1 \oplus \chi_3$. Assume now that $l \geq 5$. Let i and j be in $\mathbb{Z}/(l-1)\mathbb{Z}$ such that $i + j = k-1$. Then ρ is isomorphic to $\chi_l^i \oplus \chi_l^j$ if and only if for all prime numbers $p \neq l$ with $p \leq (l^2+l)/12$ we have $f(T_p) = p^i + p^j$ in $\mathbb{F}$.*

Proof The statements about $l=2$ and $l=3$ are proved in Théorème 3 of [Ser4] (see also Theorem 3.4 of [Edi1]). As $\mathbb{T}(1, k)$ is not zero (it has $\mathbb{F}$ as quotient), k is even.

The idea for the rest of the proof is to use suitable cuspidal eigenforms over $\mathbb{F}_l$ whose associated Galois representations give all the $\chi_l^i \oplus \chi_l^j$ with $i+j$ odd, and then to apply Proposition 2.5.18.

For a in $\mathbb{Z}$ even such that $4 \leq a \leq l-3$ or $a = l+1$, let $\overline{E}_a$ be the element of $M_a(1, \mathbb{F}_l)$ with $a_1(\overline{E}_a)=1$ and $T_p(\overline{E}_a) = (1+p^{a-1})\overline{E}_a$ for all primes p. These $\overline{E}_a$ can be obtained by reducing the Eisenstein series E_a modulo l, after multiplication by $B_k/2k$ (see Example 2.2.3). We cannot use E_{l-1} because its reduction has constant q-expansion 1, and so it cannot be normalized as we need. The $\overline{E}_a$ are eigenforms, and we have $\rho_{\overline{E}_a} = 1 \oplus \chi_l^{a-1}$. We note that the χ_l^{a-1} give all powers of χ_l except χ_l^{-1}. But we do have $1 \oplus \chi_l^{-1} = \chi_l^{-1} \otimes \rho_{\overline{E}_{l+1}}$. We also note that, for each a, $\overline{E}_a$ and $\theta^{l-1}\overline{E}_a$ give the same Galois representation. Therefore, all $\chi_l^i \oplus \chi_l^j$ with $i+j$ odd are associated with suitable cuspidal eigenforms. The proof is then finished by invoking Proposition 2.5.18. $\square$

14.3 REDUCTION TO TORSION IN JACOBIANS

Let k, $\mathbb{F}$, l, and f be as in Theorem 14.1.1, such that the representation ρ attached to f is irreducible. The following proposition is a special case of Theorem 3.4 of [Edi1].

14.3.1 Proposition *In this situation, there are a k' in $\mathbb{Z}_{>0}$, a surjective ring morphism $f': \mathbb{T}(1, k') \to \mathbb{F}$, and an i in $\mathbb{Z}/(l-1)\mathbb{Z}$ such that $2 \leq k' \leq l+1$ and $\rho \cong \rho' \otimes \chi_l^i$, where $\rho': \mathrm{Gal}(\overline{\mathbb{Q}}/\mathbb{Q}) \to \mathrm{GL}_2(\mathbb{F})$ is the Galois representation attached to f'.*

Such k', f', and i can be computed in time polynomial in k and l as follows. First, one computes the $\mathbb{F}_l$-algebras $\mathbb{F}_l \otimes \mathbb{T}(1, k')$ for $2 \le k' \le l+1$. Then, for each i in $\mathbb{Z}/(l-1)\mathbb{Z}$ and for each k', one checks if there exists an $\mathbb{F}_l$-linear map from $\mathbb{F}_l \otimes \mathbb{T}(1, k')$ to $\mathbb{F}$ that sends, for all $m \le (l^2+l)/12$ with l not dividing m, T_m to $m^{-i} f(T_m)$, and if so, if it is an algebra morphism. The previous proposition guarantees that such k', f', and i do exist. Proposition 2.5.18 guarantees that $\rho \cong \rho' \otimes \chi_l^i$.

Using standard algorithms for linear algebra over $\mathbb{Q}$, the computation of ρ is reduced to that of ρ'. By Theorem 2.5.13, ρ' is realized in $J_1(l)(\overline{\mathbb{Q}})[l]$ as the intersection of the kernels of a set of elements of $\mathbb{F}_l \otimes \mathbb{T}(l, 2)$ that can be computed in time polynomial in k and l. Hence, the proof of Theorem 14.1.1 is reduced to the case where ρ is irreducible, and is realized in $J_1(l)(\overline{\mathbb{Q}})[l]$ as described in Theorem 2.5.13.

14.4 COMPUTING THE $\mathbb{Q}(\zeta_l)$-ALGEBRA CORRESPONDING TO V

We recall the situation. The representation ρ of $\mathrm{Gal}(\overline{\mathbb{Q}}/\mathbb{Q})$ attached to the surjective ring morphism $f_k : \mathbb{T}(1, k) \to \mathbb{F}$ has image containing $\mathrm{SL}_2(\mathbb{F})$, and is also attached to a surjective ring morphism $f_2 : \mathbb{T}(l, 2) \to \mathbb{F}$. In particular, ρ is realized on the two-dimensional $\mathbb{F}$-vector space V contained in $J_1(l)(\overline{\mathbb{Q}})[l]$, that consists of all elements annihilated by $\mathrm{ker}(f_2)$. We note that $l > 5$. As in Section 8.2, we let X_l be the modular curve $X_1(5l)$ over $\mathbb{Q}$, and we embed V in the Jacobian variety J_l of X_l via pullback by the standard map from X_l to $X_1(l)$. We take a cuspidal divisor D_0 on $X_{l, \mathbb{Q}(\zeta_l)}$ as in Theorem 8.1.7. We have, for each $x \in V$, a unique effective divisor D_x of degree g_l (the genus of X_l) on $X_{l, \overline{\mathbb{Q}}}$, such that in $J_l(\overline{\mathbb{Q}})$ we have $x = [D_x - D_0]$. We write each D_x as $D_x^{\mathrm{fin}} + D_x^{\mathrm{cusp}}$, where D_x^{cusp} is supported on the cusps and D_x^{fin} is disjoint from the cusps. We write $D_x = \sum_{i=1}^{g_l} Q_{x,i}$, with $Q_{x,i}$ in $X_l(\overline{\mathbb{Q}})$, such that $D_x^{\mathrm{fin}} = \sum_{i=1}^{d_x} Q_{x,i}$.

Theorem 12.10.7 says that we have an analytic description of V inside $J_l(\mathbb{C})$, and, that for every embedding $\sigma : \mathbb{Q}(\zeta_l) \to \mathbb{C}$, complex approximations $D_{\sigma, x}$ of the D_x can be computed in polynomial time in $\#\mathbb{F}$ and the required accuracy (the number of digits on the right of the decimal point). Each such approximation is given as a sum of g_l points $Q_{\sigma, x, i}$ in $X_l(\mathbb{C})$, the numbering of which by $i \in \{1, \ldots, g_l\}$ is completely arbitrary, that is, unrelated to each other when σ varies. Similarly, complex approximations of all $b_l(Q_{x,i})$ or $(1/b_l)(Q_{x,i})$ (one of which has absolute value < 2) and of all $x_l'(Q_{x,i})$ or $(1/x_l')(Q_{x,i})$ can be computed in polynomial time in #$\mathbb{F}$ and the required accuracy. We denote such approximations by $b_l(Q_{\sigma,x,i})$, etc.

We compute such approximations, and also of $j(Q_{x,i})$ or $(1/j)(Q_{x,i})$, for all σ, x and i, with accuracy a sufficiently large absolute constant times $l^{15}\cdot(\#\mathbb{F})^6$. Here and in the rest of this section, we will use the O-notation without making the implied "absolute" constants explicit.

Using these approximations we will first decide for which (σ, x, i) the point $Q_{\sigma,x,i}$ approximates a cusp. For $x = 0$ we have $D_x = D_0$, and so all $Q_{0,i}$ are cusps. Recall that the cusps are precisely the poles of the rational function j. Hence a necessary condition for $Q_{\sigma,x,i}$ to approximate a cusp is that $(1/j)(Q_{\sigma,x,i})$ is small. Let x in V be nonzero, and i in $\{1, \ldots, g_l\}$ such that $j(Q_{x,i}) \neq 0$. By Proposition 11.7.1, the expression for j in b in Proposition 8.2.8, and Lemma 4.2.3, we have $h((1/j)(Q_{x,i})) = O(l^{12})$. The degree of $(1/j)(Q_{x,i})$ over $\mathbb{Q}$ is at most $l^2\cdot(\#\mathbb{F})^2$. By Lemma 4.2.5, we have, for all $\sigma\colon \overline{\mathbb{Q}} \to \mathbb{C}$, that if $(1/j)(Q_{x,i})$ is not zero, then

$$|\sigma((1/j)(Q_{x,i}))| \geq \exp(-O(l^{14}(\#\mathbb{F})^2)).$$

We conclude that the $Q_{\sigma,x,i}$ for which

$$|\sigma((1/j)(Q_{\sigma,x,i}))| < \exp(-O(l^{14}(\#\mathbb{F})^2))$$

are the ones that approximate cusps. This gives us the correct value of the integers d_x, and, after renumbering the $Q_{\sigma,x,i}$, approximations $D_{\sigma,x}^{\text{fin}}$ of the D_x^{fin} such that $D_{\sigma,x}^{\text{fin}} = \sum_{i=1}^{d_x} Q_{\sigma,x,i}$.

The next step is to get an integer n with $0 \leq n \leq l^4\cdot(\#\mathbb{F})^4$ such that the function $f_l := b_l + nx_l'$ separates the various $Q_{x,i}$ with $x \in V$ and $i \leq d_x$ that are distinct. We do this using just one embedding σ of $\mathbb{Q}(\zeta_l)$ in $\mathbb{C}$. For x in V and $i \leq d_x$ we have, by Proposition 11.7.1,

$$h((b_l(Q_{x,i}), x_l'(Q_{x,i}))) = O(l^{12}).$$

For x and y in V, the field over which they are both defined has degree at most $l\cdot(\#\mathbb{F})^3$ over $\mathbb{Q}$. By Lemma 4.2.5, we conclude that $Q_{\sigma,x,i}$ and $Q_{\sigma,y,j}$

approximate the same point if and only if

$$|b_l(Q_{\sigma,x,i}) - b_l(Q_{\sigma,y,j})| < \exp(-O(l^{13}\cdot(\#\mathbb{F})^3))$$

and

$$|x_l'(Q_{\sigma,x,i}) - x_l'(Q_{\sigma,y,j})| < \exp(-O(l^{13}\cdot(\#\mathbb{F})^3)).$$

We observe that the required approximations can indeed be computed within the required time because the height bounds from Proposition 11.7.1 imply that $|b_l(Q_{\sigma,x,i})| < \exp(O(l^{14}\cdot(\#\mathbb{F})^2)))$, that is, also on the left of the decimal point there are not too many digits. An integer n as above does not give a suitable f_l if and only if there are $Q_{\sigma,x,i}$ and $Q_{\sigma,y,j}$ that approximate different points and still $|f_l(Q_{\sigma,x,i}) - f_l(Q_{\sigma,y,j})| < \exp(-O(l^{13}\cdot(\#\mathbb{F})^3))$. Trying the possible n one by one until we have a suitable one gives us a function f_l as desired.

Now that we have our function f_l, we continue, as explained at the end of Section 8.2, by computing an integer m with $0 \le m \le l^2\cdot(\#\mathbb{F})^4$ such that the function

$$a_m: V \to \overline{\mathbb{Q}}, \quad x \mapsto \prod_{i=1}^{d_x}(m - f_l(Q_{x,i}))$$

is injective, and hence a generator of the $\mathbb{Q}(\zeta_l)$-algebra $A_{\mathbb{Q}(\zeta_l)}$ corresponding to V. As in the previous step, we do this using just one embedding σ of $\mathbb{Q}(\zeta_l)$ in $\mathbb{C}$. We estimate the loss in accuracy in computing a product $\prod_{i \le d_x}(m - f_l(Q_{\sigma,x,i}))$. For x in V and $i \le d_x$ we have seen that $h(f_l(Q_{x,i})) = O(l^{12})$. Hence, in the product, we have

$$|f_l(Q_{\sigma,x,i})| \le \exp(O(l^{14}\cdot(\#\mathbb{F})^2)).$$

As there are at most l^2 factors, the loss of accuracy is at most $O(l^{16}\cdot(\#\mathbb{F})^2)$ digits. Hence, from our approximations $f_l(Q_{\sigma,x,i})$ we get approximations $a_{\sigma,m}(x)$ of the $a_m(x)$ with accuracy $O(l^{15}\cdot(\#\mathbb{F})^3)$. For all candidates m and all x in V, we have that $h(a_m(x)) = O(l^{14})$, and the degree of $a_m(x)$ over $\mathbb{Q}$ is at most $(\#\mathbb{F})^2$. We conclude that a candidate m is not suitable if and only if there are distinct x and y in V with

$$|a_{\sigma,m}(x) - a_{\sigma,m}(y)| < \exp(O(-l^{14}\cdot(\#\mathbb{F})^2)).$$

Trying one by one gives us a suitable m.

We denote by $a = a_{D_0, f_l, m}$ the generator of $A_{\mathbb{Q}(\zeta_l)}$ that we obtained by finding suitable n and m. We will now compute the minimal polynomial of a over $\mathbb{Q}(\zeta_l)$:

$$(14.4.1) \qquad P = \prod_{x \in V} (T - a(x)), \qquad P = \sum P_j T^j, \qquad P_j \in \mathbb{Q}(\zeta_l).$$

As $h(a(x)) = O(l^{14})$, and, for j in $\{0, \ldots, \#V\}$, P_j is, up to a sign, an elementary symmetric polynomial in the $a(x)$, Lemma 4.2.4 gives that $h(P_j) = O(l^{14} \cdot (\#\mathbb{F})^2)$. We write the P_j in the $\mathbb{Q}$-basis $(1, \zeta_l, \ldots, \zeta_l^{l-2})$ of $\mathbb{Q}(\zeta_l)$:

$$P_j = \sum_{i < l-1} P_{j,i} \zeta_l^i, \qquad P_{j,i} \in \mathbb{Q}.$$

Then, for each j, $(P_{j,0}, \ldots, P_{j,l-2})$ is the unique solution in $\overline{\mathbb{Q}}$ of the system of linear equations, indexed by the σ in $\mathrm{Gal}(\mathbb{Q}(\zeta_l)/\mathbb{Q})$:

$$\sum_{i < l-1} \sigma(\zeta_l^i) P_{j,i} = \sigma(P_j).$$

Applying Lemma 4.2.6 and Cramer's rule gives

$$(14.4.2) \qquad h(P_{j,i}) \leq l \log l + l \cdot h(P_j) = O(l^{15} \cdot (\#\mathbb{F})^2).$$

So, in order to deduce the $P_{j,i}$ from approximations as in Proposition 3.2, the accuracy we need is $O(l^{15} \cdot (\#\mathbb{F})^2)$. We estimate the loss of accuracy in the evaluation of the product in (14.4.1). We know that $h(a(x)) = O(l^{14})$, and that the degree of $a(x)$ over $\mathbb{Q}$ is $(\#\mathbb{F})^2$. Therefore, for all embeddings σ of $\mathbb{Q}(\zeta_l)$ into $\mathbb{C}$, we have $|a_\sigma(x)| = \exp(O(l^{14} \cdot (\#\mathbb{F})^2))$. As there are $(\#\mathbb{F})^2$ factors in (14.4.1), the loss of accuracy is at most $O(l^{14} \cdot (\#\mathbb{F})^4))$. We conclude that our approximations $P_\sigma = \prod_{x \in V} (T - a_\sigma(x))$ at all σ are accurate enough to get good enough approximations of the $P_{j,i}$ so that Proposition 3.2 gives us the exact values of the $P_{j,i}$. So, finally, we know $A_{\mathbb{Q}(\zeta_l)}$ explicitly as

$$A_{\mathbb{Q}(\zeta_l)} = \mathbb{Q}(\zeta_l)[T]/(P) = \bigoplus_{0 \leq i < (\#\mathbb{F})^2} \mathbb{Q}(\zeta_l) \cdot T^i.$$

We remark that the definition of a directly implies that $a(0) = 1$. Hence P has a factor $T-1$. Under our assumption that the image of ρ contains $\mathrm{SL}_2(\mathbb{F})$ the polynomial $P/(T-1)$ is irreducible over $\mathbb{Q}(\zeta_l)$.

14.5 COMPUTING THE VECTOR SPACE STRUCTURE

The addition map $+: V \times V \to V$ corresponds to a morphism of $\mathbb{Q}$-algebras $+^*: A \to A \otimes_{\mathbb{Q}} A$, called co-addition. We will now explain how to compute the co-addition over $\mathbb{Q}(\zeta_l)$, that is, the morphism of $\mathbb{Q}(\zeta_l)$-algebras,

$$+^*: A_{\mathbb{Q}(\zeta_l)} \longrightarrow A_{\mathbb{Q}(\zeta_l)} \otimes_{\mathbb{Q}(\zeta_l)} A_{\mathbb{Q}(\zeta_l)} = \mathbb{Q}(\zeta_l)[U, V]/(P(U), P(V)).$$

To give this morphism is equivalent to giving the image of our generator a of the previous section. This image can be written uniquely as a polynomial in U and V of degree less than $(\#\mathbb{F})^2$ in each variable. Hence, there are unique $\mu_{i,j}$ in $\mathbb{Q}(\zeta_l)$, for i and j in $\{0, \dots, (\#\mathbb{F})^2-1\}$, such that for all x and y in V we have, in $\overline{\mathbb{Q}}$,

$$(14.5.1) \qquad a(x + y) = \sum_{i,j} \mu_{i,j} \cdot a(x)^i a(y)^j.$$

We view (14.5.1) as an inhomogeneous system of $(\#\mathbb{F})^4$ linear equations in the $\mu_{i,j}$. Then our bound $h(a(x)) = O(l^{14})$, together with Cramer's rule and Lemma 4.2.6, gives $h(\mu_{i,j}) = O(l^{14} \cdot (\#\mathbb{F})^6)$. Writing $\mu_{i,j}$ as a $\mathbb{Q}$-linear combination $\sum_{k<l-1} \mu_{i,j,k} \zeta_l^k$, we have $h(\mu_{i,j,k}) = O(l^{15} \cdot (\#\mathbb{F})^6)$. Hence our approximations $a_\sigma(x)$ are sufficiently precise to deduce the exact values of the $\mu_{i,j,k}$.

We also want to compute the multiplication map $\mathbb{F} \times V \to V$. That is, for each λ in $\mathbb{F}$ we want to know the map $(\lambda \cdot)^*$ from $A_{\mathbb{Q}(\zeta_l)}$ to itself that it induces. For $\lambda = 0$ this is the map that sends a to 1. Let now λ be in $\mathbb{F}^\times$. Then there are unique α_i in $\mathbb{Q}(\zeta_l)$, for i in $\{0, \dots, (\#\mathbb{F})^2-1\}$ such that $(\lambda \cdot)^*(a) = \sum_i \alpha_i \cdot a^i$. These α_i uniquely determined by the following equalities in $\overline{\mathbb{Q}}$, for all x in V:

$$(14.5.2) \qquad a(\lambda \cdot x) = \sum_i \alpha_i \cdot a(x)^i.$$

Arguments as above for the addition show that our approximations $a_\sigma(x)$ allow us to get the exact values of the α_i.

14.6 DESCENT TO $\mathbb{Q}$

At this moment, we finally have to pay the price for working with a divisor D_0 on $X_{l,\mathbb{Q}(\zeta_l)}$ and not on X_l itself. We have $A_{\mathbb{Q}(\zeta_l)} = \mathbb{Q}(\zeta_l) \otimes A$, hence we have a semilinear action of $\mathrm{Gal}(\mathbb{Q}(\zeta_l)/\mathbb{Q})$ on the $\mathbb{Q}(\zeta_l)$-algebra $A_{\mathbb{Q}(\zeta_l)}$: for τ in $\mathrm{Gal}(\mathbb{Q}(\zeta_l)/\mathbb{Q})$, λ in $\mathbb{Q}(\zeta_l)$, and x in $A_{\mathbb{Q}(\zeta_l)}$ we have $\tau(\lambda x) = \tau(\lambda) \cdot \tau(x)$. The $\mathbb{Q}$-algebra A is precisely the subset of $A_{\mathbb{Q}(\zeta_l)}$ of elements of that are fixed by this action.

In order to understand what the action of $\mathrm{Gal}(\mathbb{Q}(\zeta_l)/\mathbb{Q})$ does with our generator a of $A_{\mathbb{Q}(\zeta_l)}$, we must include the divisor D_0 in its notation: we will write a_{D_0} for it. Then, for each τ in $\mathrm{Gal}(\mathbb{Q}(\zeta_l)/\mathbb{Q})$ we have $\tau(a_{D_0}) = a_{\tau D_0}$, where $a_{\tau D_0}$ is defined as a, but with the divisor D_0 replaced by τD_0. For each $\sigma \colon \mathbb{Q}(\zeta_l) \to \mathbb{C}$, and for each x in $V \subset J_l(\mathbb{C})$, we have the approximations $a_{\sigma,D_0}(x)$ of $a_{D_0}(x)$, and $a_{\sigma\tau,D_0}(x)$ of $a_{\tau D_0}(x)$.

We take a generator τ of $\mathrm{Gal}(\mathbb{Q}(\zeta_l)/\mathbb{Q}) = \mathbb{F}_l^\times$. There are unique c_i in $\mathbb{Q}(\zeta_l)$, for i in $\{0, \ldots, (\#\mathbb{F})^2 - 1\}$, such that $\tau(a) = \sum_i c_i \cdot a^i$. These c_i are uniquely determined by the following system of equalities in $\overline{\mathbb{Q}}$, indexed by the x in V:

$$(14.6.1) \qquad a_{\tau D_0}(x) = \sum_{0 \le i < (\#\mathbb{F})^2} c_i \cdot a_{D_0}(x)^i.$$

Lemma 4.2.6 implies that for all i we have $h(c_i) = O(l^{14} \cdot (\#\mathbb{F})^6)$. Hence, writing, as usual, $c_i = \sum_{j < l-1} c_{i,j} \zeta_l^j$, we have $h(c_{i,j}) = O(l^{15} \cdot (\#\mathbb{F})^6)$. We conclude that our approximations $a_{\sigma,D_0}(x)$ of $a_{D_0}(x)$, and $a_{\sigma\tau,D_0}(x)$ of $a_{\tau D_0}(x)$, for all x and σ, are sufficiently accurate to get the exact values of the c_i.

Linear algebra over $\mathbb{Q}$ gives us then A, in terms of a $\mathbb{Q}$-basis with multiplication table, and with the maps $+^* \colon A \to A \otimes A$ and $(\lambda \cdot)^* \colon A \to A$ that correspond to the $\mathbb{F}$-vector space structure on V.

14.7 EXTRACTING THE GALOIS REPRESENTATION

We finish our computation of the Galois representation ρ as indicated in Chapter 3. We view $V \times V$ as $\mathrm{Hom}_{\mathbb{F}}(\mathbb{F}^2, V)$. This gives a right-action by $\mathrm{GL}_2(\mathbb{F})$ on $V \times V$, hence a left-action on $A \otimes A$. This action can be expressed in the co-addition and the $\mathbb{F}^\times$-action. We let B be the $\mathbb{Q}$-algebra corresponding to the subset $\text{Isom}_\mathbb{F}(\mathbb{F}^2, V)$ of $\text{Hom}_\mathbb{F}(\mathbb{F}^2, V)$. To compute B, as a factor of $A \otimes A$, we compute its idempotent, that is, the element of $A \otimes A$ that is 1 on $\text{Isom}_\mathbb{F}(\mathbb{F}^2, V)$ and 0 on its complement, as follows. In $A_{\mathbb{Q}(\zeta_l)} = \mathbb{Q}(\zeta_l)[T]/((T-1)P_l)$ we have the idempotent $a_1 = P_l/P_l(1)$, which, as a function on V, is the characteristic function of $\{0\}$. Then a_1 is an element of A. Let $a_2 = 1 - a_1$ in A. Then a_2 is the characteristic function of $V - \{0\}$. We let a_3 be the element of $A \otimes A$ obtained by taking the product of the $g \cdot (a_2 \otimes 1)$, where g ranges through $\text{GL}_2(\mathbb{F})$. Then a_3 is the characteristic function of $\text{Isom}_\mathbb{F}(\mathbb{F}^2, V)$. We compute $B = (A \otimes A)/(1 - a_3)$ by linear algebra over $\mathbb{Q}$, in terms of a basis with a multiplication table, and with the $\text{GL}_2(\mathbb{F})$-action.

We factor the algebra B as a product of fields, using a polynomial time factoring algorithm over $\mathbb{Q}$ (see [Le-Le-Lo] and [Len2]). Each factor K of B then gives us an explicit realization of ρ, as explained in Chapter 3: let $G \subset \text{GL}_2(\mathbb{F})$ be the stabilizer of a chosen factor K; then $G = \text{Gal}(K/\mathbb{Q})$ and the inclusion *is* a representation from G to $\text{GL}_2(\mathbb{F})$. This finishes the proof of Theorem 14.1.1. $\square$

14.7.1 Remark In the factorization of the algebra B above, our assumption that $\text{im}\rho$ contains $\text{SL}_2(\mathbb{F})$ implies that the idempotents lie in the subalgebra B^G of invariants by the subgroup G of $\text{GL}_2(\mathbb{F})$ consisting of the g with $\det(g)$ a $(k-1)$th power in $\mathbb{F}_l^\times$. This subalgebra is a product of copies of $\mathbb{Q}$.

If $\mathbb{F} = \mathbb{F}_l$, factoring B can be avoided by twisting ρ by a suitable power of χ_l. Indeed, $\rho' := \rho \otimes \chi_l^{1-k/2}$ has image $\text{GL}_2(\mathbb{F}_l)$, and ρ can then be obtained as $\rho' \otimes \chi_l^{k/2-1}$.

14.8 A PROBABILISTIC VARIANT

In this section we give a probabilistic Las Vegas type algorithm, based on the results of Chapter 13, that computes the representation ρ as in Theorem 14.1.1, in probabilistic running time polynomial in k and $\#\mathbb{F}$. A nice feature of computations over finite fields is that there is no loss of accuracy, as in the previous sections where computations with complex numbers were used. On the other hand, information obtained modulo varying primes is a bit harder to combine, and here the point of view of Galois theory that we have taken, relating sets with Galois action to algebras, is very convenient.

Let $f: \mathbb{T}(1, k) \to \mathbb{F}$ be as in Theorem 14.1.1, as well as l and the representation $\rho: \mathrm{Gal}(\overline{\mathbb{Q}}/\mathbb{Q}) \to \mathrm{GL}_2(\mathbb{F})$. Sections 14.2 and 14.3 apply without any change. So we can now put ourselves in the situation as at the beginning of Section 14.4. Then imρ contains $\mathrm{SL}_2(\mathbb{F})$, ρ is realized on a two-dimensional $\mathbb{F}$-vector space in $J_l(\overline{\mathbb{Q}})[l]$, and for each x in V there is a unique effective divisor D_x of degree g_l on $X_{l,\overline{\mathbb{Q}}}$ such that $x = [D_x - D_0]$ in $J_l(\overline{\mathbb{Q}})$. For each x in V, we write $D_x = D_x^{\mathrm{fin}} + D_x^{\mathrm{cusp}}$ as before, with $D_x^{\mathrm{fin}} = \sum_{i=1}^{d_x} Q_{x,i}$ and $D_x = \sum_{i=1}^{g_l} Q_{x,i}$.

Using Theorem 13.7.3, we try to compute the reductions $D_{x,\mathbb{F}_q}$ over suitable extensions of the residue fields of $\mathbb{Z}[\zeta_l]$ at successive prime numbers p up to $c \cdot l^{15} \cdot (\#\mathbb{F})^6$, with c a suitable absolute constant, skipping 5 and l. These fields $\mathbb{F}_q$ have degree at most $l \cdot (\#\mathbb{F})^3$ over their prime field $\mathbb{F}_p$. If some $D_{x,\mathbb{F}_q}$ is not unique, this will be detected by our computations, and we throw the corresponding prime p away. By Theorem 11.7.10, at most $O(l^{12} \cdot (\#\mathbb{F})^2)$ primes p are thrown away. For the V-good primes $p \leq B$, with V-good defined as in Theorem 11.7.10, we then have computed all $D_{x,\mathbb{F}_q}$.

We split each such $D_{x,\mathbb{F}_q}$ in a cuspidal part $D_{x,\mathbb{F}_q}^{\mathrm{cusp}}$ and a noncuspidal part $D_{x,\mathbb{F}_q}^{\mathrm{fin}}$. By Theorem 11.7.10, there are at most $O(l^{12} \cdot (\#\mathbb{F})^2)$ primes p where at some $\mathbb{F}_q$, the sum $\sum_{x \in V} \deg D_{x,\mathbb{F}_q}^{\mathrm{fin}}$ is less than $\sum_{x \in V} d_x$. (In fact, as imρ contains $\mathrm{SL}_2(\mathbb{F})$, all d_x for $x \neq 0$ are equal, but our argument does not need this.) This means that we have computed the unordered list of d_x. We discard the primes p where for some $\mathbb{F}_q$ the sum $\sum_{x \in V} \deg D_{x,\mathbb{F}_q}^{\mathrm{fin}}$ is less than $\sum_{x \in V} d_x$. For the remaining primes, we have computed the $D_{x,\mathbb{F}_q}^{\mathrm{fin}}$ for all x in V.

We want an integer n such that $f_l := b_l + nx_l'$ separates the D_x^{fin}. Let p be the smallest V-good prime with $p > l^4 \cdot (\#\mathbb{F})^4$. We compute the $D_{x,\mathbb{F}_q}^{\mathrm{fin}}$ over some $\mathbb{F}_q$ at p. We view the effective divisor $D_{\mathbb{F}_q} := \sum_{x \in V} D_{x,\mathbb{F}_q}^{\mathrm{fin}}$ as a closed subscheme of $X_{l,\mathbb{F}_q}$. Then f_l as above embeds $D_{\mathbb{F}_q}$ into $\mathbb{A}_{\mathbb{F}_q}^1$ if and only if it is injective on the geometric points of $D_{\mathbb{F}_q}$ and has nonzero derivative at the multiple points of $D_{\mathbb{F}_q}$. As the degree of $D_{\mathbb{F}_q}$ is at most $l^2 \cdot (\#\mathbb{F})^2$, this excludes at most $l^4 \cdot (\#\mathbb{F})^4$ elements of $\mathbb{F}_p$, and the algorithms of Chapter 13 let us compute these in polynomial time. We choose n in

$\{0, \ldots, l^4 \cdot (\#\mathbb{F})^4\}$ such that f_l embeds $D_{\mathbb{F}_q}$ into $\mathbb{A}^1_{\mathbb{F}_q}$. Then, by Nakayama's lemma, f_l embeds $\sum_{x \in V} D_x^{\mathrm{fin}}$ into $\mathbb{A}^1_{\overline{\mathbb{Q}}}$, and hence separates the D_x^{fin}.

The next step is to compute an integer m in $\{0, \ldots, l^2 \cdot (\#\mathbb{F})^4\}$ such that the function

$$a_m : V \to \overline{\mathbb{Q}}, \quad x \mapsto \prod_{i=1}^{d_x} (m - f_l(Q_{x,i}))$$

is injective, and hence a generator of the $\mathbb{Q}(\zeta_l)$-algebra $A_{\mathbb{Q}(\zeta_l)}$ corresponding to V. Let p and $\mathbb{F}_q$ be as in the preceding paragraph. For $x \in V$, write $D_{x,\mathbb{F}_q}^{\mathrm{fin}} = \sum_{i=1}^{d_x} Q_{x,i,\mathbb{F}_q}$. We take m in $\{0, \ldots, l^2 \cdot (\#\mathbb{F})^4\}$ such that the elements $\prod_{i=1}^{d_x} (m - f_l(Q_{x,i,\overline{\mathbb{F}}_q}))$ of $\mathbb{F}_q$, for $x \in V$, are all distinct. Then $a := a_m$ has the desired property.

The minimal polynomial P in $\mathbb{Q}(\zeta_l)[T]$ of a over $\mathbb{Q}(\zeta_l)$ is given as

$$P = \prod_{x \in V} (T - a(x)), \quad P = \sum_j P_j T^j, \quad P_j \in \mathbb{Q}(\zeta_l).$$

We have seen in (14.4.2) that, when writing $P_j = \sum_{i < l-1} P_{j,i} \zeta_l^i$, we have $h(P_{j,i}) = O(l^{15} \cdot (\#\mathbb{F})^2)$. Our construction shows that all $P_{j,i}$ are integral at all primes p that are V-good. From the $D_{x,\mathbb{F}_q}^{\mathrm{fin}}$ that we have computed, we get the images of the $P_{j,i}$ in $\mathbb{F}_p$, for all p that are V-good, up to $c \cdot l^{15} \cdot (\#\mathbb{F})^2$. Proposition 3.7 then gives us the $P_{j,i}$. So, at this point, we have computed the $\mathbb{Q}(\zeta_l)$-algebra $A_{\mathbb{Q}(\zeta_l)}$.

To compute the $\mathbb{F}$-vector space structure, we proceed as in Section 14.5. For each V-good prime p up to our bound $c \cdot l^{15} \cdot (\#\mathbb{F})^6$, we compute at each $\mathbb{F}_q$ over p the images in $\mathbb{F}_q$ of the $\mu_{i,j}$ from the linear system (14.5.1), over $\mathbb{F}_q$, and then the $\mu_{i,j,k}$ in $\mathbb{F}_p$. Proposition 3.7 and the height bound $h(\mu_i, j, k) = O(l^{15} \cdot (\#\mathbb{F})^6)$ give us the $\mu_{i,j,k}$. The α_i as in (14.5.2) can be computed in similarly, as well as the c_i as in (14.6.1). From here on, the computation proceeds as in Section 14.7.

Chapter Fifteen

Computing coefficients of modular forms

by Bas Edixhoven

In this chapter we apply our main result on the computation of Galois representations attached to modular forms of level one to the computation of coefficients of modular forms. In Section 15.1 we treat the case of the discriminant modular form, that is, the computation of Ramanujan's τ-function at primes. In Section 15.2 we deal with the more general case of forms of level one and arbitrary weight k, reformulated as the computation of Hecke operators T_n as $\mathbb{Z}$-linear combinations of the T_i with $i < k/12$. In Section 15.3 we give an application to theta functions of even, unimodular positive definite quadratic forms over $\mathbb{Z}$.

15.1 COMPUTING $\tau(p)$ IN TIME POLYNOMIAL IN $\log p$

We recall that Ramanujan's τ-function is defined by the following identity of formal power series with integer coefficients:

$$x \prod_{n \geq 1} (1 - x^n)^{24} = \sum_{n \geq 1} \tau(n) x^n.$$

15.1.1 Theorem *There is a deterministic algorithm that on input a prime number p gives $\tau(p)$, in running time polynomial in $\log p$.*

Proof Deligne has proved in [Del1] and [Del2] that for all prime numbers p we have $|\tau(p)| < 2p^{11/2}$. Therefore, it suffices to compute $\tau(p) \bmod l$ for all primes $l < x$, if the product of these l is at least $4p^{11/2}$. Analytic number theory (see, for example [Ten], I.2.6, Corollary 10.1) tells us that we can take $x = O(\log p)$, hence the proof is reduced to showing that there is a deterministic algorithm that computes $\tau(p)$ mod l for prime numbers p and l in time polynomial in $\log p$ and l. Of course, the slightly weaker but much more elementary bound $|\tau(n)| = O(n^6)$ in [Miy], Corollary 2.1.6 also suffices for our purposes.

We take an algorithm as in Theorem 14.1.1, and we apply it with $k = 12$. We have $\mathbb{T}(1, 12) = \mathbb{Z}$, and for each n in $\mathbb{Z}_{\geq 1}$, the element T_n of $\mathbb{T}(1, k)$ is the integer $\tau(n)$. We must now show that from the output of the algorithm we can compute $\tau(p)$ mod l deterministically, in time polynomial in $\log p$ and l. For l prime, we let $\rho_l \colon \mathrm{Gal}(K_l/\mathbb{Q}) \hookrightarrow \mathrm{GL}_2(\mathbb{F}_l)$ denote the Galois representation attached to Δ. As the discriminant of the ring of integers A of K_l is a power of l, Theorem 1.4 of [Bu-Le] gives the existence of a deterministic polynomial time algorithm that, given K_l, produces A, given by a $\mathbb{Z}$-basis. The maximal order A is preserved by the action of $\mathrm{Gal}(K_l/\mathbb{Q})$. Then $\mathrm{Gal}(K_l/\mathbb{Q})$ acts on the étale $\mathbb{F}_p$-algebra $\overline{A} := A/pA$, and $\mathrm{Hom}(\overline{A}, \overline{\mathbb{F}}_p)$ is a $\mathrm{Gal}(K_l/\mathbb{Q})$-torsor. Moreover, $\overline{A}$ is the product of its finitely many residue fields $\overline{A}/m$, where m ranges through the maximal ideals of $\overline{A}$.

We let Frob denote the absolute Frobenius endomorphism of $\overline{A}$; it sends a to a^p, it is an automorphism, and it induces the absolute Frobenius automorphism on each of the residue fields. The matrix of Frob can be computed in time polynomial in l and $\log p$.

The Frobenius element σ_m attached to a maximal ideal m of $\overline{A}$ is the unique element σ of $\mathrm{Gal}(K_l/\mathbb{Q})$ that fixes m and induces the absolute Frobenius on $\overline{A}/m$. For varying m, the σ_m form the Frobenius conjugacy class (at p) in $\mathrm{Gal}(K_l/\mathbb{Q})$.

For each σ in $\mathrm{Gal}(K_l/\mathbb{Q})$, we let $\overline{A}_\sigma$ be the quotient of $\overline{A}$ by the ideal generated by the image of Frob $- \sigma$. Such an $\overline{A}_\sigma$ can be computed in polynomial time. The σ in the Frobenius conjugacy class are precisely those σ for which $\overline{A}_\sigma$ is nonzero. We can try the σ one by one until we have found a σ in the Frobenius conjugacy class. Then we apply the map ρ_l given by Theorem 14.1.1 to it, and get an element $\rho_l(\sigma)$ of $\mathrm{GL}_2(\mathbb{F}_l)$. The trace of $\rho_l(\sigma)$ is then $\tau(p)$ mod l. $\qquad\qquad\square$

15.2 COMPUTING T_n FOR LARGE n AND LARGE WEIGHT

In this section we prove the following two results, for which we first recall some notation. For positive integers k and N we have defined, in
Section 2.4, $\mathbb{T}(N,k)$ as the $\mathbb{Z}$-algebra in $\mathrm{End}_{\mathbb{C}}(S_k(\Gamma_1(N)))$ generated by
the Hecke operators T_n ($n \geq 1$) and the $\langle a \rangle$ (a in $(\mathbb{Z}/N\mathbb{Z})^\times$). By Theorem 2.5.11, $\mathbb{T}(N,k)$ is generated as $\mathbb{Z}$-module by the Hecke operators T_i
with $1 \leq i \leq k \cdot [\mathrm{SL}_2(\mathbb{Z}) : \Gamma_1(N)]/12$. Just before (2.4.9) we have defined $M_k(\Gamma_1(N), \mathbb{Z})$ as the sub $\mathbb{Z}$-module of $M_k(\Gamma_1(N))$ consisting of the
f with all $a_i(f)$ in $\mathbb{Z}$. By the *generalized Riemann hypothesis (GRH) for
a number field K* we mean the statement that every complex zero of the
Dedekind zeta funtion of K with real part between 0 and 1 has real part $1/2$.

15.2.1 Theorem *One can compute, on input two positive integers k and n,
and the factorization of n into prime factors, the element T_n of the Hecke
algebra $\mathbb{T}(1,k)$, by computing the Galois representations attached to sufficiently many maximal ideals of $\mathbb{T}(1,k)$ as in Theorem 14.1.1. The computation gives T_n as $\mathbb{Z}$-linear combination of the T_i with $i \leq k/12$. This
algorithm is deterministic. For fixed k, it has running time polynomial in
$\log n$. If GRH holds for all number fields that occur as quotient of some
$\mathbb{Q} \otimes \mathbb{T}(1,k)$, then the algorithm has running time polynomial in k and $\log n$.*

15.2.2 Corollary *Assume GRH. There exists a deterministic algorithm that
on input the weight $k \geq 0$, the coefficients $a_i(f)$ with $0 \leq i \leq k/12$ of a
modular form f in $M_k(\mathrm{SL}_2(\mathbb{Z}), \mathbb{Z})$, and a positive integer n together with
its factorization into primes, computes $a_n(f)$ in running time polynomial in
k, $\log n$, and the maximum of the $\log(1 + |a_i(f)|)$ with $i \leq k/12$.*

The principle of the proof of Theorem 15.2.1 is first to reduce the computation of T_n to that of the T_p for the primes p that divide n, using the
identities implicit in (2.2.12). For p prime, T_p is computed from its images
in sufficiently many residue fields $\mathbb{T}(1,k)/m$, using the LLL-algorithm for
lattice reduction. Such an image is computed as the trace of a Frobenius
element at p of the Galois representation ρ_m attached to m. The Galois
representation is computed as in Theorem 14.1.1. The problem in doing
all this is to keep the residue fields small, because the computation of ρ_m

takes time polynomial in k and $\#(\mathbb{T}(1,k)/m)$. Here it makes a big difference whether we assume GRH. If we assume GRH, there are sufficiently many m with $\mathbb{T}(1,k)/m$ of size polynomial in k and $\log p$ so that T_p can be reconstructed from its images in these $\mathbb{T}(1,k)/m$. If we do not assume GRH, then we cannot rule out the possibility that all "small" primes (small in terms of the discriminant of $\mathbb{T}(1,k)$) are completely inert in $\mathbb{T}(1,k)$. Before we give the proof of Theorem 15.2.1, where the details of the algorithm are given, and the analysis of the running time, we state and prove some preliminary results. Corollary 15.2.2 will be deduced from Theorem 15.2.1, using some elementary properties of the Eisenstein series E_k and the interpretation of the $\mathbb{Z}$-module $S_k(\mathrm{SL}_2(\mathbb{Z}),\mathbb{Z})$ as the $\mathbb{Z}$-dual of $\mathbb{T}(1,k)$.

We start with a simple result that is well known.

15.2.3 Proposition *Let $k \geq 4$ be an even integer. Let $n = (k-14)/12$ if $k \equiv 2 \mod 12$, and $n = \lfloor k/12 \rfloor$ otherwise. Then $\mathbb{T}(1,k)$ is free of rank n as $\mathbb{Z}$-module, and the T_i with $1 \leq i \leq n$ form a $\mathbb{Z}$-basis for $\mathbb{T}$.*

Proof   Theorem 2.5.11 tells us that $\mathbb{T} := \mathbb{T}(1,k)$ is generated as $\mathbb{Z}$-module by the T_i with $1 \leq i \leq k/12$, but we will not use this. What we do use is that the pairing

$$S_k(\mathrm{SL}_2(\mathbb{Z}),\mathbb{Z}) \times \mathbb{T} \longrightarrow \mathbb{Z} \quad (f,t) \mapsto a_1(tf)$$

is perfect (see (2.4.9)). What me must show is that then $S_k(\mathrm{SL}_2(\mathbb{Z}),\mathbb{Z})$ is free of rank n as $\mathbb{Z}$-module, and that the maps $a_i \colon f \mapsto a_i(f)$, with $1 \leq i \leq n$, form a $\mathbb{Z}$-basis of the dual of $S_k(\mathrm{SL}_2(\mathbb{Z}),\mathbb{Z})$. As Δ is q times a unit in $\mathbb{Z}[[q]]$, we have

$$S_k(\mathrm{SL}_2(\mathbb{Z}),\mathbb{Z}) = \Delta \cdot M_{k-12}(\mathrm{SL}_2(\mathbb{Z}),\mathbb{Z}).$$

According to Swinnerton-Dyer ([Swi], Section 3):

$$(15.2.4) \qquad M(\mathrm{SL}_2(\mathbb{Z}),\mathbb{Z}) = \mathbb{Z}[E_4,\Delta] \oplus E_6 \cdot \mathbb{Z}[E_4,\Delta],$$

with E_4 and Δ algebraically independent, and $E_6^2 = E_4^3 - 1728 \cdot \Delta$. It follows that $S_k(\mathrm{SL}_2(\mathbb{Z}),\mathbb{Z})$ is free of rank n. Suitable monomials in E_4, Δ and E_6 show that $(a_1,\ldots,a_n)$ is a $\mathbb{Z}$-basis of $S_k(\mathrm{SL}_2(\mathbb{Z}),\mathbb{Z})^\vee$. $\square$

We note that we do not absolutely need the previous proposition, because from the set of generators T_i with $1 \le i \le k/12$ as given by Theorem 2.5.11 one can also compute a $\mathbb{Z}$-basis in time polynomial in k, and this is what one will probably do in the case of arbitrary level and weight. But in this case of level one we have chosen to be more explicit.

We consider the Hecke algebras $\mathbb{T}(1, k)$ as lattices in the $\mathbb{R}$-algebras $\mathbb{T}(1, k)_{\mathbb{R}} := \mathbb{R} \otimes \mathbb{T}(1, k)$. As all T_m are selfadjoint as operators on $S_k(\mathrm{SL}_2(\mathbb{Z}))$ with respect to the Petersson inner product (see Section 2.2), all their eigenvalues are real. As the level is 1, all normalized eigenforms are newforms; let us write them as $f_1, \dots, f_n$, and view them as ring morphisms $f_i \colon \mathbb{T}(1, k) \to \mathbb{R}$. Then the map

$$f \colon \mathbb{T}(1, k)_{\mathbb{R}} \longrightarrow \mathbb{R}^n, \quad t \mapsto (f_1(t), \dots, f_n(t))$$

is an isomorphism of $\mathbb{R}$-algebras. The trace form on the $\mathbb{Z}$-algebra $\mathbb{T}(1, k)$ induces the trace form of the $\mathbb{R}$-algebra $\mathbb{R}^n$, that is, the standard inner product on $\mathbb{R}^n$. We equip $\mathbb{T}(1, k)_{\mathbb{R}}$ with the standard volume form, that is, the one for which a unit cube has volume 1. Our first goal is now to get a bound for the absolute discriminant of $\mathbb{T}(1, k)$, or, equivalently, for the volume of the quotient $\mathbb{T}(1, k)_{\mathbb{R}} / \mathbb{T}(1, k)$.

15.2.5 Proposition *Let $k \ge 12$ be an even integer. Then we have*

$$\log \mathrm{Vol}\,(\mathbb{T}(1, k)_{\mathbb{R}} / \mathbb{T}(1, k)) = \frac{1}{2} \log |\mathrm{discr}\,\mathbb{T}(1, k)| \le \frac{k^2}{24} \log k.$$

Proof As $T_1, \dots, T_n$ is a $\mathbb{Z}$-basis of $\mathbb{T}(1, k)$, we have

$$\mathrm{Vol}(\mathbb{T}(1, k)_{\mathbb{R}} / \mathbb{T}(1, k)) = |\det(f(T_1), \dots, f(T_n))|.$$

Using Deligne's bound $|a_p(f_j)| \le 2p^{(k-1)/2}$ of [Del2] as we used the Weil bounds in the proof of Lemma 11.1.1, we get

(15.2.6)
$$|(f(T_i))_j| = |a_i(f_j)| \le \sigma_0(i) i^{(k-1)/2}$$
$$\le 2 {\cdot} i^{1/2} i^{(k-1)/2} = 2 {\cdot} i^{k/2}.$$

Hence the square of the length of $f(T_i)$ is at most $4ni^k$. This implies

$$|\det(f(T_1), \dots, f(T_n))| \le \prod_{i=1}^{n} (2 {\cdot} i^{k/2} {\cdot} \sqrt{n}\,) = 2^n {\cdot} (n!)^{k/2} {\cdot} n^{n/2}.$$

Hence:

$$\log \mathrm{Vol}(\mathbb{T}(1,k)_{\mathbb{R}}/\mathbb{T}(1,k)) \leq n \log 2 + \frac{k}{2} \log(n!) + \frac{n}{2} \log n$$

$$\leq n \log 2 + \frac{kn}{2} \log n - \frac{kn}{2} + \frac{k}{2} + \frac{n}{2} \log n,$$

where we have used that $\log n! \leq n \log n - n + 1$. Simple estimates for $k > 24$ and direct checks in the remaining cases give the result (the only nontrivial case being $k = 24$, in which case the discriminant equals $2^6 3^2 144169$).   □

The next ingredient to be used in our proof of Theorem 15.2.1 comes from analytic number theory: an *effective* prime number theorem for number fields, under the assumption of GRH. Effectivity means that for all real numbers x in a specified interval such as $(2, \infty)$ an estimate for the number of prime ideals of norm at most x of an arbitrary number field K is given, whereas the usual prime number theorem is a different asymptotic statement for each K.

15.2.7 Theorem (Weinberger) *For K a number field and x a real number, let $\pi(x, K)$ denote the number of maximal ideals m of the ring of integers O_K of K with $\#(O_K/m) \leq x$. For $x > 2$ in $\mathbb{R}$ let $\mathrm{li}\, x = \int_2^x (1/\log y) dy$. Then there exists c_1 in $\mathbb{R}$ such that for every number field K for which GRH holds, and for every $x > 2$, one has*

$$|\pi(x, K) - \mathrm{li}\, x| \leq c_1 \sqrt{x} \log \left(|\mathrm{discr}(O_K) x^{\dim_{\mathbb{Q}} K}| \right).$$

Weinberger states this result in [Wei4] (it is the lemma on p. 181, and says that it is proved in more generality in [Wei3] (probably on p. 328, just after (4.4))).

15.2.8 Corollary *There exist c_2 and c_3 in $\mathbb{R}$ such that for every number field K for which GRH holds and for every x in $\mathbb{R}$ such that*

$$x > c_2 \cdot (\log |\mathrm{discr}(O_K)|)^2 \cdot (\log(1 + \log |\mathrm{discr}(O_K)|))^2 \quad \text{and}$$
$$x > c_3 \cdot (\dim_{\mathbb{Q}} K)^2 \cdot (1 + \log \dim_{\mathbb{Q}} K)^4,$$

we have

$$\pi(x, K) \geq \frac{1}{2} \frac{x}{\log x}.$$

Proof Let K be a number field for which GRH holds, and let $x \in \mathbb{R}_{>2}$. We write d_K for $|\mathrm{discr}(O_K)|$ and n_K for $\dim_{\mathbb{Q}} K$. Then Theorem 15.2.7 says that

$$
\begin{aligned}
\pi(x, K) &\geq \mathrm{li}(x) - c_1 x^{1/2} \log(d_K x^{n_K}) \\
&\geq \frac{x-2}{\log x} - c_1 x^{1/2} \log(d_K x^{n_K}) \\
&= \frac{x}{\log x}\left(1 - \frac{2}{x} - \frac{c_1 (\log x)(\log d_K)}{x^{1/2}} - \frac{c_1 n_K (\log x)^2}{x^{1/2}}\right).
\end{aligned}
$$

In order to estimate the last term of the previous line, one uses the substitutions $x = y^4$ and $y = n_K^{1/2} z$. For the last but one term, one uses $x = y^2$ and $y = 2c_1 (\log d_K) z$ if $d_K \neq 1$. $\qquad \square$

In order to find an element of $\mathbb{T}(1, k)$ from sufficiently many of its images modulo maximal ideals, we will need a lower bound on the length of a shortest nonzero vector in the intersection of the maximal ideals. The following lemma gives a general, well-known lower bound in the context of orders in number fields.

15.2.9 Lemma *Let K be a number field, n its dimension as $\mathbb{Q}$-vector space, $A \subset K$ an order, that is, a subring of finite index in the ring of integers of K, and $I \subset A$ a nonzero ideal. We equip $K_{\mathbb{R}} := \mathbb{R} \otimes K$ with the inner product induced from the standard inner product on $\mathbb{C}^n$, where $K_{\mathbb{R}}$ is embedded in $\mathbb{C}^n$ via all distinct $\sigma_i \colon K \to \mathbb{C}$. We consider A and I as a lattices in $K_{\mathbb{R}}$. Then we have*

$$
\mu_1(I) \geq \sqrt{n} \cdot \#(A/I)^{1/n},
$$

where $\mu_1(I)$ is the length of a shortest element of $I - \{0\}$.

Proof Let $x \in I$ be nonzero. Then $A \cdot x \subset I$, and we have:

$$
|x_1| \cdots |x_n| = |N_{K/\mathbb{Q}}(x)| = \#(A/A \cdot x) \geq \#(A/I).
$$

The inequality between geometric and arithmetic mean gives:

$$
\left(|x_1|^2 \cdots |x_n|^2\right)^{1/n} \leq \frac{|x_1|^2 + \cdots + |x_n|^2}{n} = \frac{\|x\|^2}{n}.
$$

Combining the last two inequalities gives

$$\|x\| \geq \sqrt{n} \cdot |N_{K/\mathbb{Q}}(x)|^{1/n} \geq \#(A/I)^{1/n},$$

which finishes the proof. □

The next proposition summarizes the standard approach for using the LLL-algorithm for the "closest or nearest vector problem." For more details on lattice reduction we refer to [Le-Le-Lo], [Len3], and [Coh], Section 2.6.

15.2.10 Proposition *Let $n \geq 0$ and let L be a free $\mathbb{Z}$-module of finite rank n, equipped with a positive symmetric bilinear form $b: L \times L \to \mathbb{Z}$. We view L as a lattice in the $\mathbb{R}$-vector space $L_{\mathbb{R}}$ equipped with the inner product given by b, and for x in $L_{\mathbb{R}}$ we put $\|x\| := (b(x,x))^{1/2}$. Let L' be a submodule of finite index of L, and let $\mu_1(L')$ be the length of a shortest nonzero element of L'. Let t be an element of L such that*

$$\|t\| < 2^{-(n+1)/2} \mu_1(L').$$

Let $e = (e_1, \ldots, e_n)$ be an "LLL-reduced basis" of L': if $e^* = (e_1^*, \ldots, e_n^*)$ denotes the orthogonal $\mathbb{R}$-basis of $L_{\mathbb{R}}$ obtained from e by letting e_i^* be the orthogonal projection of e_i to the orthogonal complement of the subspace of $L_{\mathbb{R}}$ generated by $\{e_j \mid j < i\}$ (that is, by the Gram-Schmidt orthogonalization process), and $\mu_{i,j} := b(e_i, e_j^*)/b(e_j^*, e_j^*)$, then we have

$$|\mu_{i,j}| \leq \frac{1}{2} \quad \text{for } 1 \leq j < i \leq n, \text{ and}$$

$$\|e_i^*\|^2 \geq \left(\frac{3}{4} - \mu_{i,i-1}^2\right) \|e_{i-1}^*\|^2 \quad \text{for } 1 < i \leq n.$$

Then t is the shortest element of $t + L'$, and for any x in $t + L'$ we recover t as follows:

- *put $x_n := x$;*

- *for i going down from n to 1 let $x_{i-1} := x_i - [b(x_i, e_i^*)/b(e_i^*, e_i^*)]e_i$, where, for y in $\mathbb{Q}$, $[y]$ denotes the largest of the (one or two) integers nearest to y;*

- *then $t = x_0$.*

Proof We claim that the orthogonal block

$$B := \left\{ \sum_i \lambda_i e_i^* \mid -\frac{1}{2} \le \lambda_i < \frac{1}{2} \right\}$$

is a fundamental domain for L' acting on $L_{\mathbb{R}}$ by translations. Indeed, for x in $L_{\mathbb{R}}$ and x_i as above, x_0 is in B and $x - x_0$ is in L', and moreover, B and $L_{\mathbb{R}}/L'$ have the same volume, namely, $\prod_i \|e_i^*\|$. By the defining properties of an LLL-reduced basis we have, for i in $\{2, \dots, n\}$, that $\|e_i^*\|^2 \ge (1/2)\|e_{i-1}^*\|^2$. Also, as $e_1^* = e_1$, we have $\|e_1^*\| \ge \mu_1(L')$. It follows that for all i

$$\|e_i^*\| \ge 2^{-(i-1)/2}\|e_1\| \ge 2^{-(i-1)/2}\mu_1(L'),$$

and, in particular

$$\|e_i^*\| \ge 2^{-(n-1)/2}\mu_1(L').$$

Hence, for any x in $L_{\mathbb{R}}$ not in B we have

$$\|x\| \ge \min_i \|e_i^*\|/2 \ge 2^{-(n+1)/2}\mu_1(L').$$

$\square$

We can now finally prove Theorem 15.2.1. We split it into three parts: description of the algorithm, proof of its correctness, and running time analysis.

Proof [of Theorem 15.2.1]

Description of the algorithm

Let k and n be given, with the factorization of $n = \prod_p p^{v_p(n)}$ in prime factors. Let $\mathbb{T} := \mathbb{T}(1, k)$, and let r be its rank. Proposition 15.2.3 tells us that $r \le k/12$ and that $(T_1, \dots, T_r)$ is a $\mathbb{Z}$-basis of $\mathbb{T}$. Using modular symbols algorithms (see Chapter 8 of [Ste2]), one computes the $\mathbb{Z}$-algebra structure of $\mathbb{T}$, that is, one computes $T_i T_j$ as linear combinations of the T_l, with $1 \le i, j, l \le r$. The identity (2.2.12) gives us

$$T_n = \prod_{p|n} T_{p^{v_p(n)}}, \quad T_{p^i} = T_p T_{p^{i-1}} - p^{k-1}T_{p^{i-2}},$$

where p is prime and $i \geq 2$. Hence the computation of T_n is reduced to that of the T_p for the primes p dividing n.

Let now p be a prime dividing n. Using a factorization algorithm as in [Le-Le-Lo], we factor the $\mathbb{Q}$-algebra $\mathbb{T}_{\mathbb{Q}} = \prod_i K_i$ as a product of fields. In fact, in all cases that we know of, the number of factors K_i is at most one (see [Fa-Ja]). For each i, let A_i be the image of $\mathbb{T}$ in K_i. The A_i are computed as $\mathbb{Z}$-algebra, and the surjections $\mathbb{T} \twoheadrightarrow A_i$ are described by the images of the T_j for $j \leq r$. These morphisms embed $\mathbb{T}$ into $\prod_i A_i$. The computation of T_p is reduced to that of its images in the A_i.

Let K be one of the factors K_i, let A be A_i, let $n_K := \dim_{\mathbb{Q}} K$, and let $d_A := |\mathrm{discr}(A)|$. We define

(15.2.11) $$B_A := c \cdot ((\log d_A)^2 + n_K^2 + n_K \cdot k \cdot \log p)$$

for c a suitable absolute constant. For all primes l with

$$6(k-1) < l < B_A \cdot (\log B_A)^4 ,$$

we compute the maximal ideals of A/lA, and order them by their norms: $\#(A/m_1 A) \leq \#(A/m_2 A)$, etc. Then we take j minimal such that:

(15.2.12) $$\#(A/m_1 \cdots m_j) > \left(2^{(n_K+1)/2} \cdot 2 \cdot p^{(k-1)/2} \right)^{n_K} .$$

For each of the m_i with $i \leq j$, we compute the Galois representation ρ_{m_i} as in Theorem 14.1.1, and a Frobenius element $\rho_{m_i}(\mathrm{Frob}_p)$ at p as described in Section 15.1. Then we have the images of T_p in all $\mathbb{T}/m_i$ for $1 \leq i \leq j$.

We compute a $\mathbb{Z}$-basis $e' = (e'_1, \ldots, e'_{n_K})$ of $m_1 \cdots m_j$, starting from the $\mathbb{Z}$-basis of A that is part of the description of A, adding the congruence condition modulo the m_i one by one. The LLL-algorithm gives us an LLL-reduced basis e of $m_1 \cdots m_j$.

Let $\overline{T_p}$ denote the image of T_p in $\prod_{i \leq j} A/m_i$. We compute a preimage T'_p in A of $\overline{T_p}$, adding the congruence conditions one by one. We compute T_p itself with the algorithm of Proposition 15.2.10.

Correctness of the algorithm

We will now show that the computation works, and that it gives the correct result, if the constant c in (15.2.11) is large enough.

The first thing we have to show is that there are distinct maximal ideals $m_1, \ldots, m_j$ of A, whose residue characteristics are between $6(k-1)$ and $B_A \cdot (\log B_A)^4$, not equal to p, and such that (15.2.12) holds. We note that for each l in that range there is at least one maximal ideal m in A/lA, and that $\#(A/m)$ is at least l. According to Corollary 10.1 in Section I.2.6 of [Ten], for sufficiently large real numbers x, the sum $\sum_{l \leq x} \log(l)$ is at least $x/2$. So, indeed, taking all l between $6(k-1)$ and a constant times $n_K \cdot k \cdot \log p$ implies that the required m_i exist. We observe that the choice of B_A is sufficiently large for this.

The second point where an argument is needed is that the ρ_{m_i} are either reducible or have image containing $\mathrm{SL}_2(A/m_i)$, so that they can be computed by the algorithm of Theorem 14.1.1. But this is guaranteed by Theorem 2.5.20.

The third point is that in the situation where we invoke the algorithm of Proposition 15.2.10, the assumptions of that proposition hold. That means that we must check that $\|T_p\|$ is strictly less than $2^{-(n_K+1)/2} \cdot \mu_1(m_1 \cdots m_j)$. By Deligne's bound of [Del2], we know that $\|T_p\| \leq \sqrt{n_K} \cdot 2 \cdot p^{(k-1)/2}$. Lemma 15.2.9 says that $\mu_1(m_1 \cdots m_j) \geq \sqrt{n_K} \cdot (\#(A/m_1 \cdots m_j))^{1/n_K}$. Hence the required inequality follows from (15.2.12). This finishes the proof of the correctness of the algorithm.

Running time analysis

We show the two claims on the running time. We also indicate at the appropriate places that the algorithm is deterministic.

Let k, and $n = \prod_p p^{v_p(n)}$ be given. The computation of the Hecke algebra $\mathbb{T} := \mathbb{T}(1, k)$, that is, of the products $T_i T_j$ as $\mathbb{Z}$-linear combinations of the T_l, with $1 \leq i, j, l \leq r$, using modular symbols as in Chapter 8 of [Ste2], is deterministic and is done in time polynomial in k. Multiplication of two elements of $\mathbb{T}$ can be done in time polynomial in k and the maximum of the heights of the coordinates of the elements with respect to the $\mathbb{Z}$-basis $T = (T_1, \ldots, T_r)$. Lemma 4.2.7 and (15.2.6) give that the heights of the coordinates of T_m (with $m \in \mathbb{Z}_{>0}$) with respect to T are $O(k^3(\log(km)))$. Hence the computation of T_n from the T_p for p dividing n is done in time polynomial in k and $\log n$. The number of primes dividing n is at most $\log_2 n$, and each of them is at most n. Let p be a prime dividing n.

Factorization as in [Le-Le-Lo] is a deterministic polynomial time algorithm, hence we do get the factorization $\mathbb{T}_{\mathbb{Q}} = \prod_i K_i$ and the surjections $\mathbb{T} \to A_i \subset K_i$ in time polynomial in k.

The number of factors K_i is at most $r \leq k/12$. Let A be one of the A_i. As the morphism $\mathbb{T} \to \prod_i A_i$ is injective, we have

$$\prod_i \mathrm{discr}(A_i) \leq \mathrm{discr}(\mathbb{T}) \leq (k^2 \log k)/12.$$

Therefore, the number B_A as in (15.2.11) is $O(k^2 \cdot ((k \log k)^2 + \log n))$.

For l a prime number, the computation of the maximal ideals of A/lA can be done deterministically in time polynomial in l and n_K as follows (compare with Algorithm 3.4.10 of [Coh], "Berlekamp for small primes"). Take $i \in \mathbb{Z}_{\geq 1}$ such that $l^i \geq n_K$. The product I of the maximal ideals of A/lA is the subspace of x with $x^{l^i} = 0$. This can be computed with linear algebra over $\mathbb{F}_l$. One computes, again with linear algebra, the subalgebra $(A/lA)'$ of elements x with $x^l = x$. This algebra is isomorphic to $\mathbb{F}_l^d$ for some d. If $d > 1$, then for any x in $(A/lA)'$ that is not in $\mathbb{F}_l$ there is an a in $\mathbb{F}_l$ such that $(x - a)^{l-1}$ is a nontrivial idempotent in $(A/lA)'$. Repeating this procedure splits $(A/lA)'$ completely. The maximal ideal corresponding to an elementary idempotent ι of $(A/lA)'$ is then $(1 - \iota)I$.

We conclude that the computation of all maximal ideals of the A/lA, for all primes l with $6(k-1) < l < B_A \cdot (\log B_A)^4$ can be done deterministically in time polynomial in k and $\log n$. We can order them by their norms during their computation. Then we have $m_1, \ldots, m_j$ such that (15.2.12) holds. As $\#(A/m_i) \geq 2$ for all i, we have

$$j \leq n_K \cdot ((n_K + 3)/2 + (k - 1)(\log_2 p)/2) = O(k^2 \log n).$$

Let m be one of the m_i, where $i \leq j$. Then $\#(A/m) \leq l^{n_K}$, where l is the largest prime with $l < B_A \cdot (\log B_A)^4$. A rough estimate gives that there is an absolute constant such that

$$\#(A/mA) \leq (c \cdot k^5 \log n)^{k/6}.$$

The running time for computing the Galois representation ρ_m, is therefore bounded by a fixed power of $(c \cdot k^5 \log n)^{k/6}$. For fixed k, this is of polynomial size in $\log n$. Assuming GRH for K, we get a much smaller estimate for $\#(A/m)$ as follows. We use the fact that if a is in $\mathbb{R}_{>e}$, and $x > 2a \log a$, then $x/\log x > a$ (to prove this, use that $x \mapsto x/\log x$ is increasing on $\mathbb{R}_{>e}$, and that $e/\log e > 2$). We let $x = B_A \cdot (\log B_A)^4$. Then x satisfies the two hypotheses of Corollary 15.2.8 (assuming the constant c of (15.2.11) large enough with respect to c_2 and c_3), and therefore (here we use GRH!) $\pi(x, K) \geq x/2 \log x$. But then $\pi(x, K) \geq B_A/2$, that is, there are at least $B_A/2$ maximal ideals of O_K with $\#(O_K/m) \leq x$. Now O_K/A is finite, of order at most $d_A^{1/2}$, hence there are at most $(1/2) \cdot \log_2 d_A$ distinct primes dividing $\#(O_K/A)$, and hence at most $(n_K/2) \cdot \log_2 d_A$ maximal ideals of A where A and O_K differ. Similarly, there are at most $6(k-1)n_K$ maximal ideals of A with residue characteristic at most $6(k-1)$. The definition of B_A in equation (15.2.11) implies that

$$\frac{x}{2 \log x} - \frac{n_K}{2 \log_2 d_A} - 6k n_K > n_K \left(\frac{n_K + 1}{2} + 1 + \frac{k-1}{2} \log_2 p \right),$$

assuming the absolute constant c large enough. Let I be the product of the maximal ideals m in A with $\#(A/m) \leq x$ and with residue characteristic at least $6(k-1)$. Using that for all m containing I we have $\#(A/m) \geq 2$, we get

$$\#(A/I) > \left(2^{(n_K+1)/2} \cdot 2 \cdot p^{(k-1)/2} \right)^{n_K}.$$

The definition of j, see (15.2.12), implies that for every $i \leq j$ we have $\#(A/m_i) \leq x$. As we have already noticed above, x is of size polynomial in k and $\log n$. Summarizing: if we assume GRH for K, then running time for the computation of ρ_m is polynomial in k and $\log n$.

The computation of a Frobenius element $\rho_m(\text{Frob}_p)$, given ρ_m, is deterministic and has running time polynomial in k and $\log n$; see the proof of Theorem 15.1.1.

At this point we have computed the images of T_p in the A/m_i, for all $i \leq j$. The computation of a $\mathbb{Z}$-basis e' of $m_1 \cdots m_j$ can be done in time polynomial in k and $\log n$. One starts with the basis of A that is part of its description, and computes successively $\mathbb{Z}$-bases of m_1, $m_1 m_2$, and so on. At the ith step, the maximum of the absolute values of the coordinates of the elements of our temporary basis with respect to the basis of A gets at most

$n_K l_i$ times larger, where l_i is the characteristic of A/m_i. It follows that the absolute values of the coordinates of the e_i' with respect to the basis of A are at most $n_K^j \cdot \#(A/m_1 \cdots m_j)$, which is of size polynomial in k and $\log n$.

The LLL-algorithm gives us an LLL-reduced basis e of $m_1 \cdots m_j$ in time polynomial in k and $\log n$, see [Le-Le-Lo], [Len3], or [Coh], Section 2.6.

A preimage T_p' in A of the image $\overline{T_p}$ of T_p in $\prod_{i \leq j} A/m_i$ can be computed in time polynomial in k and $\log n$ as follows. One lifts the image of T_p in A/m_1 to an element $T_{p,1}$ of A, with small coordinates with respect to the $\mathbb{Z}$-basis of A, then one adjusts $T_{p,1}$ with an element of m_1 with small coordinates with respect to the $\mathbb{Z}$-basis that was already computed, to get the correct image in A/m_2, and so on. The size of the coordinates of T_p' with respect to the $\mathbb{Z}$-basis of A is polynomial in k and $\log n$.

The algorithm of Proposition 15.2.10 then computes T_p for us, in time polynomial in k and $\log n$. □

Proof [of Corollary 15.2.2] Let $k \geq 0$, and let f be in $M_k(\mathrm{SL}_2(\mathbb{Z}), \mathbb{Z})$. We are given the integers k and the $a_i(f)$ for $0 \leq i \leq k/12$, and a positive integer n together with its factorization into primes. We are to compute $a_n(f)$.

If $M_k(\mathrm{SL}_2(\mathbb{Z}), \mathbb{Z})$ is zero, then so is f, and we have $a_n(f) = 0$. If $k = 0$, then f is a constant power series, hence $a_n(f) = 0$. So, we may and do assume that $k \geq 4$, and that k is even.

Let E_k be the Eisenstein form of weight k and level one normalized as in Example 2.2.3. Note that the coefficients $a_i(E_k)$ are rational numbers, not necessarily integers. Let $g := f - a_0(f) \cdot E_k$ in $S_k(\mathrm{SL}_2(\mathbb{Z}), \mathbb{Q})$. Then we have

$$f = a_0(f) \cdot E_k + g, \quad \text{hence} \quad a_n(f) = a_0(f) a_n(E_k) + a_n(g).$$

As the factorization of n into primes is given, we can compute the coefficient:

$$a_n(E_k) = -\frac{2k}{B_k} \sigma_{k-1}(n) = -\frac{2k}{B_k} \prod_{p|n} \left(1 + p^{k-1} + \cdots + p^{(k-1)v_p(n)}\right)$$

in time polynomial in k and $\log n$. Hence it remains to compute $a_n(g)$.

Let $\mathbb{T}$ denote the Hecke algebra $\mathbb{T}(1, k)$ acting on $S_k(\mathrm{SL}_2(\mathbb{Z}), \mathbb{Z})$. Recall that its rank is at most $k/12$. Via the perfect pairing from (2.4.9),

$$S_k(\mathrm{SL}_2(\mathbb{Z}), \mathbb{Z}) \times \mathbb{T} \longrightarrow \mathbb{Z} \quad (h, t) \mapsto a_1(th),$$

we view g as the element in $\mathrm{Hom}_{\mathbb{Z}\text{-Mod}}(\mathbb{T}, \mathbb{Q})$ that sends t to $a_1(tg)$. We then have, by (2.2.13),

$$a_r(g) = g(T_r) \quad \text{for all } r \geq 1.$$

The T_i with $1 \leq i \leq \mathrm{rank}(\mathbb{T})$ form a $\mathbb{Z}$-basis for $\mathbb{T}$, by Proposition 15.2.3. The products $T_i T_j$ of these basis elements can be computed in time polynomial in k.

The coefficients $a_i(E_k)$ for $0 \leq i \leq k/12$ can be computed in time polynomial in k from the standard formulas as given in Example 2.2.3. The coefficients $a_i(f)$ for $0 \leq i \leq k/12$ are given. Hence the $g(T_i) = a_i(g)$ with $1 \leq i \leq \mathrm{rank}(\mathbb{T})$ are computed in time polynomial in k and the maximum of the $\log(1 + |a_i(f)|)$ with $0 \leq i \leq k/12$. Theorem 15.2.1 says that T_n can be computed as a $\mathbb{Z}$-linear combination of the T_i with $1 \leq i \leq \mathrm{rank}(\mathbb{T})$, in time polynomial in k and $\log n$. Applying g, viewed as element of $\mathrm{Hom}_{\mathbb{Z}\text{-Mod}}(\mathbb{T}, \mathbb{Q})$, to this linear combination gives $a_n(g) = g(T_n)$, in the required time. $\qquad\qquad\square$

15.3 AN APPLICATION TO QUADRATIC FORMS

In this section we apply our results on the computation of coefficients of modular forms to theta functions of positive definite quadratic forms over $\mathbb{Z}$ that are even and unimodular. According to Corollary 2.3.6, such a theta function is a modular form on $\mathrm{SL}_2(\mathbb{Z})$, of weight equal to half the dimension of the space of the quadratic form.

In view of Example 2.3.10, Theorem 15.1.1 has the following consequence.

15.3.1 Theorem *There is a deterministic algorithm that on input an integer $m > 0$ together with its factorization into primes gives the representation number $r_L(m)$, where L is the Leech lattice, in running time polynomial in $\log m$.*

Let us now turn to arbitrary even unimodular lattices. We can then apply
Corollary 15.2.2 to prove the following result.

15.3.2 Theorem *Assume GRH. There is a deterministic algorithm that on
input the rank n_L and the integers $r_L(2i)$ for $1 \leq i \leq n_L/24$ of an even unimodular lattice (L, b), and an integer $m > 0$ together with its factorization
into primes, computes $r_L(m)$ in running time polynomial in n_L and $\log m$.*

Proof Let n_L and the $r_L(i)$ for $1 \leq i \leq n_L/24$ be given. As L is even and
unimodular, n_L is even (in fact, it is a multiple of 8, see [Ser5], Chapter VII,
Section 6). Let $k := n_L/2$. Then θ_L is in $M_k(\mathrm{SL}_2(\mathbb{Z}), \mathbb{Z})$. The coefficients
$a_i(\theta_L) = r_L(2i)$ with $0 \leq i \leq k/12$ are given to us. The fact that the open
balls in $L_{\mathbb{R}}$ centered at the $x \in L$ and with radius $2^{-1/2}$ do not overlap gives
that for i with $0 \leq i \leq k/12$ we have $\log(1 + r_L(2i)) \leq k \log(1 + k/6)$.
Corollary 15.2.2 then means that $r_L(m) = a_{m/2}(\theta_L)$ can be computed in
the required time. $\square$

Let us point out that Theorem 15.3.2 can be applied to the orthogonal direct
sum (L, b) of even unimodular lattices (L_i, b_i) for which the $r_{L_i}(j)$ can be
computed in time polynomial in n_{L_i} and j. Indeed,

$$\theta_{L_1} \cdots \theta_{L_r} = \sum_{x \in L} q^{(b_1(x_1, x_1) + \cdots + b_r(x_r, x_r))/2} = \sum_{x \in L} q^{b(x, x)/2} = \theta_L.$$

The coefficients $a_i(\theta_L)$ with $0 \leq i \leq n_L/24$ can be computed in time
polynomial in n_L by computing the product of the images of the θ_{L_i} in the
truncated power series ring $\mathbb{Z}[[q]]/(q^{n_L/24+1})$.

Epilogue

Theorems 14.1.1 and 15.2.1 will certainly be generalized to spaces of cusp forms of arbitrarily varying level and weight. This has already been done for the probabilistic variant of Theorem 14.1.1, in the case of square-free levels (and of level two times a square-free number for reasons that will become clear below) by Peter Bruin in his PhD thesis (Leiden, September 2010), see [Bru]. We describe some details and some applications of Bruin's work, and a perspective on point counting outside the context of modular forms. Deterministic generalizations of the two theorems mentioned above will lead to deterministic applications.

Computation of Galois representations

Peter Bruin gives, in [Bru], a probabilistic algorithm that on input positive integers k and N, with N square-free, and a surjective morphism $f : \mathbb{T}(2N, k) \to \mathbb{F}$ to a finite field, computes ρ_f with probabilistic running time polynomial in k, N and $\#\mathbb{F}$. An important new ingredient in his method, suggested by Couveignes, is to be more flexible concerning the choice of the divisor D_0 of Section 8.2. His algorithm computes, for each x in V, the smallest integer d_x such that $h^0(X_1(2N)_{\overline{\mathbb{Q}}}, \mathcal{L}_x(d_x \cdot \infty)) = 1$, and then represents x by the unique effective divisor D_x of degree d_x on $X_1(2N)_{\overline{\mathbb{Q}}}$ such that $x = [D_x - d_x \cdot \infty]$ in $J_1(2N)(\overline{\mathbb{Q}})$. This generalizes the probabilistic variant of Theorem 14.1.1 that is described in Section 14.8.

Computing coefficients of modular forms

Using the methods of Chapter 15, Peter Bruin will obtain a probabilistic algorithm that, assuming GRH, on input positive integers k, N, and n, with N square-free, together with the factorization of n into prime factors, computes the element T_n of the Hecke algebra $\mathbb{T}(2N, k)$ in probabilistic running time polynomial in k, N and $\log n$.

Lattices, sums of squares

This last result, applied to theta functions of lattices as in Section 15.3, will have the following remarkable consequence that, assuming GRH, there is a probabilistic algorithm that on input positive integers k and n, together with the factorization of n into prime factors, computes the number

$$r_{\mathbb{Z}^{2k}}(n) = \#\{x \in \mathbb{Z}^{2k} \mid x_1^2 + \cdots + x_{2k}^2 = n\}$$

in time polynomial in k and $\log n$.

For some small values of k there are well-known explicit formulas for the representation numbers $r_{\mathbb{Z}^{2k}}(n)$, owing their existence to the fact that there are no non-CM cuspidal eigenforms on $\Gamma_1(4)$ of weight k. In order to give these formulas, let $\chi \colon \mathbb{Z} \to \mathbb{C}$ be the map obtained from the character $(\mathbb{Z}/4\mathbb{Z})^\times = \{1, -1\} \subset \mathbb{C}^\times$, by extending it by zero to $\mathbb{Z}/4\mathbb{Z}$ and composing the result with the reduction map $\mathbb{Z} \to \mathbb{Z}/4\mathbb{Z}$. Then we have:

$$r_{\mathbb{Z}^2}(n) = 4 \sum_{d \mid n} \chi(d),$$

$$r_{\mathbb{Z}^4}(n) = 8 \sum_{2 \nmid d \mid n} d + 16 \sum_{2 \nmid d \mid (n/2)} d,$$

$$r_{\mathbb{Z}^6}(n) = 16 \sum_{d \mid n} \chi\left(\frac{n}{d}\right) d^2 - 4 \sum_{d \mid n} \chi(d) d^2,$$

$$r_{\mathbb{Z}^8}(n) = 16 \sum_{d \mid n} d^3 - 32 \sum_{d \mid (n/2)} d^3 + 256 \sum_{d \mid (n/4)} d^3,$$

$$r_{\mathbb{Z}^{10}}(n) = \frac{4}{5} \sum_{d \mid n} \chi(d) d^4 + \frac{64}{5} \sum_{d \mid n} \chi\left(\frac{n}{d}\right) d^4 + \frac{8}{5} \sum_{d \in \mathbb{Z}[i], |d|^2 = n} d^4.$$

For the history of these formulas, featuring, among others, Fermat, Euler, Lagrange, Legendre, Gauss, Jacobi, and Liouville, we refer the reader to [Mil2] and to Chapter 20 of [Ha-Wr].

In her master's thesis (Leiden, June 2010), see [Var], Ila Varma proves that there are no other values of k for which the theta function $\theta_{\mathbb{Z}^{2k}} = \theta_{\mathbb{Z}}^{2k}$, where $\mathbb{Z}^{2k}$ is equipped with the standard inner product, is a linear combination of Eisenstein series and cusp forms coming from Hecke characters.

We conclude that, even in the absence of formulas as above, one will be able to compute the numbers $r_{\mathbb{Z}^{2k}}(n)$ as fast as if such formulas existed. Often, theta functions are considered to be modular forms of which the coefficients are easy to compute, and hence useful for computing Hecke operators and coefficients of eigenforms. However, it seems that for coefficients $a_n(f)$ with n large, this will be the other way around, from now on.

Point counting on modular curves

Another consequence of Peter Bruin's results mentioned above is that, again assuming GRH, there will be a probabilistic algorithm that on input a positive square-free number n and a finite field $\mathbb{F}_q$, computes the number of $\mathbb{F}_q$-rational points of $X_1(n)$ in time polynomial in n and $\log q$. Indeed, this is a matter of computing the element T_p (where p is the prime dividing q) in the Hecke algebra $\mathbb{T}(n, 2)$ acting on the space $S_2(\Gamma_1(n), \mathbb{Z})$.

Point counting in a more general context

The methods we have used in this book can also be tried outside the context of modular forms. Let us consider, for example, a smooth surface S in $\mathbb{P}^3_{\mathbb{Q}}$ of degree at least 5, say. Letting l be a prime number, one has the cohomology groups $\mathrm{H}^i(S_{\overline{\mathbb{Q}},\mathrm{et}}, \mathbb{F}_l)$ for $0 \leq i \leq 4$, being finite dimensional $\mathbb{F}_l$-vector spaces with $\mathrm{Gal}(\overline{\mathbb{Q}}/\mathbb{Q})$-action. It seems reasonable to suspect that, again, there is an algorithm that on input a prime l computes these cohomology groups, with their $\mathrm{Gal}(\overline{\mathbb{Q}}/\mathbb{Q})$-action, in time polynomial in l. Once such an algorithm is known, one also has an algorithm that, on input a prime p of good reduction of S, gives the number $\#S(\mathbb{F}_p)$ of $\mathbb{F}_p$-valued points of S in time polynomial in $\log p$. This result would be of interest because the known p-adic algorithms for finding such numbers have running time exponential in $\log p$.

In this case, we choose a Lefschetz fibration from a blow-up of S to $\mathbb{P}^1_{\mathbb{Q}}$, and use the derived direct images of the constant sheaf $\mathbb{F}_{l,S}$ under this fibration to compute the étale cohomology of $S_{\overline{\mathbb{Q}}}$ with $\mathbb{F}_l$-coefficients. The most complicated contribution then comes from the first derived image $\mathcal{F}_l$, which is a locally constant sheaf of $\mathbb{F}_l$-vector spaces of dimension r, say, with r independent of l, on the open part U of $\mathbb{P}^1_{\mathbb{Q}}$ over which the fibration is smooth. This open part U is independent of l as well, and it is the analog of the open part of the j-line over which all modular curves are unramified.

For each l let $V_l := \underline{\mathrm{Isom}}_U(\mathbb{F}_l^r, \mathcal{F}_l)$. These V_l play the role of the covers $X_1(l)$ of the j-line, as, by definition, they trivialize the sheaves $\mathcal{F}_l$. Each cover $V_l \to U$ is finite Galois with group $G = \mathrm{GL}_r(\mathbb{F}_l)$, and $\mathrm{H}^1(U_{\overline{\mathbb{Q}},\mathrm{et}}, \mathcal{F}_l)$ is closely related to $\mathrm{H}^1(V_{l,\overline{\mathbb{Q}},\mathrm{et}}, \mathbb{F}_l^r)$ which sits in the l-torsion of the Jacobian variety of the smooth projective model $\overline{V_l}$ of V_l. It is our hope that methods as in this book (height bounds, approximations) can show that there is a polynomial algorithm for computing these cohomology groups.

In [Ed-dJ-Sc] a first step in this program is taken, by proving, in the function field case, that, for varying l, the height of $\overline{V_l}$ is bounded by a fixed power of l.

Modular forms of half integral weight

Much to our regret, we have nothing to say about modular forms of half integer weight. Nevertheless, it would be very interesting to be able to compute coefficients of such modular forms, as they encode interesting arithmetic information, such as class numbers.

Bibliography

[Ab-Ul] A. Abbes and E. Ullmo. *Comparaison des métriques d'Arakelov et de Poincaré sur $X_0(N)$.* Duke Math. J. 80 (1995), no. 2, 295–307.

[Abr] D. Abramovich. *A linear lower bound on the gonality of modular curves.* Internat. Math. Res. Notices 30 (1996), 1005–1011.

[Ad-Hu] L.M. Adleman and M-D. Huang. *Counting points on curves and abelian varieties over finite fields.* J. Symbolic Comput. 32 (2001), no. 3, 171–189.

[Ara] S.Y. Arakelov. *An intersection theory for divisors on an arithmetic surface,* Math. USSR Izvestija 8 (1974), 1167–1180.

[Asa] T. Asai. *On the Fourier coefficients of automorphic forms at various cusps and some applications to Rankin's convolution.* J. Math. Soc. Japan 28 (1976) no. 1, 48–61.

[At-Li] A.O.L. Atkin and W.W. Li. *Twists of newforms and pseudo-eigenvalues of W-operators.* Invent. Math. 48 (1978), 221–243.

[Ba-Ch] E. Bach and D. Charles. *The hardness of computing an eigenform.* In *Computational arithmetic geometry,* 9–15, Contemp. Math. 463, Amer. Math. Soc., Providence, 2008.

[Ba-Ki] J. Basmaji and I. Kiming. *A table of A_5-fields.* Chapter II in: *On Artin's conjecture for odd 2-dimensional representations,* 37–46, 122–141, Lecture Notes in Math. 1585, Springer, Berlin, 1994.

[Ba-Ne] P. Bayer and J. Neukirch. *On automorphic forms and Hodge theory.* Math. Ann. 257 (1981), no. 2, 137–155.

[Bak] M. H. Baker. *Torsion points on modular curves*. Ph.D. thesis, University of California, Berkeley (1999).

[Beh] K. Behrend. *The Lefschetz trace formula for algebraic stacks*. Invent. Math. 112 (1993), no. 1, 127–149.

[Ber] J. Bergström. *Cohomology of moduli spaces of curves of genus three via point counts*. J. Reine Angew. Math. 622 (2008), 155–187.

[Bir] B.J. Birch. *How the number of points of an elliptic curve over a fixed prime field varies*. J. London Math. Soc. 43 (1968), 57–60.

[Bla] D. Blasius. *Modular forms and abelian varieties*. Séminaire de Théorie des Nombres, Paris, 1989–1990, 23–29, Progr. Math. 102, Birkhäuser Boston, Boston, MA, 1992.

[Bl-Sh-Sm] L. Blum, M. Shub, and S. Smale. *On a theory of computation and complexity over the real numbers: NP-completeness, recursive functions, and universal machines*. Bull. Amer. Math. Soc. 21(1), 1989.

[Boa] R. Ph. Boas. *Entire functions*, Academic Press, New York, 1954.

[Bo-Ca] J-F. Boutot and H. Carayol. *Uniformisation p-adique des courbes de Shimura: les théorèmes de Cerednik et de Drinfeld*. Courbes modulaires et courbes de Shimura (Orsay, 1987–1988). Astérisque No. 196–197 (1991), 7, 45–158 (1992).

[Bo-Gu] E. Bombieri, W. Gubler. *Heights in Diophantine geometry*. New Mathematical Monographs 4. Cambridge University Press, Cambridge, 2006.

[Bo-Le-Ri] N. Boston, H.W. Lenstra and K.A. Ribet. *Quotients of group rings arising from two-dimensional representations*. C.R. Acad. Sci. Paris, t. 312, Série I, 323–328 (1991).

[Bo-Lu-Ra] S. Bosch, W. Lütkebohmert and M. Raynaud. *Néron models*. Springer Verlag, Ergebnisse 3, 21, Springer, Berlin, 1990.

[Bos1] J. Bosman. *A polynomial with Galois group* $SL_2(\mathbb{F}_{16})$. The LMS
Journal of Computation and Mathematics 10 (2007), 378–388. Available on arXiv:math/0701442, and on
`http://www.lms.ac.uk/jcm/10/lms2007-024/`

[Bos2] J. Bosman. *Explicit computations with modular Galois representations*. Ph.D. thesis, Universiteit Leiden, December 2008. Available on
`https://openaccess.leidenuniv.nl/`

[Br-Ku] R.P. Brent and H.T. Kung. *Fast algorithms for manipulating formal power series*. J. Assoc. Comput. Mach. 25 (1978), no. 4, 581–595.

[Bru] P.J. Bruin. *Modular curves, Arakelov theory, algorithmic applications*. Ph.D. thesis, Universiteit Leiden, 2010. Available on
`http://www.math.leidenuniv.nl/en/theses/PhD/`

[Bu-Le] J.A. Buchmann and H.W. Lenstra. *Approximating rings of integers in number fields*. J. Théor. Nombres Bordeaux 6 (1994), no. 2, 221–260.

[Buc1] J.D. Buckholtz. *Extremal problems for sums of powers of complex numbers*. Acta Math. Hung. 17 (1967), 147–153.

[Buc2] J.D. Buckholtz. *Sums of powers of complex numbers*. J. Math. Anal. Appl. 17 (1967), 269–279.

[Ca-De-Ve] W. Castryck, J. Denef, F. Vercauteren. *Computing zeta functions of nondegenerate curves*. IMRP Int. Math. Res. Pap. 2006, Art. ID 72017.

[Ca-Fr] *Algebraic number theory*. Proceedings of the instructional conference held at the University of Sussex, Brighton, September 1–17, 1965. Edited by J.W.S. Cassels and A. Fröhlich. Reprint of the 1967 original. Academic Press. Harcourt Brace Jovanovich, London, 1986.

[Ca-Ko-Lu] R. Carls, D. Kohel and D. Lubicz. *Higher-dimensional 3-adic CM construction*. J. Algebra 319 (2008), no. 3, 971–1006.

[Ca-Lu] R. Carls and D. Lubicz. *A p-adic quasi-quadratic time point counting algorithm.* Int. Math. Res. Not. IMRN (2009), no. 4, 698–735.

[Car] H. Carayol. *Sur les représentations l-adiques associées aux formes modulaires de Hilbert.* Annales Scientifiques de l'École Normale Supérieure Sér. 4, 19 (1986), no. 3, 409–468.

[Cart] M.L. Cartwright. *Integral functions,* Cambridge University Press, London, 1956.

[Cas] E. Casas-Alvero. *Singularities of plane curves.* London Mathematical Society Lecture Note Series 276. Cambridge University Press, London, 2000.

[Ch] A. Chambert-Loir. *Compter (rapidement) le nombre de solutions d'équations dans les corps finis.* Séminaire Bourbaki 2006–2007. Astérisque No. 317 (2008), Exp. No. 968, vii, 39–90.

[Co-Di] H. Cohen and F. Diaz y Diaz. *A polynomial reduction algorithm.* Sém. Th. Nombres Bordeaux (Série 2) 3 (1991), 351–360.

[Co-Ed] R.F. Coleman and S.J. Edixhoven. *On the semi-simplicity of the U_p-operator on modular forms.* Math. Ann. 310 (1998), no. 1, 119–127.

[Co-Ku] H. Cohn and A. Kumar. *The densest lattice in twenty-four dimensions.* Electron. Res. Announc. Amer. Math. Soc. 10 (2004), 58–67 (electronic).

[Co-Si] *Arithmetic geometry,* Papers from the conference held at the University of Connecticut, Storrs, Connecticut, July 30–August 10, 1984. Edited by Gary Cornell and Joseph H. Silverman. Springer, Berlin, 1986.

[Coh] H. Cohen. *A course in computational algebraic number theory.* Graduate Texts in Mathematics, 138. Springer, Berlin, 1993.

[Conr] B. Conrad. *Modular forms and the Ramanujan conjecture.* Cambridge Studies in Advanced Mathematics. Cambridge University Press, in preparation.

[Conw] J.H. Conway. *A characterisation of Leech's lattice.* Invent. Math. 7 (1969), 137–142.

[Cou] J-M. Couveignes. *Jacobiens, jacobiennes et stabilité numérique.* Groupes de Galois arithmétiques et différentiels, 91–125, Sémin. Congr., 13, Soc. Math. France, Paris, 2006.

[Cou2] J-M. Couveignes. *Linearizing torsion classes in the Picard group of algebraic curves over finite fields.* Journal of Algebra 321 (2009), 2085–2118.

[Cre] J.E. Cremona. *Algorithms for modular elliptic curves.* Cambridge University Press, London, 1997.

[Cu-Re] C.W. Curtis and I. Reiner. *Representation theory of finite groups and associative algebras.* Pure and Applied Mathematics 11, Wiley, New York, 1962.

[De-Ra] P. Deligne and M. Rapoport. *Les schémas de modules des courbes elliptiques.* In Modular Functions of One Variable II. Springer Lecture Notes in Mathematics 349, Springer, Berlin, 1973.

[De-Se] P. Deligne and J-P. Serre. *Formes modulaires de poids* 1. Ann. Sci. École Norm. Sup. (4) 7 (1974), 507–530.

[De-Ve] J. Denef and F. Vercauteren. *Computing zeta functions of hyperelliptic curves over finite fields of characteristic 2.* Advances in cryptology—CRYPTO 2002, 369–384, Lecture Notes in Comput. Sci. 2442, Springer, Berlin, 2002.

[Del1] P. Deligne. *Formes modulaires et représentations l-adiques.* Séminaire Bourbaki, 355, Février 1969.

[Del2] P. Deligne *La conjecture de Weil. I.* Inst. Hautes Études Sci. Publ. Math. 43 (1974), 273–307.

[Dem] V.A. Demjanenko. *An estimate of the remainder term in Tate's formula.* (Russian) Mat. Zametki 3 (1968), 271–278.

[Di-Im] F. Diamond and J. Im. *Modular forms and modular curves.* Seminar on Fermat's Last Theorem (Toronto, ON, 1993–1994), 39–133, CMS Conf. Proc. 17, Amer. Math. Soc., Providence, RI, 1995.

[Di-Sh] F. Diamond and J. Shurman. *A first course in modular forms.* GTM 228, Springer, Berlin, 2005.

[Dic1] M. Dickinson. *On the modularity of certain 2-adic Galois representations.* Duke Math. J. 109 (2001), no. 2, 319–382.

[Dic2] L.E. Dickson. *Linear groups: With an exposition of the Galois field theory.* With an introduction by W. Magnus. Dover, New York, 1958.

[Die] C. Diem. *On arithmetic and the discrete logarithm problem in class groups of curves.* Habilitation thesis, Leipzig, 2008.

[Dri] V.G. Drinfeld. *Two theorems on modular curves.* Funkcional. Anal. i Priložen. 7 (1973), no. 2, 83–84.

[Ed-dJ-Sc] S.J. Edixhoven, R.S. de Jong, J. Schepers. *Covers of surfaces with fixed branch locus.* Internat. J. Math. 21 (2010), 1–16.

[Ed-Ev] S.J. Edixhoven and J-H. Evertse. *Diophantine approximation and abelian varieties.* Lecture Notes in Mathematics 1566, Springer, Berlin, 1993, 2nd printing 1997.

[Edi1] S.J. Edixhoven. *The weight in Serre's conjectures on modular forms.* Invent. Math. 109 (1992), 563–594.

[Edi2] S.J. Edixhoven. *Rational elliptic curves are modular (after Breuil, Conrad, Diamond and Taylor).* Séminaire Bourbaki, 1999–2000. Astérisque No. 276 (2002), 161–188.

[Edi3] S.J. Edixhoven. *Point counting after Kedlaya.* Syllabus for the graduate course "Mathematics of Cryptology," at the Lorentz Center in Leiden, September 2003. Available at:
www.math.leidenuniv.nl/~edix/
oww/mathofcrypt/carls_edixhoven/kedlaya.pdf

[Elk] R. Elkik. *Fonctions de Green, volumes de Faltings. Application aux surfaces arithmétiques.* Chapter III of [Szp].

[Eng] A. Enge. *Elliptic curves and their applications to cryptography, an introduction.* Kluwer Academic, New York, 1999.

[Fa-Ja] D.W. Farmer and K. James. *The irreducibility of some level 1 Hecke polynomials.* Math. Comp. 71 (2002), no. 239, 1263–1270.

[Fa-Jo] G. Faltings, B.W. Jordan. *Crystalline cohomology and* $GL(2, Q)$. Israel J. Math. 90 (1995), no. 1-3, 1–66.

[Fa-Kr] H.M. Farkas and I. Kra. *Riemann Surfaces.* Springer Graduate Texts in Mathematics 71, second edition, Springer, Berlin, 1991.

[Fal1] G. Faltings, *Calculus on arithmetic surfaces,* Ann. of Math. 119 (1984), 387–424.

[Fal2] G. Faltings. *Lectures on the arithmetic Riemann-Roch theorem.* Notes taken by Shouwu Zhang. Annals of Mathematics Studies 127. Princeton University Press, Princeton, 1992.

[Fis] T. Fisher. *On 5 and 7 descents for elliptic curves.* Available on the author's home page:
`http://www.dpmms.cam.ac.uk/~taf1000/`

[Fo-Ga-Ha] M. Fouquet, P. Gaudry and R. Harley. *An extension of Satoh's algorithm and its implementation.* J. Ramanujan Math. Soc. 15 (2000), no. 4, 281–318.

[For] O. Forster, *Riemannsche Flächen.* Springer, Berlin, 1977.

[Fr-Ki] E. Freitag and R. Kiehl. *Étale cohomology and the Weil conjecture.* Translated from the German by Betty S. Waterhouse and William C. Waterhouse. With an historical introduction by J.A. Dieudonné. Ergebnisse der Mathematik und ihrer Grenzgebiete 3, 13, Springer, Berlin, 1988.

[Fr-Ru] G. Frey and H-G. Rück. *A remark concerning m-divisibility and the discrete logarithm in the divisor class group of curves.* Mathematics of computation 62 (1994), no. 206, 865–874.

[Fre]   G. Frey and M. Müller. *Arithmetic of modular curves and applica-tions.* In "On Artin's conjecture for odd 2-dimensional representa-tions," Lecture Notes in Math. 1585, Springer, Berlin, 1994.

[Ga-Ge] J. von zur Gathen and J. Gerhard. *Modern Computer Algebra.* Cambridge University Press, London, 1999.

[Ga-Gu] P. Gaudry and N. Gürel. *An extension of Kedlaya's point-counting algorithm to super-elliptic curves.* In C. Boyd (ed.), Advances in Cryptology — ASIACRYPT 2001. Lecture Notes in Computer Sci-ence 1807, Springer, Berlin, 2000, 19–34.

[Ge-Kl] K. Geissler and J. Klüners. *Galois group computation for rational polynomials.* J. Symbolic Comput. 30 (2000), 653–674.

[Ger1] R. Gerkmann. *Relative rigid cohomology and deformation of hy-persurfaces.* Int. Math. Res. Pap. IMRP 2007, no. 1, Art. ID rpm003, 67 pp.

[Ger2] R. Gerkmann. *Relative rigid cohomology and point counting on families of elliptic curves.* J. Ramanujan Math. Soc. 23 (2008), no. 1, 1–31.

[Gor]   D. Gorenstein. *An arithmetic theory of adjoint plane curves.* Trans. Amer. Math. Soc. 72 (1952), 414–436.

[Gro]   B.H. Gross. *A tameness criterion for Galois representations associ-ated to modular forms (mod p).* Duke Math. J. 61 (1990), No. 2, 445–517.

[Ha-Wr] G.H. Hardy and E.M. Wright. *An introduction to the theory of numbers.* Fifth edition. Clarendon Press, New York, 1979.

[Hac]   G. Haché. *Computation in algebraic function fields for effective con-struction of algebraic-geometric codes.* In "Proceedings of the 11th International Symposium on Applied Algebra, Algebraic Algorithms and Error-Correcting Codes," pages 262–278, 1995.

[Hart] R. Hartshorne. *Algebraic geometry.* Graduate Texts in Mathemat-ics 52. Springer, New York, 1977.

[Harv] D. Harvey. *Kedlaya's algorithm in larger characteristic.* Int. Math. Res. Not. IMRN 2007, no. 22, Art. ID rnm095, 29 pp.

[Hav] G. Havas, B.S. Majewski, and K.R. Matthews. *Extended gcd and Hermite normal form algorithms via lattice basis reduction.* Experimental Mathematics 7 (1998), 125–136.

[Hay] B. Hayes. *A lucid interval.* American Scientist 91 (2003) no. 6, 484–488.

[Hen] P. Henrici. *Applied and Computational Complex Analysis Volume I.* Wiley Classics Library, Wiley, New York, 1974.

[Hig] N.J. Higham. *Accuracy and stability of numerical algorithms,* second edition. Society for Industrial and Applied Mathematics, Philadelphia, 2002.

[Hor] L. Hörmander. *Linear partial differential operators.* Die Grundlehren der mathematischen Wissenschaften in Einzeldarstellungen 116, Springer, Berlin, 1976.

[Hri] P. Hriljac. *Heights and Arakelov's intersection theory.* Amer. J. Math. 107 (1985), no. 1, 193–218.

[Hub] H. Hubrechts. *Point counting in families of hyperelliptic curves.* Found. Comput. Math. 8 (2008), no. 1, 137–169.

[Igu] J. I. Igusa, *Theta functions.* Grundlehren der Math. Wissenschaften 194, Springer, Berlin, 1972.

[Jo-Ke] B. Jordan and B. Kelly. *The vanishing of the Ramanujan Tau function.* Preprint, 1999.

[Jo-Kr] J. Jorgenson and J. Kramer. *Bounds on canonical Green's functions.* Compos. Math. 142 (2006), no. 3, 679–700.

[Joc] N. Jochnowitz. *A study of the local components of the Hecke algebra mod l.* Trans. Amer. Math. Soc. 270 (1982), no. 1, 253–267.

[Ka-Lu] G. Kato and S. Lubkin. *Zeta matrices of elliptic curves.* Journal of Number Theory 15 (1982), 318–330.

[Ka-Ma] N.M. Katz, B. Mazur. *Arithmetic moduli of elliptic curves*. Annals of Mathematics Studies 108. Princeton University Press, Princeton, 1985.

[Kal] W. van der Kallen. *Complexity of the Havas, Majewski, Matthews LLL Hermite normal form algorithm*. arXiv:math/9812130v1, 2008.

[Ked] K. Kedlaya. *Counting points on hyper-elliptic curves using Monsky-Washnitzer cohomology*. J. Ramanujan Math. Soc. 16 (2001), no. 4, 323–338.

[Ke-Um] K.S. Kedlaya and C. Umans. *Fast modular composition in any characteristic*. Foundations of Computer Science FOCS, 2008.

[Kha] C. Khare. *Serre's modularity conjecture: the level one case*. Duke Math. J. 134 (2006), 557–589.

[Kh-Wi1] C. Khare and J-P. Wintenberger. *Serre's modularity conjecture (I)*. Invent. Math. 178 (2009), no. 3, 485–504.

[Kh-Wi2] C. Khare and J-P. Wintenberger. *Serre's modularity conjecture (II)*. Invent. Math. 178 (2009), no. 3, 505–586.

[Ki-Ve] I. Kiming and H. Verrill. *On modular mod l Galois representations with exceptional images*. J. Number Theory 110 (2005), no. 2, 236–266.

[Ki-Wi] L.J.P. Kilford and G. Wiese. *On the failure of the Gorenstein property for Hecke algebras of prime weight*. Experiment. Math. 17 (2008), no. 1, 37–52.

[Kis1] M. Kisin. *Moduli of finite flat group schemes, and modularity*. Ann. of Math. (2) 170 (2009), no. 3, 1085–1180.

[Kis2] M. Kisin. *Modularity of 2-adic Barsotti-Tate representations*. Invent. Math. 178 (2009), no. 3, 587–634.

[La-Wa1] A.G.B. Lauder and D. Wan. *Counting points on varieties over finite fields of small characteristic*. In "Algorithmic number theory: lattices, number fields, curves and cryptography," J.P. Buhler and P.

Stevenhagen (ed.), Math. Sci. Res. Inst. Publ. 44, Cambridge University Press, Cambridge, 2008.

[La-Wa2] A.G.B. Lauder and D. Wan. *Computing zeta functions of Artin-Schreier curves over finite fields.* LMS J. Comput. Math. 5 (2002), 34–55 (electronic).

[Lan1] S. Landau. *Factoring polynomials over algebraic number fields.* SIAM J. Comput. 14 (1985), no. 1, 184–195.

[Lan2] S. Lang. *Algebraic groups over finite fields.* Amer. J. Math. 78 (1956), 555–563.

[Lan3] S. Lang. *Abelian varieties*, volume 7 of "Interscience Tracts in Pure and Applied Mathematics." Interscience Publishers, 1959.

[Lan4] S. Lang. *Introduction to modular forms.* Springer, Berlin, 1976.

[Lan5] S. Lang. *Elliptic curves: Diophantine analysis.* Grundlehren der mathematischen Wissenschaften 231, Springer, New York, 1978.

[Lan6] S. Lang. *Algebraic number theory.* Second edition. Graduate Texts in Mathematics 110. Springer, New York, 1994.

[Lau1] A.G.B. Lauder. *Computing zeta functions of Kummer curves via multiplicative characters.* Found. Comput. Math. 3 (2003), no. 3, 273–295.

[Lau2] A.G.B. Lauder. *Deformation theory and the computation of zeta functions.* Proc. London Math. Soc. 88 Part 3 (2004), 565–602.

[Le-Le-Lo] A.K. Lenstra, H.W. Lenstra, and L. Lovász. *Factoring polynomials with rational coefficients.* Math. Ann. 261 (1982), no. 4, 515–534.

[Le-Lu] R. Lercier and D. Lubicz. *A quasi quadratric time algorithm for hyperelliptic curve point counting.* Ramanujan J. 12 (2006), no. 3, 399–423.

[Lee1] J. Leech. *Some sphere packings in higher space.* Canad. J. Math. 16 (1964), 657–682.

[Lee2] J. Leech. *Notes on sphere packings.* Canad. J. Math. 19 (1967), 251–267.

[Leh] D.H. Lehmer. *The vanishing of Ramanujan's function* $\tau(n)$ Duke Math. J. 10 (1947), 429–433.

[Len1] A.K. Lenstra. *Factoring polynomials over algebraic number fields.* Computer algebra (London, 1983), 245–254, Lecture Notes in Comput. Sci. 162, Springer, Berlin, 1983.

[Len2] H.W. Lenstra. *Algorithms in algebraic number theory.* Bull. Amer. Math. Soc. (N.S.) 26 (1992), no. 2, 211–244.

[Len3] H.W. Lenstra. *Lattices.* In "Algorithmic number theory: lattices, number fields, curves and cryptography," 127–181, Math. Sci. Res. Inst. Publ. 44, Cambridge University Press, Cambridge, 2008.

[Lic] S. Lichtenbaum. *Duality theorems for curves over p-adic fields.* Invent. Math. 7, 120–136, 1969.

[Ma-Ri] B. Mazur and K.A. Ribet. *Two-dimensional representations in the arithmetic of modular curves.* Courbes modulaires et courbes de Shimura (Orsay, 1987/1988). Astérisque No. 196–197, (1991), 6, 215–255 (1992).

[Magma] W. Bosma, J.J. Cannon and C.E. Playoust. *The magma algebra system I: the user language.* J. Symbolic Comput. 24 (1997), no. 3/4, 235–265.

[Man1] Y. Manin. *Parabolic points and zeta function of modular curves.* Math. USSR Izvestija 6 (1972), no. 1, 19–64.

[Me-Ok-Va] A. Menezes, T. Okamoto, and S. Vanstone. *Reducing elliptic curve logarithms to logarithms in a finite field.* IEEE Trans. Inf. Theory, IT 39 (1993), no. 5, 1639–1646.

[Mer] L. Merel. *Universal Fourier expansions of modular forms.* In "On Artin's conjecture for odd 2-dimensional representations," Lecture Notes in Math. 1585, Springer, Berlin, 1994, 59–94.

[Mil1] J.S. Milne. *Étale cohomology*. Princeton Mathematical Series 33. Princeton University Press, Princeton, 1980.

[Mil2] S.C. Milne. *Infinite families of exact sums of squares formulas, Jacobi elliptic functions, continued fractions, and Schur functions*. Ramanujan J. 6 (2002), no. 1, 7–149.

[Miy] T. Miyake. *Modular forms*. Translated from the 1976 Japanese original by Yoshitaka Maeda. Reprint of the first 1989 English edition. Springer Monographs in Mathematics, Springer, Berlin, 2006.

[Mo-Ta] H. Moon and Y. Taguchi *Refinement of Tate's discriminant bound and non-existence theorems for mod p Galois representations*. Documenta Math. Extra Volume Kato (2003), 641–654.

[Mor1] B.M. Moret. *The theory of computation*. Addison-Wesley, Reading, 1998.

[Mor2] L. Moret-Bailly. *Métriques permises*. Chapter II of [Szp].

[Mor3] L. Moret-Bailly. *La formule de Noether pour les surfaces arithmétiques*. Invent. Math. 98 (1989), 491–498.

[Mu-S-T] V. Müller, A. Stein, and C. Thiel. *Computing discrete logarithms in real quadratic function fields of large genus*. Math. Comp. 68 (1999), 807–822.

[Nie] H-V. Niemeier. *Definite quadratische Formen der Dimension 24 und Diskriminante 1*. J. Number Theory 5 (1973), 142–178.

[Pan] V.Y. Pan. *Approximating complex polynomial zeros: Modified Weyl's quadtree construction and improved newton's iteration*. J. Complexity 16 (2000), no. 1, 213–264.

[Pap] C.H. Papadimitriou. *Computational complexity*. Addison Wesley, Reading, 1967.

[Par] P. Parent. *Bornes effectives pour la torsion des courbes elliptiques sur les corps de nombres*. J. Reine Angew. Math. 506 (1999), 85–116.

[PARI] C. Batut, K. Belabas, D. Bernardi, H. Cohen, and M. Olivier. *User's guide to PARI/GP (version 2.3.1)*. http://pari.math.u-bordeaux.fr.

[Pil] J. Pila. *Frobenius maps of abelian varieties and finding roots of unity in finite fields*. Math. Comp. 55 (1990), no. 192, 745–763.

[Po-Za] M. Pohst, H. Zassenhaus. *Algorithmic algebraic number theory*. Cambridge University Press, Cambridge, 1989.

[Poo1] B. Poonen. *Gonality of modular curves in characteristic p*. Math. Res. Lett. 14 (2007) no. 4, 691–701.

[Poo2] B. Poonen. *Computational aspects of curves of genus at least 2*. Lecture Notes in Computer Science 1122, Springer, Berlin, 1996, 283–306.

[Rib1] K. Ribet. *On l-adic representations attached to modular forms*. Invent. Math. 28 (1975), 245–275.

[Rib2] K. Ribet. *Galois representations attached to eigenforms with Nebentypus*. Modular functions of one variable, V (Proc. Second Internat. Conf., Univ. Bonn, Bonn, 1976), 17–51. Lecture Notes in Math. 601, Springer, Berlin, 1977.

[Rib3] K. Ribet. *On l-adic representations attached to modular forms. II*. Glasgow Math. J. 27 (1985), 185–194.

[Rib4] K. Ribet. *Images of semistable Galois representations*. Olga Taussky-Todd: in memoriam. Pacific J. Math. Special Issue (1997), 277–297.

[SAGE] SAGE Mathematics Software, http://www.sagemath.org/

[Sai1] T. Saito. *Modular forms and p-adic Hodge theory*. Invent. Math. 129 (1997), no. 3, 607–620.

[Sai2] T. Saito. *Hilbert modular forms and p-adic Hodge theory*. Compos. Math. 145 (2009), no. 5, 1081–1113.

[Sat]   T. Satoh. *The canonical lift of an ordinary elliptic curve over a finite field and its point counting.* J. Ramanujan Math. Soc. 15 (2000), no. 4, 247–270.

[Sch1] A.J. Scholl. *Motives for modular forms.* Invent. Math. 100 (1990), no. 2, 419–430.

[Sch2] R.J. Schoof. *Elliptic curves over finite fields and the computation of square roots mod p.* Math. Comp. 44 (1985), no. 170, 483–494.

[Sch3] R.J. Schoof. *Counting points on elliptic curves over finite fields.* Les Dix-huitièmes Journées Arithmétiques (Bordeaux, 1993). J. Théor. Nombres Bordeaux 7 (1995), no. 1, 219–254.

[Ser1] J-P. Serre. *Groupes algébriques et corps de classes.* Hermann, Paris, 1959.

[Ser2] J-P. Serre. *Une interprétation des congruences relatives à la fonction τ de Ramanujan.* 1969 Séminaire Delange-Pisot-Poitou: 1967–1968, Théorie des Nombres, Fasc. 1, Exp. 14.

[Ser3] J-P. Serre. *Propriétés galoisiennes des points d'ordre fini des courbes elliptiques.* Invent. Math. 15 (1972), no. 4, 259–331.

[Ser4] J-P. Serre. *Valeurs propres des opérateurs de Hecke modulo l.* Journées arithmétiques Bordeaux, Astérisque 24–25, 109–117 (1975). (Oeuvres 104).

[Ser5] J-P. Serre. *Cours d'arithmétique.* Deuxième édition revue et corrigée. Le Mathématicien 2. Presses Universitaires de France, Paris, 1977.

[Ser6] J-P. Serre. *Modular forms of weight one and Galois representations.* Algebraic number fields: L-functions and Galois properties (A. Frölich, ed.), Academic Press, London, 1977, 193–268.

[Ser7] J-P. Serre. *Représentations linéaires des groupes finis.* Third revised edition. Hermann, Paris, 1978.

[Ser8] J-P. Serre. *Sur la lacunarité des puissances de η*. Glasgow Math. J. 27 (1985), 203–221.

[Ser9] J-P. Serre. *Sur les représentations modulaires de degré 2 de* Gal($\overline{\mathbb{Q}}/\mathbb{Q}$). Duke Math. J. 54 (1987) no. 1, 179–230.

[Ser10] J-P. Serre, *Lectures on the Mordell-Weil theorem*, Asp. Math. E15, Vieweg, 1989.

[SGA4] A. Grothendieck. *Théorie des topos et cohomologie étale des schémas*. Séminaire de Géométrie Algébrique du Bois-Marie 1963–1964 (SGA 4). Dirigé par M. Artin, A. Grothendieck, et J.L. Verdier. Avec la collaboration de N. Bourbaki, P. Deligne et B. Saint-Donat. Lecture Notes in Mathematics 269, 270, and 305. Springer, Berlin, 1972, 1973.

[SGA4.5] P. Deligne. *Cohomologie étale*. Séminaire de Géométrie Algébrique du Bois-Marie SGA $4\frac{1}{2}$. Avec la collaboration de J.F. Boutot, A. Grothendieck, L. Illusie et J.L. Verdier. Lecture Notes in Mathematics 569. Springer, Berlin, 1977.

[SGA5] *Cohomologie l-adique et fonctions L*. Séminaire de Géometrie Algébrique du Bois-Marie 1965–1966 (SGA 5). Edité par Luc Illusie. Lecture Notes in Mathematics 589. Springer, Berlin, 1977.

[SGA7] P. Deligne. *Groupes de monodromie en géométrie algébrique. II.* Séminaire de Géométrie Algébrique du Bois-Marie 1967–1969 (SGA 7 II). Dirigé par P. Deligne et N. Katz. Lecture Notes in Mathematics 340. Springer, Berlin, 1973.

[Shi1] G. Shimura. *Introduction to the arithmetic theory of automorphic functions*. Iwanami Shoten and Princeton University Press, Princeton, 1971.

[Shi2] G. Shimura. *On the periods of modular forms*. Math. Ann. 229 (1977) no. 3, 211–221.

[Sho] V.V. Shokurov. *Shimura integrals of cusp forms*. Math. USSR Izvestija 16 (1981) no. 3, 603–646.

[Sil] J. Silverman. *The difference between the Weil height and the canonical height on elliptic curves.* Math. Comp. 55 (1990), no. 192, 723–
743.

[Sou] C. Soulé. *Lectures on Arakelov geometry.* With the collaboration of
D. Abramovich, J.-F. Burnol and J. Kramer. Cambridge Studies in
Advanced Mathematics 33. Cambridge University Press, Cambirdge,
1992.

[Ste1] W.A. Stein *Explicit approaches to modular abelian varieties.* Ph.D.
thesis, University of California, Berkeley (2000).

[Ste2] W.A. Stein. *Modular forms, a computational approach.* With an ap-
pendix by Paul E. Gunnells. Graduate Studies in Mathematics, 79.
American Mathematical Society, Providence, RI, 2007. xvi+268 pp.
ISBN: 978-0-8218-3960-7; 0-8218-3960-8.

[Stu] J. Sturm. *On the congruence of modular forms.* Number Theory (New
York, 1984–1985), 275–280, Lecture Notes in Mathematics 1240,
Springer, 1987.

[Suz] M. Suzuki. *Group Theory I.* Grundlehren der mathematischen Wissenschaften 247, Springer, New York, 1982.

[Swi] H. P. F. Swinnerton-Dyer. *On l-adic representations and congruences
for coefficients of modular forms.* Modular functions of one variable,
III (Proc. Internat. Summer School, Univ. Antwerp, 1972), pp. 1–55.
Lecture Notes in Math. 350, Springer, Berlin, 1973.

[Szp] L. Szpiro. *Séminaire sur les pinceaux arithmétiques: la conjecture
de Mordell.* Astérisque No. 127 (1985), Société Mathématique de
France, 1990.

[Tat1] J. Tate. *Endomorphisms of abelian varieties over finite fields.* Invent.
Math. 2 (1966), 134–144.

[Tat2] J. Tate and P. Deligne. *Courbes elliptiques: formulaire.* In B. Birch
and W. Kuyk eds. *Modular Functions of One Variable IV.* Springer,
Berlin, 1975.

[Ten] G. Tenenbaum. *Introduction to analytic and probabilistic number theory*. Cambridge Studies in Adv. Math. 46. Cambridge University Press, Cambridge, 1995.

[Var] I. Varma. *Sums of Squares, Modular Forms, and Hecke Characters*. Masters thesis, Universiteit Leiden, June 2010. Available on
http://www.math.leidenuniv.nl/en/theses/Master/

[Vel] J. Vélu. *Isogénies entre courbes elliptiques*. Comptes Rendus de l'Académie de Sciences de Paris, Série A, 273 (1971), 238–241.

[Vol] E.J. Volcheck. *Computing in the jacobian of a plane algebraic curve*. In "Algorithmic number theory, ANTS I," Lecture Notes in Computer Science 877, 221–233. Springer, Berlin, 1994.

[Wei1] A. Weil. *Über die Bestimmung Dirichletscher Reihen durch Funktionalgleichungen*. Math. Ann. 168 (1967), 149–156.

[Wei2] A. Weil. *Basic number theory*. Reprint of the second (1973) edition. Classics in Math. Springer, Berlin, 1995.

[Wei3] P.J. Weinberger. *On Euclidean rings of algebraic integers*. Analytic number theory (Proc. Sympos. Pure Math., Vol. XXIV, St. Louis Univ., St. Louis, Mo., 1972), pp. 321–332. Amer. Math. Soc., Providence, 1973.

[Wei4] P.J. Weinberger. *Finding the number of factors of a polynomial*. J. Algorithms 5 (1984), no. 2, 180–186.

[Wie1] G. Wiese. *On the faithfulness of parabolic cohomology as a Hecke module over a finite field*. J. Reine Angew. Math. 606 (2007), 79–103.

[Wie2] G. Wiese. *Multiplicities of Galois representations of weight one*. With an appendix by Niko Naumann. Algebra Number Theory 1 (2007), no. 1, 67–85.

[Wit] E. Witt. *Collected papers. Gesammelte Abhandlungen*. With an essay by Günter Harder on Witt vectors. Edited and with a preface in English and German by Ina Kersten. Springer, Berlin, 1998.

[Zim] H. Zimmer. *On the difference of the Weil height and the Néron-Tate height.* Math. Z. 147 (1976), no. 1, 35–51.

Index

Words,

Milton Keynes UK
Ingram Content Group UK Ltd.
UKHW020739180824
447095UK00005B/337